U0622505

何中虎◎主编

中麦895选育与主要特性解析

中国农业出版社

农村读物出版社

北　京

序

2003 年 5 月，作为中国农业科学主管科研的副院长，我参加了作物科学研究所和棉花研究所育种合作协议签署仪式并为安阳育种站揭牌，这标志着中国农业科学院作物科学研究所小麦育种的重点正式转向黄淮麦区。在历任院所领导的大力支持下，在庄巧生先生的指导下，双方成功共建安阳育种站，面向黄淮麦区合作开展小麦育种工作。在何中虎研究员的带领下，组建了跨两个研究所、育种与分子标记和品质检测密切结合的育种团队，近 20 年取得了很好效果。2009 年至今，先后育成 12 个通过国家和省级审定的新品种，其中高产耐热的中麦 895、优质高产兼顾的中麦 578 表现最为突出。

安阳育种站将矮秆抗倒、耐热和优质作为育种的主攻方向，将田间落黄特性、高千粒重和年度间的稳定性作为耐热性选择的主要指标，育成的中麦 895 实现了矮秆高产、耐热性与抗病性的良好结合，耐热性和面条馒头加工品质显著优于大面积推广的周麦 16，2012 年通过黄淮南片国家审定。中麦 895 先后被选为河南、安徽、陕西的主推品种，曾为全国第三大小麦品种，累计推广近 5 000 万亩；还成为黄淮麦区高产耐热育种的骨干亲本，为小麦生产和育种作出重要贡献。"耐热高产优质小麦新品种中麦 895 的选育与应用"获得了 2019 年农业农村部中华农业科技奖科研成果一等奖。

在加强品种选育和推广的同时，团队还对中麦 895 的主要特性做了系统深入研究。主要创新性包括三个方面：（1）用建立起的高通量田间表型鉴定技术分析表明，中麦 895 携带 4 个高千粒重和灌浆速率 QTL，这是其耐高温和高产广适的主要原因；（2）携带 1 个全生育期抗条锈病主效 QTL、2 个成株期抗白粉病 QTL 和 1 个抗赤霉病 QTL，综合抗病性好；（3）含低分子量麦谷蛋白亚基 Glu-A3d 等优质基因，酚酸含量高且稳定，营养价值高。上述意见是否成立有待读者评说。

在中国农业科学院棉花研究所和作物科学研究所领导、全国同行与推广部门的大力支持和帮助下，安阳育种站的工作取得了显著进展。目前正在大面积推广 2020 年审定的中麦 578，该品种实现了优质强筋、高产、抗病抗逆、广泛适应性的良好结合，初步解决了优质高产的矛盾，订单种植快速发展。期望他们继续努力，为小麦育种和生产做出更大的贡献。

刘　旭

2023 年 10 月

前　　言

20世纪90年代，由于北部冬麦区小麦面积大幅度下降，中国农业科学院作物科学研究所便将育种工作重点逐步转向黄淮麦区。2003年，在时任所长万建民院士和喻树迅院士的大力支持下，中国农业科学院作物科学研究所和棉花研究所签署了长期育种合作协议，共建安阳育种站，面向黄淮南片麦区联合开展小麦育种工作。我们组建了跨两个研究所、一体化运行、多学科合作的育种团队，取得了良好效果。在团队首席科学家领导下，安阳的育种工作具体由闫俊研究员和张勇研究员共同负责，主要参加人员包括王洪森研究员（返聘至2007年）、徐春英农艺师（2005—2015年）、闫俊良农艺师和田宇兵助理农艺师，其中张勇和田宇兵是作物科学研究所派到安阳工作的全职人员，近几年徐开杰副研究员和张赵星助理研究员的加入使育种队伍进一步壮大。位于北京的品质实验室和分子标记实验室负责安阳育种材料的品质测试和标记检测工作。

根据生产发展要求，鉴于气候变化的影响越来越大，我们确定了新的育种思路，并逐步形成了一套新的做法。安阳育种站与团队的石家庄育种站、山东农业科学院作物所刘建军研究员领导的育种组、泰安农科院吴科研究员领导的育种组建立了优良材料交换与穿梭选择机制；分离世代在田间加强了抗倒性和抗热性选择，在室内加强了品质测试；为了实现高产与广适性的有效结合，新品系初级产量比较试验和产量比较试验同时在河北、山东、河南、安徽、陕西的10多个地点进行；新品系的抗病性鉴定分别在淮安和武汉（赤霉病）、周口（纹枯病）、杨凌和成都（条锈病）进行。期望育成的新品种在黄淮南片大面积推广的同时，还能在黄淮北片和新疆扩大利用，为小麦生产作出应有贡献。

经过同事们的不懈努力，安阳育种站育成并推广了以中麦895为代表的耐热高产广适新品种，优质强筋高产新品种中麦578正在黄淮南片和北片及新疆大面积推广，这标志着中国农业科学院培育的新品种开始在黄淮麦区发挥重要作用。中麦895实现了高产潜力、耐热性与抗病性、优良面条品质及广适性的良好结合，于2012年通过黄淮南片国家审定，成为河南、安徽、陕西的主推品种，年最大推广面积1 062万亩，曾为全国第三大小麦品种，累计推广近5000万亩；还成为黄淮麦区高产耐热育种的骨干亲本，为小麦育种和生产作出重要贡献。"耐热高产优质小麦新品种中麦895的选育与应用"有幸获得2019年农业农村部中华农业科技奖科研成果一等奖。

在加强品种推广的同时，我们还专门培育了中麦871/中麦895重组自交系群体和扬麦16/中麦895双单倍体群体，研究并建立了高通量田间表型鉴定平台，采用育种、栽培、生理、谷物化学与基因组学相结合的方法，对中麦895的株型、耐热性和粒重、水肥利用效率、抗病性、加工品质与营养特性等做了较为系统深入的研究。主要创新性可归纳为三点：

(1) 中麦895携带矮秆基因 *Rht2* 与 *Rht24*、2个控制分蘖角的主效 QTL，株型矮、紧凑，旗叶小且直立；含4个高千粒重与灌浆速率 QTL 和5个抗旱基因，根系活力好，叶功能期长，灌浆速率快，千粒重高且稳定，为耐高温和高产稳产提供保障；(2) 携带1个全生育期抗条锈病主效 QTL、2个成株期抗白粉病 QTL 和1个抗赤霉病 QTL，综合抗病性好，赤霉病病穗率低；(3) 含低分子量麦谷蛋白亚基 Glu-A3d 等优质基因，面团筋力中等，颜色亮黄，面条和馒头品质优良；酚酸含量高且稳定，营养价值高。

为了便于虚心向各部门同行学习，全面提高我们的育种和研究水平，同时也向主管部门和资助机构汇报工作进展，我们挑选了学术性与应用价值兼顾的25篇论文编成《中麦895选育与主要特性解析》，包括耐热性和产量性状、根系与肥料利用效率、抗病性与黑胚、加工品质与营养特性共4部分。这些论文已在国内外发表，为了减少工作量，文章基本保持原貌，只对文字和个别差错之处做了补充、修改。

安阳育种站的工作一直得到中国农业科学院、棉花研究所和作物科学研究所历任领导的大力支持和帮助，先后获得863计划、948重大国际合作项目（2003—2015）、国家重点研发计划、中国农业科学院创新工程等资助，在此一并表示感谢。除安阳育种站的工作人员外，夏先春研究员、肖永贵副研究员、郝元峰研究员、曹双和研究员、张艳研究员等和许多研究生为中麦895主要特性解析付出了不懈努力。特别感谢庄巧生院士、李振声院士、程顺和院士、喻树迅院士、万建民院士、赵振东院士、李付广研究员、郑天存研究员、郭天财教授、茹振钢教授等的长期鼓励和指导。感谢李鸣博士在文件转换、文字核对和内容审读等方面做了不可或缺的工作。还要感谢中国农业出版社杨天桥编审为本书出版所做的努力。特别感谢刘旭院士在百忙中为本书作序，我们将继续努力，为小麦育种做出更大贡献。

谨将此书献给国际杰出小麦育种家、2014年世界粮食奖获得者 Sanjaya Rajaram 博士，他因患新冠肺炎不幸于2021年2月17日在墨西哥逝世，享年78岁。Raj 是我和阎俊共同的导师，还是我的密切朋友，在关键时刻总能给予建设性指导和无私帮助。我于1990年5月下旬陪同 Raj、Hans Braun（2003—2020年任 CIMMYT 小麦项目主任）和 Ravi Singh（2005年至今任 CIMMYT 卓越科学家）到安阳育种站考察小麦育种工作，1990年11月至1993年9月我在 Raj 领导的小麦育种团队做博士后，阎俊于1992年3月至1993年6月跟 Raj 做访问学者。Raj 对推动中国与 CIMMYT 合作、CIMMYT 中国办事处建立与发展等发挥了关键作用。2017年 Raj 应邀参加中国农业科学院60周年庆祝活动并再次访问安阳育种站，与我们一起观摩小麦新品种中麦578。2021年3月我们为《麦类作物学报》撰写了"追忆国际卓越小麦育种家拉贾拉姆"一文，也编入此书。Raj 热爱小麦、追求卓越、尊重同事的精神将永远激励我们做好小麦研究工作。

由于时间较短，加上作者水平有限，疏漏、错误之处在所难免，敬请指正。

何中虎

2023年10月

目　　录

加工品质与营养特性 ·· 297

追忆国际卓越小麦育种家拉贾拉姆

邢清清[1]　王馨[2,3]　Hans Joachim Braun[3]　Marten van Ginkel[4]　何中虎[2,3]

1 北京理工大学国际交流合作处，北京 100081；
2 中国农业科学院作物科学研究所，北京 100081；
3 国际玉米小麦改良中心，墨西哥 56100；
4 国际干旱地区农业研究中心，黎巴嫩

摘要：拉贾拉姆博士从事小麦育种研究 50 多年，在布劳格博士创立的国际小麦育种基础上，实现了 CIMMYT 小麦育种全球化，主要贡献包括四个方面：（1）大规模开展冬春杂交，实现产量、抗性和适应性重大进展；（2）将慢锈性理念系统融入小麦育种项目，成功培育并在全世界推广高产慢锈性品种；（3）耐酸性土壤和耐旱育种取得显著进展；（4）先后育成 481 个高产抗病优质小麦新品种，在 51 个发展中国家年推广种植 6,000 万公顷；他还为发展中国家和 CIMMYT、ICARDA 等国际组织的人才培养做出了杰出贡献。

关键词：拉贾拉姆；小麦育种；冬春杂交；慢锈性

Remembering the Life and Legacy of Sanjaya Rajaram

Xing Qingqing[1]　Wang Xin[2,3]　Hans Joachim Braun[3]　Marten van Ginkel[4]　He Zhonghu[2,3]

1 International Office，Beijing Institute of Technology；
2 Crop Science Institute，Chinese Academy of Agricultural Sciences；
3 International Maize and Wheat Improvement Center（CIMMYT）；
4 International Center for Agricultural Research in the Dry Area（ICARDA）

Abstract：Dr. Rajaram was engaged in international wheat breeding for more than 50 years. Based on Dr. Borlaug's international wheat breeding achievements，Dr. Rajaram made further major contributions in four aspects by leading CIMMYT.（1）Launched a large scale of spring×winter wheat crossing，and breakthrough was achieved in enhancing yield potential with a wide range of adaptability and resistance to major diseases；（2）Employed slow-rusting concept in wheat breeding program and varieties with durable resistance were later developed and used by farmers worldwide；（3）Developed breeding tools and screening methods for tolerance to acid soil and drought stress；（4）Released 481 varieties in 51 countries during 1976-2001，with annual sowing acreage of 60 million ha in developing countries. He also

原文发表在《麦类作物学报》，2021，41（6）：653-657。

made outstanding contributions in training scientists both in less developed countries as well as those in international organizations such as CIMMYT and ICARDA.

Keywords：Sanjaya Rajaram；Wheat breeding；Spring and winter cross；Slow rusting

20 世纪 60 年代兴起的绿色革命在为解决全球粮食问题作出重要贡献的同时，使得一批甘于奉献的育种家将追求科技进步和服务发展中国家的人道主义精神传播到国际农业界。历史上最具影响力的小麦育种家之一，世界粮食奖得主、国际玉米小麦改良中心（Centro Internacional de Mejoramientode Maizy Trigo，CIMMYT）小麦项目第二代学科带头人桑贾亚·拉贾拉姆（Sanjaya Rajaram，同事同行都亲切叫他 Raj，中文简称为拉贾）博士因培育高产抗病广适新品种，为提高世界小麦生产水平，减少全球饥荒做出了杰出贡献，也为中国小麦育种和人才培养做出了重要贡献。1972 年以来，拉贾拉姆先后担任 CIMMYT 小麦育种课题组主持人、小麦项目主任及国际干旱地区农业研究中心基因资源项目主任等职务。1976—2001 年，他先后培育了 481 个高产抗病优质广适小麦新品种，在 51 个国家推广种植 6,000 万公顷。他培育的品种使小麦绿色革命的辉煌从 1980 年代延续到 21 世纪初，使全球小麦总产量增加了 2 亿多吨。除此之外，拉贾拉姆引领的小麦育种研究突破了意识形态和国界差异。由于他领导得力，获得全球同行信任，显著扩大了全球小麦协作网络，使得小麦品种资源的收集、试验和交流渠道更加畅通，加速了高产小麦品种培育和全球推广，使全球小麦产量的提高速度领先于人口增长。

拉贾拉姆的贡献被世界各国广泛认可，他的成功经历被誉为印度农村穷苦孩子成长为世界农业巨人的传奇故事。1992—2015 年，他获得了印度、墨西哥、中国、土耳其、美国、英国、澳大利亚等 10 多个国家的 40 多个奖项和荣誉称号。其中代表性的奖项包括获得 2014 年世界粮食奖，2015 年被授予印度海外人士的最高荣誉——普拉瓦西·巴拉蒂亚·萨姆曼奖，中国政府友谊奖，由美国农学会和美国作物学会共同颁发的会长奖，伊朗颁发的赫瓦里兹米国际奖，危地马拉颁发的奎萨尔勋章。2021 年 2 月 17 日，拉贾拉姆因罹患新冠肺炎在墨西哥不幸逝世，享年 78 岁。CIMMYT 为纪念他为世界小麦育种做出的杰出贡献，将在墨西哥托卢卡的小麦实验站更名为"桑贾亚·拉贾拉姆实验站"（另一实验站于 2012 年更名为诺曼·布劳格实验站）。本文谨回顾拉贾拉姆博士的

生平、育种成就、对中国小麦育种的贡献及其高尚品格，以纪念这位世界小麦育种先驱。

1. 生平概况

拉贾拉姆 1943 年出生并成长于印度北方邦瓦拉纳西附近一个小村庄的普通家庭，是村里为数不多受教育的孩子之一。他勤奋学习，通过奖学金资助在印度戈勒克布尔的大学获得农学学士学位；之后在新德里的印度国家农业研究所师从著名农学家史瓦密纳坦（M. S. Swaminathan）等人，获得植物遗传育种学硕士学位；在悉尼大学获得植物育种学博士学位之后，他在印度进行了六个月的博士后研究。1969 年，拉贾拉姆加入 CIMMYT，开始在诺贝尔和平奖获得者、"绿色革命之父"诺曼·布劳格的团队中工作，先是进行博士后研究，两年后晋升为遗传学家岗位。1972 年拉贾拉姆年仅 29 岁时就被布劳格任命为 CIMMYT 小麦育种课题组负责人。1996 年，拉贾拉姆升任为 CIMMYT 卓越科学家（50 年来共 8 人），并被任命为国际小麦项目主任。

拉贾拉姆带领 CIMMYT 团队进行高产抗病小麦新品种研究，领导了国际小麦生产的第二次绿色革命，一直持续到 2003 年。他从 2005 年开始担任国际干旱地区农业研究中心（International Center for Agricultural Research in the Dry Areas，ICARDA）生物多样性和综合基因管理项目主任，于 2008 年正式退休。退休后，他担任 CIMMYT 和 ICARDA 特别科学顾问，并成功创办了小麦育种公司。近年来，拉贾拉姆博士还担任墨西哥西北部亚克流域基金会主席，通过非营利基金会植树造林并用本地树木重新造林，从而改善墨西哥索诺拉州的环境。

2. 国际育种贡献

2.1　育种方法创新

拉贾拉姆从 1969 年到 2021 年一直从事小麦育种工作，在布劳格工作的基础上，他善于开拓创新，实现了 CIMMYT 小麦育种全球化，他对育种技术和方法的主要贡献可概括为以下四个方面。一是通过冬春

杂交大幅度提高产量和适应性。他是大规模开展小麦冬春杂交育种的先驱，通过国际合作实现了小麦产量、抗病性和适应性的突破。二是成功培育持久抗性品种。国际上有关持久抗性的学术研究始于20世纪50年代，但缺乏有目的的育种应用。拉贾拉姆利用中等或微效慢锈的兼抗基因如 $Lr34/Yr18/Sr57/Pm38$，成功培育兼抗型成株抗性或持久抗性品种（即一个基因兼抗三种锈病和白粉病等，且抗性不因小种变异而丧失），减少小种特异性主效抗性基因使用，培育的"慢锈性"小麦品种 Pavon 76 等在世界各地推广。三是建立了小麦抗逆育种方法。通过与巴西和美国合作，形成了品种耐酸性土壤筛选技术，明确其遗传基因，将耐酸性土壤特性转入高产小麦品种中，提高了产量潜力和稳定性；通过与西亚、北非和ICARDA合作以及开发利用人工合成小麦等，显著提高了耐旱性和适应性。四是开发国际小麦育种信息系统（International Wheat Information System）。该系统不仅方便全球遗传育种信息交流，还扩大了全球小麦评价测试系统，有助于筛选产量潜力高、抗病性和抗逆性优良的小麦新品种，实现了从CIMMYT种质资源库-小麦育种课题组-国际试验圃的标准化和信息化管理。他与同事和学生合作发表学术论文419篇，其中119篇在国际期刊发表。

2.2 品种国际影响

据CIMMYT资料，1976—2001年间，拉贾拉姆先后培育了481个高产抗病优质广适的新小麦品种。通过冬春杂交培育的高产广适组合 Veery（组合代号为CM33027，KVZ/BUH/KAL/BB，CIMMYT只命名组合，不审定或命名品种）在30个国家共审定了60个品种，表现高产抗病节肥，叶子持绿，耐热节水，适应性非常广泛，在水浇地和旱地表现都特别优异。1985年该组合在全世界年种植面积1,000万～2,500万公顷，其中巴基斯坦的PAK 81约600万公顷，墨西哥80%的小麦面积为Veery组合品种。他用Veery做亲本育成的PBW-343（即Attila姊妹系）于1995年在印度审定，年最大种植面积超过700万公顷，大面积比对照增产10%～15%，使印度小麦总产量提高了20%～25%。Veery及其后代被誉为小麦的第二次绿色革命。又如他培育的第一个品种Pavon 76兼抗条锈、叶锈、秆锈和白粉病、且面包加工品质优质，还是节水的慢锈性品种，在澳大利亚等12个国家审定推广，曾是巴基斯坦和墨西哥的主栽

品种。Bobwhite作为小麦转基因标准型品种，对生物技术发展产生了巨大影响，并获得生产专利，在9个国家审定推广。拉贾拉姆培育的小麦品种（包括做亲本育成的）在全世界发展中国家年种植约6,000万公顷（约2,000万公顷为CIMMYT品种，4,000万公顷做亲本育成品种），仅南亚就约3,000万公顷。另据保守估计，在发达国家如澳大利亚和美国的种植面积也多达2,500多万公顷。

2.3 ICARDA和自创公司

拉贾拉姆2005年开始担任位于叙利亚的ICARDA生物多样性和综合基因项目主任，2008年正式退休后，继续担任ICARDA高级科学顾问到2013年。在ICARDA工作期间，他带领团队建立了冬小麦育种课题组，优化完善了大麦育种课题组，显著提升了ICARDA的育种水平。他的育种团队与西亚、北非地区的同事和ICARDA合作，培育了在干旱条件下表现优良的小麦品种，为改善非洲、中东和中亚干旱地区农民的生活提供了品种技术支撑。

拉贾拉姆酷爱小麦育种工作，他说小麦育种是他唯一的爱好。他退休后自筹资金成立并运营了资源种子国际公司，继续致力于小麦品种改良的研究应用。公司虽然条件简陋而且规模很小，仅有12人，在土卢卡和墨西哥西北部仅各有5公顷试验地（每年约200个杂交组合），但育成的品种表现特别突出，公司的所有育种材料和品系及其试验资料于2014年被先正达收购。

拉贾拉姆在CIMMYT、ICARDA和自创的资源种子国际公司三个阶段都取得了巨大成功，国际同行说他走到哪里就把成功带到哪里，被誉为小麦育种奇才。他目标明确，敢于面对工作挑战，在不同岗位都善于学习和合作，总能找到实现目标的科学可行方法和具体实施办法。

3. 对中国的贡献

3.1 建立抗赤霉病穿梭育种项目

拉贾拉姆负责建立了中国-CIMMYT抗赤霉病穿梭育种国际合作项目。赤霉病一直是我国长江流域的主要病害。20世纪80年代初正值改革开发的初期，他意识到提高中国小麦育种水平和国际化的战略重要性，与中国合作不仅有利于中国，而且将推动中国小麦种质资源独特性状研究利用，从而提高全球小麦产

量。他提出了中国-CIMMYT小麦抗赤霉病穿梭育种项目建议，旨在将CIMMYT小麦的高产潜力与赤霉病抗病性相结合，为中国培育优良品种。1988年，中国-CIMMYT小麦穿梭育种合作协议正式签署，主要内容包括种质交换与双边穿梭育种，人员培训与互访，信息交流。中方参加单位包括中国农业科学院、江苏省农业科学院、四川省农业科学院和黑龙江省农业科学院，另外河北农业科学院、河南农业科学院、山东农业科学院、陕西省农业科学院等与CIMMYT总部和CIMMYT土耳其冬麦项目种质交流与人员培训及互访，春麦区的新疆、宁夏、甘肃、内蒙古和湖北、云南等省区农业科学院持续获得CIMMYT各类国际圃，并参加人员培训。CIMMYT种质对提高我国小麦的产量、抗条锈病和加工品质发挥了重要作用。

除此之外，全世界的科学家都通过中国-CIMMYT抗赤霉病穿梭育种项目受益。我国向CIM-MYT提供了500多份主栽品种和高代品系，对提高CIMMYT小麦的抗赤霉病、抗印度黑穗病和改进灌浆特性发挥了关键作用，宁麦号、武汉号、上海号等品系不仅在CIMMYT成功利用，还在美国、加拿大等国抗赤霉病育种中发挥了核心作用。

3.2　成立CIMMYT中国办事处

在拉贾拉姆的大力推动下，CIMMYT中国办事处于1997年正式成立。经过20多年的发展，与中国农业科学院等国内科研院校深入合作，合作建立了农业农村部-CIMMYT玉米小麦联合实验室、CIMMYT-河南小麦玉米联合研究中心和CIMMYT国际小麦赤霉病鉴定试验站，并派6名科学家在华全职工作，其中1人被评为CIMMYT卓越科学家，1人入选外专千人计划，1人入选国家百千万人才工程，1人获国家自然科学基金优青项目，2人入选外国高端人才（A类）。7名国内知名农业专家或管理人员担任CIMMYT理事，现有3名中国人在CIM-MYT总部任高级科学家。

CIMMYT中国办事处的成立直接推动中国小麦育种研究深入发展。双方合作建立了兼抗型成株抗性育种新方法，发掘验证50个基因特异性标记，并在美国、澳大利亚等22个国家和CIMMYT、ICARDA等广泛应用，合作发表论文300多篇。从20世纪80年代中期至今，国内引进各类小麦种质9万多份次，包括育成品种（系）、农家种、近缘种、遗传群体和

近等基因系。将筛选的15,000份可能有利用价值的优异资源交国家和地方种质库长期保存。国内用CI-MMYT亲本育成小麦品种300多个，累计推广6.5亿亩，其中邯6172、绵农4号、克丰6号和宁春4号等分别成为黄淮地区、四川、黑龙江和西北春麦区的主栽品种，用人工合成小麦育成的川麦42成为西南地区的突破性品种，8个品种先后获得国家科技进步奖，山东农业科学院和中国农业科学院获2007年国际农业磋商组织（Consultative Group on International Agricultural Research，CGIAR）合作奖。举办国际会议等30多次，累计参会人数超过5,000人次。CI-MMYT获2016年度中华人民共和国国际科学技术合作奖，8名CIMMYT科学家获中国政府"友谊奖"，8人获省（自治区）政府颁发的国际合作奖。

3.3　小麦育种人才培养

拉贾拉姆十分重视对中国科研人员的培养和能力建设。从20世纪80年代中期至2003年，CIMMYT先后资助了50名中国科学家参加6-8个月的中长期培训班，136人次参与小麦穿梭育种项目，15人次参加合作研究项目，50余人次参加国际会议。到目前为止，赴CIMMYT小麦项目参观学习的中国学者和管理人员超过1,000人次，350名中国科研人员赴CIMMYT总部学习、参加培训或合作研究，学成归国后85人晋升为研究员或教授，5人晋升为国内农业科研院校高层管理人员。联合培养研究生48人，其中2人获珍妮·布劳格女性科学家奖。拉贾拉姆先后20次访问我国，为中国与CIMMYT搭建起了合作桥梁，实现了双方在小麦研究、国际合作、人才培养等方面的互利共赢。

4. 品格楷模

拉贾拉姆深受发展中国家同行的爱戴和尊敬。CIMMYT的使命是为发展中国家服务，建立并保持互相尊重的合作关系至关重要。他为人随和谦逊，正直诚恳，乐于助人，与发展中国家的同行和行政领导建立了良好关系，这是他取得非凡成就的重要原因。他在百忙之中认真友好接待来自世界各国的访问人员，无论是高级官员、资深科学家，还是一般访客，他总是一视同仁，尊重和关注每一位客人。在工作繁忙的情况下，他常常牺牲周末，邀请他们共同品尝墨西哥佳肴；还为同行颁发奖励或者证书，以表彰他们

的勤奋工作。在 CIMMYT 和 ICARDA 工作期间，拉贾拉姆非常重视国家育种项目的人员培养，培训了来自 90 个发展中国家的 700 多名科学家，指导了 22 名硕士和博士研究生的学位论文，帮助世界各地的科研团队在具有世界最高影响力水平的层面上开展工作。

拉贾拉姆以身作则且具有远见卓识。他是 CIMMYT 尊重和帮助员工与同事的楷模，信任是他管理和使用员工的基本原则。他用切实的行动而不是豪言壮语激励着一代代科学家发挥最大潜力，突破自我，寻求卓越。从 1980 年到 2020 年期间，CIMMYT 和 ICARDA 的骨干育种家都是他培养的。在重要科学问题上，他坚持真理，与高级管理层就全球战略问题进行过多次辩论，也曾经过深思熟虑之后，直接向伙伴国家的国家领导人提出过坚定的反对意见和建议。

拉贾拉姆爱护贤才且善于用人。在人才使用过程中给予年轻科学家充分的成长空间，又时时在田间给予指导，教他们与小麦对话。他总对年轻人说 "I am always available（我永远有时间）"。他擅长与经验丰富的工作人员一起为科学研究设置新挑战。CIMMYT 中国办事处的成长和发展也得益于他的直接指导和启迪。

拉贾拉姆博士虽然离开了我们，但他高尚的品格和科学精神永远影响国际小麦育种界！

❖ 参考文献

Dubin H J，Braun H-J，Singh R，Kohli M，2021. In Memoriam——SANJAYA RAJARAM，CSA News，in press.

Govindan V，2015. MR. GOLDEN GRAIN The Life and Work of the MAHARAJA of WHEAT Dr. Sanjaya Rajaram，Bloomsbury Publishing India Pvt. Ltd. ，New Delhi，India.

Jikun Huang，Cheng Xiang and Yanqing Wang，2015. The impact of CIMMYT wheat germplasm on wheat productivity in China. Mexico，D. F. ：CGIAR Research Program on Wheat.

Jikun Huang，Cheng Xiang and Yanqing Wang，2016. Hidden value of CGIAR training program for national research capacity，a case study of CIMMYT's impact on China's wheat R&D productivity. Mexico，D. F. ：CGIAR Research Program on Wheat.

中麦 895 高产稳产优质特性遗传解析

张　勇[1]　阎　俊[2]　肖永贵[1]　郝元峰[1]　张　艳[1]　徐开杰[2]　曹双河[1]　田宇兵[1]

李思敏[1]　闫俊良[2]　张赵星[2]　陈新民[1]　王德森[1]　夏先春[1]　何中虎[1,3]

[1] 中国农业科学院作物科学研究所，北京 100081；

[2] 中国农业科学院棉花研究所，河南安阳 455000；

[3] CMMYT 中国办事处，北京 100081

摘要：解析中麦 895 高产潜力、广泛适应性、综合抗性及其优良品质性状机理，有助于为新品种培育提供理论和方法指导。用中麦 871/中麦 895 重组自交系和扬麦 16/中麦 895 双单倍体 2 个群体的 QTL 定位研究和区域试验、生产试验示范资料对中麦 895 遗传特性及其生产中表现进行了分析。中麦 895 具有高产、稳产、抗病、优质 4 个方面优良特性：（1）携带 2 个控制分蘖角度和 5 个控制旗叶夹角主效 QTL，株型紧凑，旗叶小且直立，为单位面积穗数多（640 个·m^{-2}）提供保障；携带矮秆基因 *Rht2* 和 *Rht24*，植株矮（75 cm），茎秆弹性好，抗倒伏能力强。（2）含 4 个高千粒重与灌浆速率 QTL 和 5 个抗旱基因，春化基因组成 *vrn-A1x*、*Vrn-B1a* 和 *vrn-D1w*，灌浆速率快，千粒重高（48 g）高且稳定，根系活力好，水肥利用效率高，叶功能期长，耐高温和耐晚播能力强。（3）携带 1 个全生育期抗条锈病主效 QTL，2 个成株期抗白粉病 QTL 和 1 个抗赤霉病 QTL，综合抗病性好，中感条锈和白粉病，赤霉病病穗率低。（4）含低分子量麦谷蛋白亚基 *Glu-A3d* 等优质基因，面团筋力中等，颜色亮黄，面条和馒头品质优良；酚酸含量高（748 μg·g^{-1}）且稳定，营养价值高。本研究为黄淮麦区高产广适新品种培育提供了重要经验和理论支撑。

关键词：普通小麦；中麦 895；高产潜力；广泛适应性；抗病性

Characterization of Wheat Cultivar Zhongmai 895 with High Yield Potential，Broad Adaptability，and Good Quality

ZHANG Yong[1]，YAN Jun[2]，XIAO YongGui[1]，HAO YuanFeng[1]，ZHANG Yan[1]，

XU KaiJie[2]，CAO ShuangHe[1]，TIAN YuBin[1]，LI SiMin[1]，YAN JunLiang[2]，

ZHANG ZhaoXing[2]，CHEN XinMin[1]，WANG DeSen[1]，

XIA XianChun[1]，HE ZhongHu[1,3]

[1] Institute of Crop Sciences，Chinese Academy of Agricultural Sciences，Beijing 100081；

原文发表在《中国农业科学》，2021，54（15）：3158-3167。

基金项目：中国农业科学院重大任务（CAAS-ZDRW202002）、中国农业科学院创新工程

联系方式：张勇，E-mail：zhangyong05@caas.cn。通信作者何中虎，E-mail：zhhecaas@163.com

[2] Institute of Cotton Research, Chinese Academy of Agricultural Sciences, Anyang 455000, Henan;
[3] CIMMYT-China Office, Beijing 100081

Abstract: Characterization of leading cultivars will provide crucially important information for cultivar development. The objective of this research is to characterize high yield potential, broad adaptation, good disease resistance and stress tolerance, as well as good quality in wheat cultivar Zhongmai 895. The dataset of two populations of Zhongmai 871/Zhongmai 895 recombinant inbred lines and Yangmai 16/Zhongmai 895 doubled haploids, as well as regional yield and pilot trail related with Zhongmai 895 were used in this research. The high yield potential of Zhongmai 895 was largely due to the increased spike number which could easily reach $640 \cdot m^{-2}$, guaranteed by erect plant type with small leaves, and short plant height around 75 cm related to the outstanding lodging resistance, on the basis of two QTL for tiller angle, five QTL for leaf angle, and combination of semidwarfing genes *Rht2* and *Rht24*. It was characterized with high and stable thousand grain weight (48 g) related to the fast grain filling rate which provided excellent tolerance to high temperature during grain filling period, contributing to the perfact performance in late sowing environment, together with the contribution from high water and fertilizer use efficiency, high activity of root system, and slow leaf senescence, on the basis of four QTL for high thousand grain weight and grain filling rate, five genes for drought tolerance with vernalization gene combination of *vrn-A1x*, *Vrn-B1a*, and *vrn-D1w*. The performance of the broad adaptability was further contributed by the good resistance to stripe rust, powdery mildew and Furisium head blight in the production, due to the presence of one major QTL for stripe rust resistance, two QTL for slow powdery mildew, and one new QTL for Furisium head blight resistance. Zhongmai 895 was one of the few cultivars performing good dual qualities for Chinese dry white noodles and steamed bread with stable medium gluten strength, bright and yellow flour colour across environments in the Yellow-Huai River Valleys Winter Wheat Region, with *Glu-A3d*, and performed high nutrition quality with stable and high phenolic acid concentration of $748 \mu g \cdot g^{-1}$. This research provides very important information and experience for developing new cultivars with high yield potential and broad adaptability.

Keywords: common wheat; Zhongmai 895; yield potential; broad adaptability; disease resistance

0 引言

高产、稳产、抗病抗逆、优质小麦新品种的培育和推广对保障中国粮食特别是居民的口粮安全意义重大。黄淮冬麦区南片是中国第一大小麦产区,包括河南省(除信阳市和南阳南部部分地区以外)平原灌区、江苏和安徽两省沿淮及淮河以北地区、陕西省关中灌区,播种面积和总产分别约占全国的 38% 和 43%。该区处于南方和北方麦区的过渡地带,随着全球气候变化加剧,小麦生育期间冬春干旱、春季低温霜冻、灌浆中后期倒伏、干热风等灾害频发,小麦—玉米轮作及秸秆还田和旋耕面积扩大等因素使赤霉病、根腐病和茎基腐病等有持续加重发生的趋势。小麦育种的主要目标包括高产、抗病、抗逆、优质,在冬春两季应具有较好的抗寒性、抗旱性,灌浆期应具有耐高温与抗倒伏能力,并对条锈、叶锈、白粉、纹枯等常发病害和赤霉病、根腐病和茎基腐病表现中感或以上的抗性,生产上迫切需要高产稳产、矮秆抗倒、耐热抗病且面条和馒头品质较好的新品种[1-2]。

自 2000 年以来,笔者逐步将育种工作的重点由北部冬麦区转向黄淮麦区,与位于河南省安阳市中国农业科学院棉花研究所合作,建立了面向黄淮麦区的小麦育种试验站。针对该麦区的上述目标,建立了高产优质耐热育种技术体系,并集成应用于株型、粒重和品质改良,育成的中麦 895 株型紧凑、旗叶小且直立、粒重高且稳定,表现高产稳产、水肥高效、抗逆抗病、面条和馒头品质优良,2012 年通过国家审定(国审麦 2012010),已成为河南北中部、安徽北部和陕西关中地区的主栽品种。本研究采用育种、生理、栽培、谷物化学与分子标记相结合的方法，对中麦 895 的上述优良特性进行了较为全面的解析，目的是为培育高产稳产新品种提供理论支撑和经验。

1　高产特性解析

中麦 895 的高产潜力主要表现在以下 3 个方面：（1）区域试验中比对照平均增产 4.6%；51 点次 47 点增产，增产点率 92%（表 1）。（2）河南省和陕西省多地将中麦 895 作为高产创建主要品种，曾 2 次创造陕西省最高单产纪录，高达 11 730 kg·hm⁻²（表 2）。（3）大面积生产示范表现突出，容易达到 8 950 kg·hm⁻²。如陕西省泾阳县 2013 年和三原县 2013 年及 2014 年 6.6 hm² 平均单产分别为 9 411、9 429 和 10 086 kg·hm⁻²，三原县 2014 年 666 hm² 平均单产 8 950 kg·hm⁻²。

1.1　历史品种分析

从区域试验结果来看，中麦 895 株型紧凑、叶片较小且直立，穗数平均较对照周麦 18 多 11.9%；株高平均 74.3 cm，抗倒伏能力较强；后期叶功能期长，千粒重高，平均 48.3 g（表 1）。为进一步分析中麦 895 的高产机制，2013—2015 年在周口市和郑州市两点研究了黄淮南片 20 世纪 50 年代至今代表性历史品种产量及产量因子、生理性状的变化规律，中麦 895 产量居 26 份参试品种首位，显著高于 20 世纪 90 年代之前育成的所有品种及之后育成的豫麦 13，高达 8 906 kg·hm⁻²，且其生物量、花期叶面积指数、千粒重均显著高于豫麦 13、豫麦 49、百农 AK58、周麦 18 及其母本周麦 16 等品种（表 3）[3]，而 20 世纪 90 年代后育成的其他半冬性品种间产量差异不显著。豫麦 13 为该区 20 世纪 90 年代突破性主栽品种，豫麦 49 和周麦 18 分别为 2000—2005 年和 2007 年至今的国家区域试验对照品种，百农 AK58 是近 15 年黄淮南片推广面积最大的主栽品种。生物学产量和千粒重分别较上述 4 个主栽品种高 10% 和 20%，是中麦 895 实现高产的主要原因。

1.2　株型性状遗传解析

TIAN 等[4]定位了株高基因 *Rht24*，可使株高平均降低 7.0 cm，并成功开发了 3 个基因特异性标记。分子标记检测表明，携带 2 个矮秆基因 *Rht2* 和 *Rht24*，是中麦 895 表现矮秆的遗传基础[4]；中麦 895 的母本周麦 16 也含有这两个矮秆基因，其父本荔垦 4 号含有 *Rht2*。

以中麦 871/中麦 895 重组自交系（RIL）为材料，通过 3 年 8 个环境试验，定位了 2 个来自中麦 895 的株型紧凑 QTL——*QTA. caas-1AL* 和 *QTA. caas-5DL*，可解释表型变异的 9.0% 和 26.2%；标记检测表明，这两个 QTL 均来自荔垦 4 号；已将效应较大的 QTL 命名为 *TaTAC1-5D*，位于 *QTA. caas-5DL* 标记区间，开发了相关 KASP 标记，验证了 *TaTAC1-5D* 对分蘖角度的影响[5]。以扬麦 16/中麦 895 双单倍体（doubled haploid，DH）群体为材料，通过 2 年 4 个环境试验，定位了 5 个来自中麦 895 的叶型直立 QTL，可解释表型变异的 5.2%—9.9%[6]。株型紧凑、旗叶夹角较小、叶型直立是中麦 895 单位面积穗数较多的主要原因。

1.3　灌浆相关特性分析

为进一步明确中麦 895 千粒重高且稳定的机制，选取黄淮麦区 14 份代表性主栽品种和苗头品系进行分析，中麦 895 的粒重和各时期灌浆速率均显著高于其双亲及黄淮北片和南片对照品种济麦 22 和周麦 18（图 1）[7]。选取北方冬麦区 13 份代表性主栽品种，2013—2015 年安阳市、石家庄市和衡水市 3 点大田正常温度和花后 14 d 塑料大棚覆盖热处理试验表明，热处理灌浆中后期平均最高温度 42.1℃（比大田高 2.1℃）条件下，中麦 895 的千粒重和平均灌浆速率与我国知名耐热品种京冬 8 号相当（表 4），两者均显著高于河北省区域试验对照品种衡 4399；而京冬 8 号和衡 4399 的耐热性居北方冬麦区主栽品种前列[8-9]。因此，在多种环境下均能保持快的灌浆速率，是中麦 895 千粒重和产量潜力较高的主要原因之一。

表 1 中麦 895 黄淮南片区域试验产量表现

Table 1 Yield performance of Zhongmai 895 in the Southern part of the Yellow and Huai River Valleys Winter Wheat regional trials

类型 Trialtype	年份 Year	品种 Cultivar	株高 PH (cm)	穗数 SN (m^{-2})	穗粒数 KNS	千粒重 TKW (g)	产量 GY (kg·hm^{-2})	±CK (%)	增产点次 YIS
区域试验 RYT	2010—2011	中麦 895 Zhongmai 895	71b	678a	29.8b	47.1a	8 820a	5.1	17 (19#)
		周麦 18 Zhoumai 18 (CK)	74a	603b	34.3a	44.7b	8 392b		
	2011—2012	中麦 895 Zhongmai 895	75b	651a	29.7b	45.8a	7 590a	4.4	16 (17)
		周麦 18 Zhoumai 18 (CK)	79a	575b	32.3a	44.9b	7 275b		
生产试验 RPT	2011—2012	中麦 895 Zhongmai 895	75b	648a	29.3b	48.0a	7 665a	4.3	14 (15)
		周麦 18 Zhoumai 18 (CK)	77a	569b	32.0a	44.9b	7 350b		
平均 Average		中麦 895 Zhongmai 895	74.3b	641a	29.2b	48.3a	7 973a	4.6	
		周麦 18 Zhoumai 18 (CK)	78.0a	573b	32.5a	46.0b	7 561b		

RYT：区域试验；RPT：生产试验；PH：株高；SN：穗数；KNS：穗粒数；TKW：千粒重；GY：产量；YIS：增产点次。不同字母表示在 $P=0.05$ 水平差异显著。#：试验点总数。下同

RYT：Regional yield trial；RPT：Regional pilot trial；PH：Plant height；SN：Spike number；KNS：Kernel number per spike；TKW：Thousand kernel weight；GY：Grain yield；YIS：Yield increase site. Different letters are significantly different at $P=0.05$. #：Total site of the yield trial. The same as below

表 2 中麦 895 黄淮南片高产创建产量表现

Table 2 Yield performance of Zhongmai 895 in farmer fields in the Southern part of the Yellow and Huai River Valleys Winter Wheat Region

地点 Location	年份 Year	面积 Area (hm^2)	产量 Yield (kg·hm^{-2})	备注
陕西省三原县 Sanyuan，Shaanxi	2012	0.30	10562	实产，本省最高单产纪录 Harvest yield，provincial record
陕西省三原县 Sanyuan，Shaanxi	2014	0.20	10862	实产，本省最高单产纪录 Harvest yield，provincial record
河南省滑县 Huaxian，Henan	2015	0.20	11714	实产 Harvest yield
河南省新乡县 Xinxiang，Henan	2016	0.34	11730	实产 Harvest yield

表 3 黄淮南片部分代表性品种产量性状表现

Table 3 Yield and yield components of representative cultivars in the Southern part of the Yellow and Huai River Valleys Winter Wheat Region

品种 Cultivar	株高 PH (cm)	生物量 Biomass (kg·hm^{-2})	花期叶面积指数 LAI	穗数 SN	穗粒数 KNS	千粒重 TKW (g)	产量 GY (kg·hm^{-2})	收获指数 HI
豫麦 13 Yumai 13	84.6a	15 302bc	4.92b	757a	37b	41.6c	7 806bc	0.38ab
豫麦 49 号 Yumai 49	77.7bc	15 281bc	4.59cd	645bc	36bc	48.0b	8 390b	0.39a
周麦 16 Zhoumai 16	73.5c	15 765b	4.35d	574c	40a	54.1a	8 433b	0.38ab
周麦 18 Zhoumai 18	79.2b	14 572c	4.74bc	602bc	34c	45.7bc	8 546b	0.35c
百农 AK58 Bainong AK58	69.3d	15 290bc	4.83bc	630bc	36bc	46.8bc	7 563c	0.38a
中麦 895 Zhongmai 895	77.9bc	17 367a	5.27a	586c	37b	56.0a	8 906a	0.37b

LAI：花期叶面积指数；HI：收获指数。数据来自 GAO 等[3]

LAI：Leaf area index；HI：Harvest index. Data from GAO et al [3]

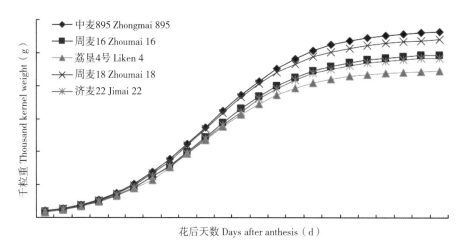

图1 黄淮麦区部分代表性水地主栽品种籽粒灌浆曲线

Fig. 1 Grain filling rate curves of representative leading cultivars in the Yellow and Huai River Valleys Winter Wheat Region

表4 大田正常温度与大棚增温热处理条件下中麦 895 与抗热对照品种千粒重和平均灌浆速率比较

Table 4 TKW and average grain-filling rate of Zhongmai 895 and check under normal and heat stress environments

品种 Cultivar	大田 Normal			热处理 Heat stress		
	千粒重 TKW (g)	平均灌浆速率	Va (g·1000-kernel⁻¹·d⁻¹)	千粒重 TKW (g)	平均灌浆速率	Va (g·1000-kernel⁻¹·d⁻¹)
中麦 895 Zhongmai 895	51.0a		1.27a	46.7a		1.15a
京冬 8 Jingdong 8	50.0a		1.26a	48.7a		1.21a
衡 4399 Heng 4399	44.0b		1.05b	38.5b		0.97b

Va：平均灌浆速率。数据来自韩利明等[8]和苗永杰[9]

Va：Average grainfilling rate. Data from HAN et al[8] and MIAO [9]

多年多点观察表明，中麦 895 籽粒饱满，灌浆速率快，粒重高且稳定。以中麦 871/中麦 895 RIL 群体为材料，通过 3 年 10 个环境试验，在 1AL、2BS、3AL 和 5B 染色体定位到同时与高千粒重和灌浆速率相关且正向效应均来自中麦 895 的稳定 QTL，可解释表型变异的 4.4%—17.3% 和 4.6%—13.0%，相关 KASP 标记得到验证[10]；1AL、2BS 和 5B 上 QTL 均来自荔垦 4 号，3AL 上 QTL 来自周麦 16。进一步分析发现，1AL 上控制千粒重、粒宽和平均灌浆速率的 QTL 在不同遗传背景中效应稳定且对穗粒数影响较小，在育种中具有重要利用价值，与其紧密连锁的 KASP 标记可用于分子标记辅助育种。

在建立表型鉴定平台的基础上，将高光谱遥感技术快速有效用于生物量、叶面积指数、叶绿素含量等参数分析[11]。以扬麦 16/中麦 895 DH 群体为材料，通过 2 年 2 点试验，定位了 4 个来自中麦 895 的持绿性相关 QTL，可解释表型变异的 7.2%—20.3%，其中 1 个位于 4DS 染色体上的 QTL 同时与叶片持绿性和产量相关，并参与调控幼苗干重及根苗比，表明根系生物量主效 QTL 同时参与调控灌浆期叶绿素含量，而灌浆中后期较高的叶绿素含量和花期叶面积指数以及较低的冠层温度，为提高中麦 895 灌浆速率和延缓叶片衰老提供了重要保障[12]，进一步检测表明，4DS 上 QTL 来自荔垦 4 号。

由此可见，灌浆速率快、叶功能期长、粒重和生物量高是中麦 895 高产的重要原因。

2 稳产特性解析

中麦 895 的稳产特性主要表现为水肥利用效率高，耐晚播能力强，综合抗病性好。为进一步了解其对水肥的反应程度并明确其机理，2016—2018 年在安阳市和新乡市设置 W1（全生育期不灌溉）、W2（越冬水）、W3（越冬水＋拔节孕穗水）和 W4（越冬水＋拔节孕穗水＋灌浆水）共 4 个梯度灌溉处理试验，并在每个处理施用基肥过磷酸钙（750 kg·hm⁻²）和硫酸钾（180 kg·hm⁻²）基础上，设置 T1（基肥纯氮 0 kg）、T2（基肥纯氮 120 kg）、T3（基肥 120 kg＋拔节期追施纯氮 60 kg）、T4（基肥纯氮 120 kg＋拔节期追施纯氮 120 kg）共 4 个梯度氮肥处理试验，对中麦 895 与百农 AK58 进行系统比较。

2.1 水分利用效率

在全生育期灌溉 1 水、2 水和 3 水条件下，中麦 895 的穗粒数、千粒重、产量和水分利用效率均高于百农 AK58[12]。随着灌溉次数的增加，中麦 895 在少于 3 水条件下的水分利用效率呈显著增加趋势，而其 3 水条件下的水分利用效率显著低于 2 水；百农 AK58 的水分利用效率则呈持续下降趋势。这与 2 个品种的每平方米穗数、穗粒数、千粒重的变化情况密切相关，且这 2 个品种在 2 水和 3 水条件下的每平方米穗数、穗粒数、千粒重和产量差异不显著，但均显著高于 0 水和 1 水。与百农 AK58 相比，随着灌溉次数的增加，中麦 895 的穗粒数和千粒重增加较快，这是其产量和水分利用效率高的主要原因（表 5）[12]。

分子标记检测表明，中麦 895 含 CWI-4A（Hap-4A-C）、CWI-5D（Hap-5D-C）、TaMoc-A1（Hap-H）、TaSST-A2（SST-A2a）和 TaSST-D1（SST-D1a）共 5 个抗旱基因，为其节水特性奠定基础；CWI-4A（Hap-4A-C）和 TaMoc-A1（Hap-H）来自荔垦 4 号，周麦 16 和荔垦 4 号均携带 CWI-5D（Hap-5D-C）、TaSST-A2（SST-A2a）和 TaSST-D1（SST-D1a）。在正常自然降水条件下，2014 年安徽省阜阳市颍上县全生育期未浇水 68 hm² 连片种植示范田平均产量 9 655 kg·hm⁻²；2015 年蓝田县三里镇旱地全生育期喷灌 2 次 666 hm² 实打验收平均产量 8 325 kg·hm⁻²，创该省小麦旱作节水攻关高产记录；在仅浇 1 水条件下，2016 年河南省新乡县小冀镇全生育期仅浇 1 水 166 hm² 连片种植示范田实收平均产量 8 700 kg·hm⁻²，比当地主栽品种百农 AK58 增产 11.6%。上述示范结果进一步验证了中麦 895 的高水分利用效率特性。

2.2 肥料利用效率

在所有氮肥处理条件下，中麦 895 的穗数、千粒重、产量和氮肥利用效率均高于百农 AK58；且随着氮肥施用量的增加，2 个品种的氮肥农学效率、吸收效率和利用效率均呈显著下降趋势，中麦 895 的穗粒数不断降低，千粒重和产量先快速增加，在氮肥施用量为 180 kg·hm⁻² 时达到最高，之后显著降低。粒重显著增加是中麦 895 取得高产的主要原因（表 6）[12]。

在上述分析基础上，采用溶液培养法，在不同氮磷梯度处理条件下，扬麦 16/中麦 895 DH 群体的苗期根系形态、结构及其生物量分析表明，中麦 895 苗期根系在高氮或低磷环境下根长较长、根直径及根表面积较大[12]，其田间表现根系活力强、氮肥吸收利用效率高。以该 DH 群体为材料，定位 3 个与氮调控相关且正向效应来自中麦 895 的稳定 QTL，分别位于 2BS、4DS 和 7BL 染色体，可解释根干重和根尖数、幼苗干重和根苗比、幼苗干重和根表面积表型变异的 8.8% 和 8.9%、5.3% 和 19.0%、5.1% 和 7.8%；定位 6 个与磷调控相关且正向效应来自中麦 895 的稳定 QTL，分别位于 3AS、4BS、6BL、6DS、7AS 和 7AL 染色体，可解释幼苗及根系干重和根苗总重、根表面积和根苗比、根长和幼苗干重、根表面积和幼苗干重、根长和根表面积、根长和根尖数表型变异的 6.0% 和 7.7%、8.1% 和 17.7%、6.8% 和 8.6%、4.9% 和 8.4%、10.8% 和 11.9%、4.6% 和 6.2%[12]。

2.3 耐晚播特性

自 2012 年通过黄淮南片国家审定以来，中麦 895 累计推广面积约 300 万 hm²，已成为河南、安徽和陕西省的主栽品种，并在江苏北部大面积示范推广，其中年最大面积 70.8 万 hm²，目前推广面积居全国第 3 位，表现广泛适应性，这与其春化基因组成 vrn-A1x、Vrn-B1a 和 vrn-D1w 密切相关，为其耐晚播特性奠定基础。2017—2018 年在播种期遭遇连阴雨、较常年推迟播种 2 个月、越冬期持续低温、晚春低温霜冻严重条件下，河南省洛阳市宜阳县 66 hm² 示范方平均产量 5 520 kg·hm⁻²，创造了当地晚播条件下大面积高产典型，为实现小麦—玉米周年高产水平奠定基础。

表 5 不同灌溉次数对中麦 895 和百农 AK58 产量及水分利用效率相关性状的影响

Table 5 Effect of irrigation times on yield and water use efficiency related traits of Zhongmai 895 and Bainong AK58

灌溉处理 Irrigation	穗数 SN (m^{-2})	穗粒数 KNS	千粒重 TKW (g)	产量 Yield (kg · hm^{-2})	水分利用效率 WUE (kg · mm^{-1})
W1	410c[#]/382c[S]	26.7c/28.7c	45.9c/44.5c	4 365c/4 775c	14.7c/16.1a
W2	460b/433b	30.1b/29.9bc	49.5b/45.8bc	5 806b/5 666b	16.2b/15.9ab
W3	541a/519a	33.0a/31.4ab	51.6ab/47.1ab	7 115a/6 565a	17.1a/15.7ab
W4	565a/566a	33.2a/32.3a	52.1a/48.5a	7 481a/6 995a	15.7bc/14.7b

W1：全生育期不灌溉；W2：越冬水；W3：越冬水+拔节孕穗水；W4：越冬水+拔节孕穗水+灌浆水（单次灌溉水量 60 mm，基肥纯氮、过磷酸钙、硫酸钾分别为 180、750 和 180 kg · hm^{-2}）。WUE：水分利用效率。[#]：中麦 895；[s]：百农 AK58

W1：No irrigation；W2：Overwintering irrigation；W3：Overwintering and jointing irrigation；W4：Overwintering, jointing, and grain filling irrigation （1 time irrigation 60 mm water；base fertilizer 180 kg · hm^{-2} N, 750 kg · hm^{-2} $CaH_4P_2O_8$, and 180 kg · hm^{-2} K_2SO_4）. WUE：Water use efficiency. [#]：Zhongmai895；[s]：Bainong AK58

表 6 不同氮肥梯度对中麦 895 和百农 AK58 产量及氮肥利用效率相关性状的影响

Table 6 Effect of N fertilizer levels on yield and nitrogen use efficiency related traits of Zhongmai 895 and Bainong AK58

氮肥处理 Nitrogen	平米穗数 SN (m^{-2})	穗粒数 KNS	千粒重 TKW (g)	产量 Yield (kg · hm^{-2})	氮肥农学效率 NAE (kg · kg^{-1})	氮肥吸收效率 NUE (%)	氮肥利用效率 NUtE (kg · kg^{-1})
T1	558a/512a	29.9a/33.9a	45.3b/43.7b	6 220c/6 008b	0/0	0/0	0/0
T2	574a/522a	28.8ab/31.3ab	53.1a/48.4a	6 536b/6 062b	5.4/4.9	0.24/0.21	27.2/24.4
T3	552a/482b	27.4bc/30.3bc	54.1a/49.6a	6 918a/6 186ab	3.8/3.2	0.17/0.15	19.2/16.6
T4	496b/471b	26.0c/26.0c	51.3a/49.4a	6 634b/6 370a	2.8/2.6	0.14/0.12	13.8/13.2

T1：基肥过磷酸钙、硫酸钾分别为 750 和 180 kg · hm^{-2}；T2：基肥过磷酸钙、硫酸钾、纯氮分别为 750、180 和 120 kg · hm^{-2}；T3：基肥过磷酸钙、硫酸钾、纯氮分别为 750、180 和 120 kg · hm^{-2}，拔节期追施纯氮 60 kg · hm^{-2}；T4：基肥过磷酸钙、硫酸钾、纯氮分别为 750、180、120 kg · hm^{-2}，拔节期追施纯氮 120 kg · hm^{-2}。NAE：氮肥农学效率；NUE：氮肥吸收效率；NUtE：氮肥利用效率

T1：Base fertilizer 750 kg · hm^{-2} $CaH_4P_2O_8$ and 180 kg · hm^{-2} K_2SO_4；T2：Base fertilizer 750 kg · hm^{-2} $CaH_4P_2O_8$, 180 kg · hm^{-2} K_2SO_4, and 120 kg · hm^{-2} N；T3：Base fertilizer 750 kg · hm^{-2} $CaH_4P_2O_8$, 180 kg · hm^{-2} K_2SO_4, and 120 kg · hm^{-2} N, and topdressing fertilizer 60 kg · hm^{-2} N at jointing stage；T4：Base fertilizer 750 kg · hm^{-2} $CaH_4P_2O_8$, 180 kg · hm^{-2} K_2SO_4, and 120 kg · hm^{-2} N, and topdressing fertilizer 120 kg · hm^{-2} N at jointing stage. NAE：N agronomy efficiency；NUE：N uptake efficiency；NUtE：N utilization efficiency

2.4 综合抗病性好

黄淮南片区域试验抗病性接种鉴定表明，中麦 895 中感条锈和白粉病，赤霉病病穗率低。为进一步了解其在生产中的抗病性，于 2013—2015 年将中麦 895 分别种植于四川省成都市和安徽省濉溪县，在自然发病较重条件下鉴定条锈和白粉病抗性，确认其中抗条锈和白粉病；2012、2013 和 2016 年黄淮南片赤霉病重发年份对河南省安阳市、开封市和南阳市中麦 895、百农 AK58 及国家区域试验对照品种周麦 18 的赤霉病病穗率调查表明，中麦 895 的 3 年多点平均病穗率仅为百农 AK58 的 49.9%，与周麦 18 相当（个人资料）。

为进一步分析中麦 895 的抗病分子机制，以扬麦 16/中麦 895 DH 群体为材料，通过 3 个环境，定位了 1 个正向效应来自中麦 895 且稳定的条锈病抗性新

QTL，位于 2AL 染色体，可解释表型变异的 54.3%，已开发紧密连锁 CAPS 标记 Yr-2AL-CAPS[13]；该抗病 QTL 来自其父本荔垦 4 号。

通过 2 年 5 个环境接种鉴定试验，定位了 4 个正向效应来自中麦 895 且稳定的白粉病成株抗性 QTL，分别位于 2DL、4BS、6BL 和 7BS 染色体，可解释表型变异的 7.2%—15.2%，其中 6BL 和 7BS 可能是新的白粉病抗性位点[14]；连锁 KASP 标记检测表明，6BL 和 7BS 上 QTL 分别来自周麦 16 和荔垦 4 号。

通过 2 年 5 个环境试验，定位 2 个正向效应来自中麦 895 且稳定的赤霉病抗性 QTL，分别位于 4AS 和 5AL 染色体，可解释表型变异的 5.6% 和 9.0%[15]；QFhb.caas-5AL 连锁标记 InDel _ AX-89588684 检测表明，该抗病 QTL 来自其父本荔垦 4 号。

从以上分析可知，中感条锈和白粉病、赤霉病病

穗率低为中麦895的高产稳产广适特性提供了保障。

3 优质特性解析

3.1 面条和馒头兼用品质优

多年多点品质测试表明，中麦895平均籽粒蛋白质含量中等（12.2％，14％湿基），籽粒硬度（65.7，SKCS指数）和出粉率（平均为73.5％）较高，吸水率、形成时间、稳定时间、拉伸面积、延展性、最大抗延阻力分别为61.1％、2.9 min、4.3 min、40 cm²、161 mm和163 BU，属于硬质中筋类型[16-17]。中麦895面粉和面片L*值较高、a*值和b*值较低，颜色亮黄；食味和黏弹性、光滑性、软硬度等性状均显著好于其母本周麦16，与黄淮南片优质品种郑麦366和优质面条商业对照雪花粉相近（表7）。分子标记检测表明，中麦895的高低分子量麦谷蛋白亚基组成分别为1、7＋9、2＋12、Glu-A3d、Glu-B3j，并携带基因 Ppo-A1a、Ppo-D1a、TaPod-A1a 和 TaZds-A1b、TaZds-D1b、PSY-A1a 和 PSY-B1c；Ppo-D1a、TaPod-A1a 和 PSY-B1c 均来自其父本荔垦4号，这是其多酚氧化酶活性中等、过氧化物酶活性较低、面粉和面片颜色亮黄、面条品质优的主要原因。中麦895是23个在生产中加工品质能稳定达到优质中筋小麦标准的品种之一（万富世，个人通讯）。

中麦895馒头品质较好，主要表现为体积大、比容高，结构与雪花粉相近，外形、表面光滑性、色泽、压缩张弛性和总分均明显好于其母本周麦16（表7）。在同一环境下，中麦895的面条和馒头品质均较好，显著优于周麦16，这与农业农村部谷物品质监督检验测试中心（北京）的结果一致。

表 7 中麦895和对照品种的面条和馒头加工品质比较

Table 7 Comparison of dry white Chinese noodle and steamed bread qualities of Zhongmai 895 and the Check

类型 Type	特性 Trait	中麦895 Zhongmai 895	周麦16 Zhoumai 16	郑麦366 Zhengmai 366	雪花粉 Xuehua
面条 Noodle	色泽 Color（15）	9.2	9.0	9.6	10.5
	表面状况 Appearance（10）	6.3	6.5	6.8	7.0
	软硬度 Firmness（20）	14.3	12.5	13.8	14.0
	黏弹性 Viscoelasticity（30）	21.4	18.4	21.4	21.0
	光滑性 Smoothness（15）	11.4	9.2	9.8	10.5
	食味 Flavor（10）	8.3	7.8	7.8	7.0
	总分 Total score（100）	70.8	63.3	68.9	70.0
馒头 Steamed bread	比容 Specific volume（20）	18	11.8	16.3	15.0
	外形 Shape（10）	6.0	4.3	6.8	8.0
	表面光滑 Smoothness（10）	7.5	5.0	7.3	8.0
	表面色泽 Skin color（10）	4.5	4.0	6.8	10.0
	压缩张弛性 Stress relaxation（35）	29.0	21.5	29.0	35.0
	结构 Structure（15）	11.8	7.9	12.4	12.0
	总分 Total score（100）	76.8	54.4	78.4	88.0

资料来源于孔欣欣等[16]和赵德辉等[17] Data from KONG et al[16] and ZHAO et al[17]

3.2 营养价值高

酚酸可提高抗氧化能力，通过保护细胞免受损伤，显著降低糖尿病、心脑血管和直肠癌等疾病的发病率[18-19]。2008—2010年对37份北方冬麦区代表性品种自由酚酸、结合酚酸以及总酚酸含量分析表明，中麦895总酚酸含量2年度均较高（748 μg·g⁻¹），可在生产中加以利用，有利于保障人体健康[20]。

4 育种经验分享与工作设想

4.1 组合选配成功经验

黄淮冬麦区南片是中国最重要的主产麦区，自1950年至今，该麦区小麦产量年遗传增益约为0.60％—1.05％[3,21-22]，其中1970—1993年产量年提高72.1 kg·hm⁻²，单位面积穗数、生物学产量和收获指数的显著提高及株高的显著降低是获得较高产量潜力遗传增益的主要原因[21]；1981—2008 年得益于千粒重和收获指数的显著提高，产量年提高 51.3 kg·hm^{-2}[22]。GAO 等[3]进一步研究表明，1950—2012 年该麦区产量年提高 57.5 kg·hm^{-2}，单位面积穗数和千粒重的显著提高发挥了关键作用，同时还与株高的显著降低及生物学产量、收获指数和叶绿素含量的有效改良有关。值得注意的是，在 20 世纪 90 年代后育成的 11 个半冬性品种中，中麦 895 的产量显著高于豫麦 13，而其他品种与豫麦 13 的产量差异均不显著。

河南省周口市农业科学院育成的周麦系列品种因具有产量潜力高、穗大、结实性好、穗粒数多等优点，已成为当地的当家品种和骨干亲本，其中，以周麦 16 表现较为突出，综合农艺性状好，株高 75 cm 左右，株型较紧凑，中抗条锈和白粉病，籽粒较大。但该品种存在以下几个缺点：（1）单位面积穗数偏少；（2）籽粒灌浆速率偏慢，后期不耐高温，年度间饱满度不稳定；（3）中感纹枯病，高感赤霉病，后期遇雨易穗发芽；（4）籽粒品质一般，面食品加工品质差。因此，通过适当增加单位面积穗数，延长叶功能期，提高籽粒灌浆速率，并改良其赤霉病抗性水平，改善加工品质，以满足国内对面条和馒头等主要面食品品质的要求，是进一步提高其产量潜力和适应性的重要途径。荔垦 4 号是陕西省大荔县农垦中心育成的高代品系，突出特点是灌浆速率快，成熟落黄好，籽粒饱满，千粒重 43 g 左右，综合抗病性较好，且对赤霉病有一定抗性，面条品质优良。因此，笔者于 2002 年组配周麦 16/荔垦 4 号杂交组合，针对黄淮南片对高产稳产、矮秆抗倒、抗病抗热特性的要求，育种早代通过晚播和灌浆期塑料大棚增温增加耐热性选择压力，田间接种条锈菌优势混合小种，重点对株型、穗数、籽粒灌浆和条锈病抗性进行选择，中选单株要求矮秆、株型紧凑、穗数多、抗病、旗叶小且衰老慢、落黄好，考种时注重籽粒大小、饱满度和外观品质。系谱法选择至 F$_5$ 代，中选 18 个株系，千粒重介于 45.5—56.5 g，对照品种周麦 18 则为 46.8 g。

2007—2008 年初级产量比较试验时，将中选 18 个株系同时在河南和陕西省多个鉴定点种植，田间对其繁茂性、穗部性状、落黄、综合抗病性和产量进行筛选，并于收获后进行考种和蛋白质含量、籽粒硬度、和面时间等加工品质参数测试，获得 5 个优异品系；于 2008—2009 年分别种植于安阳、焦作、许昌、

周口、驻马店和徐州、阜南、咸阳、济南、高邑、临汾，通过多点产量比较试验，进一步对其抗寒性、粒重稳定性、综合抗病性、产量水平和适应性进行鉴定，获得 2 个优异品系，分别命名 08CA95 和 08CA75（即中麦 895 和中麦 875）。中麦 895 表现株型紧凑，叶片直立且旗叶夹角小，矮秆抗倒伏，熟期比对照周麦 18 略早，穗数多，籽粒大且粒重稳定，中感条锈和白粉病，田间赤霉病病穗率低，在黄淮南片预试中，产量比对照周麦 18 平均增产 8.1%，列 50 个参试品种第 1 位；由于在 2010—2011 年第 1 年区域试验中比对照周麦 18 平均增产 5.1%，表现优异，继续参加该组区域试验，并同步生产试验，于 2012 年通过国家审定（表 1）。中麦 875 表现株型适中，株高和抗倒性中等，熟期比周麦 18 略晚，穗数中等，穗大，籽粒大且粒重稳定，中感条锈和白粉病，于 2014 年通过河南省审定；由于其田间表现赤霉病病穗率低，之后推荐参加湖北省鄂北组和中作联合体黄淮南片区域试验，并分别于 2019 和 2020 年通过湖北省和国家黄淮南片审定；从而基本实现了预期育种目标。

4.2 中麦 895 亲本利用方向

中麦 895 试验示范和生产中表现穗数多、灌浆速率快、耐热性突出，粒重高且稳定，高产稳产，中感条锈和白粉病、赤霉病病穗率低，面条品质优；但遗憾的是，还存在以下 4 个方面问题：（1）在组合选配之初，倒春寒和纹枯病在安阳市的育种环境基本没有发生，使得这两个品种的抗倒春寒能力和纹枯病抗性中等；（2）没有收集到周麦系列品种的穗发芽抗性资料，安阳市也不适合对材料的穗发芽抗性进行鉴定筛选，使得这两个品种的抗穗发芽能力一般；（3）在世代材料考种时没有对中麦 895 和中麦 875 的黑胚进行严格筛选，致使其籽粒外观品质表现虽然较好，但仍存在一定程度的黑胚率；（4）在高世代鉴定时没有设置高肥力或高密度试验，未对其抗倒伏性进行严格把关，致使这两个品种在产量水平达到 11 250 kg·hm^{-2} 以上时，存在一定倒伏风险，且中麦 875 的茎秆强度还略弱于中麦 895；上述 4 个因素对中麦 895 的进一步推广应用带来一定影响，这在育种中利用中麦 895 作亲本时应加以注意。

笔者已经定位了中麦 895 相关株型、粒重和灌浆速率、条锈和赤霉病等病害抗性 QTL，并发掘了相应紧密连锁分子标记，在保留中麦 895 灌浆速率快、

耐热性突出基础上，选用黄淮北片国审品种山农17为父本，与中麦895选系杂交，以进一步改良其抗寒性特别是抗倒春寒能力和抗倒伏性以及综合抗病性，增强其后代育成品系的适应区域，其中山农17具有发育稳健、冬春抗寒性突出，抗倒伏能力强，综合抗病性好等优点。杂交组合经系谱法选育至 F_3 后，将中选单株种于高邑进行抗寒性和综合农艺性状选择，并将高代品系种在北京进行抗寒性鉴定，育成中麦30，表现高产稳产广适、冬季和春季抗寒性好、抗倒伏能力强、穗数中等、结实性好（平均穗粒数37粒）、千粒重46 g左右、籽粒黑胚率较低、外观商品性好、纹枯病较轻，综合抗病性较好。该品种已于2020和2021年通过黄淮南片和北片国家审定，完成北部冬麦区试验程序，有望于2021年底报国家审定，成为第1个通过上述3个麦区国家审定的品种。此外，还以中麦895作为耐热骨干亲本，进一步改良黄淮麦区品种，先后育成中麦5215和GY16004，均表现籽粒大且饱满，灌浆速率快，已完成黄淮南片试验程序，有望于2021年底报国家审定。

4.3　高产优质广适新品种培育初步设想

在气候变化日益加剧和赤霉病等重大病害频发、重发面积不断扩大的背景下，在保证较高的产量水平基础上，既要提高肥水利用效率，降低其投入，还要改善品质，以提高产业竞争力[2]。总之，高产广适、水肥高效、抗病抗逆新品种培育至关重要，在保障产量的基础上，提高育成品种的品质。就育种技术和方法而言，用于小麦品质和抗病性的分子育种技术已经成熟并在国内外大范围应用[23]，全基因组选择在水稻等作物中已开始取得阶段性进展[24]，但对于产量、水肥利用效率、抗寒、抗旱、耐热等性状来说，由于表型鉴定的复杂性，近期内发掘效应较大、育种家可用的基因并开发其功能标记的难度相当大，高产高效、广适抗逆小麦新品种的培育仍需主要依靠常规育种技术。因此，实际育种过程应重点关注以下4个环节：（1）在充分了解亲本主要优缺点的基础上，抓好组合配置；（2）育种早代 F_2—F_4 针对所在麦区的主要育种目标，进行抗寒、耐热、抗旱和抗病等性状的选择与鉴定，需特别注意材料鉴定的可靠性；（3）分离世代材料针对抗病性和品质等目标性状，及时采用 *Yr18*、*Glu-D1d* 等育种可用的功能标记，结合分子标记和表型进行鉴定；（4）高世代材料加大多点鉴定力度，注重广泛适应性、抗病性、抗逆性和产量潜力

筛选。

考虑到黄淮麦区小麦播种总面积和产量均占全国小麦的60％以上，除在安阳市和高邑县与当地合作建立育种站外，增加了新乡育种站，并在原有鉴定点基础上，分别在黄淮北片的石家庄市和济宁市、淄博市，黄淮南片的盐城市、淮安市和宿州市、阜阳市，以及新疆泽普县新增了鉴定点，形成了较完善的品种比较试验网络，这是近年来育种工作取得较快进展的重要原因。自中麦895育成以来，在黄淮南片和北片陆续审定了中麦875、中麦30、中麦578和中麦23、中麦30、中麦578等品种，并有望育成一批适应区域覆盖黄淮南片和北片的高产高效、优质广适新品种，为全国小麦生产和育种技术发展做出贡献。

❀ 参考文献
References

[1] 庄巧生. 中国小麦品种改良及系谱分析. 北京：中国农业出版社，2003.
ZHUANG Q S. Chinese Wheat Breeding and Pedigree Analysis. Beijing：China Agriculture Press，2003. （in Chinese）

[2] 何中虎，庄巧生，程顺和，于振文，赵振东，刘旭. 中国小麦产业发展与科技进步. 农学学报，2018，8（1）：99-106.
HE Z H, ZHUANG Q S, CHENG S H, YU Z W, ZHAO Z D, LIU X. Wheat production and technology improvement in China. Journal of Agriculture，2018，8（1）：99-106. （in Chinese）

[3] GAO F M, MA D Y, YIN G H, RASHIEED A, DONG Y, XIAO Y G, XIA X C, WU X X, HE Z H. Genetic progress in grain yield and physiological traits in Chinese wheat cultivars of southern Yellow and Huai Valley Winter Wheat Zone since 1950. Crop Science，2017，57：760-773.

[4] TIAN X, WEN W, XIE L, FU L, XU D, FU C, WANG D, CHEN X, XIA X, CHEN Q, HE Z, CAO S. Molecular mapping of reduced plant height gene Rht24 in bread wheat. Frontiers in Plant Sciences，2017，8：1379.

[5] ZHAO D H, YANG L, XU K J, CAO S H, TIAN Y B, YAN J, HE Z H, XIA X C, SONG X Y, ZHANG Y. Identification and validation of genetic loci for tiller angle in bread wheat. Theoretical and Applied Genetics，2020，133：3037-3047.

［6］ XU K J，ZHANG Y，TIAN Y B，YAN J L，ZHANG Z X，XIAO Y G，XIA X C，HE Z H，YAN J. QTL mapping for flag leaf angle in common wheat. Euphytica，2020（in Press）.

［7］ 苗永杰，阎俊，赵德辉，田宇兵，闫俊良，夏先春，张勇，何中虎. 黄淮麦区小麦主栽品种粒重与籽粒灌浆特性的关系. 作物学报，2018，44（2）：252-259.
MIAO Y J，YAN J，ZHAO D H，TIAN Y B，YAN J L，XIA X C，ZHANG Y，HE Z H. Relationship between grain filling parameters and grain weight in leading wheat cultivars in the Yellow and Huai Rivers Valley. Acta Agronomic Sinica，2018，44（2）：252-259.（in Chinese）

［8］ 韩利明，张勇，彭惠茹，乔文臣，何明琦，王洪刚，曲延英，何中虎. 从产量和品质性状的变化分析北方冬麦区小麦品种抗热性. 作物学报，2010，36（9）：1538-1546.
HAN L M，ZHANG Y，PENG H R，QIAO W C，HE M Q，WANG H G，QU Y Y，HE Z H. Analysis of heat resistance for cultivars from North China Winter Wheat Region by yield and quality traits. Acta Agronomic Sinica，2010，36（9）：1538-1546.（in Chinese）

［9］ 苗永杰. 高温胁迫对小麦籽粒灌浆特性及主要品质性状的影响［D］. 北京：中国农业科学院，2016.
MIAO Y J. Effect of heat stress on grain filling and major quality traits of common wheat［D］. Beijing：Chinese Academy of Agricultural Sciences，2016.（in Chinese）

［10］ YANG L，ZHAO D H，MENG Z L，XU K J，YAN J，XIA X C，CAO S H，TIAN Y B，HE Z H，ZHANG Y. Rapid QTL mapping for grain yield-related traits in bread wheat via SNP-based selective genotyping. Theoretical and Applied Genetics，2020，133：857-872.

［11］ YANG M，HASSAN M A，XU K，ZHENG C，RASHEED A，ZHANG Y，JIN X，XIA X，XIAO Y，HE Z. Assessment of water and nitrogen use efficiencies trough UAV-based multispectral phenotyping in winter wheat. Frontiers in Plant Science，2020，11：927.

［12］ 杨梦娇. 冬小麦中麦 895 节水抗旱相关生理机制及遗传特性研究［D］. 北京：中国农业科学院，2019.
YANG M J. Study on water saving and drought resistance related physiology mechanisms and genetic research of winter wheat Zhongmai 895［D］. Beijing：Xinjiang Agricultural University，2019.（in Chinese）

［13］ 朱展望. 利用全基因组连锁分析和关联分析定位小麦赤霉病抗性基因及分子标记开发［D］. 北京：中国农业科学院，2020.
ZHU Z W. Genome-wide linkage and association mapping of resistance genes to fusarium head blight and development of molecular markers in wheat［D］. Beijing：Chinese Academy of Agricultural Sciences，2020.（in Chinese）

［14］ XU X T，ZHU Z W，JIA O L，WANG F J，WANG J P，ZHANG Y L，FU C，FU L P，BAI G H，XIA X C，HAO Y F，HE Z H. Mapping of QTL for partial resistance to powdery mildew in two Chinese common wheat cultivars. Euphytica，2020，216：3.

［15］ ZHU Z，XU X，FU L，WANG F，DONG Y，FANG Z，WANG W，CHEN Y，GAO C，HE Z，XIA X，HAO Y. Molecular mapping of QTL for Fusarium head blight resistance in a doubled haploid population of Chinese bread wheat. Plant Disease，2020.（in Press）

［16］ 孔欣欣，张艳，赵德辉，夏先春，王春平，何中虎. 北方冬麦区新育成优质小麦品种面条品质相关性状分析. 作物学报，2016，42（8）：1143-1159.
KONG X X，ZHANG Y，ZHAO D H，XIA X C，WANG C P，HE Z H. Noodle quality evaluation of new wheat cultivars from northern China winter wheat regions. Acta Agronomic Sinica，2016，42（8）：1143-1159.（in Chinese）

［17］ 赵德辉，张勇，王德森，黄玲，陈新民，肖永贵，阎俊，张艳，何中虎. 北方冬麦区新育成优质品种的面包和馒头品质性状. 作物学报，2018，44（5）：697-705.
ZHAO D H，ZHANG Y，WANG D S，HUANG L，CHEN X M，XIAO Y G，YAN J，ZHANG Y，HE Z H. Pan bread and steamed bread qualities of novel-released cultivars in Northern Winter Wheat Region of China. Acta Agronomic Sinica，2018，44（5）：697-705.（in Chinese）

［18］ JACOBS D R，MEYER K A，KUSHI L H，FOLSOM A R. Whole grain intake may reduce risk of coronary heart disease death in postmenopausal women：The Iowa Women's Health Study. American Journal of Clinical Nutrition，1998，68：248-257.

［19］ MEYER K A，KUSHI L H，JACOB D J，SLAVIN J，SELLERS T A，FOLSOM A R. Carbohydrates，dietary fiber，incident type 2 diabetes mellitus in older women. American Journal of Clinical Nutrition，2000，71：921-930.

[20] ZHANG Y，WANG L，YAO Y，YAN J，HE Z H. Phenolic acid profiles of Chinese wheat cultivars. Journal of Cereal Science，2012，56：629-635.

[21] ZHOU Y，HE Z H，SUI X X，XIA X C，ZHANG X K，ZHANG G S. Genetic improvement of grain yield and associated traits in the Northern China Winter Wheat Region from 1960 to 2000. Crop Science，2007，47：245-253.

[22] ZHENG T C，ZHANG X K，YIN G H. WANG L N，HAN Y L，CHEN L，HUANG F，TANG J W，XIA X C，HE Z H. Genetic gains in grain yield，net photosynthesis and stomatal conductance achieved in Henan Province of China between 1981 and 2008. Field Crops Research，2011，122：225-233.

[23] LIU Y N，HE Z H，APPES R，XIA X C. Functional markers in wheat：Current status and future prospects. Theoretical and Applied Genecics，2012，125：1-10.

[24] CUI Y R，LI R D，LI G W，ZHANG F，ZHU T T，ZHANG Q F，ALI J，LI Z K，XU S Z. Hybrid breeding of rice via genomic selection. Plant Biotechnology Journal，2020，18：57-67.

耐热性和产量性状

黄淮麦区小麦主栽品种粒重与籽粒灌浆特性的关系

苗永杰[1]　阎　俊[2]　赵德辉[1]　田宇兵[1]　闫俊良[2]　夏先春[1]
张　勇[1,*]　何中虎[1,3]

[1] 中国农业科学院作物科学研究所/国家小麦改良中心，北京 100081；[2] 中国农业科学院棉花研究所，河南安阳 455000；[3] 国际玉米小麦改良中心（CIMMYT）中国办事处，北京 100081；
* 通信作者，E-mail：Zhangyong05@caas.cn.

摘要：研究粒重与籽粒灌浆特性的关系对提高小麦产量潜力和稳定性具有重要意义。采用 Logistic 方程，对 2012—2015 年连续 3 年度种植在河南安阳的 14 份黄淮麦区主栽品种和苗头品系的粒重及其籽粒灌浆特性研究表明，粒重和灌浆速率参数主要受基因型控制，灌浆持续时间主要受环境影响。不同粒重类型品种间平均灌浆速率、最大灌浆速率和各时期灌浆速率均存在显著差异，表现为高粒重＞中等粒重＞低粒重，灌浆持续时间则差异不显著。灌浆速率，特别是快增期灌浆速率快慢是造成品种间粒重差异的主要原因。粒重与所有灌浆速率参数均呈显著正相关（$P<0.001$），与快增期灌浆速率和平均灌浆速率的相关系数分别为 0.97 和 0.90，与灌浆持续时间相关不显著。建议采用平均灌浆速率对相关性状进行基因定位，以进一步改良黄淮麦区小麦品种的粒重。

关键词：普通小麦；粒重；品种粒重类型；平均灌浆速率

Relationship between Grain Filling Parameters and Grain Weight in Leading Wheat Cultivars in the Yellow and Huai Rivers Valley

MIAO Yong-Jie[1]，YAN Jun[2]，ZHAO De-Hui[1]，TIAN Yu-Bing[1]，
YAN Jun-Liang[1]，XIA Xian-Chun[1]，ZHANG Yong[1,*]，and HE Zhong-Hu[1,3]

[1] Institute of Crop Science / National Wheat Improvement Center, Chinese Academy of Agricultural Sciences (CAAS)，Beijing 100081，China；[2] Institute of Cotton Research, CAAS，Anyang 455000，Henan，China；[3] CIMMYT-China Office, c/o CAAS，Beijing 100081，China；* Corresponding author

Abstract：The knowledge on relationship between grain weight and grain-filling parameters is important for yield potential and stability improvement of common wheat. Logistic equation was used for fitting

原文发表在《作物学报》，2018，44（2）：260-267.

the grain-filling dataset from 14 leading cultivars and advanced lines, sown at Anyang, Henan province in three successive seasons from 2012 to 2015. The results showed that grain weight and all grain-filling rate (GFR) related parameters were mainly influenced by genotype, while grain-filling period related parameters were mainly influenced by environment. There was significant difference on all GFR parameters including the average and the highest GFR, and those in the three periods among cultivar groups based on grain weight, showing a trend of high-grain-weight cultivar > medium-grain-weight cultivar > low-grain-weight cultivar, whereas there was no significant difference for grain-filling period related parameters among the cultivar groups. GFR, especially in the fast increase period, was the major factor that made the significant difference of grain weight among cultivars. Positive correlations between grain weight and all GFR related parameters were observed ($P < 0.001$), with the coefficients of 0.97 for GFR in the fast increase period and 0.90 for average GFR, whereas no significant correlations were found between grain weight and grain-filling period related parameters. Therefore, average GFR was proposed to be used in quantitative trait loci mapping to improve grain weight of wheat in the Yellow and Huai Rivers Valley.

Keywords: common wheat; grain weight; cultivar group based on grain weight; average grain-filling rate

黄淮麦区是我国最重要的小麦产区，播种面积和总产分别约占全国的 55% 和 60%，对保障粮食安全至关重要[1]。过去 60 余年，黄淮麦区小麦产量年遗传增益约为 0.48%～1.05%，其构成因素中千粒重年遗传增益较大，为 0.35%～0.51%[2]。虽然近 15 年产量遗传增益放缓，但粒重仍持续提高[3]，表明粒重改良是该麦区产量显著提高的关键因素。粒重主要受基因型控制，并受环境显著影响，粒重在产量构成因素中的遗传力最高[4-5]。已定位了大量粒重相关 QTL，位于 2A、4D、5B、6B、7B 和 7D 染色体上的 QTL 效应较大，单个位点可解释 4.8%～28.0% 的表型变异[6-8]。其中，$TaCwi$[9]、$TaGW2$[10]、$TaSus2$[11]、$TaCKX6$[12]、$TaSAPl$[13]、$TaGS1a$[14]、$TaGS-D1$[15]、$TaGASR7-A1$[16]、$TaGS5-3A$[17]、$6-SFT-A2$[18] 等多个相关基因已被克隆，可解释 4.8%～14.6% 的表型变异。

粒重由籽粒灌浆速率和持续时间决定[19-23]。灌浆速率主要受基因型控制，灌浆持续时间主要由特定地区的气候和耕作栽培制度决定[23-26]。有关灌浆速率和持续时间对粒重的贡献尚无定论，多数研究认为，在灌浆期偏短的地区，灌浆速率对粒重的贡献大于持续时间[26-30]。因此，育种工作的重点是提高灌浆速率。

灌浆速率是决定我国北部冬麦区小麦品种粒重最重要的参数[29]；黄淮麦区水地和旱地品种籽粒灌浆特性存在显著差异，旱地品种的灌浆速率高于水地品

种，但其籽粒灌浆时间较短[30]。这些研究多集中于描述具体品种的灌浆特性，对不同粒重类型品种间的籽粒灌浆特性可能因为工作量大而缺乏系列研究，且很少涉及黄淮麦区水地主栽品种的粒重与籽粒灌浆特性之间的关系。本研究通过分析品种间籽粒灌浆特性的差异，明确粒重与籽粒灌浆特征参数的关系，揭示灌浆速率和持续时间对粒重的相对重要性，旨在为黄淮麦区小麦品种的粒重改良提供理论依据。

1 材料与方法

1.1 品种及田间设计

选用黄淮麦区 14 份水地主栽品种和苗头品系（表 1），于 2012—2015 年度种植在河南安阳。良星 99 和周麦 18 分别是黄淮北片和南片冬麦区国家水地组区域试验的对照品种。济麦 22 和矮抗 58 是黄淮麦区近 10 年累计推广面积最大的两个主栽品种。郑麦 366、周麦 16、良星 66、周麦 27、存麦 1 号、中麦 895 和中麦 875 均为当前生产上的主推品种，中麦 871 和中麦 140 是本课题组新育成的苗头品系。荔垦 4 号是中麦 895、中麦 875 和中麦 871 的父本。

采用随机区组设计，2 次重复。小区面积 6 m²，6 行区，4 m 行长，行距 20 cm。试验地前茬玉米收获后秸秆还田，播种前底施复合肥（N、P_2O_5、K_2O 比例 17∶1∶17）750 kg·hm⁻² 和尿素 75 kg·hm⁻²，深翻。10 月 10 日前后播种，每公顷基本苗

210 万株。分别于越冬期、返青期、孕穗期和灌浆期灌溉 4 次，并结合越冬水和孕穗水分别追施尿素 112.5 kg·hm^{-2} 和 150 kg·hm^{-2}。返青期化学除草一次，抽穗扬花期"一喷三防"，其他管理措施同当地大田生产。

1.2 田间调查指标及方法

开花至成熟期间的天数即为灌浆持续时间。于开花初期从每个小区选取同一天开花，穗型、长势和大小基本一致，无病虫害的单茎 200 个挂牌标记。从花后第 6 天开始，每隔 6 d 取样一次，直至成熟收获。每次各小区取 15 穗，105℃杀青 15 min，65℃烘干 24 h，调查粒数，并用千分之一天平称重，计算千粒重。

以花后相对灌浆时间（x）为自变量，粒重（y）为因变量，采用 Logistic 方程 $y = k/(1+ae^{-bx})$ 对不同品种的籽粒灌浆进程进行拟合，绘制籽粒灌浆速率曲线，其中 a 和 b 为品种参数，k 为理论粒重。

采用一阶和二阶求导，计算最大灌浆速率（R_{max}，mg·grain^{-1}·d^{-1}）及其到达时间（T_{max}，

d）、渐增期持续时间（T_1，d）和灌浆速率（R_1，mg·grain^{-1}·d^{-1}）及其增重（W_1，g）、快增期持续时间（T_2，d）和灌浆速率（R_2，mg·grain^{-1}·d^{-1}）及其增重（W_2，g）、缓增期持续时间（T_3，d）和灌浆速率（R_3，mg·grain^{-1}·d^{-1}）及其增重（W_3，g）、平均灌浆持续时间（T，d）和灌浆速率（R_a，mg·grain^{-1}·d^{-1}）等参数。气压、温度、湿度、降水量、风速和日照等主要气象资料来自中国气象数据网（http：//cdc.cma.gov.cn/）河南安阳站"中国地面气候资料日值数据集"。

1.3 统计分析

采用 Statistical Analysis System（SAS9.2）软件，调用 PROC NLIN 程序进行灌浆数据的 Logistic 方程拟合。调用 PROC CLUSTER 程序，按 Ward 最小平方和法，依据品种粒重进行聚类分析。调用 PROC GLM 模型，粒重类型为固定效应，类内品种、年份内重复及其相关互作为随机效应，进行方差分析。调用 PROC CORR 程序进行相关分析。

表 1　14 份参试品种名称、系谱及其审定年份
Table 1　Names, pedigrees and released years of the 14 cultivars investigated

编号 Code	品种 Cultivar	系谱 Pedigree	审定年份 Year released
1	周麦 16 Zhoumai 16	豫麦 21/周 8425B Yumai 21/Zhou 8425B	2002
2	周麦 18 Zhoumai 18	内乡 185/豫麦 21 Neixiang 185/Yumai 21	2004
3	良星 99 Liangxing 99	济 91102/鲁麦 14//PH 85-16 Ji 91102/Lumai 14//PH 85-16	2004
4	矮抗 58 Aikang 58	周麦 11//豫麦 49/郑州 8960 Zhoumai 11//Yumai 49/Zhengzhou 8960	2005
5	郑麦 366 Zhengmai 366	豫麦 2/百泉 3199 Yumai 2/Baiquan 3199	2005
6	济麦 22 Jimai 22	935024/935106	2006
7	良星 66 Liangxing 66	济 91102/济麦 19 Ji 91102/Jimai 19	2008
8	周麦 27 Zhoumai 27	周麦 16/矮抗 58 Zhoumai 16/Aikang 58	2011
9	丰德存麦 1 号 Fengdec unmai 1	周 9811/矮抗 58 Zhou 9811/Aikang 58	2011
10	中麦 895 Zhongmai 895	周麦 16/荚垦 4 号 Zhoumai 16/Liken 4	2012
11	中麦 875 Zhongmai 875	周麦 16/荚垦 4 号 Zhoumai 16/Liken 4	2014
12	中麦 871 Zhongmai 871	周麦 16/荚垦 4 号 Zhoumai 16/Liken 4	/
13	中麦 140 Zhongmai 140	良星 99/矮抗 58 Liangxing 99/Aikang 58	/
14	荚垦 4 号 Liken 4	未知 Unknown	/

"/"表示未审定。"/" denotes that the lines has not been released yet.

2　结果与分析

2.1　籽粒灌浆特征参数基本统计量分析

千粒重及所有灌浆特征参数的变异范围均较大

（表 2）。参试品种的千粒重介于 41.8～53.8 g 之间，平均和最大灌浆速率分别为 1.15～1.50 mg·grain^{-1}·d^{-1} 和 1.94～2.63 mg·grain^{-1}·d^{-1}，渐增期、快增期和缓增期灌浆速率分别为 0.78～1.02、1.70～2.31 和 0.69～1.10 mg·grain^{-1}·d^{-1}，相应

时期的籽粒增重量分别为 9.2～12.1、25.2～33.3 和 9.2～12.1 mg·grain^{-1}，上述参数的品种间变异范围均大于年份。品种间灌浆持续时间、最大灌浆速率到达时间和渐增期持续时间的变异范围分别为 36.3～40.3、18.0～19.8 和 11.3～12.4 d，品种间变异范围均小于年份。品种间快增期和缓增期持续时间的变异范围分别为 12.7～15.5 和 11.1～13.9 d，品种间变异范围与年份相当。

表 2　14 份参试品种千粒重和籽粒灌浆特征参数均值及其变异范围

Table 2　Mean and variation of thousand-grain weight（TGW）and grain-filling parameters of the 14 cultivars investigated

灌浆参数 Grain-filling parameter	均值 Mean	总变异区间 Range of total variation	品种间变异区间 Range of cultivar variation	年份间变异 Range of year variation
千粒重 TGW（g）	47.2	38.3—59.1	41.8—53.8	45.2—51.1
平均灌浆持续时间 T（d）	38.4	34.0—43.0	36.3—40.3	36.6—41.6
平均灌浆速率 R_a（mg·grain^{-1}·d^{-1}）	1.28	1.07—1.57	1.15—1.50	1.25—1.31
最大灌浆速率到达时间 T_{max}（d）	19.0	14.9—22.2	18.0—19.8	16.8—21.0
最大灌浆速率 R_{max}（mg·grain^{-1}·d^{-1}）	2.30	1.72—2.87	1.94—2.63	2.23—2.38
渐增期持续时间 T_1（d）	11.9	9.2—14.2	11.3—12.4	10.4—13.2
渐增期灌浆速率 R_1（mg·grain^{-1}·d^{-1}）	0.88	0.72—1.09	0.78—1.02	0.84—0.93
渐增期增重量 W_1（mg·grain^{-1}）	10.4	8.0—13.6	9.2—12.1	9.7—11.5
快增期持续时间 T_2（d）	14.2	9.9—18.3	12.7—15.5	12.7—15.7
快增期灌浆速率 R_2（mg·grain^{-1}·d^{-1}）	2.02	1.51—2.52	1.70—2.31	1.96—2.08
快增期增重量 W_2（mg·grain^{-1}）	28.5	21.9—37.2	25.2—33.3	26.4—31.4
缓增期持续时间 T_3（d）	12.3	9.0—15.9	11.1—13.9	10.8—13.4
缓增期灌浆速率 R_3（mg·grain^{-1}·d^{-1}）	0.86	0.53—1.26	0.69—1.10	0.73—0.95
缓增期增重量 W_3（mg·grain^{-1}）	10.4	8.0—13.6	9.2—12.1	9.7—11.5

TGW：thousand-grain weight；T：average grain-filling-period；R_a：average grain-filling rate；T_{max}：days reaching the maximum grain-filling rate；R_{max}：maximum grain-filling rate；T_1：grain-filling pyramid period；R_1：grain-filling rate in T_1；W_1：grain weight accumulated in T_1；T_2：grain-filling fast increase period；R_2：grain-filling rate in T_2；W_2：grain weight accumulated in T_2；T_3：grain-filling slow increase period；R_3：grain-filling rate in T_3；W_3：grain weight accumulated in T_3.

2.2　籽粒灌浆特征参数方差分析

千粒重、所有籽粒灌浆速率及各时期籽粒增重量参数均受基因型和年份效应的显著影响，灌浆持续时间和缓增期持续时间的年份和基因型×年份效应、渐增期和快增期持续时间的基因型和年份效应显著（表 3）。千粒重、所有籽粒灌浆速率及各时期增重量的基因型效应较大，其次为年份效应，且基因型效应远大于基因型×年份效应，表明粒重、灌浆速率和增重量各参数主要受基因型控制。所有灌浆持续时间的年份效应较大，其次为基因型效应，且基因型效应均大于基因型×年份效应，表明灌浆持续时间主要受环境影响。

按千粒重将 14 个品种聚为 3 类，高粒重品种包括周麦 18、中麦 895 和中麦 875，中等粒重品种包括周麦 16、良星 99、济麦 22、存麦 1 号和中麦 140，低粒重品种包括矮抗 58、良星 66、郑麦 366、周麦 27、中麦 871 和荔垦 4。千粒重、最大灌浆速率到达时间、所有籽粒灌浆速率及其增重量参数均受粒重类型及类内基因型效应的显著影响，渐增期和快增期持续时间的类内基因型效应显著，灌浆持续时间、缓增期持续时间及其灌浆速率的类内基因型×年份效应显著。粒重的类型效应远大于类内基因型效应，类型×年份及类内基因型×年份效应不显著。所有灌浆速率参数及各时期籽粒增重量的类型效应均大于类内基因型效应，灌浆持续时间相关参数的类型效应不显著，表明不同粒重类型品种间的灌浆速率及各时期籽粒增重量均存在显著差异，灌浆持续时间则差异不显著。

表 3　14 份参试品种干粒重和籽粒灌浆特征参数方差分析

Table 3　Analysis of variance for TGW and grain-filling parameters of the 14 cultivars

变异来源 Source of variance	自由度 df	千粒重 TGW	平均灌浆 Average grain-filling		最大灌浆 The maximum grain-filling		渐增期 Grain weight accumulated in T_1			快增期 Grain weight accumulated in T_2			缓增期 Grain weight accumulated in T_3		
			T	R_a	T_{max}	R_{max}	T_1	R_1	W_1	T_2	R_2	W_2	T_3	R_3	W_3
基因型 Genotype (G)	13	1 225.4***	66.9	1.13***	30.1***	2.95***	9.62*	0.42***	73.8***	49.5***	2.27***	551.0***	49.4	1.13**	73.8***
类型 Cluster (C)	2	1 110.6***	17.1	1.02***	10.8***	2.16***	4.13	0.32***	70.4***	6.4	1.66**	525.5***	10.4	0.89**	70.4***
类内基因型 (G_c)	11	114.9*	49.8	0.11*	19.3***	0.79***	5.49*	0.10**	3.4*	43.1***	0.61***	25.4**	39.0	0.24*	3.4*
年份 Year (Y)	2	608.3***	408.8***	0.05**	235.5***	0.26***	100.77***	0.12***	48.8***	114.7***	0.20*	363.9***	102.6*	0.77**	48.7***
基因型×年份 G×Y	26	98.0	58.3***	0.06	7.0	0.36	3.94	0.04	2.8	14.9	0.28	21.3	48.2*	0.25*	2.8
类型×年份 C×Y	4	3.3	10.9	0.01	1.2	0.11	0.39	0.01	0.1	4.5	0.08	0.5	3.8	0.04	0.1
类内基因型×年份 G_c×Y	22	94.7	47.4***	0.05	5.8	0.25	3.55	0.03	2.8	10.3	0.20	20.8	44.5***	0.21*	2.8
年份内重复 Rep (Y)	3	0.5	0.1	0.00	0.2	0.01	0.17	0.00	0.1	0.1	0.01	0.4	0.2	0.00	0.1
误差 Error	39	98	0.1	0.06	7.4	0.37	5.20	0.03	4.1	14.5	0.29	30.8	17.0	0.16	4.1
基因型/基因型×年份 G/G×Y		12.5	1.1	11.0	4.3	8.2	2.4	13.8	1.5	3.3	8.1	1.5	1.0	4.5	1.5

TGW: 千粒重；T: 平均灌浆持续时间；R_a: 平均灌浆速率；T_{max}: 最大灌浆速率到达时间；R_1: 渐增期灌浆速率；T_1: 渐增期持续时间；R_2: 快增期灌浆速率；T_2: 快增期持续时间；W_2: 快增期增重；W_1: 渐增期增重；T_3: 缓增期持续时间；R_3: 缓增期灌浆速率；W_3: 缓增期增重。*、**和***分别表示在 0.05、0.01 和 0.001 概率水平显著。

TGW: thousand-grain weight; T: average grain-filling period; R_a: average grain-filling rate; R_{max}: maximum grain-filling rate; T_{max}: days reaching the maximum grain-filling rate; T_1: grain-filling fast increase period; R_2: grain-filling rate in T_2; W_2: grain weight accumulated in T_2; R_1: grain-filling pyramid period; T_1: grain-filling slow increase period; R_3: grain-filling rate in T_3; W_3: grain weight accumulated in T_3. *, **, and ***indicate significant at the 0.05, 0.01, and 0.001 probability levels, respectively.

2.3　粒重类型间籽粒灌浆特征参数分析

所有品种的籽粒饱满度均较好，表明灌浆正常。Logistic 方程决定系数（R^2）均在 0.99 以上，说明拟合方程可以有效描述籽粒灌浆进程。不同粒重类型品种灌浆进程均呈"S"型变化曲线，灌浆速率变化趋势呈单峰曲线，表现为慢—快—慢的特征（图 1），不同粒重类型品种间平均和最大灌浆速率及渐增期、快增期和缓增期灌浆速率及其籽粒增重量均存在显著

差异，表现为高粒重＞中等粒重＞低粒重，灌浆持续时间则差异不显著（表 4）。渐增期、快增期和缓增期持续时间分别约占整个灌浆期的 31.2%、36.9% 和 31.9%，其籽粒增重量则分别约占粒重的 21.1%、57.8% 和 21.1%。与渐增期和缓增期相比，快增期高、中、低粒重类型品种间灌浆速率和籽粒增重量差异较大。高粒重和中等粒重类型品种的最大灌浆速率到达时间显著晚于低粒重类型品种，其他灌浆持续时间相关参数类型间差异均不显著。

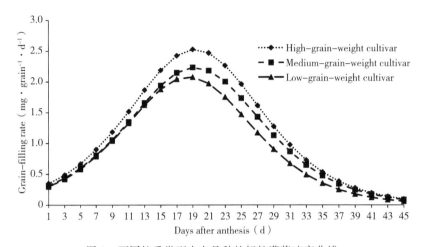

图 1　不同粒重类型小麦品种的籽粒灌浆速率曲线

Fig. 1 Grain-filling rate curve of different grain-weight wheat cultivars

表 4　不同粒重类型品种的籽粒灌浆特征参数
Table 4　Grain-filling parameters in different grain-weight cultivars

籽粒灌浆特征参数 Grain-filling parameter	高粒重品种 High-grain-weight cultivar	中等粒重品种 Medium-grain-weight cultivar	低粒重品种 Low-grain-weight cultivar
千粒重 TGW（g）	56.5 a	50.0 b	45.1 c
平均灌浆持续时间 T（d）	38.2 a	39.0 a	38.0 a
平均灌浆速率 R_a（mg·grain⁻¹·d⁻¹）	1.48 a	1.28 b	1.19 c
最大灌浆速率 R_{max}（mg·grain⁻¹·d⁻¹）	2.59 a	2.29 b	2.16 c
最大灌浆速率到达时间 T_{max}（d）	19.4 a	19.3 a	18.5 b
渐增期持续时间 T_1（d）	12.1 a	12.0 a	11.6 a
渐增期灌浆速率 R_1（mg·grain⁻¹·d⁻¹）	0.99 a	0.88 b	0.82 c
渐增期增重量 W_1（mg·grain⁻¹）	11.9 a	10.6 b	9.5 c
快增期持续时间 T_2（d）	14.1 a	14.4 a	13.9 a
快增期灌浆速率 R_2（mg·grain⁻¹·d⁻¹）	2.27 a	2.00 b	1.89 c
快增期增重量 W_2（mg·grain⁻¹）	32.6 a	28.9 b	26.1 c
缓增期持续时间 T_3（d）	11.7 a	12.5 a	12.5 a
缓增期灌浆速率 R_3（mg·grain⁻¹·d⁻¹）	1.05 a	0.85 b	0.78 c
缓增期增重量 W_3（mg·grain⁻¹）	11.9 a	10.6 b	9.5 c

TGW：thousand-grain weight；T：average grain-filling period；R_a：average grain-filling rate；T_{max}：days reaching the maximum grain-filling rate；R_{max}：maximum grain-filling-rate；T_1：grain-filling pyramid period；R_1：grain-filling-rate in T_1；W_1：grain weight accumulated in T_1；T_2：grain-filling fast increase period；R_2：grain-filling rate in T_2；W_2：grain weight accumulated in T_2；T_3：grain-filling slow increase period；R_3：grain-filling rate in T_3；W_3：grain weight accumulated in T_3. Different letters after each parameter measurements indicate significant difference among cultivar groups at $P<0.05$.

2.4 千粒重与籽粒灌浆特征参数的相关性

千粒重与各时期籽粒灌浆速率均呈显著正相关（$P<0.001$），其中与快增期灌浆速率的相关程度最密切（$r=0.97$），与平均灌浆速率的相关程度次之（$r=0.90$）（图2和表5）。千粒重与各时期籽粒增重量均呈显著正相关（$r=0.97$，$P<0.001$），与灌浆持续时间参数相关均不显著（表5）。

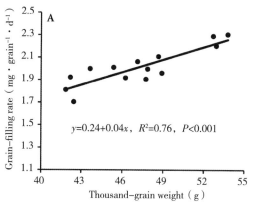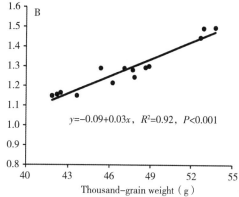

图 2 千粒重与灌浆速率的线性回归

Fig. 2 Linear regressions between thousand-grain weight and grain-filling rate

A：快增期灌浆速率；B：平均灌浆速率。

A：grain-filling ratein the fast increase period；B：average grain-filling rate.

表 5 千粒重与籽粒灌浆特征参数的相关系数
Table 5 Correlation coefficients between thousand-grain weight and grain-filling parameters

参数 Parameter	T	R_a	T_{max}	R_{max}	T_1	R_1	W_1	T_2	R_2	W_2	T_3	R_3	W_3
TGW	0.03	0.90***	0.42	0.87***	0.46	0.85***	0.97***	0.23	0.97***	0.97***	−0.40	0.86***	0.97***
T		−0.13	0.67**	−0.21	0.55*	−0.04	0.13	0.56*	−0.21	0.13	0.36	−0.08	0.13
R_a			0.40	0.87***	0.43	0.96***	0.97***	0.23	0.87***	0.97***	−0.57*	0.94***	0.97***
T_{max}				0.07	0.80***	0.38	0.57*	0.85***	0.07	0.57*	−0.42	0.57*	0.57*
R_{max}					0.37	0.81***	0.82***	−0.22	0.97***	0.82***	−0.19	0.67**	0.82***
T_1						0.30	0.57*	0.37	0.37	0.57*	−0.17	0.46	0.57*
R_1							0.95***	0.32	0.81**	0.95***	−0.51	0.91***	0.95***
W_1								0.38	0.82***	0.99***	−0.47	0.92***	0.99***
T_2									−0.22	0.38	−0.51	0.48	0.38
R_2										0.82***	−0.19	0.67**	0.82***
W_2											−0.48	0.92***	0.99***
T_3												−0.78**	−0.47
R_3													0.92***

TGW：千粒重；T：平均灌浆持续时间；R_a：平均灌浆速率；T_{max}：最大灌浆速率到达时间；R_{max}：最大灌浆速率；T_1：渐增期持续时间；R_1：渐增期灌浆速率；W_1：渐增期增重；T_2：快增期持续时间；R_2：快增期灌浆速率；W_2：快增期增重；T_3：缓增期持续时间；R_3：缓增期灌浆速率；W_3：缓增期增重。*、**和***分别表示在0.05、0.01和0.001概率水平显著相关。

TGW：thousand-grain weight；T：average grain-filling period；R_a：average grain-filling rate；T_{max}：days reaching the maximum grain-filling rate；R_{max}：maximum grain-filling rate；T_1：grain-filling pyramid period；R_1：grain-filling rate in T_1；W_1：grain weight accumulated in T_1；T_2：grain-filling fast increase period；R_2：grain-filling rate in T_2；W_2：grain weight accumulated in T_2；T_3：grain-filling slow increase period；R_3：grain-filling rate in T_3；W_3：grain weight accumulated in T_3. *, **, and *** indicate significant correlation at the 0.05, 0.01, and 0.001 probability levels, respectively.

3　讨论

黄淮冬麦区小麦灌浆期经常遇到高温、干旱和干热风的影响，使籽粒不饱满，粒重下降，造成产量损失。大量研究表明，粒重主要取决于灌浆速率，而灌浆持续时间对粒重的贡献相对较小[6,8,12,20-21,27]。本试验结果与此一致，黄淮麦区小麦品种的千粒重与各时期籽粒灌浆速率均呈显著正相关，其中与快增期灌浆速率相关最密切，且千粒重与各时期增重量均呈显著正相关，与灌浆持续时间各参数相关不显著。渐增期、快增期和缓增期的持续时间分别约占灌浆期31.2%、36.9%和31.9%，籽粒增重量则分别约占粒重的21.1%、57.8%和21.1%，进一步表明快增期籽粒增重量对粒重影响较大。由此可见，灌浆速率，特别是快增期灌浆速率，对粒重的形成具有重要贡献。

粒重与平均灌浆速率呈显著正相关（$r=0.90$，$P<0.001$）。由于灌浆速率测定繁琐，利用分离群体对该性状进行基因定位难度很大。鉴于灌浆速率对粒重的重要性，可以根据粒重与灌浆持续时间计算平均灌浆速率，仅需记录开花期、成熟期和粒重，对相关性状进行基因定位，以进一步解析粒重的遗传机制。王瑞霞等[31]利用平均灌浆速率将粒重相关QTL定位于1B、2A和3B染色体，可分别解释5.8%~20.8%的表型变异。因此，进一步明确TaCwi[9]、TaGW2[10]等已克隆的粒重基因与籽粒灌浆速率的关系，将有助于这些基因在育种中的有效利用，进一步改良黄淮麦区小麦品种的籽粒灌浆特性。本研究还表明，来自同一组合的3个品种中麦895、中麦875和中麦871在粒重和籽粒灌浆速率方面存在显著差异，中麦895和中麦875的粒重和各时期灌浆速率均显著高于其姊妹系中麦871及其双亲周麦16和荔垦4号。周麦16籽粒较大，但灌浆速率偏慢，常年饱满度一般，粒重并不高，约46g；荔垦4号成熟落黄好，籽粒饱满。我们当时组配周麦16和荔垦4号组合是期望通过提高周麦16的灌浆速率来改良其粒重，以进一步提高产量。根据我们多年多点观察结果，中麦895和中麦875的籽粒饱满度好，灌浆速率快，粒重高且稳定，这说明在进行粒重改良时，杂交组合亲本选配和后代选择至关重要，后代选择应主要考察品种的田间落黄特性及其收获后籽粒的大小和饱满度。我们正在对中麦895的快速灌浆和高粒重特性进行研究，期望发掘相关基因和分子标记，以辅助改良粒重。

4　结论

黄淮麦区小麦粒重和灌浆速率各参数主要受基因型控制，灌浆持续时间主要受环境影响。不同粒重类型品种间灌浆速率存在显著差异，表现为高粒重＞中等粒重＞低粒重。灌浆速率，特别是快增期灌浆速率的差异是导致品种间粒重高低的主要因素。

◈ 参考文献
References

[1] 茹振钢，冯素伟，李淦. 黄淮麦区小麦品种的高产潜力与实现途径. 中国农业科学，2015，48：3388-3393.
Ru Z G, Feng S W, Li G. High yield potential and effective ways of wheat in Yellow and Huai River Valley Facultative Winter Wheat Region. Sci Agric Sin, 2015, 48: 3388-3393. (in Chinese with English abstract)

[2] Zhou Y, He Z H, Sui X X, Xia X C, Zhang X K, Zhang G S. Genetic improvement of grain yield and associated traits in the Northern China Winter Wheat Region from 1960 to 2000. Crop Sci, 2007, 47: 245-253.

[3] Gao F M, Ma D Y, Yin G H, Rashieed A, Dong Y, Xiao Y G, Xia X C, Wu X X, He Z H. Genetic progress in grain yield and physiological traits in Chinese wheat cultivars of southern Yellow and Huai Valley Winter Wheat Zone since 1950. Crop Sci, 2017, 57: 760-773.

[4] 肖世和，何中虎. 小麦产量潜力和品质的改良. 见：庄巧生. 中国小麦品种改良及系谱分析. 北京：中国农业出版社，2003. pp 497-542.
Xiao S H, He Z H. Wheat yield and end use quality improvement in China. In: Zhuang Q S, ed. Chinese Wheat Improvement and Pedigree Analysis. Beijing: China Agriculture Press, 2003. pp 497-542. (in Chinese)

[5] Wang Y Q, Hao C Y, Zheng J, Ge H M, Zhou Y, Ma Z Q, Zhang X Y. A haplotype block associated with thousand kernel weight on chromosome 5DS in common wheat (Triticum aestivum L.), J Integr Plant Biol, 2015, 57: 662-672.

[6] Huang X Q, Coster H, M. Ganal W, Röder M

S. Advanced backcross QTL analysis for the identification of quantitative trait loci alleles from wild relatives of wheat (*Triticum aestivum* L.). *Theor Appl Genet*, 2003, 106: 1379-1389.

[7] Elouafi I, Nachit M M. A genetic linkage map of the Durum × *Triticum dicoccoides* backcross population based on SSRs and AFLP markers, and QTL analysis for milling traits. *Theor Appl Genet*, 2004, 108: 401-413.

[8] Quarrie S A, Steed A, Calestani C, Semikhodskii A, Lebreton C, Chinoy C, Steele N, Pljevljakusić D, Waterman E. A high-density genetic map of hexaploid wheat (*Triticum aestivum* L.) from the cross Chinese Spring× SQ1 and its use to compare QTLs for grain yield across a range of environments. *Theor Appl Genet*, 2005, 110: 865-880.

[9] Ma D Y, Yan J, He Z H, Wu L, Xia X C. Characterization of a cell wall invertase gene *TaCwi-A1* on common wheat chromosome 2A and development of functional markers. *Mol Breed*, 2010, 29: 43-52.

[10] Su Z Q, Hao C Y, Wang L F, Dong Y C, Zhang X Y. Identification and development of a functional marker of *TaGW2* associated with grain weight in bread wheat (*Triticum aestivum* L.). *Theor Appl Genet*, 2011, 122: 211-223.

[11] Jiang Q Y, Hou J, Hao C Y, Wang L F, Ge H M, Dong Y S. Zhang X Y. The wheat (*T. aestivum*) sucrose synthase 2 gene (*TaSus2*) active in endosperm development is associated with yield traits. *Funct Integr Genomics*, 2011, 11: 49-61.

[12] Zhang L, Zhao Y L, Gao L F, Zhao G Y, Zhou R H, Zhang B S, Jia J Z, *TaCKX6-D1*, the ortholog of rice *OsCKX1*, is associated with grain weight in hexaploid wheat. *New Phytol*, 2012, 195: 574-584.

[13] Chang J Z, Zhang J N, Mao X G, Li A, Jia J Z, Jing R L. Polymorphism of *TaSAP1-A1* and its association with agronomic traits in wheat. *Planta*, 2013. 237: 1495-1508.

[14] Guo Y, Sun J, Zhang G, Wang Y, Kong F, Zhao Y, Li S. Haplotype, molecular marker and phenotype effects associated with mineral nutrient and grain size traits of *TaGS1a* in wheat. *Field Crops Res*, 2013, 154: 119-125.

[15] Zhang Y J, Liu J D, Xia X C, He Z H. *TaGS-D1*, an ortholog of rice *OsGS3*, is associated with grain weight and grain length in common wheat. *Mol Breed*, 2014, 34: 1097-1107.

[16] Dong L, Wang F, Liu T, Dong Z, Li A, Jing R, Mao L, Li Y, Liu X, Zhang K, Wang D. Natural variation of *TaGASR7-A1* affects grain length in common wheat under multiple cultivation conditions. *Mol Breed*, 2014, 34: 937-947.

[17] Ma L, Li T, Hao C, Wang Y, Chen X, Zhang X. *TaGS5-3A*, a grain size gene selected during wheat improvement for larger kernel and yield. *Plant Biotech J*, 2016, 14: 1269-1280.

[18] Yue A Q, Li A, Mao X G, Chang X P, Li R Z, Jing R L. Identification and development of a functional marker from *6-SFT-A2* associated with grain weight in wheat. *Mol Breed*, 2015, 35: 63.

[19] Dngid S D, Brule-Babel A L. Rate and duration of grain filling in five spring wheat (*Triticum aestivum* L.) genotypes. *Can J Plant Sci*, 1994, 74: 681-686.

[20] Saini H S, Westgate M E. Reproductive development in grain crops during drought. *Adv Agron*, 1999, 68: 59-96.

[21] Zahedi M, Jenner C F. Analysis of effects in wheat of high temperature on grain filling attributes estimated from mathematical models of grain filling. *J Agric Sci*, 2003, 141: 203-212.

[22] Yang J C, Zhang J H. Grain filling of cereals under soil drying. *New Phytol*, 2006, 169: 223-236.

[23] Kamaluddin, Singh R M, Abdin M Z, Khan M A, Alam T, Kham S, Prasad L C, Joshi A K. Inheritance of grain filling duration in spring wheat (*Triticum aestivum* L. em Thell). *J Plant Biol*, 2007, 50: 504-507.

[24] Wong L S L, Baker R J. Selection for time to maturity in spring wheat. *Crop Sci*, 1986, 26: 1171-1175.

[25] Talbert L E, Lanning S P, Murphy R L, Martin J M. Grain fill duration in twelve hard red spring wheat crosses: genetic variation and association with other agronomic traits. *Crop Sci*, 2001, 41: 1390-1395.

[26] 吴晓丽, 汤永禄, 李朝苏, 吴春, 黄钢, 马蓉. 四川盆地小麦籽粒的灌浆特性. 作物学报, 2014, 40: 337-345.
Wu X L, Tang Y L, Li C S, Wu C, Huang G, Ma R. Characteristics of grain filling in wheat growing in Sichuan basin. *Acta Agron Sin*, 2014, 40: 337-345. (in Chinese with English abstract)

[27] Dias A S, Lidon F C. Evaluation of grain filling rate and duration in bread and durum wheat, under heat stress after anthesis. *J Agron Crop Sci*, 2009, 195:

137-147.

[28] Motzo R, Giunta F, Pruneddu G. The response of rate and duration of grain filling to long-term selection for yield in Italian durum wheats. *Crop Pasture Sci*, 2010, 61: 162-169.

[29] 曾浙荣, 庞家智, 周桂英, 赵双宁, 曹梅林. 我国北部冬麦区小麦品种籽粒灌浆特性的研究. 作物学报, 1996, 22: 720-728.
Zeng Z R, Pang J Z, Zhou G Y, Zhao S N, Cao M L. Grain filling properties of winter wheat varieties in northern part of China. *Acta Agron Sin*, 1996, 22: 720-728. (in Chinese with English abstract)

[30] 吴少辉, 段国辉, 高海涛, 张学品, 温红霞, 余四平, 马飞. 黄淮麦区水、旱生态型小麦籽粒灌浆进程研究. 麦类作物学报, 2009, 29: 1015-1021.

Wu S H, Duan G H, Gao H T, Zhang X P, Wen H X, Yu S P, Ma F. Research on wheat grain filling process of water and dryland ecological types of wheat in Huang-Huai area. *J Triticeae Crops*, 2009, 29: 1015-1021. (in Chinese with English abstract)

[31] 王瑞霞, 张秀英, 伍玲, 王瑞, 海林, 闫长生, 游光霞, 肖世和. 不同生态环境条件下小麦籽粒灌浆速率及千粒重 QTL 分析. 作物学报, 2008, 34: 1750-1756.
Wang R X, Zhang X Y, Wu L, Wang R, Hai L, Yan C S, You G X, Xiao S H. QTL mapping for grain-filling rate and thousand-grain weight in different ecological environments in wheat. *Acta Agron Sin*, 2008, 34: 1750-1756. (in Chinese with English abstract)

Genetic progress in grain yield and physiological traits in Chinese wheat cultivars of southern Yellow and Huai Valley since 1950

Fengmei Gao, Dongyun Ma, Guihong Yin, Awais Rasheed, Yan Dong, Yonggui Xiao, Xianchun Xia, Xiaoxia Wu, and Zhonghu He[*]

F. Gao, A. Rasheed, Y. Dong, Y. Xiao, X. Xia, and Z. He, Institute of Crop Science, National Wheat Improvement Center, Chinese Academy of Agricultural Sciences (CAAS), 12 Zhongguancun S. St., Beijing 100081, China

F. Gao, Key Laboratory of Soybean Biology, Soybean Research Institute, Ministry of Education, Northeast Agricultural Univ., Harbin, China; F. Gao, Keshan Agricultural Research Institute, Heilongjiang Academy of Agricultural Sciences, Keshan 161600, Heilongjiang, China

D. Ma, National Wheat Engineering Research Center, Henan Agricultural Univ., 2 Nongye Rd., Zhengzhou 450000, Henan Province, China; G. Yin, Zhoukou Academy of Agricultural Sciences, Zhoukou 466000, China; X. Wu, Key Lab. of Soybean Biology, China

A. Rasheed and Z. He, International Maize and Wheat Improvement Center (CIMMYT) China Office, c/o CAAS, 12 Zhongguancun S. St., Beijing 100081, China. F. Gao and D. Ma contributed equally to this work.

Abstract: Understanding the key characteristics associated with genetic gains achieved through breeding is essential for improving yield-limiting factors and designing future breeding strategies in bread wheat (*Triticum aestivum* L.) cultivars. The objective of the present study was to investigate the genetic progress in yield-related and physiological traits in cultivars released from 1950 to 2012 for irrigated conditions in the southern Yellow and Huai Valleys Winter Wheat Zone. Field trials including 26 leading cultivars from 1950 to the present time were conducted at Zhengzhou and Zhoukou in Henan Province, during the 2013 − 2014 and 2014 − 2015 cropping seasons, providing data from four environments. Grain yield (GY) was significantly increased by the linear rate of 57.5 kg ha^{-1} yr^{-1} or 0.70% ($R^2 = 0.66$, $P < 0.01$) and significantly correlated with increased thousand-kernel weight (TKW) ($r = 0.48$, $P < 0.05$), spike number m^{-2} ($r = 0.44$, $P < 0.05$), kernels m^{-2} ($r = 0.56$, $P < 0.01$), aboveground biomass (AGBM) ($r = 0.80$, $P < 0.01$), harvest index (HI) ($r = 0.84$, $P < 0.01$), water-soluble carbohydrate at 10 d postanthesis (WSC-10) ($r = 0.80$, $P < 0.01$), and reduced plant height (PH) ($r = -0.85$, $P < 0.01$). There was no significant change in kernel number per spike, heading date, normalized difference in vegetation index at anthesis and at 10 d postanthesis, leaf area index at anthesis and at 10 d postanthesis, and canopy temperature depression at anthesis during the past 60 yr. Soil plant analysis development (SPAD) estimates of chlorophyll content at 10 d postanthesis

Published in Crop Sci., 2017, 57: 760-773.

* Corresponding author (zhhecaas@163.com).

(Chl-10) increased with year of release and were significantly correlated with GY ($r = 0.69$, $P < 0.01$), PH ($r = -0.76$, $P < 0.01$), AGBM ($r = 0.52$, $P < 0.01$), HI ($r = 0.71$, $P < 0.01$), and WSC-10 ($r = 0.73$, $P < 0.01$). Cultivars conferring *Rht-D1b* and *Rht-D1b* + *Rht8c* showed increased GY, TKW, AGBM, HI, WSC-10, and Chl-10. Stem water solubility content can be used as a selection criterion for further improving yield potential.

Abbreviations: AGBM, aboveground biomass; Chl, SPAD estimates of chlorophyll content; Chl-10, SPAD estimates of chlorophyll content at 10 d postanthesis; Chl-A, SPAD estimates of chlorophyll content at anthesis; CTD, canopy temperature depression; CTD-10, canopy temperature depression at 10 d postanthesis; CTD-A, canopy temperature depression at anthesis; DPA, days postanthesis; GY, grain yield; HD, heading date; HI, harvest index; KASP, kompetitive allele specific polymerase chain reaction; KN, kernels m^{-2}; KNS, kernel number per spike; LAI, leaf area index; LAI-10, leaf area index at 10 d postanthesis; LAI-A, leaf area index at anthesis; NDVI, normalized difference in vegetation index; NDVI-10, normalized difference in vegetation index at 10 d postanthesis; NDVI-A, normalized difference in vegetation index at anthesis; PH, plant height; SN, spike number m^{-2}; SPAD, soil plant analysis development; TKW, thousand-kernel weight; WSC-10, stem water-soluble carbohydrate at 10 d postanthesis; YHVWWZ, the Yellow and Huai Valleys Winter Wheat Zone.

China is the largest wheat (*Triticum aestivum* L.) producer and consumer in the world, and bread wheat is the third leading cereal crop after maize (*Zea mays* L.) and rice (*Oryza sativa* L.) in China, with a planting area of about 24 million ha, an average yield of 4762 kg ha^{-1}, and annual production of around 115 million Mg in 2008 (He et al., 2010). Chinese wheat production is facing significant challenges, due to reducing arable land area, increased occurrence of diseases, especially Fusarium head blight (mainly caused by *Fusarium graminearum* Schwabe), increased water scarcity, and greater fluctuations in temperature and water supply caused by global climate change. Therefore, it is a primary breeding objective to improve grain yield (GY) in China, although processing quality has also become very important. The Yellow and Huai River Valleys Winter Wheat Zone (YHVW-WZ) is the most important wheat-producing region in China, accounting for ~65% of national wheat production (He et al., 2010). The southern YHVWWZ, including most of Henan Province, the central Shanxi Plain, and northern parts of Anhui and Jiangsu Provinces, is the largest wheat production area of around 8 million ha and production of around 40 million Mg annually. It is an irrigated area with an intensive annual double-crop wheat-maize system. The major wheat-breeding objectives in this zone are to improve yield potential and processing quality for pan bread and noodle, tolerance to low temperatures in winter and high temperatures during grain filling, and resistance to Fusarium head blight and powdery mildew [caused by *Blumeria graminis* (DC) Speer f. sp. *tritici* emend. É. J. Marchal].

Several studies have examined progress in GY in historical Chinese wheat cultivars. The annual GY genetic gains were 0.48 and 1.23% yr^{-1} in 57 wheat cultivars released from 1960 to 2000 in the northern China Winter Wheat Region, including the Northern China Plain Winter Wheat Zone and the YHVWWZ (Zhou et al., 2007b), respectively, 0.60% in 18 leading cultivars released between 1981 and 2008 in Henan Province (Zheng et al., 2011), and 0.85% yr^{-1} for 15 cultivars released from 1969 to 2006 in the northern part of the YHVWWZ (Xiao et al., 2012). All above studies indicated that significant genetic progress has been achieved in improving the yield potential of Chinese winter wheat cultivars. Genetic gains in wheat yield potential were also reported in other countries, such as in North America (Graybosch and Peterson, 2010), Finland (Peltonen-Sainio et al., 2009), Russia (Morgounov et al., 2010), Australia

(Sadras and Lawson, 2011), the United Kingdom (Shearman et al., 2005; Mackay et al., 2011; Clarke et al., 2012), Spain (Sanchez-Garcia et al., 2013), Canada (Hucl et al., 2015) and Mexico (Manès et al., 2012; Aisawi et al., 2015). However, other studies showed no or very small genetic gains in yield potential over time (Graybosch and Peterson, 2010; Matus et al., 2012; Beche et al., 2014).

Understanding the key yield components and morpho logical and physiological traits associated with genetic gains in yield potential is important for identifying yield-limiting factors and for future design of breeding strategies. Thou sand-kernel weight (TKW) (Zhou et al., 2007a; Zheng et al., 2011; Sanchez-Garcia et al., 2013; Aisawi et al., 2015) and kernels m^{-2} (KN) (Donmez et al., 2001; Royo et al., 2007; Xiao et al., 2012; Sanchez-Garcia et al., 2013) were reported as major factors contributing to increased grain yield, whereas several other studies showed no (Brancourt-Hulmel et al., 2003; Zhou et al., 2007b) or even negative changes in TKW (Siddique et al., 1989; Royo et al., 2007). Spike number m^{-2} (SN) was attributed to increased number of spikes per plant (Royo et al., 2007). Most studies showed that increased GY was significantly associated with increased harvest index (HI) and reduced plant height (PH) (Canevara et al., 1994). Decreased PH not only enhanced lodging resistance and transport of assimilates to the repro ductive organs due to reduced size of the vegetative organs (Ehdaie et al., 2006; Álvaro et al., 2008) but also increased HI (Sayre et al., 1997; Brancourt-Hulmel et al., 2003).

Although little genetic change in aboveground biomass (AGBM) was observed in some studies (Tian et al., 2011; Zheng et al., 2011; Sanchez-Garcia et al., 2013), significant genetic change was reported in others (Donmez et al., 2001; Shearman et al., 2005; Xiao et al., 2012; Aisawi et al., 2015). Water-soluble carbohydrate (WSC) stored in wheat stems and leaf sheaths make significant contributions to final grain weight and kernel size (Salem et al., 2007; Dreccer et al., 2009); stem WSC at the grain-filling stage showed significant genetic change over time and positive correlations with grain yield (Shearman et al., 2005; Foulkes et al., 2007). Photosyn thesis is the primary source of grain yield and Chl (soil plant analysis development [SPAD] estimates of chlorophyll con tent) of the flag leaf is considered an important determinant of photosynthesis (Gladun and Karpov, 1993; Chen et al., 1995; Makino, 2011). Therefore, increasing Chl in the flag leaf during grain filling can raise yield when other genetic factors are not altered (Long et al., 2006b). However, most investi gations reported little genetic change in Chl of the flag leaf, except for Xiao et al. (2012). High CTD (canopy temperature depression) was significantly and positively associated with wheat yield (Pinto et al., 2010; Kumar et al., 2013) and dem onstrated to be a useful selection criterion for improving grain yield (Reynolds et al., 2001; Ayeneh et al., 2002). Fischer et al. (1998) showed that genetic gain in grain yield was associated with CTD at the grain filling stage for CIMMYT semidwarf cultivars released from 1962 to 1988. Grain yield increased along with leaf area index (LAI) between seedling stage and at anthesis (Acreche et al., 2009). Significant genetic changes in LAI at anthesis were observed in winter cultivars from Shandong Province (Xiao et al., 2012).

Genetic gains in yield and related traits in wheat cultivars in the southern YHVWWZ was reported in our previous study (Zheng et al., 2011). The yield potential was 4704.5 kg ha^{-1} for the cultivar 'Bima 1' released in 1951 and rose up to 8906 kg ha^{-1} for 'Zhongmai 895' in 2012 in the southern YHVWWZ. Although Zheng et al. (2011) studied the prog ress in GY of wheat cultivars released from 1981 to 2008 in this region, the genetic progress in cultivars before 1981 and after 2008 has not been reported. Furthermore, only 18 culti vars released from 1981 to 2008 were investigated, and some important physiological traits such as stem WSC, Chl, CTD, normalized difference in vegetation index (ND-VI), and LAI at the filling stage were not evaluated. Wheat improvement started in 1940 in this region, and a comprehensive study of yield potential of leading cultivars from 1950 until the present time will provide useful information for planning future breeding strategies for the region. Such information will also be useful for the international wheat community due to unique early maturity required for wheat-maize double cropping, the germplasm used in breeding, and food types consumed in China. Thus, 26 leading cultivars released in the southern YHVWWZ from 1951 to 2012 were used to dissect the genetic gains in yield-related and physiological traits. The objectives were (i) to determine the genetic changes in grain yield and yield-related and physiological traits among those cultivars, and (ii) to identify the relationship between yield and yield-related components and physiological traits.

MATERIALS AND METHODS

Plant Materials and Field Trials

Twenty-six leading cultivars, released in the southern YHVWWZ from 1951 to 2012, were evaluated in this study (Supplemental Table S1). The field trials were conducted at Zhengzhou ($34°44'$ N, $113°42'$ E) and Zhoukou ($33°37'$ N, $114°38'$ E) in Henan Province during the 2013—2014 and 2014—2015 cropping seasons. Zhengzhou and Zhoukou represent regions with the highest wheat yields in this region. The trial was planted on 11 (Zhoukou) and 12 (Zhengzhou) Oct. 2013 and on 9 (Zhoukou) and 10 (Zhengzhou) Oct. 2014 in completely randomized blocks with three replicates at each location. Each plot consisted of six rows of 7.15-m length and 1.20-m width, with 250 seeds m^{-2}. Field management followed local normal practices and allowed achievement of high yield potential. The soils at both locations are sandy clay, highly organic, and slightly alkaline. Before planting, ammonium phosphate (170 kg ha^{-1} of P, 67 kg ha^{-1} of N), urea (13 kg ha^{-1} of N), and potassium chloride (75 kg ha^{-1} of K) were applied. An additional 120 kg ha^{-1}

of N was top dressed at stem elongation. Following local practices, three irrigations were provided (pre-winter, mid-March, and mid-April) to ensure the achievement of high yield potential. Nets with a mesh size of approximately $20×20$ cm were used to prevent cultivars from lodging. Weeds were cleared by hand. Powdery mildew and stripe rust (caused by *Puccinia striiformis* Westend.) were controlled by fungicides (Triadimefon [Dupont Co. Ltd., Shanghai] at a rate of 0.2 mL L^{-1}) at booting and anthesis stage. Aphid damage was prevented by use of insecticides (Cyhalothrin [Nanjing Kingsun Bio-tech Co. Ltd., Nanjing] at 3.3 mL L^{-1} and Imidacloprid [Karegreen Biotechnology Co. Ltd., Hangzhou] at a rate of 0.07 mL L^{-1}) between heading and mid-grain filling. The monthly average temperatures in January at Zhengzhou and Zhoukou were -0.1 and $0.3℃$, respectively, while the highest temperatures in May were 33 and $32℃$, respectively. The yearly average rainfalls were 640.5 and 689 mm at Zhengzhou and Zhoukou, respectively.

Plant Measurement and Observation

Heading date (HD) was recorded at 50% spike emergence. Nor-malized difference in vegetation index (NDVI), Chl, canopy temperature, and LAI were measured at anthesis and at 10 d postanthesis (DPA) in each plot. The NDVI was measured by scanning with a portable spectroradiometer (GreenSeeker, Ntech Industries, Inc, Ukiah, CA). The Chl was scored as the average of six flag leaves per plot using a chlorophyll meter SPAD-502 (Inolta, Japan). Canopy temperature was measured between 11:00 AM and 1:00 PM in fine windless and cloudless days at anthesis and at 10 DPA, respectively, using a portable infrared thermometer (Mikron M90 Series, Mikron Infrared Instrument Co., Inc., Oakland, NJ) according to the CIMMYT Physiological Breeding Field Guide (Pask et al., 2012). Canopy temperature depression was calculated as the difference between the air temperature at the time of measurement and canopy temperature to account for fluctuations throughout the measurement period. Leaf area indices were calcu lated from incident and transmitted radiation measurements at noon under sunny days (Calderini et al., 1997) using the AccuPAR ceptometer (Decagon Devices LP-80, Pullman, WA). Because there are several days of differences for flowering dates among cultivars, CTD and other physiological traits were measured on the same calendar day in the middle of a given stage of development.

Ten stems with the same heading date in each plot were cut from the soil surface to about 20 cm above the ground at 10 DPA and were used to measure stem WSC contents by near-infrared spectroscopy (Wang et al., 2011). Plant height was measured from the ground to the tip of the spike excluding awns at the late grain-filling stage. Kernel number per pike (KNS) was calculated as the mean of 30 randomly selected spikes in each plot. At physiological maturity, two 1m sections (subsamples) in the middle of each plot were cut with a sickle at soil level. The SN was scored using the subsamples and then transformed to spike numbers m^{-2}. The KN was calculated by KNS×SN. The four central rows in each plot, excluding the subsamples, were hand harvested, threshed, dried, and weighed to obtain grain yield at a 14% moisture level. The subsamples were used to record the dry weight of AGBM to determine HI. After harvest, TKW was measured by weighing duplicates of 500 kernels from each plot. Grain yield was determined as the weight of grain harvested per unit area (kg ha^{-1}). There was no cold damage at any location.

KASP Marker Test

The 1B. 1R translocation, dwarfing, and photoperiod genes were determined using gene-specific markers by the KASP method (kompetitive allele specific polymerase chain reaction; Rasheed et al., 2016). The primer sequences and more detailed allele information are provided in Supplemental Table S2.

The Master mix included 2 μL of 50 to 100 ng μL^{-1} template DNA, 2.5 μL of 2 × KASP master mix, 0.07 μL of KASP assay mix, and 2.5 × μL of water. Polymerase chain reaction was performed in a 384-well format (S1000, Thermal Cycler, USA) by the following procedure: hot start at 95℃ for 15 min, followed by 10 touchdown cycles (95℃ for 20 s, touchdown at 65℃ initially and decreasing at −1℃ per cycle for 25 s), followed by 30 additional cycles of annealing (95℃ for 10 s, 57℃ for 60 s).

Statistical Analysis

Means and standard errors (SE) were determined using PROC MEANS. Analysis of variance (ANOVA) was performed using PROC MIXED in the Statistical Analysis System (SAS Institute, 2000) for all traits, with cultivars as fixed effects and environments, cultivar× environment interactions, and replication nested in environment effects as random. Differences among cultivars were tested using Tukey's test. Genotypic least square means were used for subsequent analysis. Pearson's linear correlation coefficients were calculated by PROC CORR. The equations $\gamma_i = a + bx_i + u$ [1] or $\gamma = b/GY_{2011-2012} \times 100$ [2] were used to estimate absolute (GY gains in kg ha^{-1} yr^{-1}) or relative (% GY gain yr^{-1}) genetic gains in GY and related traits, where γ_i is the estimated mean GY of cultivar i in each trial, x_i is the year in highly significant, and GY increased with a genetic gain which cultivar i was released, and the intercept of equations was estimated by a, while slope b measured absolute grain yield gains [1] and u represents the residual error; γ is relative rates of yield gain (% yr^{-1}), and GY$_{2011-2012}$ is the averaged grain yield of cultivars released in 2011 to 2012 (Grassini et al., 2013; Fischer, 2015).

RESULTS

Analyses of variance based on averaged data across four environments were conducted for GY, HD, PH, KNS, SN, KN, TKW, AGBM, HI, stem water-soluble carbohydrate at 10 DPA (WSC-10), SPAD estimates of chlorophyll content at anthesis (Chl-A), SPAD estimates of chlorophyll content at 10 DPA (Chl-10), NDVI at anthesis (NDVI-A), NDVI at 10 DPA (NDVI-10), CTD at anthesis (CTD-A), CTD at 10 DPA (CTD-10), LAI at anthesis (LAI-A), and LAI at 10 DPA (LAI-10). Most of the traits were significantly affected by cultivar, environments, and cultivar×environment interactions (Table 1). Experimental error variances were homogeneous ($P>0.05$) over environments using Bartlett's test for all traits. Significant cultivar effects and cultivar×environment interactions were observed for all traits except for CTD-A and CTD-10.

Grain Yield

Grain yields were significantly different across the 26 cultivars, ranging from 4632.8 kg ha^{-1} for 'Bima 4' released in 1956 to 8906 kg ha^{-1} for Zhongmai 895 released in 2012 (Table 2). A linear regression equation showed that the relationship between yield and year of release was of 0.70% or 57.5 kg ha^{-1} ($R^2=0.66$, $P<0.01$) from 1951 to 2012 (Table 2, Fig. 1A). Bima 1 marked the first mile-stone cultivar in the history of Chinese wheat breeding; it was the first improved cultivar developed for commercial production through a cross between a local Chinese landrace and an introduction from the United States. Bima 1 was characterized by a large spike, high TKW, good yield potential, white grain, early maturity, stripe rust resistance, and wide adaptability. It was sown in almost every part of the YHVWWZ, with an annual growing area that peaked at 6 million ha in 1959. Bima 4 was less broadly adapted, but its maximum seeding area reached 1.1 million ha in 1960 (He et al., 2010). 'Funo' and 'Abbondanza' were Italian cultivars introduced into China in 1956 and were widely grown in the YHVWWZ from 1959 to 1963, occupying about 1.1 and 0.5 million ha yr^{-1}, respectively, due to high yield potential, lodging tolerance, and high rust resistance. 'Zhoumai 27' is becoming a leading cultivar in this zone due to its high yield potential, large spikes, and outstanding processing quality. Zhongmai 895, developed by the Chinese Academy of Agricultural Sciences (CAAS) in 2012, has high yield, large spikes, semidwarf PH, lodging tolerance, multiple disease resistance, and acceptable quality for noodles and steamed bread.

Table 1. Mean square from a combined analysis of variance for yield and physiological traits of 26 winter wheat genotypes across four environments.

Source of variation†	Cultivar	Environment	Replicate	Cultivar×environment	Error
df	25	3	2	75	206
GY	18 677 605**	468 265	38 917	1 507 844**	522 902
HD	39.6**	0**	11.7	0**	0.4
PH	4 414.6**	4 383.2**	16.8	246.1**	19.5
KNS	112.8**	131.7**	23.3	22.6**	11.2
SN	67 187**	332 648**	14 071	21 342**	8 497
KN	86 503 705**	382 607 227**	2 726 468 754	38 066 930**	14 561 619
TKW	323.6**	3 302.0**	7.9	37.9**	10.8
AGBM	26 716 670**	2 209 481 754**	4 170 022	17 221 517**	6 676 492
HI	0.03**	0.1**	0.0004	0.004**	0.002
WSC-10	94.9**	455.0**	12.3	8.6**	2.7
Chl-A	75.9**	233.5**	32.4**	8.9**	5.5
Chl-10	62.3**	264.0**	11.6	8.8	8.9
NDVI-A	0.005*	3.7**	0.07	0.006**	0.003
NDVI-10	0.003**	4.1**	0.003*	0.002**	0.001
CTD-A	1.6	357.2**	8.8*	1.2	2.3
CTD-10	0.8	446.1**	43.9**	1.6	1.5
LAI-A	2.8**	11.9**	0.5	2.6**	0.8
LAI-10	0.6**	170.8**	7.2**	1.3**	0.2

* Significant at the 0.05 probability level.

** Significant at the 0.01 probability level.

† GY, grain yield; HD, heading date; PH, plant height; KNS, kernel number per spike; SN, spike number m^{-2}; KN, kernels m^{-2}; TKW, thousand-kernel weight; AGBM, aboveground biomass; HI, harvest index; WSC-10, water-soluble carbohydrate at 10 d postanthesis; Chl-A, SPAD (soil plant analysis development) estimates of chlorophyll content at anthesis; Chl-10, SPAD estimates of chlorophyll content at 10 d postanthesis; CTD-A, canopy temperature depression at anthesis; CTD-10, canopy temperature depression at 10 d postanthesis; NDVI-A, normalized difference in vegetation index at anthesis; NDVI-10, normalized difference in vegetation index at 10 d postanthesis; LAI-A, leaf area index at anthesis; LAI-10, leaf area index at 10 d postanthesis.

Table 2. Yield components and morphological traits of 26 winter wheat cultivars averaged for four environments, genetic gain ($\%$ yr^{-1}), and correlation with year of release (R^2).

Cultivar[†]	Trait[‡]						
	GY kg ha^{-1}	HD[§] d	PH cm	KNS	SN	KN	TKW g
Bima 1	4 704. 5 k	102 a	130. 5 a	38cdefg	501ijk	18 842ih	38. 6 hi
Neixiang 5	4 633. 1 k	97 hi	106. 5 d	38efgh	475jk	21 493cdefgh	45. 6 def
Funo	5 408ijk	99fg	117. 8 c	36ghigk	525ghigk	22 430cdefgh	41. 5gh
Abbondanza	5 712. 4hij	100 b	115. 2 c	41abc	535ghijk	22 490cdefgh	35. 4 j
Zhengzhou 3	6 141. 8ghi	98gh	121. 4 b	38cdefg	457 k	19 382ghi	44. 7ef
St1472/506	7 525. 4bcdef	98gh	100. 6 e	44 ab	556efghij	18 286 j	38. 2i
Bima 4	5 100. 2jk	97 hi	128. 2 a	36efghi	507hijk	18 008 j	47. 2 de
Bainong 3217	6 262. 7ghi	99fg	116. 6 c	33jkl	677bc	22 280cdefgh	42. 8fg
Yumai 2	7 415. 6cdef	98gh	84. 5 g	38cdefgh	749 ab	20 500efghi	41. 0ghi
Yumai 7	7 635. 4bcdef	96ij	79. 7hij	37efgh	529ghijk	27 559 ab	45. 7 def
Yumai 13	7 806. 4bcde	98gh	84. 6 g	37efghi	757 a	23 156cdef	41. 6gh
Yumai 18	8 083. 3abdce	95jk	82. 3ghi	36efghi	647 dc	24 556bcd	42. 8fg
Yumai 21	8 401. 4abcd	100 cd	70. 0mn	38efgh	654 dc	28 262 a	45. 8 def
Yumai 34	7 317. 1 def	96ij	83. 0ghi	33 kl	614cdefg	20 121fghi	52. 0bc
Yumai 49	8 389. 6abcd	99fg	77. 7 kl	36efghi	645cde	23 505cdef	48. 0 d
Yumai 54	7 401. 5cdef	98gh	83. 3gh	41bcd	577defghi	23 260cdef	39. 4 hi
Zhengmai 9023	7 070. 2efg	94 k	85. 4 g	32 l	636cdef	19 854fghi	47. 2 de
Zhoumai 16	8 432. 9abc	100bc	73. 5lm	40cde	574defghi	20 096fghi	54. 1 ab
Xinmai 18	7 576bcdef	102 a	95. 7 f	39cdef	571defghi	24 055cde	39. 5 hi
Zhoumai 18	8 546. 4 ab	97 hi	79. 2ij	34ijkl	602cdefg	17 976ghi	45. 7 def
Yanzhan 4110	6 611. 7fgh	98gh	79. 1ij	41bc	550fghij	21 493defghi	45. 9 de
Aikang 58	7 562. 7bcdef	98gh	69. 3 n	36ghigk	630cdef	22 587cdefg	46. 8 de
Zhoumai 22	8 069. 5abcde	99ef	79. 3hij	39cdef	577defghi	22 966cdefg	51. 2 c
Zhoumai 27	8 541. 4 ab	99fg	74. 6 kl	44 a	593defghi	22 589cdefg	45. 2 def
Zhongmai 895	8 906 a	100 de	77. 9jk	37efghi	586defghi	22 599cdefg	56. 0 a
Zhongmai 875	7 125. 1efg	98gh	73. 3lm	35hijkl	654 dc	25 825abc	53. 4abc
Mean	7 168	98	91. 1	38	591	28 262	45. 2
SEM¶	723. 1	0. 6	4. 4	3. 4	92. 2	3816	3. 3
Absolute genetic gains	57. 5	0. 01	−0. 93	−0. 01	2	64	0. 18
Genetic gains (%)	0. 70**	0	−1. 24**	0. 04	0. 31*	0. 27*	0. 35**
R^2	0. 66	0. 02	0. 71	0. 01	0. 24	0. 19	0. 37

* Significant at the 0. 05 probability level.
** Significant at the 0. 01 probability level.
† Cultivars ranked according to year of release.
‡ For trait abbreviations see Table 1.
§ Days from 1 January.
¶ SEM, standard error of the means.

Yield Components and Other Agronomic Traits

There was nochange in HD with year of release. Significant genetic changes were found in PH, which decreased linearly at −1. 24% yr^{-1} or 0. 93 cm yr^{-1} ($R^2 = 0. 71$, $P < 0. 01$) (Table 2, Fig. 1C), ranging from 69. 3 cm for 'Aikang 58' to 130. 5 cm for Bima 1.

Significant differences were also observed for SN,

KN, and TKW among cultivars. Mean SN was 591, ranging from 457 for 'Zhengzhou 3' to 757 for 'Yumai 13', with a genetic gain of 0.31% yr^{-1} or 2 yr^{-1} ($R^2 = 0.24$, $P < 0.05$). Mean KN was 22, 137, ranging from 17, 676 for Zhengzhou 3 to 28, 244 for 'Yumai 2', with a genetic gain of 0.27% yr^{-1} or 64 yr^{-1} ($R^2 = 0.16$, $P < 0.05$) (Table 2). Mean TKW was 45.2 g, ranging from 35.4 g for Funo to 56.0 g for Zhongmai 895, with a linear genetic gain of 0.35% yr^{-1} or 0.18 g yr^{-1} ($R^2 = 0.37$, $P < 0.01$) (Table 2, Fig. 1B). There was no significant genetic gain in KNS during the past 60 yr.

Physiological Traits

There were significant changes in AGBM, HI, stem WSC-10, Chl-A, Chl-10, and CTD-A, whereas no genetic changes were observed for NDVI-A, NDVI-10, CTD-A, and LAI-10 (Table 3). Aboveground biomass significantly increased by 0.39% yr^{-1} or 62.6 kg ha^{-1} ($R^2 = 0.53$, $P < 0.01$), ranging from 12, 702 kg ha^{-1} for 'Bainong 3217' to 17, 367 kg ha^{-1} for Zhongmai 895 (Table 3, Fig. 1D). Harvest index increased at an average genetic gain of 0.51% yr^{-1} ($R^2 = 0.48$, $P < 0.01$) (Table 3, Fig. 1E), ranging from 0.24 for Abbondanza to 0.42 for 'Yumai 7' and Zhoumai 27. Stem WSC-10 showed the largest genetic gain among yield-related physiological traits and increased significantly and linearly by 0.81% yr^{-1} or 0.13 mg g^{-1} ($R^2 = 0.66$, $P < 0.01$), ranging from 8.5 mg g^{-1} for Abbondanza to 18.1 mg g^{-1} for Aikang 58 (Table 3, Fig. 1F). The traits Chl-A, Chl-10, and CTD-A increased at 0.15, 0.17, and 0.12% yr^{-1}, respectively. Mean Chl-A, Chl-10 and CTD-A were 50.9 units, 52.3 units, and 5.0℃, ranging from 45.0 to 56.1 units, 46.7 to 55.4 units, and 4.5 to 5.1℃, respectively. The LAI-A decreased at −0.21% yr^{-1}.

Correlations among Grain Yield, Agronomic, and Physiological Traits

Grain yield was significantly and positively correlated with SN ($r = 0.44$, $P < 0.01$), KN ($r = 0.56$, $P < 0.01$), TKW ($r = 0.48$, $P < 0.01$), AGBM ($r = 0.80$, $P < 0.01$), HI ($r = 0.84$, $P < 0.01$), and stem WSC-10 ($r = 0.80$, $P < 0.01$) and negatively correlated with PH ($r = -0.85$, $P < 0.01$) (Table 4, Fig. 2), indicating that GY improvement in Chinese wheat cultivars was mainly contributed by increased TKW, SN, KN, AGBM, HI, and WSC-A and reduced PH.

Plant height wassignificantly and negatively correlated with AGBM ($r = -0.77$, $P < 0.01$), HI ($r = -0.87$, $P < 0.01$), and stem WSC-10 ($r = -0.91$, $P < 0.01$), revealing that reduced PH contributed to the increases in AGBM, HI, and stem WSC-10. Spike number exhibited positive correlations with AGBM ($r = 0.53$, $P < 0.01$), HI ($r = 0.46$, $P < 0.01$), and stem WSC-10 ($r = 0.60$, $P < 0.01$) and a negative correlation with PH ($r = -0.54$, $P < 0.01$). Kernel number showed significant and positive correlations with SN ($r = 0.78$, $P < 0.01$), AGBM ($r = 0.65$, $P < 0.01$), HI ($r = 0.42$, $P < 0.05$), WSC-10 ($r = 0.50$, $P < 0.01$), Chl-A ($r = 0.44$, $P < 0.05$), and Chl-10 ($r = 0.41$, $P < 0.05$) and a negative correlation with PH ($r = -0.54$, $P < 0.01$). Significant and positive correlations were observed between TKW and AGBM ($r = 0.47$, $P < 0.01$), HI ($r = 0.55$, $P < 0.01$), and WSC-10 ($r = 0.53$, $P < 0.01$). Thousand-kernel weight showed a significantly negative correlation with PH ($r = -0.50$, $P < 0.01$). Aboveground biomass was positively correlated with WSC-10 ($r = 0.65$, $P < 0.01$). The highest positive correlation was observed between HI and WSC-10 ($r = 0.91$, $P < 0.01$).

Grain yield, AGBM, HI, and stem WSC-10 were significantly and positively correlated with Chl-A ($r = 0.63$, 0.43, 0.63, and 0.60, respectively; $P < 0.01$) and Chl-10 ($r = 0.69$, 0.52, 0.71, and 0.73, respectively; $P < 0.01$) (Table 4; Fig. 3A, 3C, and 3D). Plant height showed significant negative correlations with Chl-A ($r = -0.68$, $P < 0.01$) and Chl-10 ($r = -0.76$, $P < 0.01$) (Table 4, Fig. 3B) and positive correlations with LAI-A ($r = 0.54$, $P < 0.01$). Significant negative correlations were also observed between HI and CTD-A ($r = -0.50$, $P < 0.01$), WSC-10 and LAI-A ($r = -0.55$, $P < 0.01$), respectively.

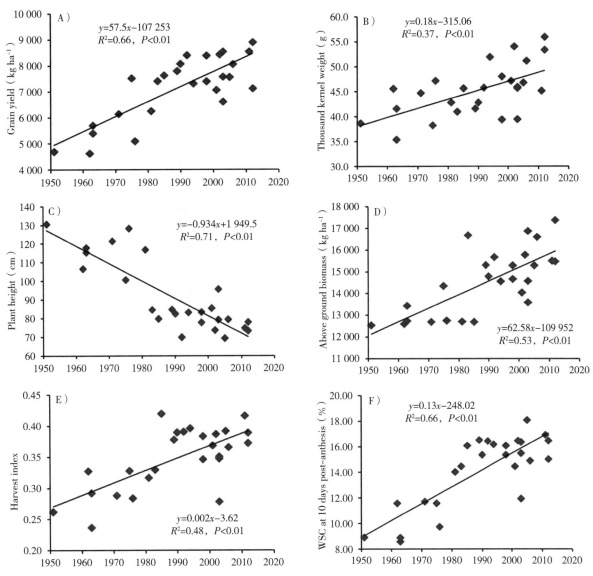

Fig. 1. Fitted linear curves on year of release for (A) grain yield, (B) thousand kernel weight, (C) plant height, (D) aboveground biomass, (E) harvest index, and (F) water-soluble carbohydrate (WSC) at 10 d post-anthesis. The fitted data were for environmental means for 26 cultivars released from 1951 to 2012.

Table 3. Means of physiological traits for 26 winter wheat cultivars grown in four environments, genetic gain ($\%$ yr^{-1}), and correlation with year of release (R^2).

Cultivar[†]	Trait[‡]										
	AGBM kg ha^{-1}	HI	WSC-10 mg g^{-1}	Chl-A	Chl-10	NDVI-A	NDVI-10	CTD-A ℃	CTD-10	LAI-A	LAI-10
Bima 1	12 531 d	0. 26 kl	8. 9 g	45. 0 l	46. 7 j	0. 59bcd	0. 55abcdef	6. 8 a	5. 0 a	4. 67defgh	3. 42 de
Neixiang 5	12 581 d	0. 32efghi	11. 6 f	51. 7fghij	50. 6efgh	0. 59bcd	0. 52efg	6. 2 a	4. 8 a	4. 73defgh	3. 63abcde
Funo	12 723 d	0. 24 l	8. 5 g	47. 6 k	47. 8ij	0. 62abcd	0. 57 ab	6. 6 a	4. 9 a	5. 39bcdefg	4. 00 a
Abbondanza	13 427 cd	0. 29hijk	8. 9 g	50. 6hij	49. 7ghi	0. 62abcd	0. 54bcdef	6. 4 a	4. 8 a	6. 26 a	3. 90abc
Zhengzhou 3	12 673 d	0. 30ghijk	11. 7 f	50. 1ij	50. 9defgh	0. 59bcd	0. 53cdefg	6. 6 a	4. 5 a	5. 50abcde	3. 98 ab

（continued）

Cultivar†	Trait‡										
	AGBM kg ha^{-1}	HI	WSC-10 mg g^{-1}	Chl-A	Chl-10	NDVI-A	NDVI-10	CTD-A ℃	CTD-10	LAI-A	LAI-10
St1472/506	14 346bcd	0. 33defg	11. 6 f	52. 5defgh	51. 4cdefgh	0. 59bcd	0. 52fg	6. 1 a	5. 0 a	6. 00 ab	3. 80abcd
Bima 4	12 738 d	0. 29ijk	9. 7 g	49. 5 j	50. 9defgh	0. 63abc	0. 52efg	6. 0 a	4. 9 a	5. 41bcdef	3. 30ef
Bainong 3217	12 702 d	0. 31fghij	14. 0 e	47. 5 k	47. 7ij	0. 64 ab	0. 54bcdefg	6. 9 a	5. 4 a	5. 70abc	3. 82abcd
Yumai 2	16 671 ab	0. 33defgh	14. 4 e	52. 1efghi	50. 3fghi	0. 62abcd	0. 55abcde	6. 9 a	5. 1 a	4. 67defgh	3. 70abcde
Yumai 7	12 683 d	0. 42 a	16. 1bcd	56. 1 a	53. 2abcdef	0. 58bcd	0. 52efg	6. 0 a	4. 9 a	5. 28bcdefg	3. 74abcd
Yumai 13	15 302abc	0. 38abc	16. 5bc	53. 1cdefg	52. 3bcdefg	0. 58 dc	0. 53efg	6. 1 a	4. 8 a	4. 92cdefgh	3. 72abcde
Yumai 18	14 775bcd	0. 40 ab	15. 4bcde	52. 6defgh	50. 3fghi	0. 57 d	0. 52efg	6. 3 a	5. 1 a	4. 74defgh	3. 04 f
Yumai 21	15 660abc	0. 39abc	16. 4bcd	52. 6defgh	51. 2defgh	0. 65 a	0. 56abc	6. 8 a	4. 9 a	4. 64efgh	3. 70abcde
Yumai 34	14 549bcd	0. 39 ab	16. 2bcd	51. 2ghij	48. 8hij	0. 62abcd	0. 50 g	6. 0 a	4. 8 a	5. 39bcdefg	3. 64abcde
Yumai 49	15 281abc	0. 39abc	16. 1bcd	53. 8bcdef	53. 6abcd	0. 60abcd	0. 57 a	6. 0 a	4. 8 a	4. 59fgh	3. 69abcde
Yumai 54	14 666bcd	0. 35cdef	15. 4bcde	55. 9 ab	55. 0 ab	0. 62abcd	0. 52efg	6. 3 a	5. 1 a	4. 98cdefgh	3. 67abcde
Zhengmai 9023	14 038 cd	0. 37bcd	14. 4 e	51. 1ghij	49. 8ghi	0. 60abcd	0. 54bcdefg	6. 1 a	5. 5 a	5. 4abcdef	3. 49cde
Zhoumai 16	15 765abc	0. 38abc	16. 5bcd	54. 8abc	54. 4 ab	0. 61abcd	0. 53cdefg	6. 3 a	5. 3 a	4. 35 h	3. 62abcde
Xinmai 18	13 579 cd	0. 28jk	11. 9 f	51. 1ghij	51. 5cdefgh	0. 62abcd	0. 55abcdef	6. 8 a	4. 9 a	5. 55abcd	3. 55bcde
Zhoumai 18	14 572bcd	0. 35cdef	16. 3bcd	51. 5ghij	52. 7abcdef	0. 60abcd	0. 53defg	6. 0 a	5. 0 a	4. 74defgh	3. 81abcd
Yanzhan 4110	16 866 ab	0. 36bcde	15. 5bcde	51. 8efghi	53. 3abcde	0. 62abcd	0. 55abcdef	6. 3 a	5. 0 a	4. 79defgh	3. 93 ab
Aikang 58	15 290abc	0. 38abc	18. 1 a	53. 9bcde	53. 5abcd	0. 61abcd	0. 56abcd	6. 0 a	4. 9 a	4. 83cdefgh	3. 81abcd
Zhoumai 22	16 593 ab	0. 37bcd	14. 9 de	52. 0efghi	54. 2abc	0. 61abcd	0. 55abcdef	6. 7 a	5. 3 a	4. 81cdefgh	4. 05 a
Zhoumai 27	15 504abc	0. 42 a	16. 9 ab	54. 5abcd	55. 4 a	0. 59bcd	0. 54bcdef	6. 2 a	5. 6 a	4. 79defgh	3. 91abc
Zhongmai 895	17 367 a	0. 37bcd	15. 0cde	51. 1ghij	50. 6efgh	0. 63abc	0. 54abcdef	6. 4 a	5. 1 a	5. 27bcdefg	3. 98 ab
Zhongmai 875	15 468abc	0. 39 ab	16. 5bcd	50. 9ghij	51. 3cdefgh	0. 60bcd	0. 53cdefg	6. 3 a	5. 5 a	4. 50gh	3. 77abcd
Mean	14 552	0. 35	14. 13	50. 88	52. 25	0. 61	0. 54	6. 3	5. 0	5. 07	3. 72
SEM §	2 583. 9	0. 04	1. 6	2. 3	3. 0	0. 06	0. 03	1. 5	1. 2	0. 9	0. 4
Absolute genetic gains	62. 6	0. 002	0. 13	0. 08	0. 09	0. 0001	0	−0. 005	0. 007	−0. 01	0. 002
Genetic gains (%)	0. 39**	0. 51**	0. 81**	0. 15**	0. 17**	0. 02	0	−0. 07	0. 12**	−0. 21	0. 05
R^2	0. 55	0. 49	0. 67	0. 30	0. 46	0. 02	0. 05	0. 05	0. 25	0. 13	0. 02

** Significant at the 0. 01 probability level.

† Cultivars ranked according to year of release.

‡ See footnote to Table 1 for trait abbreviations.

§ SEM，standard error of the means.

Table 4. Pearson's coefficients of correlation among yield and physiological traits of 26 winter wheat cultivars based on means across four environments

Trait[†]	GY	HD	PH	KNS	SN	KN	TKW	AGBM	HI	WSC-10	Chl-A	Chl-10	NDVI-A	NDVI-10	CTD-A	CTD-10	LAI-A
HD	-0.25																
PH	-0.85**	0.26															
KNS	0.17	0.46*	0.02														
SN	0.44*	-0.20	-0.54**	-0.35													
KN	0.56**	0.10	-0.54**	0.31	0.78**												
TKW	0.48*	-0.20	-0.50**	-0.35	0.08	-0.15											
AGBM	0.80**	0.03	-0.77**	0.17	0.53**	0.65**	0.47*										
HI	0.84**	-0.44*	-0.87**	-0.08	0.46*	0.42*	0.55**	0.57**									
WSC-10	0.80**	-0.35	-0.91**	-0.17	0.60**	0.50**	0.53**	0.65**	0.91**								
Chl-A	0.63**	-0.18	-0.68**	0.43*	0.14	0.44*	0.17	0.43*	0.63**	0.60**							
Chl-10	0.69**	-0.23	-0.76**	0.34	0.17	0.41*	0.29	0.52**	0.71**	0.73**	0.90**						
NDVI-A	-0.10	0.33	0.03	-0.13	0.09	0.01	0.12	0.15	-0.20	-0.06	-0.14	-0.23					
NDVI-10	0.06	0.48	-0.09	0.01	0.16	0.18	-0.04	0.26	-0.17	-0.04	-0.12	-0.09	0.31				
CTD-A	-0.24	0.61**	0.30	0.16	0.07	0.17	-0.30	0.04	-0.50**	-0.34	-0.43	-0.42	0.32	0.44			
CTD-10	0.28	-0.07	-0.34	-0.03	0.32	0.30	0.29	0.33	0.34	0.33	0.07	0.16	0.07	-0.02	0.14		
LAI-A	-0.43*	0.003	0.54**	0.08	-0.28	-0.25	-0.43*	-0.48*	-0.45*	-0.55**	-0.24	-0.44*	0.17	-0.28	-0.02	-0.23	
LAI-10	0.23	0.25	-0.12	0.28	-0.10	0.08	0.13	0.26	-0.01	0.07	0.11	0.16	0.21	0.28	0.11	0.01	0.18

* Significant at the 0.05 Probability level.
** Significant at the 0.01 Probability level.
† For trait abbreviations see Table 1.

Dwarfing Genes and the 1B. 1R Translocation

Gene-specific marker results showed that dwarfing genes was frequently used in the improvement of wheat cultivars, whereas particular photoperiod genes were not strongly selected in wheat breeding during the past 60 yr (Supplemental Table S1). *Rht-D1b* was the predomi nant

dwarfing gene (73. 1%), followed by *Rht8c* (34. 6%), and *Rht-B1b* was present only in Zhongmai 875. Cultivars genotyped with either *Rht-B1b* + *Rht8c* or *Rht-D1b* + *Rht8c* exhibited higher GY, SN, KN, TKW, AGBM, HI, stem WSC-10, Chl-A, and Chl-10 and lower PH (29. 5, 18. 0, 7. 0, 17. 0, 17. 3, 23. 7, 38. 8, 6. 2, 7. 1, and −51. 6%, respectively) (Table 5).

Table 5. Influence of dwarfing alleles on grain yield (GY), thousand-kernel weight (TKW), spike number (SN), plant height (PH), aboveground biomass (AGBM), harvest index (HI), stem water-soluble carbohydrate at 10 d postanthesis (WSC-10), SPAD (soil plant analysis development) estimates of chlorophyll content at anthesis (Chl-A), and SPAD estimates of chlorophyll content at 10 d postanthesis (Chl-10) of 26 cultivars from the southern Yellow and Huai Valleys Wheat Zone

Genotype	Number	GY kg ha^{-1}	PH cm	SN	KN	AGBM kg ha^{-1}	HI	WSC-10 mg g^{-1}	Chl-A	Chl-10
No dwarfing gene	3	5 589 b	117. 3 a	504 b	21 122 b	12 846 b	0. 29 b	9. 8 c	48. 6 b	49. 5 b
Rht-B1b, *Rht-D1b*, *Rht8*	9	6 372 b	103. 8 a	603 ab	21 705 ab	13 641 b	0. 32 b	12. 6 b	50. 2 ab	51. 5 ab
Rht-B1b+*Rht8*, *Rht-D1b*+*Rht8*	14	7 931 a	77. 4 b	615 a	22 703a	15 533 a	0. 38 a	16. 0 a	51. 8 a	53. 3 a
Increase (%)		29. 5	−51. 6	18. 0	7. 0	17. 3	23. 7	38. 8	6. 2	7. 1

From the early 1980s to 1990s, cultivars with the 1B. 1R translocation, such as Yumai 7 and Yumai 13, predomi-nated in the southern YHVWWZ (Supplemental Table S1). All leading cultivars released from 2003 to the present retain the 1B. 1R translocation, indicating a con tinuing contribution to yield potential.

DISCUSSION

Evaluations of historical cultivars is a common approach to assess the achievements of past breeding efforts (Donmez et al., 2001; Morgounov et al., 2010; Xiao et al., 2012; Sanchez-Garcia et al., 2013; Aisawi et al., 2015). The current study inclu-ding 26 winter wheat cultivars widely grown in the southern YHVWWZ during the past 60 yr was used to estimate the genetic gain in yield and to analyze con-tributory factors through the identification of yield-re-lated and physiological traits associated with yield im-provements.

Yield Potential and Components

Yield potential increasedlinearly in cultivars released in the southern YHVWWZ from 1951 to 2012 by 0. 70% yr^{-1} or 57. 5 kg ha^{-1} yr^{-1}. This was higher than the rate of 0. 60% ha^{-1} yr^{-1} (Zheng et al., 2011) and lower than 1. 05% yr^{-1} (Zhou et al., 2007b) esti-mated by others for this region. The progress in yield potential was mostly due to high yielding cultivars, such as Bainong 3217, 'Yumai 54', Aikang 58, Zhoumai 27, and Zhongmai 895 released after 1980. Elite parents were the main reason for yield im-provement. Zhongmai 895 released in 2012 has the highest yield potential at 8906 kg ha^{-1}. It was derived from 'Zhoumai 16' × 'Liken 4', and Zhoumai 16 was developed from 'Yumai 21' × 'Zhou 8425B'. Zhou 8425B is characterized by semidwarf PH (*Rht-B1b* and *Rht-D1b*) of around 65 cm, large spikes, high TKW, and multiple disease resistances. More than 100 cultivars derived from this line have been re-leased (Xiao et al., 2011). Zhou 8425B has become a milestone core parent in the southern part of the YH-

VWWZ（He et al.，2011）. It is estimated that around 80% or at least 4 million ha of the current area in Henan Province is covered by cultivars derived from Zhou 8425B（'Zhou 78A'×'Annong 7959'）. Although these cultivars have high yield potential and broad adaptation, they tend to be prone to black point（*Alternaria alternate* Keissl.）and sharp eye spot（*Rhizoctonia cerealis* Van der Hoeven）and have lower processing qualities due to the presence of the 1B/1R translocation.

Among yield components, KNS wasnot associated with genetic progress in GY, consistent with other reports from the region（Zhou et al.，2007a; Xiao et al.，2012）. Spike number was significantly increased over time, different from previous studies（Xiao et al.，2012; Aisawi et al.，2015 ）. Normally, increased KN was derived from enhanced KNS and SN（KN=KNS×SN）（Brancourt-Hulmel et al.，2003）. Hence, significant gains for KN achieved in cultivars

over time was largely attributed to increased SN. Genetic gains in GY from 1951 to 2012 were significantly associated with increasing TKW in the present study, in agreement with previous studies（Zhou et al.，2007a; Morgounov et al.，2010; Tian et al.，2011; Zheng et al.，2011; Lopes et al.，2012; Aisawi et al.，2015）. However, several other studies found no change in TKW with year of release and that GY gains were mainly associated with KN（Sayre et al.，1997; Shearman et al.，2005; Acreche et al.，2009; Brisson et al.，2010; Xiao et al.，2012）. Kernel number（$r = 0.56$, $P < 0.01$）showed larger correlation with GY than SN（$r=0.44$, $P<0.01$）and TKW（$r = 0.48$, $P<0.01$）in the present study, indicating that KN has contributed more to increased yield potential than SN and TKW. Meanwhile, SN and TKW were also positively correlated with GY and have significant progress over time, indicating that these components have commonly contributed to increased grain yield.

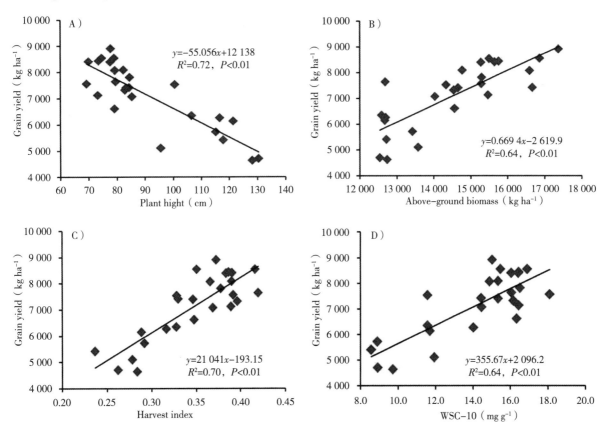

Fig. 2. Relationships between grain yield and （A）plant height, （B）aboveground biomass, （C）harvest index, and （D）water-soluble carbohydrates at 10 dpostanthesis（WSC-10）.

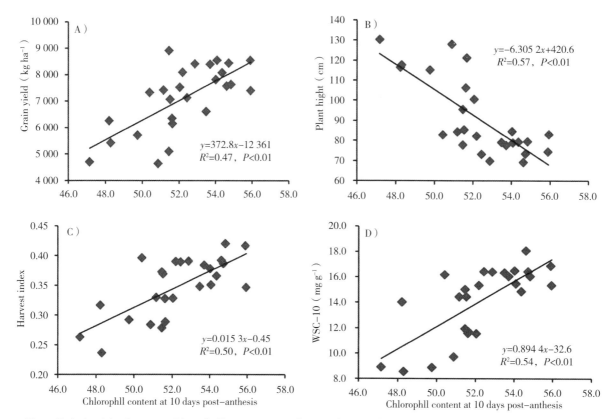

Fig. 3. Relationships between chlorophyll content at 10 dpostanthesis and (A) grain yield, (B) plant height, (C) harvest index, and (D) water-soluble carbohydrate at 10 d postanthesis (WSC-10).

Contributions of Physiological Traits

Although increases in yield potential of wheat was mainly contributed by improvements in HI (Slafer and Andrade, 1993) rather than increased AGBM (Austin et al., 1977; Brancourt-Hulmel et al., 2003; Sanchez-Garcia et al., 2013), AGBM and HI showed significant genetic gains over time and were linearly and positively correlated with GY in the present study, consistent with other reports (Xu et al., 2000; Foulkes et al., 2007; Xiao et al., 2012). The improvement in HI was highly associated with reduction in PH and increased SN. Reduced PH and increased SN increases AGBM and improves partitioning of photosynthate to the reproductive organs (Austin et al., 1980; Donmez et al., 2001; Reynolds et al., 2009). Further increasing HI, consequently, improves the final GY. The highest HI in this study was 0.42, leaving a gap of 0.20 compared with the theoretical upper limit of 0.62 (Fischer and Edmeades, 2010). It seems that increasing HI could

further improve GY in the southern YHVWWZ. However, it is difficult to attain a HI above 0.5 in wheat, and future yield improvement should rely on achieving greater AGBM while at least maintaining present HI levels.

Considering changes in resource-capture traits during anthesis and postanthesis, stem WSC-10 showed a significant, linear, and positive correlation with GY, consistent with previous reports in the United Kingdom (Foulkes et al., 1998; Shearman et al., 2005). However, Aisawi et al. (2015) reported that genetic progress in GY in northwest Mexico was not associated with increased stem WSC accumulation or stem WSC remobilization. At preanthesis, WSC accumulation in crop stems was mainly partitioned to structural stem dry matter (Beed et al., 2007), whereas WSC in stem and leaf sheaths postanthesis were mainly partitioned to the grain sink (Reynolds et al., 2009; Foulkes et al., 2011). To examine those observations in more detail, WSC in stems at anthesis, at 10, and at 20 DPA were simultaneously measured in the 2014—2015 season in the present study. The results indicated significant gains in stem WSC at all three measures over time, and these were positively correlated with GY. Stem WSC in stems at anthesis (0.94%, $P < 0.01$) showed the highest genetic progress, followed at 10 DPA (0.93%, $P < 0.01$) and at 20 DPA (0.74%, $P < 0.01$) (Supplemental Table S3). How ever, the highest correlation ($r = 0.74$, $P < 0.01$) between stem WSC and GY, and slope (0.17 mg g^{-1}) was observed at 10 DPA (Supplemental Table S4). Similar changes in WSC were reported for winter wheats released in the UK from 1972 to 1995 (Shearman et al., 2005). Although stem WSC during the late grain-filling period was consistently decreasing, much of this carbohydrate can be remobilized to grains (Foulkes et al., 2007; Xue et al., 2009). Therefore, increasing the stem WSC in leaf sheaths and stems during the grain-filling period should potentially increase grain yield. A high broad-sense heritability ($h^2 = 0.90$) for stem WSC accumulation was reported by Ruuska et al. (2006), and a gene (*TaSST-D1*) associated with stem WSC in bread wheat stems was cloned by Dong et al. (2016), indicating that selection of genotypes with high stem WSC is feasible and should be regarded as an indirect selection criterion for improvement of grain yield.

Parameters NDVI-A, NDVI-10, CTD-A, LAI-A, and LAI-10 did not change over time, and none were associated with genetic progress in yield potential. Although CTD-10 showed a significant genetic gain, it was not significantly associated with GY, in agreement with Xiao et al. (2012), but differing from a report for northwestern Mexico (Aisawi et al., 2015). Both Chl-A and Chl-10 showed significant genetic gains over time, and both exhibited linear and significant correlations with GY, AGBM, HI, stem WSC-10, and PH (Table 4, Fig. 3). Chlorophyll content is an important component of flag leaves, the major functional leaves participating in photosynthesis (Gladun and Karpov, 1993). Therefore, increasing

potential leaf photosynthesis could increase dry matter accumulation when other genetic factors are not altered (Long et al., 2006a). Most cultivars released in the southern YHVWWZ after 1980s have compact plant architectures, reduced PH, and semi-erect or erect flag leaves (Zhuang, 2003; He et al., 2010), which could improve canopy structure and allow greater light access to a larger proportion of the leaf area (Parry et al., 2011). Thus, potential leaf photosynthesis could be enhanced, enabling increased biomass production. Murata (1981) also found a higher potential of photosynthesis postanthesis in modern cultivars than in old cultivars, indicating that photosynthesis at the single-leaf level can be an important factor for potential biomass production. Consequently, there is a potential to increase yield in the southern YHVWWZ through improving photosynthetic capacity and efficiency at the postanthesis stage. In addition, AGBM, HI, and stem WSC-10 exhibited significant and positive correlations with KN and TKW, indicating that a high grain-filling source (or photosynthesis capacity) and large sink (KN and potential KW) simultaneously contributed to increased GY.

Dwarfing Genes and the1B. 1R Translocation

During the last 60 yr, significant progress in GY was achieved by utilization of dwarfing genes (He et al., 2010). In the present study, the significant yield increase that occurred in the early 1980s corresponded to reduction in PH. Almost all cultivars released after 1981 carried *Rht-D1b* and *Rht8c*, whereas only Zhongmai 875 carried dwarfing gene *Rht-B1b* (Supplemental Table S2). Compared with cultivars carrying *Rht-B1b*, *Rht-D1b*, or no dwarfing genes, *Rht-B1b+Rht8c* or *Rht-D1b+Rht8c* cultivars showed significantly higher GY, TKW, SN, AGBM, HI, stem WSC-10, Chl-A, and Chl-10 and lower PH, consistent with previous reports (Borojevic and Borojevic, 2005; Xiao et al., 2012). Plant height was reduced from 130.5 to 69.6 cm, mainly due to the integration of *Rht-D1b* and *Rht8c*, but other genes for reduced height, such as *QPH. caas-6A* (Li et al., 2015), were also involved. More effort is needed to identify additional genes for PH and to determine its possible relationship with GY.

Plant height exhibitedsignificant and negative correlations with GY, indicating that the reduced PH increased GY during the past 60 yr. However, PH cannot be continuously decreased because reducing PH will decrease biomass and further GY. Therefore, an ideal combination of PH and GY will be required to minimize the conflicts between increasing yield potential and lodging risk. An optimum PH can fully extract maximum yield potential (Fischer, 2007). Plant height has been stable at around 80 cm since the 1980s, within the reported optimum range (Richards, 1992; Miralles and Slafer, 1995). This might be one of reasons that GY sustainably increased with year of release in the southern YHVWWZ. However, lodging is again becoming a limiting factor for yield achievement and stability in farmer lands due to significantly increased seeding rates (300 − 375 kg ha^{-1}) being used in the conservation agriculture practices for the wheat-maize rotation system. Therefore, increased attention must be given to breeding for lodging resistance by increasing stem strength and toughness to achieve this (Xiao et al., 2015).

The current study showed that cultivars with the 1B. 1R translocation had higher GY, TKW, AGBM, HI, stem WSC-10, Chl-A, and Chl-10 and lower PH, in agreement with previous reports (Foulkes et al., 2007; Zhou et al., 2007a, 2007b; Wu et al., 2014). A positive effect of 1B. 1R on TKW was reported by Moreno-Sevilla et al. (1995). Considerable evidence suggests that AGBM increases with introduction of 1B. 1R into wheat germ-plasm (Carver and Rayburn, 1994; Villareal et al., 1998). The frequency of the 1B. 1R translocation remains high in Chinese wheat cultivars. Among the 26 cultivars studied here, most of those released between 1985 and 2012 carried the 1B. 1R translocation, especially those released from 2003 to 2012. At the same time, the 1BL. 1RS translocation was extensively used in wheat breeding programs in many other countries because of its better disease resistance (Lukaszewski, 1990; Rajaram et al., 1990). However, its negative effects on pan bread, noodle quality (Graybosch, 2001; He et al., 2005), and flour color (Zhang et al., 2009) are increasingly reducing its use.

In summary, increased TKW, SN, and KN and significantly decreased PH have contributed to the continuous genetic progress in yield potential from 1951 to 2012 in the southern YHVWWZ. Significant improvements in AGBM, HI, and stem WSC were highly associated with increased yield potential. The introduction of dwarfing genes (including *Rht-D1b* and *Rht8c*) into winter wheat germplasm at the southern YHVWWZ increased the yield potential. Therefore, the physiological traits such as AGBM, HI, and stem WSC should be considered important criteria in breeding besides selection of SN, KN, and TKW in the southern YHVWWZ. In addition, control of weeds, plant diseases, and insect pests, and utilization of disease resistant germplasm, particularly for Fusarium head blight resistance, will be important for future yield improvement and sustainability in the region.

Acknowledgements

The authors thank Prof. R. A. McIntosh, Plant Breeding Institute, University of Sydney, for review of this manuscript. This study was supported by the National Natural Science Foundation of China (31161140346) and Beijing Municipal Science and Technology Project (D151100004415003).

❖ References

Acreche, M. M., G. Briceño-Félix, J. A. M. Sánchez, and G. A. Slafer. 2009. Radiation interception and use efficiency as affected by breeding in Mediterranean wheat. Field Crops Res. 110: 91-97. doi: 10. 1016/j. fcr. 2008. 07. 005.

Aisawi, K. A. B., M. P. Reynolds, R. P. Singh, and M. J. Foulkes. 2015. The physiological basis of the genetic progress in yield potential of CIMMYT spring wheat cultivars from 1966 to 2009. Crop Sci. 55: 1749-1764. doi:

10. 2135/cropsci2014. 09. 0601.

Álvaro, F., J. Isidro, D. Villegas, L. F. Garciadelmoral, and C. Royo. 2008. Old and modern durum wheat varieties from Italy and Spain differ in main spike components. Field Crops Res. 106: 86-93. doi: 10. 1016/j. fcr. 2007. 11. 003.

Austin, R. B., J. Bingham, R. D. Blackwell, L. T. Evans, M. A. Ford, C. L. Morgan, and M. Taylor. 1980. Genetic improvements in winter wheat yields since 1900 and associated physiological changes. J. Agric. Sci. 94: 675-689. doi: 10. 1017/ S0021859600028665.

Austin, R. B., J. A. Edrich, M. A. Ford, and R. D. Blackwell. 1977. The fate of the dry matter, carbohydrates and 14 C lost from leaves and stems of wheat during grain filling. Ann. Bot. (Lond.) 41: 1309-1321.

Ayeneh, A., G. M. Van, M. P. Reynolds, and K. Ammar. 2002. Comparison of leaf, spike, peduncle, and canopy temperature depression in wheat under heat stress. Field Crops Res. 79: 173-184. doi: 10. 1016/S0378-4290 (02) 00138-7.

Beche, E., G. Benin, C. L. da Silva, L. B. Munaro, and J. A. Marchese. 2014. Genetic gain in yield and changes associated with physiological traits in Brazilian wheat during the 20th century. Eur. J. Agron. 61: 49-59. doi: 10. 1016/ j. eja. 2014. 08. 005.

Beed, F. D., N. D. Paveley, and R. Sylvester-Bradley. 2007. Predict ability of wheat growth and yield in light-limited conditions. J. Agric. Sci. 145: 63-79. doi: 10. 1017/S0021859606006678.

Borojevic, K., and K. Borojevic. 2005. Historic role of the wheat variety Akakomugi in southern and central European wheat breeding programs. Breed. Sci. 55: 253-256. doi: 10. 1270/ jsbbs. 55. 253.

Brancourt-Hulmel, M., G. Doussinault, C. Lecomte, P. Berard, B. Le Buanec, and M. Trottet. 2003. Genetic improvement of agronomic traits of winter wheat cultivars released in France from 1946 to 1992. Crop Sci. 43: 37-45. doi: 10. 2135/crop-sci2003. 3700.

Brisson, N., P. Gate, D. Gouache, G. Charmet, F. X. Oury, and F. Huard. 2010. Why are wheat yields stagnating in Europe? A comprehensive data analysis for France. Field Crops Res. 119: 201-212. doi: 10. 1016/ j. fcr. 2010. 07. 012.

Calderini, D. F., M. F. Dreccer, and G. A. Slafer. 1997. Conse quences of breeding on biomass, radiation interception and radiation-use efficiency in wheat. Field Crops Res. 52: 271-281. doi: 10. 1016/S0378-4290 (96)

03465-X.

Canevara, M. G., M. Romani, M. Corbellini, M. Perenzin, and B. Borghi. 1994. Evolutionary trends in morphological, physiological, agronomical and qualitative traits of Triticum aestivum L. cultivars bred in Italy since 1900. Eur. J. Agron. 3: 175-185. doi: 10. 1016/S1161-0301 (14) 80081-6.

Carver, B. F., and A. L. Rayburn. 1994. Comparison of related wheat stocks possessing 1B or 1RS. 1BL chromosomes: Agronomic performance. Crop Sci. 34: 1505-1510. doi: 10. 2135/cro psci1994. 0011183X003400060017x.

Chen, W. F., Z. J. Xu, and L. B. Zhang. 1995. Physiological bases of super high yield breeding in rice. (In Chinese.) Liaoning Science and Technology Publishing Company, Shenyang, China.

Clarke, S., R. Sylvester-Bradley, J. Foulkes, D. Ginsburg, O. Gaju, J. P. Werner et al. 2012. Adapting wheat to global warming or 'ERYCC' earliness and resilience for yield in a changing climate. HGCA Research and Development Rep. 496. HGCA, London.

Dong, Y., Y. Zhang, Y. G. Xiao, J. Yan, J. D. Liu, W. E. Wen et al. 2016. Cloning of TaSST genes associated with water soluble carbohydrate content in bread wheat stems and development of a functional marker. Theor. Appl. Genet. 5: 1061-1070. doi: 10. 1007/s00122-016-2683-5.

Donmez, E., R. G. Sears, J. P. Shroyer, and G. M. Paulsen. 2001. Genetic gain in yield attributes of winter wheat in the Great Plains. Crop Sci. 41: 1412-1419. doi: 10. 2135/ cropsci2001. 4151412x.

Dreccer, M. F., A. F. Herwaarden, and S. C. Chapman. 2009. Grain number and grain weight in wheat lines contrasting for stem water soluble carbohydrate concentration. Field Crops Res. 112: 43-54. doi: 10. 1016/j. fcr. 2009. 02. 006.

Ehdaie, B., G. A. Alloush, M. A. Madore, and J. G. Waines. 2006. Genotypic variation for stem reserves and mobilization in wheat: I. Postanthesis changes in internode dry matter. Crop Sci. 46: 735-746. doi: 10. 2135/crops-ci2005. 04-0033.

Fischer, R. A. 2015. Definitions and determination of crop yield, yield gaps, and of rates of change. Field Crops Res. 182: 9-18. doi: 10. 1016/j. fcr. 2014. 12. 006.

Fischer, R. A. 2007. Understanding the physiological basis of yield potential in wheat. J. Agric. Sci. 145: 99-113. doi: 10. 1017/ S0021859607006843.

Fischer, R. A., and G. O. Edmeades. 2010. Breeding and

cereal yield progress. Crop Sci. 50: S-85-S-98. doi: 10. 2135/crop-sci2009. 10. 0564.

Fischer, R. A., D. Rees, K. D. Sayre, Z. -M. Lu, A. G. Condon, and A. Larque Saavedra. 1998. Wheat yield progress associated with higher stomatal conductance and photosynthetic rate, and cooler canopies. Crop Sci. 38: 1467-1475. doi: 10. 2135/cro psci1998. 0011183X003800060011x.

Foulkes, M. J., G. A. Slafer, W. J. Davies, P. M. Berry, R. Sylvester-Bradley, P. Martre et al. 2011. Raising yield potential of wheat. III. Optimizing partitioning to grain while maintaining lodging resistance. J. Exp. Bot. 62: 469-486. doi: 10. 1093/ jxb/erq300.

Foulkes, M. J., J. W. Snape, V. J. Shearman, M. P. Reynolds, O. Gaju, and R. Sylvester-Bradley. 2007. Genetic progress in yield potential in wheat: Recent advances and future prospects. J. Agric. Sci. 145: 17-29. doi: 10. 1017/S0021859607006740.

Foulkes, M. J., J. H. Spink, R. K. Scott, and R. W. Clare. 1998. Varietal typing trials and NIAB additional character assessments. Home-Grown Cereals Authority Final Rep. 174. Vol. V. HGCA, London.

Gladun, I. V., and E. A. Karpov. 1993. Distribution of assimilates from the flag leaf of rice during the reproductive period of development. Russ. J. Plant Physiol. 40: 215-219.

Grassini, P., K. M. Eskridge, and K. G. Cassman. 2013. Distinguishing between yield advances and yield plateaus in historical crop production trends. Nat. Commun. 4: 2918. doi: 10. 1038/ ncomms3918.

Graybosch, R. A. 2001. Uneasy unions: Quality effects of rye chromatin transfers to wheat. J. Cereal Sci. 33: 3-16. doi: 10. 1006/ jcrs. 2000. 0336.

Graybosch, R. A., and C. J. Peterson. 2010. Genetic improvement in winter wheat yields in the Great Plains of North America, 1959-2008. Crop Sci. 50: 1882-1890. doi: 10. 2135/crop-sci2009. 11. 0685.

He, Z. H., L. Liu, X. C. Xia, J. J. Liu, and R. J. Pena. 2005. Composition of HMW and LMW glutenin subunits and their effects on dough properties, pan bread, and noodle quality of Chinese bread wheat. Cereal Chem. 82: 345-350. doi: 10. 1094/CC-82-0345.

He, Z. H., X. C. Xia, and A. P. A. Bonjean. 2010. Wheat improvement in China. In: Z. H. He and A. P. A. Bonjean, editors, Cereals in China. CIMMYT, D. F., Mexico. p. 51-68.

He, Z. H., X. C. Xia, X. M. Chen, and Q. S. Zhuang. 2011. Progress of wheat breeding in China and the future

perspective. (In Chinese, with English abstract.) Acta Agron. Sin. 37: 202-215. doi: 10. 3724/SP. J. 1006. 2011. 00202.

Hucl, P., C. Briggs, R. J. Graf, and R. N. Chibbar. 2015. Genetic gains in agronomic and selected end-use quality traits over a century of plant breeding of Canada western red spring wheat. Cereal Chem. 6: 537-543 doi: 10. 1094/CCHEM-02-15-0029-R.

Kumar, S., P. Kumari, U. Kumar, M. Grover, A. K. Singh, R. Singh, and R. S. Sengar. 2013. Molecular approaches for designing heat tolerant wheat. J. Plant Biochem. Biotechnol. 22: 359-371. doi: 10. 1007/s13562-013-0229-3.

Li, X. M., X. C. Xia, Y. G. Xiao, Z. H. He, D. S. Wang, R. Trethowan et al. 2015. QTL mapping for plant height and yield components in common wheat under water limited and full irrigation environments. Crop Pasture Sci. 66: 660-670. doi: 10. 1071/CP14236.

Long, S. P., E. A. Ainsworth, A. D. B. Leakey, J. Nösberger, and D. R. Ort. 2006a. Food for thought: Lower-than-expected crop yield simulation with rising CO_2 concentrations. Science 312: 1918-1921. doi: 10. 1126/science. 1114722.

Long, S. P., X. G. Zhu, S. L. Naidu, and D. R. Ort. 2006b. Can improvement in photosynthesis increase crop yields? Plant Cell Environ. 29: 315-330. doi: 10. 1111/ j. 1365-3040. 2005. 01493. x.

Lopes, M. S., M. P. Reynolds, Y. Manes, R. P. Singh, J. Crossa, and H. J. Braun. 2012. Genetic yield gains and changes in associated traits of CIMMYT spring bread wheat in a "Historic" set representing 30 years of breeding. Crop Sci. 52: 1123-1131. doi: 10. 2135/cropsci2011. 09. 0467.

Lukaszewski, A. J. 1990. Frequency of 1RS/1AL and 1RS/1BL translocations in United States wheats. Crop Sci. 30: 1151-1153. doi: 10. 2135/cropsci1990. 0011183X003000050041x.

Mackay, I., A. Horwell, J. Garner, J. White, J. McKee, and H. Philpott. 2011. Reanalyses of the historical series of UK variety trials to quantify the contributions of genetic and environmental factors to trends and variability in yield over time. Theor. Appl. Genet. 122: 225-238. doi: 10. 1007/s00122-010-1438-y.

Makino, A. 2011. Photosynthesis, grain yield, and nitrogen utilization in rice and wheat. Plant Physiol. 155: 125-129. doi: 10. 1104/pp. 110. 165076.

Manès, Y., H. F. Gomez, L. Puhl, M. Reynolds, H. J. Braun, and R. Trethowan. 2012. Genetic yield gains of the

CIMMYT international semi-arid wheat yield trials from 1994 to 2010. Crop Sci. 52: 1543-1552. doi: 10. 2135/cropsci2011. 10. 0574.

Matus, I. , M. Mellado, M. Pinares, R. Madariaga, and A. delPozo. 2012. Genetic progress in winter wheat cultivars released in Chile from 1920 to 2000. Chilean J. Agric. Res. 72: 303-308. doi: 10. 4067/S0718-58392012000300001.

Miralles, D. J. , and G. A. Slafer. 1995. Yield, biomass and yield components in dwarf, semi-dwarf and tall isogenic lines of spring wheat under recommended and late sowing dates. Plant Breed. 114: 392-396. doi: 10. 1111/j. 1439-0523. 1995. tb00818. x.

Moreno-Sevilla, B. , P. S. Baenziger, C. J. Peterson, R. A. Graybosch, and D. V. Mcvey. 1995. The 1BL/1RS translocation: Agronomic performance of derived lines from a winter wheat cross. Crop Sci. 35: 1051-1055. doi: 10. 2135/cropsci1995. 00111 83X003500040022x.

Morgounov, A. , V. Zykin, I. Belan, L. Roseeva, Yu. Zelenskiy, H. F. Gomez-Becerra et al. 2010. Genetic gains for grain yield in high latitude spring wheat grown in Western Siberia in 1900-2008. Field Crops Res. 117: 101-112. doi: 10. 1016/j. fcr. 2010. 02. 001.

Murata, Y. 1981. Dependence of the potential productivity and efficiency in solar energy utilization on leaf photosynthetic capacity in crop species. Jpn. J. Crop. Sci. 50: 223-232. doi: 10. 1626/jcs. 50. 223.

Parry, M. , M. Reynolds, M. Salvucci, C. A. Raines, P. J. Andralojc, X. G. Zhu et al. 2011. Raising yield potential of wheat. II. Increasing photosynthetic capacity and efficiency. J. Exp. Bot. 62: 453-467. doi: 10. 1093/jxb/erq304.

Pask, A. J. D. , J. Pietragalla, D. M. Mullan, and M. P. Reynolds. 2012. Physiological breeding II: A field guide to wheat phenotyping. CIMMYT, Mexico.

Peltonen-Sainio, P. , L. Jauhiainen, and I. P. Laurila. 2009. Cereal yield trends in northern European conditions: Changes in yield potential and its realisation. Field Crops Res. 110: 85-90. doi: 10. 1016/j. fcr. 2008. 07. 007.

Pinto, R. S. , M. P. Reynolds, K. L. Mathews, C. L. McIntyre, J. -J. Olivares-Villegas, and S. C. Chapman. 2010. Heat and drought adaptive QTL in a wheat population designed to minimize confounding agronomic effects. Theor. Appl. Genet. 121: 1001-1021. doi: 10. 1007/s00122-010-1351-4.

Rajaram, S. , R. Villareal, and A. Mujeeb-Kazi. 1990. The global impact of 1B/1R spring wheat. In: Agronomy abstracts. ASA, Madison, WI. p. 105.

Rasheed, A. , W. E. Wen, F. M. Gao, S. N. Zhai, H. Jin, J. D. Liu et al. 2016. Development and validation of KASP assays for functional genes underpinning key economic traits in bread wheat. Theor. Appl. Genet. 129: 1843-1860. doi: 10. 1007/ s00122-016-2743-x.

Reynolds, M. , M. J. Foulkes, G. A. Slafer, P. Berry, M. A. J. Parry, J. W. Snape, and W. J. Angus. 2009. Raising yield potential in wheat. J. Exp. Bot. 60: 1899-1918. doi: 10. 1093/jxb/erp016.

Reynolds, M. P. , J. I. Ortiz-Monasterio, and A. Mcnab. 2001. Application of physiology in wheat breeding. CIMMYT, El Batan, Mexico.

Richards, R. A. 1992. The effect of dwarfing genes in spring wheat in dry environments. I. Agronomic characteristics. Aust. J. Agric. Res. 43: 517-527. doi: 10. 1071/AR9920517.

Royo, C. , F. Álvaro, V. Martos, A. Ramdani, J. Isidro, D. Villegas, and L. F. Garciademoral. 2007. Genetic changes in durum wheat yield components and associated traits in Italian and Spanish varieties during the 20th century. Euphytica 155: 259-270. doi: 10. 1007/s10681-006-9327-9.

Ruuska, S. A. , G. J. Rebetzke, A. F. van Herwaarden, R. A. Richards, N. A. Fettell, L. Tabe, and C. L. D. Jenkins. 2006. Genotypic variation in water-soluble carbohydrate accumulation in wheat. Funct. Plant Biol. 33: 799-809. doi: 10. 1071/FP06062.

Sadras, V. O. , and C. Lawson. 2011. Genetic gain in yield and associated changes in phenotype, trait plasticity and competitive ability of South Australian wheat varieties released between 1958 and 2007. Crop Pasture Sci. 62: 533-539. doi: 10. 1071/ CP11060.

Salem, K. F. M. , M. S. Roder, and A. Borner. 2007. Identification and mapping quantitative trait loci for stem reserve mobilisation in wheat (Triticum aestivum L.). Cereal Res. Commun. 35: 1367-1374. doi: 10. 1556/CRC. 35. 2007. 3. 1.

Slafer, G. A. , and F. H. Andrade. 1993. Physiological attributes related to the generation of grain yield in bread wheat cultivars released at different eras. Field Crops Res. 31: 351-367. doi: 10. 1016/0378-4290 (93) 90073-V.

Sanchez-Garcia, M. , C. Royo, N. Aparicio, J. Martin-Sanchez, and F. Álvaro. 2013. Genetic improvement of bread wheat yield and associated traits in Spain during the 20th century. J. Agric. Sci. 151: 105-118. doi: 10. 1017/S0021859612000330.

SAS Institute. 2000. SAS user's guide: Statistics. SAS Inst. , Cary, NC.

Sayre, K. D. , S. Rajaram, and R. A. Fischer. 1997. Yield

potential progress in short bread wheat in northwest Mexico. Crop Sci. 37: 36-42. doi: 10. 2135/cropsci1997. 0011183X003700010006x.

Shearman, V. J., R. Sylvester-Bradley, R. K. Scott, and M. J. Foulkes. 2005. Physiological processes associated with wheat yield progress in the UK. Crop Sci. 45: 175-185. doi: 10. 2135/crop-sci2005. 0175.

Siddique, K. H. M., R. K. Belford, M. W. Perry, and D. Tennant. 1989. Growth, development and light interception of old and modern wheat cultivars in a Mediterranean-type environment. Aust. J. Agric. Res. 40: 473-487 doi: 10. 1071/ AR9890473.

Tian, Z. W., Q. Jing, T. Dai, D. Jiang, and W. Cao. 2011. Effects of genetic improvements on grain yield and agronomic traits of winter wheat in the Yangtze River Basin of China. Field Crops Res. 124: 417-425. doi: 10. 1016/j. fcr. 2011. 07. 012.

Villareal, R. L., O. Banuelos, A. Mujeeb-Kazi, and S. Rajaram. 1998. Agronomic performance of chromosomes 1B and 1BL. 1RS near-isolines in the spring bread wheat Seri M82. Euphytica 103: 195-202. doi: 10. 1023/A: 1018392002909.

Wang, Z. H., X. Liu, R. Li, X. P. Chang, and R. L. Jing. 2011. Development of near-infrared reflectance spectroscopy models for quantitative determination of water-soluble carbohydrate content in wheat stem and glume. Anal. Lett. 44: 2478-2490. doi: 10. 1080/00032719. 2011. 551859.

Wu, W., C. J. Li, B. L. Ma, F. Shah, Y. Liu, and Y. C. Liao. 2014. Genetic progress in wheat yield and associated traits in China since 1945 and future prospects. Euphytica 196: 155-168. doi: 10. 1007/s10681-013-1033-9.

Xiao, Y. G., J. J. Liu, H. S. Li, X. Y. Cao, X. C. Xia, and Z. H. He. 2015. Lodging resistance and yield potential of winter wheat: Effect of planting density and genotype. Front. Agric. Sci. Eng. 2: 168-178. doi: 10. 15302/J-FASE-2015061.

Xiao, Y. G., Z. G. Qian, K. Wu, J. J. Liu, X. C. Xia, W. Q. Ji, and Z. H. He. 2012. Genetic gains in grain yield and physiological traits of winter wheat in Shandong Province, China, from 1969 to 2006. Crop Sci. 52: 44-

56. doi: 10. 2135/crop-sci2011. 05. 0246.

Xiao, Y. G., G. H. Yin, H. H. Li, X. C. Xia, J. Yan, T. C. Zheng et al. 2011. Genetic diversity and genome-wide association analysis of stripe rust resistance among the core wheat parent Zhou 8425B and its derivatives. Zhong-guo Nongye Kexue (Beijing, China) 44: 3919-3929.

Xu, W., L. Hu, Z. Wu, and J. Gai. 2000. Studies on genetic improvement of yield and yield components of wheat cultivars in Mid-Shanxi area. (In Chinese, with English abstract.) ActaAgron. Sin. 26: 352-357.

Xue, G. P., C. L. McIntyre, A. R. Rattey, A. F. van Herwaarden, and R. Shorter. 2009. Use of dry matter content as a rapid and low-cost estimate for ranking genotypic differences in water-soluble carbohydrate concentrations in the stem and leaf sheath of *Triticum aestivum*. Crop Pasture Sci. 60: 51-59. doi: 10. 1071/CP08073.

Zhang, Y. L., Y. P. Wu, Y. G. Xiao, Z. H. He, Y. Zhang, J. Yan et al. 2009. QTL mapping for flour and noodlecolour components and yellow pigment content in common wheat. Euphytica 165: 435-444. doi: 10. 1007/s10681-008-9744-z.

Zheng, T. C., X. K. Zhang, G. H. Yin, L. N. Wang, Y. L. Han, L. Chen et al. 2011. Genetic gains in grain yield, net photosynthesis and stomatal conductance achieved in Henan Province of China between 1981 and 2008. Field Crops Res. 122: 225-233. doi: 10. 1016/j. fcr. 2011. 03. 015.

Zhou, Y., Z. H. He, X. X. Sui, X. C. Xia, X. K. Zhang, and G. S. Zhang. 2007a. Genetic improvement of grain yield and associated traits in the northern China winter wheat region from 1960 to 2000. Crop Sci. 47: 245-253. doi: 10. 2135/cropsci2006. 03. 0175.

Zhou, Y., H. Z. Zhu, S. B. Cai, Z. H. He, K. Zhang, X. C. Xia, and G. S. Zhang. 2007b. Genetic improvement of grain yield and associated traits in the southern China winter wheat region: 1949 to 2000. Euphytica 157: 465-473. doi: 10. 1007/s10681-007-9376-8.

Zhuang, Q. S. 2003. Chinese wheat improvement and pedigree analysis (In Chinese.) China Agricultural Press, Beijing, China.

Accuracy assessment of plant height using an unmanned aerial vehicle for quantitative genomic analysis in bread wheat

Muhammad Adeel Hassan[1], Mengjiao Yang[1,2], Luping Fu[1], Awais Rasheed[1,3,4], Bangyou Zheng[5], Xianchun Xia[1], Yonggui Xiao[1] and Zhonghu He[1,3]

[1] Institute of Crop Sciences, National Wheat Improvement Centre, Chinese Academy of Agricultural Sciences (CAAS), Beijing 100081, China.

[2] College of Agronomy, Xinjiang Agricultural University, Ürümqi 830052, China.

[3] Inter-national Maize and Wheat Improvement Centre (CIMMYT) China Office, c/o CAAS, Beijing 100081, China.

[4] Department of Plant Sciences, Quaid-i-Azam University, Islamabad 45320, Pakistan.

[5] CSIRO Agriculture and Food, Queens-land Bioscience Precinct, 306 Carmody Road, St Lucia 4067, Australia.

Abstract

Background: Plant height is an important selection target since it is associated with yield potential, stability and particularly with lodging resistance in various environments. Rapid and cost-effective estimation of plant height from airborne devices using a digital surface model can be integrated with academic research and practical wheat breeding programs. A bi-parental wheat population consisting of 198 doubled haploid lines was used for time-series assessments of progress in reaching final plant height and its accuracy was assessed by quantitative genomic analysis. UAV-based data were collected at the booting and mid-grain fill stages from two experimental sites and compared with conventional measurements to identify quantitative trait loci (QTL) underlying plant height.

Results: A significantly high correlation of $R^2 = 0.96$ with a 5.75 cm root mean square error was obtained between UAV-based plant height estimates and ground truth observations at mid-grain fill across both sites. Correlations for UAV and ground-based plant height data were also very high ($R^2 = 0.84$-0.85, and 0.80-0.83) between plant height at the booting and mid-grain fill stages, respectively. Broad sense heritabilities were 0.92 at booting and 0.90-0.91 at mid-grain fill across sites for both data sets. Two major QTL corresponding to *Rht-B1* on chromosome 4B and *Rht-D1* on chromosome 4D explained 61.3% and 64.5% of the total phenotypic variations for UAV and ground truth data, respectively. Two new and stable QTL on chromosome 6D seemingly associated with accelerated plant growth was identified at the booting stage using UAV-based data. Genomic prediction accuracy for UAV and ground-based data sets was significantly high, ranging from $r = 0.47$-0.55 using genomewide and QTL markers for plant height. However, prediction accuracy declined to $r = 0.20$-0.31 after excluding markers linked to plant height QTL.

Conclusion: This study provides a fast way to obtain time-series estimates of plant height in

Published in Plant Methods, 2019, 15 (1): 1-12.

understanding growth dynamics in bread wheat. UAV-enabled phenotyping is an effective, highthroughput and cost-effective approach to understand the genetic basis of plant height in genetic studies and practical breeding.

Keywords: Aerial surveillance, Genomic prediction, Quantitative trait loci, *Triticum aestivum*

Background

Plant height is an important agronomic trait and it was reduction in plant height that enabled the Green Revolution[1]. Although plant height has been reduced to around 75-80 cm for irrigated wheat with high yield potential, its control remains a very important aspect in breeding programs. Two major genes, *Rht1* (or *Rht-B1b*) and *Rht2* (or *Rht-D1b*) confer reduced plant height without detrimental effects on grain yield potential in varying environments[2]. *Rht* genes also have confounding effects on anther extrusion: a major trait for hybrid wheat production[3,4], resistance to Fusarium head blight (FHB)[5,6], and resistance to at least one insect pest[7]. Therefore, fine-tuning of plant height for a target environment is not only important for pure-line breeding but can also be important in hybrid wheat breeding where tallness of the male parent is required for efficient production of hybrids[8]. However, the association of *Rht-B1* and *Rht-D1* with undesirable traits, for example shortened coleoptile length, has caused wheat researcher to seek alternate dwarfing genes with less adverse effects. Recently, *Rht24* was reported as new gene for reduced plant height but affecting floral architecture and response to FHB[8,9]. It was also reported to increase kernel weight[10]. Reports of some other reduced height genes, such as *Rht4*, *Rht5*, *Rht7*, *Rht8*, *Rht9*, *Rht12*, *Rht13*, *Rht14*, *Rht16*, *Rht18*, *Rht21*, *Rht23*, and *Rht25*, also offer other possibilities for wheat improvement[11].

Marker-assisted selection based on quantitative trait loci (QTL) or functional genes can enhance the selection accuracy and ultimately increase genetic gain in each breeding cycle[12,13]. Wheat has determinate growth habit thus plant height progressively increases during vegetative growth until the reproductive stage. Conventionally, plant height is measured once, after anthesis when full height potential has been reached. Therefore, temporal characterization of plant height could provide a better understanding about the mechanism of plant growth and underlying genetics[14]. Quantitative methods, such as QTL analysis and association mapping, can give an insight about the genetic loci and genomic prediction analysis help in selection of genotypes with strong genetic basis for trait of interest[15,16].

Multi-location characterization of wheat germplasm is essential to evaluate adaptability of genotypes and patterns of G×E interaction for trait stability[17]. Field-based phenotyping tends to be laborious, with high likelihood of error and represents a major bottleneck for genome-to-phenome knowledge[18]. High throughput phenotyping platforms have higher capability for high precision, non-destructive characterization of quantitative traits[19]. Recent advances in proximal remote sensing using unmanned aerial vehicles (UAV) with RGB (red, green, blue) and multi-spectral imaging have made it possible to create high throughput, cost-effective and accurate quantitative phenotyping datasets[12,20]. UAV platforms can easily acquire multi-point data for complex traits such as biomass, normalized difference vegetation index, plant density, early emergence, rate of senescence rate, and plant height[20-25]. These platforms are low cost compared to traditional and recently advanced ground-based phenotyping platforms[25].

UAV-based plant height has been estimated using digital surface models (DSM). High correlations with ground-based reference measurements have been made for barley[21], wheat[26], poppy[27] and sorghum[28]. DSM gives information of altitude in the form of raster values. The drawbacks of previous approaches were that estimations were made of the average heights of whole canopies, including not only the heights of ears, but also the heights of lower leaves and even the elevation of bare ground patches within canopy gaps[29]. Furthermore, accurate assessment of the ground surface elevation imposes a major restriction factor data acquisition for UAV-based phenotyping of plant height in crops such as wheat with dense canopies. These limitations have made UAV-based platforms more complex and time-consuming by increasing the workload such as flights before planting and post-imaging quality control analysis[30]. This kind of data noise can adversely affect genetic analyses and genome-based selection. Previously, DSM-derived plant height data had been applied for genomic prediction in sorghum[24]. Therefore, there is a need to standardize UAV-based data for accurate and error-free characterization of plant height for quantitative genetic studies and selection of advanced lines in breeding pro-gram. To date, there is no report on the use of UAV-derived plant height data for quantitative loci analysis in wheat.

The major objectives of the present study were to (1) standardize a rapid method for plant height estimation using a UAV platform, (2) identify quantitative trait loci for plant height using UAV and ground-based measurements, and (3) assess genomic prediction accuracy for plant height in wheat.

Materials and methods

Germplasm and experimental design

A set of 198 doubled haploid (DH) lines derived from the cross Yangmai 16/Zhongmai 895 were used to evaluate a UAV-based platform for measuring plant height and its application in QTL analysis and genomic prediction. Yangmai 16 and Zhongmai 895 are elite varieties that have been widely cultivated in Yangtze River, and Yellow and Huai Valleys regions, respectively. Experiments were conducted during 2016-2017 and 2017-2018 at Xinxiang (35°18′0″N, 113°52′0″E) and Luohe (33°34′0″N, 114°2′0″E), both in Henan province. The DH lines and two parents were planted in randomized complete blocks with three replications (200 genotype × 3 replications) at each site. The size of each plot was 3.9 m² (1.3 m×3 m) with six rows at 0.30 cm spacing and the plant density was maintained at 270 plants/m². Both sites were irrigated at same developmental stages according to local agricultural practices.

Remote sensing campaign, mosaicking and DSM generation

An auto-operational DJI Inspires 1 model T600 (SZ DJI Technology Co., Shenzhen) carrying a Sequoia 4.0 camera (https://www.micasense.com/parrot-sequoia/) was used for aerial imagery. Sequoia has a 16-megapixel RGB camera and 4 monochrome sensors (NIR, Red, Green and Red-edge). Flight missions over the targeted field were controlled by flight planning software Altizure DJI version 3.6.0 (https://www.altizure.com). Images were acquired in sunny conditions from 30 m altitude while maintaining 85% forward and 85% side overlapping between images to ensure enough ground sampling distance. Pix4D Mapper(Pix4D, Lausanne, Switzerland) (https://pix4d.com/) was used for orthomosaic and DSM generation using world geographic coordinates of GCPs as previously reported by Hassan et al. [20]. Pix4D has the advantage of auto-processing in feature point matching and point cloud generation. All correspondence between overlapping images estimated from their geographical coordinates and pixels were used to detect the accuracy of matching points to minimize spaces between point clouds. The image resolution or ground sampling distance at 30 m was 2.5 cm/pixel.

DSM evaluation and plant height model (PHM) development

As wheat canopies are relatively dense at maturity, there are lower possibilities of error in detecting bare ground patches within the canopy, especially if plants densities are maintained at 270 plants/m². For more accuracy, ortho-mosaic images with Red and Green bands were used to classify the vegetation and bare ground soil[27]. Visual classification of bare soil patches and separation between plots were also done by RGB images. DSM gen-eration was based on the World Geodetic System (1984), which does not reflect the actual height of canopies. The digital terrain model (DTM) was generated through raster values of bare ground along the edges of each plot; this gave information on the altitude of the ground surface[21]. For this, polygon shapes were sketched on bare ground surfaces across the experimental area to deter-mine the lowest and highest ground elevation points in each zone, to minimise overall surface curvature using QGIS 1.18.15 (www.qgis.org). The PHM was calculated by subtracting the DTM from the DSM (Fig. 1).

$$PHM = DSM - DTM \quad (1)$$

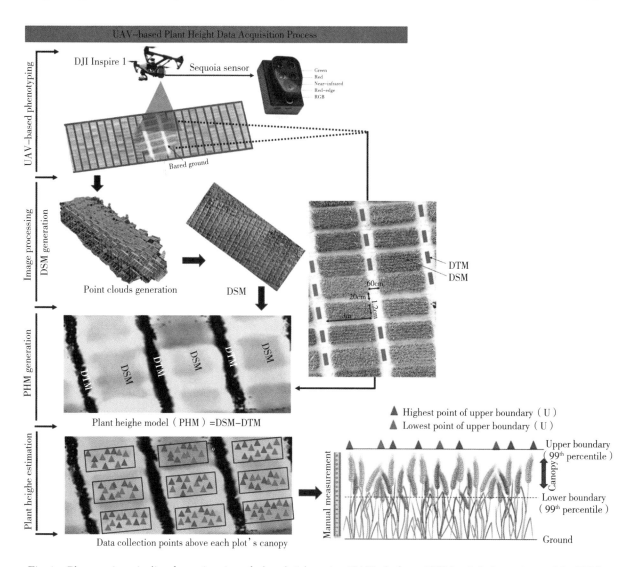

Fig. 1　Phenotyping pipeline for estimation of plant height using UAV platform. DTM, digital terrain model; DSM, digital surface model; PSM, plant surface model; UAV, unmanned aerial vehicle

Estimation and validation of UAV-based plant heights

After construction of the PHM, a workflow program reported by Hassan et al.[20] was followed for segmentation of the PHM into specific genotypes representing plots by sketching polygon shapes using QGIS 1.18.15 (www.qgis.org). For precise segmentation, ortho-mosaic images generated sequentially with DSM were used for segmentation. In order to avoid over-lapping of plants from adjacent plots, plant heights were estimated from a trimmed section of the plots to overcome expected data noise at the plot margins. UAV-based plant height was averaged from pixel values obtained at the highest and lowest points in the upper boundary of the canopy to avoid detection of low pixel values from lower canopy boundaries. The lower boundary of the canopy might include the elevation of gaps between plants and leaves. Lower and upper elevations of each plot from PHM were estimated by zonal statistics of polygon shapes using QGIS 1.18.15

(www. qgis. org). Small polygon shapes within each plot were sketched randomly to obtain upper and lower boundaries of the canopy top assuming a 10 cm difference while rejecting the extreme lower values that could not be spike height. Individual plant heights were calculated as the mean of randomly estimated upper and lower boundaries of the canopy and used for validation and statistical analysis (Fig. 1).

$$H = average \ (U+L) \qquad (2)$$

His plant height estimated from PHM, where U is the highest point and L is the lowest point of the upper boundary of the canopy at specific location.

UAV-based plant height was validated through ground-based measurements using a ruler at mid-grain fill. Plant height was averaged from 10 plants of each plot representing a DH line. A total 600 of plots were measured in 2 days at each experimental site. Average height error was calculated as the difference between ground measurements and plant height estimated from the UAV platform. The root means square error (RMSE) was also calculated along with the regression fit for validation of UAV platform measurements.

SNP genotyping, QTL analysis and genomic prediction

TheYangmai 16/Zhongmai 895 DH population and parents were genotyped at Capital Bio Corporation (Beijing, China; http: //www. capitalbio. com) using the commercially available Affymetrix wheat 660 K SNP array. Previously, this array was used for genome-wide QTL mapping studies[30-32]. IciMapping 4. 0 was used for linkage map construction using Kosambi mapping approach. Inclusive composite interval mapping-additive (ICIM-ADD) method was used for the QTL analysis at LOD threshold of 2. 5[33]. To assess the accuracy of identification of QTL from UAV-based remote sensing, we cross-validated our results with ground truth data obtained at mid-grain fill. For this, the averaged data from 2 years (2016-2017 and 2017-2018) at both experimental sites was used for quantitative genomic analysis. For temporal

assessment of genomic variation, plant height was phenotyped at booting and mid-grain fill. QTL with overlapping confidence intervals were considered to be the same. Differences between the phenotypic variances explained by QTL from both data sets were detected as validation for UAV-based QTL.

Weevaluated whether ground-based measurements can be replaced by UAV-based remote sensing for future genomic prediction of yield-related traits. For this, rrBLUP (http: //cran. rproject. org/web/packages/ rrBLUP/ index. html) was used to detect the genomic prediction accuracy of UAV-based plant height by comparison with ground-based reference data. To estimate genetic values for traits measured across environments, the following model was used for genomic best linear unbiased prediction (G-BLUP):

$$y_i = \mu_i + x_g + \varepsilon_i, \qquad (3)$$

where phenotypes are viewed as the sum of a random effect representing genomic signals (μ_i), marker effects (x_g) and a model residual (ε_i)[34].

We cross-validated UAV-based data through estimating predication accuracy by removing markers and chromosomes linked with major plant height reducing genes.

Statistical analysis

Linear regression was calculated to evaluate the relationship between UAV-based plant height and ground-based manually measured data. A mixed linear model was used to test the significance ($P \leqslant 0.05$) of variation at among DH lines, environments and effects of their interactions for both data sets by the following general model:

$$Y = X\beta + Z\mu + \varepsilon \qquad (4)$$

where Y is the response demonstrated by fixed (β) and random (μ) effects with random error (ε) and X and Z indicate fix and random effects, respectively. Furthermore, for better understanding of G× E interaction combined heritabilities across environments were calculated:

$$h^2 = \sigma_g^2 / (\sigma_g^2 + \sigma_{ge}^2/r + \sigma_\varepsilon^2/re)$$

where σ_{ge}^2 is genotype × environment interaction variance, e is number of environments and r indicates total replicates for each genotype[35]. The R packages such as "lme4" (https: //CRAN. R-project. org/package = lme4) and "car" (https: //CRAN. R-project. org/package = car) were used for all statistical analysis[36].

Results

Accuracy assessment of UAV-based plant height

Regression analysis showed high R^2 values (0.96) at both sites between UAV-based and ground-based plant height measurements at the mid-grain fill stage (Fig. 2). High correlations (R^2=0.84-0.85 and 0.80-0.83) were also obtained between booting and mid-grain fill from UAV and ground-based data sets, respectively. An accurate DTM with low error noise ranging from ± 3.5 to 4.5 cm across both sites was generated from the spaces adjacent to each plot (Fig. 3a). UAV-based single plant height was measured instead of whole canopy height through aver-aging highest points randomly detected from the canopy. Plant height was under-estimated but without probability of noise due to avoidance strategy for lower boundary of canopy and bare ground estimation. The average difference between predicted plant height from UAV and that observed from ground measurement was approximately 14.02 cm with a root mean square error (RMSE) of 5.75 cm across sites. Chances of error probability in estimation of UAV-based plant height were on average higher (15.83 cm) from plots with higher canopy elevations from ground level as compared to low elevation plots (11.08 cm) (Fig. 3b).

Phenotypic variation

The average ground measurement-based plant heights of Zhongmai 895 and Yangmai 16 were 71.11 and 85.66 cm, respectively. While UAV-based plant heights of parents and DH lines are given in Fig. 3c. Plant height showed continuous variation across the DH population and followed normal distributions at both growth stages (Fig. 2). Both UAV and ground-based data sets showed similar patterns of phenotypic variation among genotypes and G×E interaction at the mid-grain fill stages (Table 1). Significant variation (P<0.0001) among the DH lines was also observed at the booting stage from the UAV data set. The standard deviation was 20.44 cm for UAV and 18.29 cm for ground-based data in DH population. Heritabilities were very high at both developmental stages, i.e. 0.92 at booting and 0.90-0.91 at mid-grain fill for UAV and ground-based plant height, respectively (Table 1).

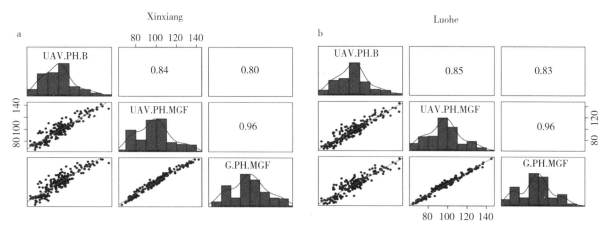

Fig. 2　Regression plots, histograms and R² values between UAV-based plant height at two developmental stages and ground measurements taken from two experiment locations (**a** Xinxiang and **b** Luohe). B, booting; G, ground; PH, plant height; MGF, mid grain fill; UAV, unmanned aerial vehicle

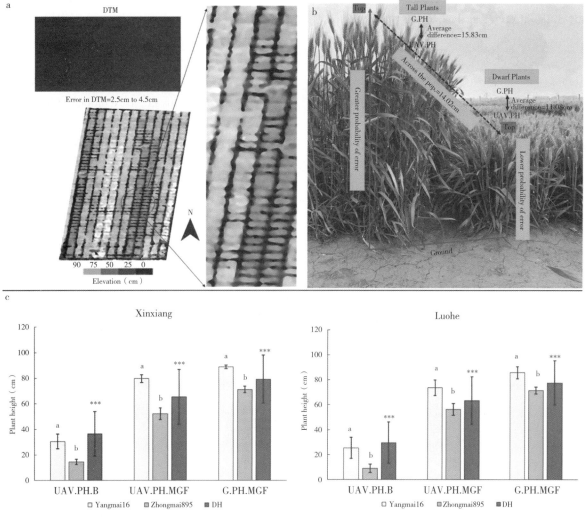

Fig. 3 **a** Estimated probability of error between UAV and ground-based data sets for plots consisted dwarf and tall plants，**b** impact of major genes on plant height across the DH population，and **c** averaged plant height of Yangmai16，Zhongmai895 and DH. Error bars indicate standard deviation；lowercase letters indicate significant difference between parent cultivars；*．**，*** indicate significant differences among DH lines. B，booting；DH，doubled haploid lines；DTM，digital terrain model；G，ground；MGF，mid-grain filling；PH，plant height；UAV，unmanned aerial vehicle

Table 1 Summary of statistics for both plant height data sets and developmental stages

	Yangmai 16 and Zhongmai 895			DH population		
	UAV. PH. B	UAV. PH. MGF	G. PH. MGF	UAV. PH. B	UAV. PH. MGF	G. PH. MGF
SD	10. 05	12. 62	8. 88	17. 26	20. 44	18. 29
G（F. value）	431. 21*	78. 01*	58. 24*	33. 99***	43. 39***	77. 24***
E（F. value）	46. 35*	0. 19	0. 54	25. 61***	27. 93***	47. 55***
G×E（F. value）	0. 06	3. 74	0. 80	0. 591.56	1. 52	
h^2	0. 92	0. 91	0. 93	0. 92	0. 90	0. 91

B，booting；MGF，mid-grain filling；PH，plant height；UAV，unmanned aerial vehicle

＊ $P < 0.05$，＊＊ $P < 0.001$ and ＊＊＊ $P < 0.0001$

Identification of QTL and their impact on phenotype

Identificationof QTL was performed using UAV-based phenotypic data collected at the booting and mid-grain fill stages and validated through ground truth data at mid-grain fill. Five QTL were identified from UAV and ground-based phenotypic data sets at both developmental stages and sites (Fig. 4a). Stable major QTL on chromosomes 4B and 4D were identified from UAV-based plant height data and were also detected with ground-based reference data across sites (Fig. 4b and Additional file 1: Table S1). These two QTL significantly reduced plant height in the DH population at both developmental stages (Fig. 3b). Genotypes of the DH lines are given in Additional file 1: Table S2. Another two QTL for plant height on chromosome 6D identified at the booting stage from UAV-based data explained 9.0-10.2% of phenotypic variation (Fig. 4c and Additional file 1: Table S1).

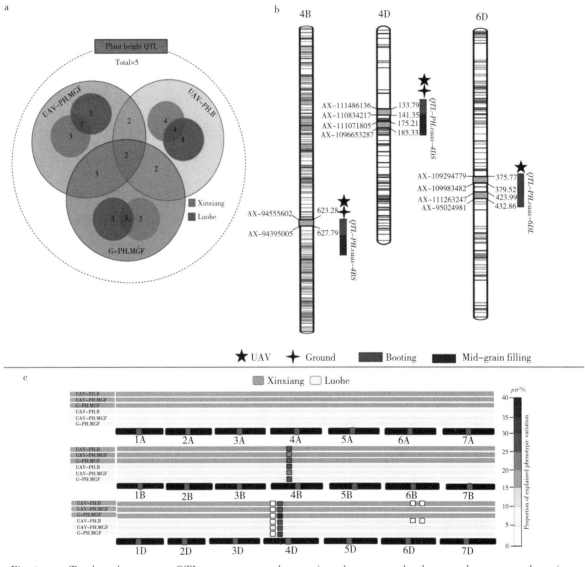

Fig. 4 **a** Total and common QTL among two phenotyping data sets, developmental stages and environments. Numbers show the QTL in data sets, developmental stages and sites, **b** location of QTL with markers, **c** Comparison of phenotypic variance explained by QTL detected from UAV and ground-based data sets at two experimental sites. Squares with different colours show proportion of phenotypic variations explained by QTL detected in particular data set and sites on chromosomes 4B, 4D and 6D. Green spots represent the centromeres. B, booting; G, ground; MGF, mid-grain fill; UAV, unmanned aerial vehicle

Validation of UAV-based QTL results

For validation of QTL predicted from UAV-based plant height data, their contribution to phenotypic variance was compared with ground truth results at mid-grain fill. The phenotypic variances explained by the QTL located on chromosomes 4B and 4D were almost the same for both data sets, i.e. 61.3% for UAV-based and 64.5% for ground-based data with very high heritabilities of 0.90 and 0.91, respectively (Fig. 4c). These two major QTL were also identified for plant height at the booting stage, explaining 73.1% of phenotypic variance for the UAV-based data. The QTL on

chromosomes 4B and 4D corresponded to reduced plant height alleles *Rht-B1b* and *Rht-D1b*, respectively. Gene-specific KASP markers (*Rht-B1_SNP* and *Rht-B1_SNP*) for *Rht-B1b* and *Rht-D1b* confirmed these results. Distributions of these alleles in the DH population are given in Fig. 5a and Additional file 1: Table S1. The QTL identified on chromosome 6D from UAV-based observations at booting showed a simi-lar trend in variation and explained 1.50 and 1.97% of the total variation in plant height at each site (Fig. 4c). Accuracy of booting stage data for plant height was vali-dated with markers for the *Rht-B1b* and *Rht-D1b* alleles (Fig. 5b).

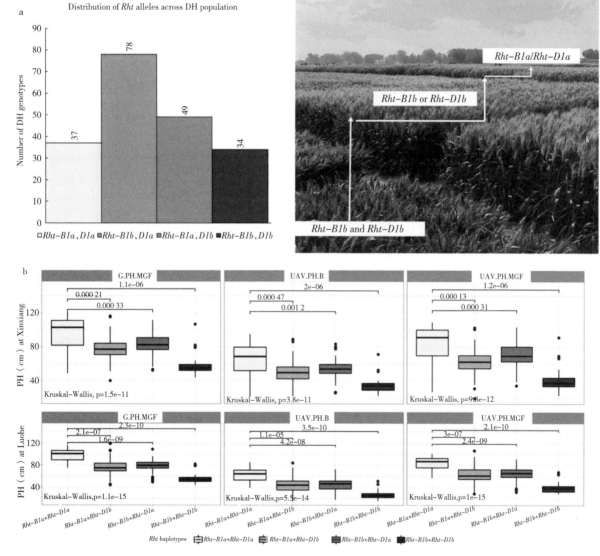

Fig. 5   **a** Distribution of *Rht* genotypes across the DH population and **b** validation of UAV-based data set through emulating impact of these alleles using UAV-based phenotype data. B, booting; G, ground; MGF, mid-grain fill; UAV, unmanned aerial vehicle

Genomic prediction accuracy of UAV-based data set

Genomic prediction accuracy was calculated through correlations between genetically estimated breeding values observed from a training population and was then tested by cross validation. Our results provided further validation of the accuracy of UAV-based plant height by showing similar trend regarding genomic prediction ability for both UAV and ground-based data. The correlations between observed and predicted genomic values for UAV and ground-based data sets ranged from $r = 0.47$- 0.53 for UAV-based plant height at mid-grain fill, but slightly lower than ground-based truth observations of and 055 across sites. Genomic prediction accuracy was higher ranging from $r = 0.56$-0.57 at booting when estimated from UAV-based plant height data. Prediction accuracy was significantly reduced to $r = 0.20$ and 0.31 when markers linked with QTL on chromosomes 4B, 4D and 6D were removed. Genomic prediction ability fell to 75% and 95% when all markers on chromosomes 4B, 4D and 6D were removed. (Fig. 6).

Discussion

Accuracy and phenotypic variations in UAV-based plantheight

UAV is a promising platform to predict time-series development of crop canopies, and further use this data to understand the genetic basis of phenotypic variation[28]. Previously some studies have been reported different workflows for the estimation of plant height using UAV platform[21,26,28] The UAV-platform requires far fewer images and less computing capacity to construct the digital elevation model compared to ground-based imaging platforms[37]. Ground-based Li-DAR technology has been reported more accurate[38], but it has some limitations such as in coving large and multi-locational trials. Aerial estimation of plant height could be error-prone due to low efficiency in pre- and post-imagery processing methods such as altitude of imaging platform, accuracy in DTM construction, and height extraction strategy from images[21,28,29].

High altitude of the UAV flight is likely to generate low pixel resolution of images casing increased data noise. UAV flights were taken at low altitude (30 m) to minimise probability of error due to low pixel numbers. DTM gives information about the elevation of the ground surface. DTM accuracy is an important factor, and low accuracy in DTM can lead to high over-or under-estimations of canopy elevation[21,27]. The precision in estimating depends on number and distribution of bare ground patches across experimental sites if the terrain is to geographically variable. In crops with dense canopies like wheat, it is difficult to generate accurate DTM from DSM images at later growth stages acquiring time-points to develop PHM from single flights. Terrain and distribution of bare ground can be handled through better experimental design and management. Our trial field was well managed with enough spacing between and along the plots to be used to estimate ground elevations across the field. DTM generated from both experimental sites at booting and mid-grain fill had low errors varying from ± 3.5 to 4.5 cm, similar to a previous report on a poppy crop[27] (Fig. 3b). It also reduced the computing load and time required for pre-planting flights to generate DTM of bare fields as done in other reports[24,27,28]. Our method also overcame the problem of data noise in height extraction from PHM due to the detection of lower parts of the canopy such as elevation of leaf from gaps between plants. Using this method, height of a single plant from a particular position of the experimental plot can be measured even in the case of a thin canopy. Higher correlations ($R^2 = 0.96$; 5.75 cm RMSE) were estimated between ground and UAV data sets at mid-grain fill. Our results were better than previous reports where correlations were slightly lower between UAV-derived plant height and reference observations (0.85-0.90) in wheat and barley[26,29] (Fig. 2). This was due to the better strategy of measuring pixel values from the highest points of the imaging to be the upper boundary of the canopy rather than mean values from the whole canopy as previously done in wheat, barley and sorghum[21,24,26]. Both data sets showed transgressive segregation among DH lines relative to the parents with significant phenotypic variation and high heritability. Moreover, high heritability and no

significant G × E allowed detection of stable quantitative loci for plant height.

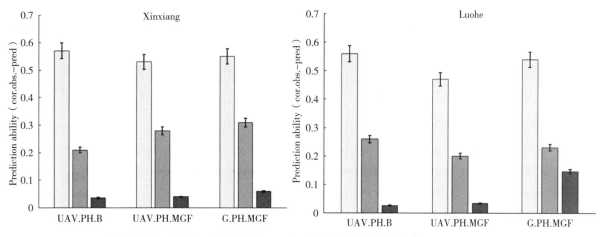

Fig. 6   Validation of genomic prediction ability of UAV-based plant height data through with and without detected QTL at booting and mid-grain fill, as well as comparison with ground truth data at both sites. B, booting; G, ground; MGF, mid-grain fill; UAV, unmanned aerial vehicle

UAV-based QTLs and their effects on phenotype

Height reducinghomoeoalleles *Rht-B1* and *Rht-D1* on the short arms of chromosomes 4B and 4D are GA-insensitive and major plummeting factor for wheat height by reduced GA response mechanism[39,40]. Plant height in wheat is a developmental trait and the genetic basis underlying for its development over time is still being unmasked from a number of potential quantitative loci[11]. *Rht-B1b* and *Rht-D1b* were already reported in parent cultivars Yangmai16 and Zhongmai 895, respectively[41]. UAV-based plant height accuracy was confirmed by identification of QTL corresponding to these *Rht* genes, high correlations between ground truth data and UAV-based data sets, and consistent identification of the same QTL in both UAV-based and ground-based datasets (Fig. 4). UAV-based phenotype data successfully verified the dynamic presence of these two major genes as previously reported by Zhang et al. [41]. Two new QTL with minor phenotypic effect of 1. 50-1. 97% was identi-fied on chromosome 6D using UAV-based booting data from both sites. QTL were also identified 6D at under heat and drought condition which help plant for adaptation without confounding agronomic effects[42]. While in our study, these QTL might be involved in seedling

vigour, but further validation is required. The QTL on chromosome 6D at booting is likely to affect the plant growth. The phenotypic validation of *Rht-B1* and *Rht-D1* on plant height measured by UAV confirmed the accuracy of this platform and proved that UAV has potential for genetic studies.

Accuracy of UAV-based QTL

In quantitative genetics, erroneous phenotypic data is a major bottleneck[19]. Probability of error in UAV-based data can influence the QTL analysis and other genomics studies. In our study, accuracy of QTL detected from both data sets was also validated from multi-location trials. The identification of chromosome 4B and 4D QTL underpinning plant height was consistent across sites (Fig. 4b and Additional file 1: Table S1). Similarly, QTL with less phenotypic variation ranging 1. 50-1. 97% at booting was also consistent at both sites confirming the accuracy of the UAV-based platform for reliable quantitative genomic analysis. The new QTL on chromosome 6D identified using UAV-based data indicated that the UAV platform was effective in detecting genetic variation. These results indicated the potential high efficiency of UAV-based remote sensing for major QTL identification as well as temporal genetic dissection of traits.

Accuracy of UAV-based data for genomic prediction

Genomic prediction is regarded as a relatively new breeding strategy to better exploit quantitative variation in crop breeding and in increasing selection accuracy by optimization of resource allocation in breeding programs[13,43]. In revolutionizing phenotyping platforms for capture of data at lower cost, accuracy for true genomic selection cannot be compromised[44]. Therefore, UAV platforms have potential to contribute in enhancement of genomic prediction accuracy cost-effectively. Rutkoski et al.[44] used UAV-based multispectral secondary traits and reported their high prediction accuracy ($r = 0.41$-0.56) for traits related to grain yield in wheat. Here we demonstrate the use of plant height data captured by a UAV-based aerial platform for high accuracy genomic selection. Similar trends in prediction ability were obtained with and without consideration of the QTL across the data sets. The prediction accuracy declined as markers linked with the QTL were excluded in both data sets. However, remaining genome-wide SNPs predicted accuracy ranged from $r = 0.20$-0.31 (Fig. 6). Our results indicated the presence of an additional gene with minor effect that was not detected in earlier QTL mapping. Our findings also indicate that the use of UAV platforms for genomic selection of quantitative traits could improve prediction ability by continuous capture of cost-effective phenotypic data from multiple environments.

Conclusions

This study describes a UAV-based method for plant height estimation in wheat and its application in quantitative genomic analysis and functional gene characterization. Traditionally, plant height is measured only once, despite the fact that progression to final plant height may differ among genotypes. Our UAV-based approach facilitates rapid, cost-effective, high-throughput capture of plant height data at different growth stages. High R^2 between UAV and ground-based data sets indicated that UAV-platforms could be used for quantitative genomic analysis. This technique can also be applied in practical breeding after adjustment of UAV data according to the average difference (in this case, 14.03 cm) calculated between UAV and ground reference observations. The potential of UAV-based high throughput plant height phenotyping not only reduces the labour costs but is also capable of providing time-lapse reproducible data from large breeding trials to identify the underlying genetics and permit genomic selection for complex traits such as plant height.

Acknowledgements

We are grateful to Prof. R. A. McIntosh, Plant Breeding Institute, University of Sydney, for reviewing this manuscript. This work was funded by the National Key Project (2016YFD0101804), the Fundamental Research Funds for the Institute Planning in Chinese Academy of Agricultural Sciences (S2018QY02), and the National Natural Science Foundation of China (31671691, 3171101265).

❖ References

[1] Griffiths S, Simmonds J, Leverington M, Wang Y, Fish L, Sayers L, et al. Meta-QTL analysis of the genetic control of crop height in elite European winter wheat germplasm. Mol Breed. 2012; 29 (1): 159-71.

[2] Wilhelm EP, Mackay IJ, Saville RJ, Korolev AV, Balfourier F, Greenland AJ, et al. Haplotype dictionary for the *Rht-1* loci in wheat. Theor Appl Genet. 2013; 126 (7): 1733-47.

[3] Royo C, Nazco R, Villegas D. The climate of the zone of origin of Medi-terranean durum wheat (*Triticum durum* Desf.) landraces affects their agronomic performance. Genet Resour Crop Evol. 2014; 61 (7): 1345-58.

[4] Würschum T, Langer SM, Longin CFH, Tucker MR, Leiser WL. A three-component system incorporating *Ppd-D1*, copy number variation at *Ppd-B1*, and numerous small-effect quantitative trait loci facilitates adaptation of heading time in winter wheat cultivars of worldwide origin. Plant Cell Environ. 2018; 41 (6): 1407-16.

[5] Lu Q, Lillemo M, Skinnes H, He X, Shi J, Ji F, et al. Anther extrusion and plant height are associated

with Type I resistance to Fusarium head blight in bread wheat line 'Shanghai-3/Catbird'. Theor Appl Genet. 2013; 126 (2): 317-34.

[6] Srinivasachary, Gosman N, Steed A, Hollins TW, Bayles R, Jennings P, et al. Semi-dwarfing *Rht-B1* and *Rht-D1* loci of wheat differ significantly in their influence on resistance to Fusarium head blight. Theor Appl Genet. 2009; 118 (4): 695-702.

[7] Emebiri LC, Tan M-K, El-Bouhssini M, Wildman O, Jighly A, Tadesse W, et al. QTL mapping identifies a major locus for resistance in wheat to Sunn pest (*Eurygaster integriceps*) feeding at the vegetative growth stage. Theor Appl Genet. 2017; 130 (2): 309-18.

[8] Boeven PH, Longin CF, Leiser WL, Kollers S, Ebmeyer E, Wurschum T. Genetic architecture of male floral traits required for hybrid wheat breeding. Theor Appl Genet. 2016; 129 (12): 2343-57.

[9] Herter CP, Ebmeyer E, Kollers S, Korzun V, Leiser WL, Würschum T, et al. *Rht24* reduces height in the winter wheat population 'Solitär Bussard' without adverse effects on Fusarium head blight infection. Theor Appl Genet. 2018; 131 (6): 1263-72.

[10] Tian X, Wen W, Xie L, Fu L, Xu D, Fu C, et al. Molecular mapping of reduced plant height gene *Rht24* in bread wheat. Front Plant Sci. 2017; 8: 1379.

[11] McIntosh RA, Dubcovsky J, Rogers WJ, Morris C, Xia XC. Catalogue of gene symbols for wheat. 2017. supplement. Downloaded from http://shigen. nig. ac. jp/wheat/komugi/genes/symbolClassList. jsp. Accessed 02 Feb 2017

[12] Araus JL, Kefauver SC, Zaman-Allah M, Olsen MS, Cairns JE. Translating high-throughput phenotyping into genetic gain. Trends Plant Sci. 2018; 23 (5): 451-66.

[13] Li H, Rasheed A, Hickey LT, He Z. Fast-forwarding genetic gain. Trends Plant Sci. 2018; 23 (3): 184-6.

[14] Wu R, Lin M. Functional mapping-how to map and study the genetic architecture of dynamic complex traits. Nat Rev Genet. 2006; 7 (3): 229-37.

[15] Huang X, Han B. Natural variations and genome-wide association studies in crop plants. Annu Rev Plant Biol. 2014; 65: 531-51.

[16] Jannink J-L. Dynamics of long-term genomic selection. Genet Sel Evol. 2010; 42 (1): 35.

[17] Chapman S, Merz T, Chan A, Jackway P, Hrabar S, Dreccer M, et al. Pheno-Copter: a low-altitude, autonomous remote-sensing robotic helicopter for high-throughput field-based phenotyping. Agronomy. 2014; 4 (2): 279.

[18] Großkinsky DK, Svensgaard J, Christensen S, Roitsch T. Plant phenomics and the need for physiological phenotyping across scales to narrow the genotype-to-phenotype knowledge gap. J Exp Bot. 2015; 66 (18): 5429-40.

[19] Furbank RT, Tester M. Phenomics-technologies to relieve the phenotyping bottleneck. Trends Plant Sci. 2011; 16 (12): 635-44.

[20] Hassan MA, Yang M, Rasheed A, Jin X, Xia X, Xiao Y, et al. Time-series multispectral indices from unmanned aerial vehicle imagery reveal senescence rate in bread wheat. Remote Sens. 2018; 10 (6): 809.

[21] Bendig J, Bolten A, Bennertz S, Broscheit J, Eichfuss S, Bareth G. Estimating biomass of barley using crop surface models (CSMs) derived from UAV-based RGB imaging. Remote Sens. 2014; 6 (11): 10395.

[22] Sankaran S, Khot LR, Espinoza CZ, Jarolmasjed S, Sathuvalli VR, Vandemark GJ, et al. Low-altitude, high-resolution aerial imaging systems for row and field crop phenotyping: a review. Eur J Agron. 2015; 70: 112-23.

[23] Shi Y, Thomasson JA, Murray SC, Pugh NA, Rooney WL, Shafian S, et al. Unmanned aerial vehicles for high-throughput phenotyping and agronomic research. PLoS ONE. 2016; 11 (7): e0159781

[24] Watanabe K, Guo W, Arai K, Takanashi H, Kajiya-Kanegae H, Kobayashi M, et al. High-throughput phenotyping of sorghum plant height using an unmanned aerial vehicle and its application to genomic prediction modeling. Front Plant Sci. 2017; 8: 421.

[25] Hassan MA, Yang M, Rasheed A, Yang G, Reynolds M, Xia X, et al. A rapid monitoring of NDVI across the wheat growth cycle for grain yield prediction using a multi-spectral UAV platform. Plant Sci. 2018. https://doi.org/10.1016/j. plantsci. 2018. 10. 022

[26] Khan Z, Chopin J, Cai J, Eichi V-R, Haefele S, Miklavcic S. Quantitative estimation of wheat phenotyping traits using ground and aerial imagery. Remote Sens. 2018; 10 (6): 950.

［27］ Iqbal F, Lucieer A, Barry K, Wells R. Poppy crop height and capsule volume estimation from a single UAS flight. Remote Sens. 2017; 9 (7): 647.

［28］ Hu P, Chapman SC, Wang X, Potgieter A, Duan T, Jordan D, et al. Estima-tion of plant height using a high throughput phenotyping platform based on un-manned aerial vehicle and self-calibration: example for sorghum breeding. Eur J Agron. 2018; 95: 24-32.

［29］ Bendig J, Yu K, Aasen H, Bolten A, Bennertz S, Broscheit J, et al. Combin-ing UAV-based plant height from crop surface models, visible, and near infrared vegetation indices for biomass monitoring in barley. Int J Appl Earth Obs Geoinf. 2015; 39: 79-87.

［30］ Condorelli GE, Maccaferri M, Newcomb M, Andrade-Sanchez P, White JW, French AN, et al. Comparative aerial and ground based high throughput phenotyping for the genetic dissection of NDVI as a proxy for drought adaptive traits in durum wheat. Front Plant Sci. 2018; 9: 893.

［31］ Jin H, Wen W, Liu J, Zhai S, Zhang Y, Yan J, et al. Genome-wide QTL map-ping for wheat processing quality parameters in a Gaocheng 8901/Zhou-mai 16 recombinant inbred line population. Front Plant Sci. 2016; 7: 1032.

［32］ Cui F, Zhang N, Fan XL, Zhang W, Zhao CH, Yang LJ, et al. Utilization of a Wheat660 K SNP array-derived high-density genetic map for high-resolu-tion mapping of a major QTL for kernel number. Sci Rep. 2017; 7 (1): 3788.

［33］ Meng L, Li H, Zhang L, Wang J. QTLIciMapping: integrated software for genetic linkage map construction and quantitative trait locus mapping in bi-parental populations. Crop J. 2015; 3 (3): 269-83.

［34］ Endelman JB. Ridge regression and other kernels for genomic selection with R package rrBLUP. Plant Ge-nome. 2011; 4 (3): 250-5.

［35］ Sehgal D, Skot L, Singh R, Srivastava RK, Das SP, Taunk J, et al. Exploring potential of pearl millet germplasm association panel for association mapping of drought tolerance traits. PLoS ONE. 2015; 10 (5): e0122165

［36］ R Core Team. R. A Language and environment for sta-tistical computing. Vienna: R Foundation for Statistical Computing; 2016.

［37］ Wang X, Singh D, Marla S, Morris G, Poland J. Field-based high-through-put phenotyping of plant height in sorghum using different sensing technolo-gies. Plant Methods. 2018; 14 (1): 53.

［38］ Madec S, Baret F, de Solan B, Thomas S, Dutartre D, Jezequel S, et al. High-throughput phenotyping of plant height: comparing unmanned aerial vehicles and ground LiDAR estimates. Front Plant Sci. 2017; 8: 2002.

［39］ Daoura BG, Chen L, Du Y, Hu Y-G. Genetic effects of dwarfing gene *Rht-5* on agronomic traits in common wheat (*Triticum aestivum* L.) and QTL analysis on its linked traits. Field Crops Res. 2014; 156: 22-9.

［40］ Gao F, Ma D, Yin G, Rasheed A, Dong Y, Xiao Y, et al. Genetic progress in grain yield and physio-logical traits in Chinese Wheat Cultivars of South-ern Yellow and Huai valley since 1950. Crop Sci. 2017; 57 (2): 760-73.

［41］ Zhang B, Shi W, Li W, Chang X, Jing R. Efficacy of pyramiding elite alleles for dynamic development of plant height in common wheat. Mol Breed. 2013; 32: 327-38.

［42］ Pinto RS, ReynoldsMP, Mathews KL, McIntyre CL, Olivares-Villegas J-J, Chapman SC. Heat and drought adaptive QTL in a wheat population designed to minimize confounding agronomic effects. Theor Appl Genet. 2010; 121 (6): 1001-21.

［43］ Werner CR, Voss-Fels KP, Miller CN, Qian W, Hua W, Guan CY, et al. Effective genomic selection in a narrow-genepool crop with low-density markers: Asian rapeseed as an example. Plant Genome 2018; 11 (2): 170084. https: // doi. org/10. 3835/plantgen-ome2017. 09. 0084

［44］ Rutkoski J, Poland J, Mondal S, Autrique E, Pérez LG, Crossa J, et al. Canopy temperature and vegeta-tion indices from high-throughput phenotyping improve accuracy of pedigree and genomic selection for grain yield in wheat. G3 Genes Genomes Genet. 2016; 6 (9): 2799-808.

Quantifying senescence in bread wheat using multispectral imaging from an unmanned aerial vehicle and QTL mapping

Muhammad Adeel Hassan,[1,†] Mengjiao Yang,[1,†] Awais Rasheed[1,2,3] Xiuling Tian,[1] Matthew Reynold,[4] Xianchun Xia,[1] Yonggui Xiao[1,]* [,‡] and Zhonghu He[1,2,‡]

[1] Institute of Crop Sciences, National Wheat Improvement Centre, Chinese Academy of Agricultural Sciences (CAAS), Beijing 100081, China

[2] International Maize and Wheat Improvement Centre (CIMMYT) China Office, c/o CAAS, Beijing 100081, China

[3] Deparment of Plant Science, Quaid-i-Azam University Islamabad 44000, Pakistan

[4] Global Wheat Program, International Maize and Wheat Improvement Centre (CIMMYT), Mexico DF 06600, Mexico

* Author for communication: xiaoyonggui@caas. cn
[†] These authors contributed equally (M. A. H. , M. Y.).
[‡] Senior authors.

M. A. H. conducted the experiments, analyzed the data, and wrote the paper under supervision of Z. H. and Y. X. M. Y. collected the ground truth data and X. T. assisted in the field management. Y. X. managed and directed the trial. X. X. , M. R. , and A. R. gave comments and suggestions to improve the manuscript.

The author responsible for distribution of materials integral to the findings presented in this article in accordance with the policy described in the Instructions for Authors (https: //academic. oup. com/pl-phys/pages/General-Instructions) is Yonggui Xiao (xiaoyonggui@caas. cn).

Abstract: Environmental stresses from climate change can alter source-sink relations during plant maturation, leading to premature senescence and decreased yields. Elucidating the genetic control of natural variations for senescence in wheat (*Triticum aestivum*) can be accelerated using recent developments in unmanned aerial vehicle (UAV) -based imaging techniques. Here, we describe the use of UAVs to quantify senescence in wheat using vegetative indices (VIs) derived from multispectral images. We detected senescence with high heritability, as well as its impact on grain yield (GY), in a doubled-haploid population and parent cultivars at various growth time points (TPs) after anthesis in the field. Selecting for slow senescence using a combination of different UAV-based VIs was more effective than using a single ground-based vegetation index. We identified 28 quantitative trait loci (QTL) for vegetative growth, senescence, and GY using a 660K single-nucleotide poly-morphism array. Seventeen of these new QTL for VIs from UAV-based multispectral imaging were mapped on chromosomes 2B, 3A, 3D, 5A, 5D, 5B, and 6D; these QTL have not been reported previously using conventional phenotyping methods. This integrated approach allowed us to identify an important, previously unreported, se-

Published in Plant Physiology, 2021: 1-14.

nescence-related locus on chromosome 5D that showed high phenotypic variation（up to 18. 1％）for all UAV-based VIs at all TPs during grain filling. This QTL was validated for slow senescence by developing kompetitive allele-specific PCR markers in a natural population. Our results suggest that UAV-based high-throughput phenotyping is advantageous for temporal assessment of the genetics underlying for senescence in wheat.

Introduction

Consumed by half the world, wheat（ *Triticum aestivum* ）is a major contributor to food security, but its production needs to increase by 3 billion tons annually by 2050 to meet global needs（ Tester and Langridge, 2010 ）. To achieve sustainable wheat production as climate change increase the severity of stresses on plants, we need to integrate new genomics and phenomics technologies in studies to more deeply dissect the mechanisms of plant stress responses to various extremes（ Araus et al. , 2018 ）. Breeding based on such cross-disciplinary knowledge that links important physiological and biochemical interactions to new genetic information will better enable us to overcome agricultural challenges（ Rasheed et al. , 2020 ）. The physiochemical, morphological, and molecular changes in plants in response to climatic fluctuations such as drought and heat（ Zhu, 2016; Sade et al. , 2017 ）can alter source-sink relationships, which affect growth and yield of plant. Two mechanisms are mainly involved in these alterations: （ 1 ）premature senescence/yellowing and（ 2 ）inhibition of photosynthesis activity by producing assimilates in source organs. This helps to de- crease consumption of resources within the source organs and increase their mobilization toward sink tissues（ Albacete et al. , 2014; Yolcu et al. , 2017 ）.

Although senescence is a genetically programmed system of plants for dynamic accumulation of nutrients to sink tissues, a "stress-induced senescence" can also be activated by external stress stimuli to tradeoff the losses in sink tissues under stress conditions（ Schippers et al. , 2015; Sade et al. , 2017 ）. Under normal conditions, the coordinated breakdown of chlorophyll is an integral process of senescence that is developmentally planned to facilitate the remobilization of nutrients from senescing tissues to newly growing organs/ tissues such as young leaves（ Lim et al. , 2007 ）. Elucidating the progress of senescence by temporal remote sensing of green biomass and chlorophyll degradation is of key importance in developing climate-resilient genotypes（ Christopher et al. , 2016 ）. However, such selection for slow senescence has been difficult and error prone in the field.

Adapting field-based advanced remote sensing technology to phenotyping physiological and biochemical traits beyond conventional traits that are assessed by eye is the future of crop breeding（ Araus and Cairns, 2014 ）. Physiological attributes such as canopy chlorophyll content（ CCC ）and photosynthesis and the impact of nutrients, water and heat stress on plant growth（ Araus et al. , 2018; Yang et al. , 2020 ）can only be quantified nondestructively by measuring variations in the spectral reflectance（ Jin et al. , 2020 ）. For example, when growing conditions are optimal, healthy plants look greener than those under stress because the absorbance maximum for chlorophyll is in the red wavelength region and green wavelengths are reflected（ Hatfield et al. , 2008 ）. Handheld active sensors can estimate chlorophyll levels and photosynthetic rates（ Pleban et al. , 2020 ）, wheat physiology can be assessed using traits derived from light reflectance such as normalized difference vegetation index（ NDVI ）to select for genotypes tolerant to abiotic stress（ Christopher et al. , 2016 ）. Operational and resolution limitations of sensors for capturing a wide range of light bands when light reflection fluctuates, however, contribute to the difficulty in resolving the complexity of senescence and underlying genetics.

The integration of new techniques for image（ RGB （ Red-Green-Blue ）, X-ray or hyper/multispectral ）-based data acquisition of traits can improve the accuracy of plant phenotyping and accelerate the molecular breeding process by increasing the rate of gene discovery（ Yang et al. 2014; Ruckelshausen and

Busemeyer, 2015; Campbell et al. 2017; Araus and Kefauver, 2018; Zhou et al. 2019). Many computer vision-based tools and imaging from various types of sensors have become vital for phenotyping with high accuracy and throughput (Mochida et al., 2018). By converting these digital measurements into useful biological knowledge (crop traits), scientists can have "smart" eyes on disease, photosynthesis, chlorophyll status, senescence, and other physiochemical properties of crops (Tardieu et al., 2017; Araus et al., 2018; Su et al., 2019; Rasheed et al., 2020). Recent deployment of unmanned aerial vehicle (UAV) platforms for aerial surveillance of crops using hyper/multispectral imaging has increased the access and capability of scientists to cover large number of experimental trials at multiple times during growth over a limited time (Jin et al., 2020; Rasheed et al., 2020). UAV-based multispectral traits derived from the visible and beyond-visible range of the light spectrum, such as NDVI, green and red edge chlorophyll indices, and normalized difference red edge have been used as smart indicators to differentiate senescence rates among genotypes exposed to drought and heat stresses (Duan et al. 2017; Hassan et al., 2018). However, no effort has been made to use this knowledge in the genetic dissection of complex traits such as senescence. Therefore, linking UAV-based VIs with high-density single-nucleotide polymorphisms (SNPs) will open a new avenue in quantitative genetic studies to discover new loci for breeding. Some genes including *NAC*, *WRKY*, *MYB*, *AP2/EREBP*, and *bZIP* have been reported to play significant roles in the regulation of age-induced senescence in many plant species such as *Arabidopsis thaliana* (Woo et al., 2013), wheat (Gregersen and Holm, 2007), and rice (Liu et al., 2008). Interestingly, these same genes are involved in stress tolerance. Most attempts to identify genes that control variation in a target trait have used data from a single time point (TP), usually at maturity or a fixed number of days from planting. Precision phenotyping to monitor plant growth over time remains a bottleneck (Furbank and Tester, 2011). Advances in UAV and computer vision are overcoming barriers to score traits at

multiple times during development (Salas Fernandez et al., 2017; York, 2018). To the best of our knowledge, no studies have yet reported on integration of VIs derived from UAV-based multispectral pixels to identify loci controlling dynamic senescence in wheat. Therefore, the aims of this study were to (1) digitally quantify senescence from heading (H) to maturation using UAV-based VIs, (2) identify highly heritable predictive traits to evaluate senescence, and (3) identify loci that contribute to variations in wheat senescence.

Results

Adaptation of field-based UAV system for multi-spectral scan of wheat physiology

We previously optimized the UAV platform for high-throughput multispectral phenotyping of the wheat canopy and predicted grain yield (GY) by monitoring seasonal growth. The accuracy of the estimates of the UAV-based traits was highly consistent with values from conventional measurements (Hassan et al., 2018; 2019a, 2019b). The multispectral scans by the UAV system could thus be used to quantify the wheat senescence and thus phenotype wheat doubled-haploid (DH) populations to identify loci controlling senescence (see Materials and methods). The average maximum focal length for each sensor was 1, 445.20 pixels (5.40 mm) to capture four multispectral images of light reflection, that is, near-infrared (NIR), red, green, and red-edge bands of varying wavelengths, with optimized ground sampling distance (GSD) of 2.23 cm (i.e. 1 pixel indicated 2.23 cm distance at ground; Figure 1D). Dense point-clouds for making 2D mosaic maps, over five images were overlapped for each pixel with an average reprojection error of 0.21 pixels across the whole data collection campaign (Figure 1E). In total, 240 GB data in the form of raw multi-spectral images were collected through aerial surveillance of 2, 400 plots during the whole phenotyping campaign consisting of 2 years and sites within 5 h flight time (Supplemental Table S1). This raw data were converted into useful pixel information through machine learning to calculate UAV-based VIs data (400 kb) to estimate the physiological status of all five growth TPs (GTPs) (see process in "Materials and methods"). The combination of multispectral pixels from different light reflectance gives a better overall estimate of the physiological status of plants. The red band is reportedly more reliable at stages before maturation due to saturation issues in detecting high chlorophyll level after canopy closure (Hatfield et al., 2008). The NIR band is best for detecting a wide range of variations in green biomass and chlorophyll, and reflectance in the green and red-edge bands ranges is also sensitive to the whole range of variations in chlorophyll and green biomass (Hatfield et al., 2008). Senescence is a complex quantitative trait, rapid, precise temporal phenotyping of each individual beyond typical physiological measurements was possible because phenotypic correlations were high (up to $r = 0.82$) between UAV-based VIs and GY compared to ground NDVI measurements (Figure 1I).

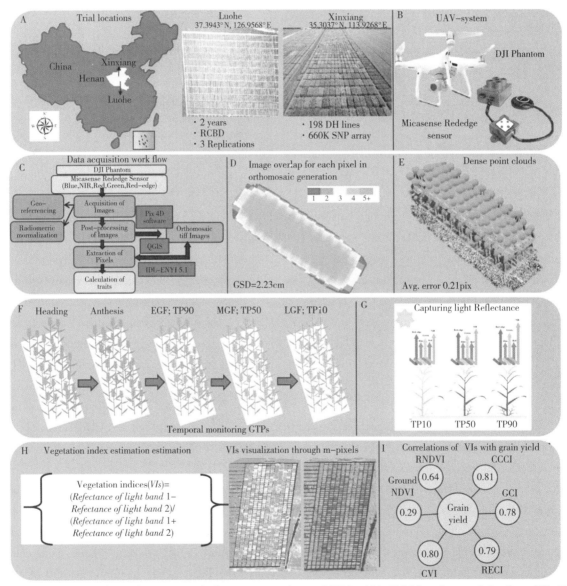

Figure 1　High-throughput phenotyping of wheat canopy using an UAV. (A) Trial locations and details, (B) UAV system, (C) data acquisition work flow, (D) results for average image overlapping to estimate each pixel in orthomosaic generation, (E) results for dense point clouds, (F) important GTPs when data were collected, (G) mechanism behind capturing different light reflectance (H) general equation for estimating value for VIs and visualization of UAV-based VIs using multispectral pixels (m-pixels) (green color indicates highest VI and red lowest VI values) and (I) correlations of VIs with GY.

Phenotypic variations in UAV-based VIs, ground NDVI, and GY

All five UAV-based VIs derived from the pixels of the 2D multispectral images obtained by aerial surveillance of field trials and the ground NDVI from the ground GreenSeeker sensor were normally distributed at all GTPs. Significant variations (at $P < 0.05$) among genotypes across environments and genotype×environment (G×E) factor were observed for five UAV-based VIs, senescence, ground NDVI at H, anthesis (Ans), early (TP90 at 10% senescence), mid (TP50 at 50% senescence), and late (TP10 at 90% senescence concluded) grain filling (GF) GTPs (Supplemental Table S2). Genotypes also varied significantly ($P < 0.05$) for GY in both environments, and the G×E interaction was also significant. The mean broad-sense heritabilities for UAV-based VIs and GY at each GTP ranged from 0.71 to 0.95 at both locations. Heritabilities for the ground NDVI data set ranged from 0.73 to 0.95 at various GTPs (Figure 2; Supplemental Table S3). As expected, variances among TPs increased postAns for all traits (Supplemental Table S2). Two UAV-based VIs, CCC index (CCCI) and red NDVI (RNDVI), were selected to visualize the physiological status of wheat canopy using images generated by different combinations of multispectral pixels for NIR, red, green, and red-edge reflectance for four GTPs (Figure 3). Visualization results of CCCI and RNDVI also illustrated a temporal decline in canopy greenness and phenotypic variations for slow senescence in some DH lines at TP50 to TP10. The temporal pattern of the decline in UAV-based VIs was not linear in most cases, indicating slow senescence in some DH lines (Figure 4). Several DH lines (DH-082, DH-116, DH-123, and DH-197) had slow-senescence phenotypes with high GY. Multispectral pixel visualization results were in accordance with the graphical results for these DH lines. Senescence in Zhongmai 895 was delayed compared to Yangmai 16 across the postAns GTPs, and transgressive segregation was observed for all five UAV-based VIs (Figure 5).

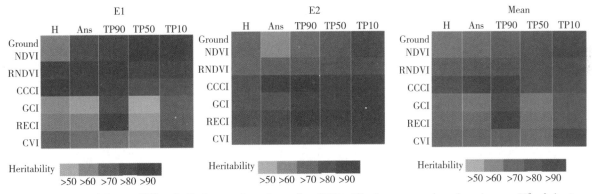

Figure 2   Heritabilities (H^2) of all phenotypic traits at five GTPs. The heat maps show broad-sense H^2 of the investigated traits in two environments E1 (Xinxiang), E2 (Luohe), and mean of both environments. The color intensity indicates high (dark) and low H^2.

Relevance of UAV-based VIs with GY assessment

All UAV-based VIs were strongly correlated with GY at all GTPs in both environments (Supplemental Table S4). For example, genotypic correlations of RNDVI, CCCI, green chlorophyll index (GCI), and red-edge chlorophyll index (RECI) with GY ranged from $r=0.22$-0.82 at TP90, $r=0.28$-0.62 at TP50, and $r=0.22$-0.69 at TP10 in E1, while $r=0.59$-0.80 at TP90, $r=0.64$-0.81 at TP50, and $r=0.61$-0.69 TP10 in E2. Genetic correlations between ground NDVI and GY were low, ranging from $r=0.31$-0.36 in E1 and from $r=0.19$-0.29 in E2 at all GTPs. A similar trend was observed in phenotypic correlations of RNDVI, CCCI, GCI RECI, and CVI with GY. Correlations for E2 were higher, ranging from 0.55 to 0.79 at TP90, 0.61 to 0.79 at TP50, and 0.58 to 0.68 at TP10 from ground NDVI compared to

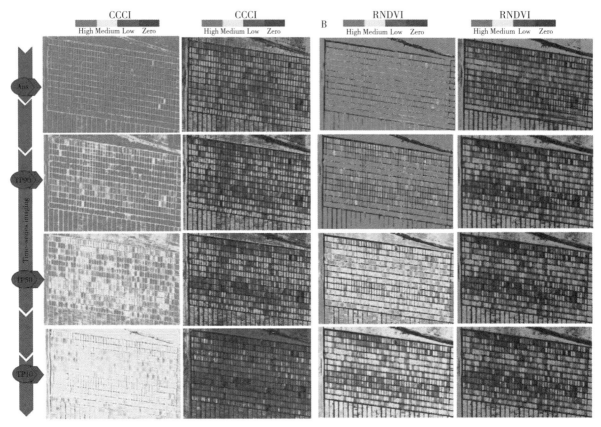

Figure 3　Temporal visualization of important VIs. A，CCCI and（B）RNDVI were derived through combining the multispectral pixels of NIR，red and green bands. Image on left and right in（A）and（B）show overall and differential illustrations of the traits among DH lines at TP90，TP50，and T10 for experimental site in Luohe，respectively. Green pixels indicate maximum VI values within the plot.

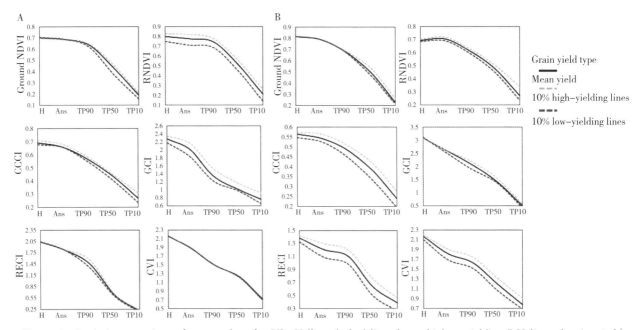

Figure 4　Logistic regressions of mean values for VIs. Yellow dashed line shows highest-yielding DH lines that is，10% higher from mean-yielding DH lines（solid black line）and the red dashed line shows lowest-yielding DH lines that is，10% lower from mean-yielding DH lines in（A）E1（Xinxiang）and（B）E2（Luohe）at five GTPs.

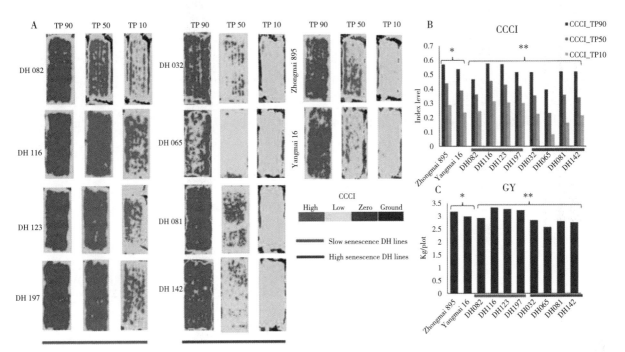

Figure 5   Assessment of senescence progress using CCCI visualization and impact on GY. A and B, CCCI was estimated for four high-yielding and four low yielding DH lines compared with parent cultivars at TP90, TP50, and TP10. Green pixels indicate maximum VI values within the plot. C, GY comparison of slow and fast senescence DH lines and parents. Statistical significance was determined by t test: $*P<0.05$, $**P<0.01$.

E1. CVI had low correlations with GY, ranging from 0.12 to 0.23 in E1.

We also examined the relevance of GTPs TP90 to TP10 for predicting GY in terms of high genetic and phenotypic correlations. To further examine the relevance of UAV-based VIs to GY, we plotted the average UAV-based VIs for the 10% higher- and 10% lower-yielding genotypes against the mean-yielding genotypes for each environment (Figure 4). Clear differences were found between the highest and lowest-yielding genotypes under both environments, with the higher-yielding genotypes having high values for UAV-based VIs. The differences appear greatest in the UAV-based VIs and ground NDVI in both environments, except for CVI in E1. The smallest differences in UAV-based VIs and ground NDVI between the high- and low-yielding groups were at TP90 when the correlation between yield and UAV-based VIs was low. These results are in accordance with the genetic correlations observed between yield and UAV-based VIs (Figure 6). UAV-based VIs of high-yielding genotypes declined slower than those of the low-yielding genotypes, but the difference in UAV-based VIs in most of the graphs decreased as senescence progressed, such that they attained full senescence at a similar time after Ans (Figure 3). Some DH lines showed a significant delay in trait decline with high-yielding effects. It is unlikely that the yield contrast between high- and low-yielding groups resulted from differences in the date of Ans. Differences between groups in the mean period from sowing to Ans were small (ranging from 3 to 4 d) and nonsignificant for all environments ($P>0.05$).

Multispectral image-based senescence was high at TP10

Senescence derivedfrom UAV-based VIs and ground NDVI at TP90, TP50, and TP10 was not linear for most of traits under both environments (Supplemental Figure S1). Among the DH lines, canopy senescence (CS) was slow from Ans to TP50 in many DH lines as illustrated in the CCCI and RNDVI multispectral pixels maps and graphical results (Figure 3; Supplemental Figure S2). However, a sudden increase was observed from TP50 to TP90. For senescence to start slowly but completing at a similar time in the slow- and the fast-senescence genotypes, senescence must progress faster at particular points, resulting in an increase in overall green leaf area and higher yield. Therefore, a majority of the DH lines reached maturity at a similar time due to the trend for fast CS at TP10. The slow CS from TP90 to TP50 provided sufficient time for the mobilization of nutrients from source to sink nutrient,

resulting in high GY. CS at TP50 was estimated to be higher for GCI and RECI in E1 compared to other UAV-based VIs and E2. The box plot in Figure 6A shows a significant deviation in UAV-based Vis from their means level across the postAns GTPs especially at TP10. This kind of natural variation in CS at TP10 can be used to detect potential loci for slow senescence and stable GY.

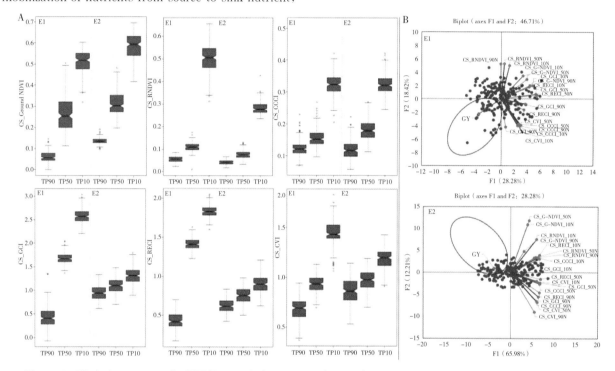

Figure 6　Variations among the DH lines and phenotypic selection for senescence. A, Boxplots show the range of phenotypic means of CS derived from ground NDVI, RNDVI, CCCI, GCI, RECI, and CVI at three GTPs for environments E1 (Xinxiang) and E2 (Luohe). Bars are indicating the upper and lower limited of the boxplots from their central line which indicates median. Dots outside of the upper and lower limited of both quantiles of boxplots are considered as outliners. B, Biplots results from PCA of CS plotted against the GY. Different line colors represent the variables, blue dots are DH lines, and red circles indicate the low-senescence DH lines with high GY. Dots near the GY vector within the red circle were selected as high-yielding with slow senescence across environments. DH lines near the CS vectors were considered as prone to rapid senescence and low GY. Data are BLUEs.

VIs-based selection of low-senescence and high-yielding genotypes

A principal component analysis (PCA)-biplot was used to examine relationships between CS calculated from UAV- based VIs and yield for selecting the low-senescence DH lines, while maintaining high yield compared to fast-senescence individuals (Figure 6B). As anticipated, high CS was negatively correlated with

GY, as discussed above, high-yielding DH lines maintained higher UAV-based VIs level across the postAns GTPs compared with the low yielding. A strong negative correlation was observed between CS derived from UAV-based VIs and GY especially at TP50 and TP10 in both environments. CS measured from CCCI and CVI were negatively correlated with those from GCI, ground NDVI, RECI, and RDVI across the postAns GTPs in E1. Vectors of CS for CC-

CI, CVI, GCI, and RECI were in a similar direction at TP90 and TP50 but slightly opposite at TP10, for which vectors were close to ground NDVI and RNDVI under E2. As Figure 6B illustrates, CS was higher at TP10 than at TP90 and TP50; CS vectors at TP10 were plotted opposite of those for TP90 and TP50 in E2. But an exception was observed in case of CCCI and CVI in E1; all of their CS vectors were in a similar direction and had high correlations with each other.

QTL were identified for senescence at individual GTPs and GY

QTL was identified for all UAV-based VIs, ground NDVI, and GY measured in multiple environments (Supplemental Table S5). From inclusive composite interval mapping- additive (ICIM-ADD) scans of the average data of 2 years (2016-2017, 2017-2018) at Xinxiang (E1) and at Luohe (E2), and BLUEs by combining E1 and E2 (i. e. analyzing data for each GTP separately), we identified 28 consistent QTL. Of these, 22 QTL were identified as senescence loci that controlled at least one of the UAV-based VIs at any post Ans GTP and detected in at least two environments (Supplemental Figure S2), and 17 QTL were identified only from UAV-based VIs. Genetic regions associated with QTL were identified on chromosomes 1B, 2B, 3A, 3D, 4B, 4D, 5A, 5B, 5D, 6A, 6D, 7A, 7B, and 7D (Supplemental Figure S3). QTL associated with yield were identified on chromosomes 2B, 4D, 5A, 5B, and 6D, coinciding with senescence QTL (QS) on 4D, 5B, and 6D and found to colocalize with QS on 2B and 5A.

Three QTL on chromosomes 1B and 2B wereassociated with CCCI, CVI, GCI, RECI, and RNDVI only during H and Ans GTPs, explaining 8. 9%-16. 4%, 5. 6%-9. 8%, and 5. 5%- 6. 8% of phenotypic variations, respectively (Supplemental Table S5). For photoperiod response (*Ppd*), vernalization response (*Vrn*), or earliness per se genes, there is little evidence of association with putative senescence loci in our DH population. But a *QTL-caas. 1B* (722. 5 Mb) on chromosome 1B that was identified at H was linked with a major photoperiod response gene *Ppd-B2* (Supplemental Table S5). Six QTL in total among chromosomes 5A, 6D, and 7B were detected for single postAns GTP, whereas five QTL in total among chromosomes 3A, 5A, and 7D were identified in more than one postAns GTPs. Eleven QTL in total among chromosomes 2B, 3D, 4B, 4D, 5B, 5D, 6D, and 7A were detected at both preAns and postAns GTPs. QTL on 4B, 4D, and 5D explained a high percentage of the phenotypic variations explained by QTL (PVE), up to 23. 4%, for traits measured at postAns GTPs compared to other QTL. Zhongmai 895 contributed a positive additive effect for identified QTL (Supplemental Table S5).

In some cases, plant height QTL were alsodetected as senescence loci (Pinto et al. , 2016; Christopher et al. , 2018). Height QTL associated with *Rht-B1* (4B) and *Rht-D1* (4D) genes, were closely colocalized with QS on chromosomes 4B and 4D in all environments (Supplemental Table S5). In our DH population, *Rht-B1b* was contributed by Yangmai 16 and *Rht-D1b* by Zhongmai 895 (Hassan et al. , 2019a, 2019b). QTL on chromosomes 2B, 3A, 3D, and 5D were detected as new QTL for senescence in wheat with 5. 2%-18. 1% of PVE. As *QTL-caas. 5D* (315. 5 Mb) showed high PVE for the senescence (Figure 7A), two SNPs in the region of this QTL at 315. 5 Mb and 320. 5 Mb with 18. 13% and 10. 1% of PVE, respectively, were converted successfully to kompetitive allele-specific PCR (KASP) markers. To check the effectiveness of this genomic region for slow senescence in wheat, 207 accessions of a natural population were genotyped with these KASP markers. Results showed significant differences in chlorophyll levels between genotypes within full and limited irrigations based on SNPs linked with *QTL-caas. 5D* (Figure 7b).

Discussion

To bridge thegenome to phenome gap, predictive traits with high heritabilities in the field condition are urgently needed (Araus and Kefauver，2018；Rasheed et al.，2020). Traits derived from UAV-based multispectral imaging have been described for precise and rapid phenotyping of wheat growth dynamics (Hassan et al.，2018). The complex mechanism underlying senescence in plants involves dynamic and diverse responses that are controlled by many loci with moderate to minor effects (Sade et al.，2017；Christopher et al.，2018). Minor-effect loci can be difficult to discover using traditional assessments of senescence-related traits. UAV-based multispectral traits can accelerate the identification of both major and minor loci and advance our understanding of the genetic architecture behind the temporal responses of plants during senescence.

We extracted five different UAV-based VIs through integrating light of various wavelengths reflected from the canopy and compared with ground truth NDVI measurements. UAV-based VIs and ground NDVI were measured at vegetative GTPs (i. e. H，Ans) and senescence GTPs (i. e. TP90，TP50，and TP10) in multiple environments (i. e. two locations in 2 years) (Figure 1，A and F). This strategy allowed us to monitor the consistency of UAV-based image pixels for calculating UAV-based VIs to predict senescence and behavior (G×E) of the genotypes in the varied environments. This approach could increase the efficiency of predicting complex physiological traits over that of traditionally limited approaches for trait acquisitions. Transgressive segregation for UAV-based VIs and yield suggests that alleles for improvements in each of these traits could be contributed by either of the parents，though they were more often con-tributed by Zongmai 895 (Supplemental Table S5). The heritability (H²) of UAV-based VIs was generally high in all environments，highlighting high genetic variances within environments. UAV-based VIs showed higher genetic and phenotypic correlations with GY compared to ground NDVI，as in our previous study on GY prediction based on multispectral traits (Hassan et al. 2018). High correlations indicate that UAV-based VIs from image pixels can be used as secondary traits to predict GY and that variations in these traits can ex-plain the underlying genetics. Multispectral pixel-based field visualization also accurately illustrated the transgressive segregations in DH lines at all postAns GTPs，as validated through selected high and low senescence DH lines. Our results suggest that field trials using multispectral visualization could speed up assessment of the overall status of genotypes. High genetic correlations between UAV-based VIs and GY especially CCCI，illustrated distinct average differences in UAV-based VIs of low- and high-yielding DH lines at senescence GTPs，especially at TP50 and T10 （Figures 3 and 4). Most of the DH lines started and completed senescence within a similar time span，with some exceptions. Thus，the low rate of senescence was not due to the duration between Ans and maturity but due to the slow degradation of chlorophyll at TP90 to TP50 in high- yielding DH lines （Christopher et al.，2016，2018). During this period，source to sink mobilization was the highest as evidenced by the retention of a high green area，resulting in a high yield (Senapati et al.，2018). TP50 to TP10 senescence was greatly accelerated due to internal and external factors with the least impact on GY in high-yield DH lines. In low- yield DH lines，the rate of senescence was rapid from right after Ans to TP10，resulting in lower yields due to the rapid degradation of chlorophyll (Senapati et al.，2018). The rate of senescence from TP90 to TP50 is more crucial in terms of yield losses and grouping the high- and low-yielding genotypes （Figure 5). But TP10 is also vital for final refilling of the sink，that could also significantly increase GY through a re-enforcement mechanism (Lim et al.，2007；Borrell et al.，2014；Christopher et al.，2016). For example，GY significantly differed among DH-082 and three other slow-senescence DH lines due to rapid senescence at TP10，whereas a significant difference for GY was also observed among four slow and four rapid senescence DH lines at TP10. We detected the same kind of phenotypic tradeoff in DH lines as reported by Christopher et al. （2016). Previous PCA-biplot results showed negative correlations between CS and GY at all postAns GTPs (Hassan et al.，2018)，whereas all postAns GTPs are similarly involved in affecting the GY (Figure 6B).

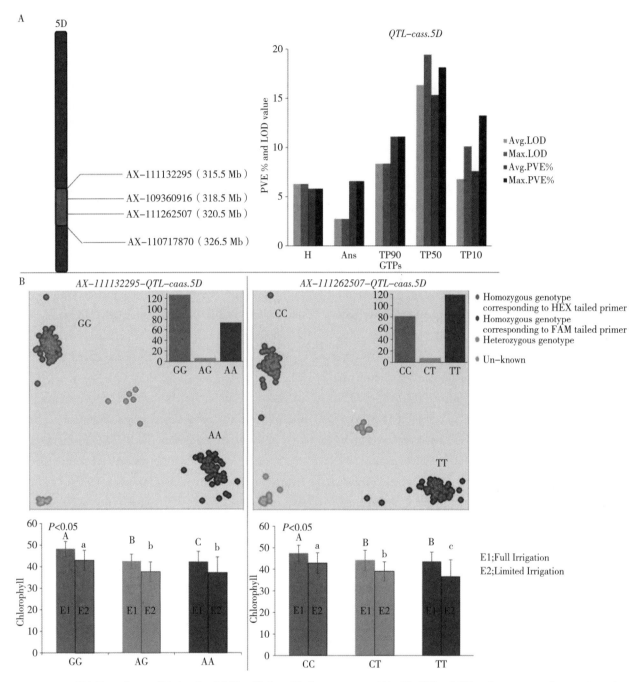

Figure 7   KASP markers validation for QS identified on 5D chromosome. (A) 5D QTL, LOD and percentage phenotypic variation explained by the QTL (PVE) across the GTPs, (B) validation of two KASP markers developed for SNPs linked with the 5D locus regarding effectiveness of this QTL for senescence using 207 genotypes of a natural population. Capital alphabets indicate variation for chlorophyll level at $P<0.05$ (by t test) among the genotypes with GG, AG, and AA SNPs within the full irrigation treatments while small alphabets indicate variation within the limited irrigation treatments. Error bars are indicating the standard deviations. Chlorophyll was estimated at MGF stage for 207 genotypes with full or limited irrigation treatments to validate the KASP markers.

Based on our results, multiple UAV-based VIs are more suitable for selecting slow senescence with high GY as compared to a single ground-based trait because CS through multi-angle data sets of light reflection plotted against GY instead of a traditional trait (ground NDVI) can provide a clear picture about the physiological performance of genotypes and increase the selection accuracy.

QTL that was identified consistently in both sites and years were considered credible, and QTL common to UAV- based VIs and ground NDVI indicated the higher accuracy and credibility of UAV-based data sets for QTL analysis. QTL detected across the GTPs were temporal loci, suggesting that the causative alleles have persistent effects at the various vegetative and senescence GTPs as reported in other time-series studies (Guo et al., 2018; Lyra et al., 2020). Clustering of the QTL for different UAV-based VIs (for biomass and chlorophyll) to the same chromosomal region indicates that a single or multiple genes at one locus affect multiple UAV-based VIs for plant physiology. Each UAV-based VI was equationally different, that is, they were derived from combinations or ratios of different spectral bands, but a surrogate of the same trait. For example, both RNDVI and CVI were proxy traits of green biomass, whereas CCCI, GCI, and RECI for chlorophyll content. Therefore, identifying QTL for any surrogate trait means that these QTL likely contributed to biomass or chlorophyll. Most of the QS detected in our study have been reported previously for stay-green or senescence-related traits (Vijayalakshmi et al., 2010; Pinto et al., 2016; Shi et al., 2017; Christopher et al., 2018). For example, Christopher et al. (2018) detected stay- green QTL on chromosomes 4B, 4D, 5B, and 7B based on analysis of a DH population in four environments. Similarly, Pinto et al. (2016) identified QTL on chromosomes 1B, 4B, 4D, 5B, 7A, and 7D in an RIL population in five heat-stress environments. In our results, loci on chromosomes 2B, 3A, 3D, and 5D were detected as new QS. *QTL-caas. 2B _ 2* (560.5Mb) on 2B temporally controlled CCCI and CVI at Ans and at TP90 and TP50 and colocalized with GY locus *QTL-caas. 2B _ 1* (567.5 Mb). A QS on chromosome 3D at 248.5 Mb also contributed to controlled stage-specific senescence, influencing chlorophyll and green biomass surrogates CCCI, RECI, and CVI from Ans to TP10. We found that the QS *QTL-caas. 5D* (315.5 Mb) identified for contributing to slow senescence was not closely linked with a major vernalization response gene (Vrn-D1) previously reported on chromosome

5D (Eagles et al., 2009). Our analysis of candidate genes showed that SNPs of the 5D locus were linked with the *TraesCS5D02G403900* gene, while the gene was annotated as encoding a glycosyltransferase (Supplemental Table S6). The glycosyltransferase protein family has been reported to be involved in stress response regulation in *A. thaliana* (Rehman et al., 2018) and rice (Shi et al., 2020). Moreover, our validation results of 5D QTL using two KASP markers indicated that it could be used to select slow-senescence genotypes. Both QTL on 4B and 4D associated with *Rht* alleles were identified as QS across the GTPs as reported previously (Pinto et al., 2016; Christopher et al., 2018), whereas GY QTL was also colocalized with senescence on 4D chromosome. GY QTL on 5A and 5B were also identified near a QS. *QTL-caas. 6D _ 3* (455.6 Mb) for CCCI, RECI and RNDVI identified at postAns GTPs also colocalized with a GY QTL on chromosome 6D. In most of these cases, alleles conferring slower senescence also conferred improved yield. In those experiments where major QTL for yield are not coincident with QTL for senescence traits, genetic factors other than senescence have a greater influence on yield (Christopher et al., 2018).

In conclusion, we found that UAV-derived VIs could be used to predict senescence and GY at postAns GTPs. More importantly, most of the traits are very difficult to measure by conventional methods and now can be assessed temporally using the low-cost UAV system. Previous QTL mapping studies have been based on conventional phenotypic data at limited growth stages, but still QTL reported in our results have good overlap with QTL previously found in several stay-green and senescence-related studies (Vijayalakshmi et al., 2010; Pinto et al., 2016; Shi et al., 2017; Christopher et al., 2018). Around 75% of the QS was detected using UAV- based VIs rather than ground NDVI. Four QTL for senescence colocalized with GY, indicating that senescence influences low- and high-yielding DH lines. We found several new QTL for senescence on chromosomes 2B, 3D, 3A, and 5D that explained the high phenotypic variations (5. 2%-18. 1%). We also demonstrated the temporal nature of the 5D QTL control at preAns and postAns GTPs. However, KASP markers based on SNPs linked with the 5D QTL can be used to select for slow senescence genotypes in breeding programs. Combining high-throughput phenotyping technology and large-scale QTL analysis not only greatly expands our knowledge of the dynamics of wheat development but also provides a new strategy for breeders to optimize plant physiology toward ideotype breeding in wheat.

Materials and methods

Germplasm and field trial

A DH population of 198 lines derived from Yangmai16/ Zhongmai895 cultivars was used to phenotype UAV-based multispectral traits to quantify senescence and discover loci that control natural variations in senescence. Yangmai16 is a spring wheat cultivar that was released in the Yangtze River area of Jiangsu Province, China in 2004. It is a popular cultivar because of its high-yield potential and stability under drought. The facultative wheat Zhongmai895 was released in the southern Yellow and Huai River valleys in 2012. It has a high yield potential with high GF rate, high nitrogen-use efficiency, and tolerance to drought and high-temperature fluctuations (Yang et al., 2020).

Field trials wereconducted during two cropping seasons (2016-2017 and 2017-2018) at two sites in Henan Province, China: Xinxiang (environment 1 [E1], 35°18′0″N, 113°52′0″E) and Luohe (E2, 33°34′0″ N, 114° 2′ 0″ E) (Figure. 1A). Both sites had experienced fluctuations in temperature, rain and daylight during maturation in both years (Supplemental Figure S3), which can decrease yield in nonresilient genotypes. The experiment was designed with 600 plots at each site, with each plot (3. 6 m² area [3 m× 1. 2 m]) having one DH line in six rows at 0. 20 m spacing. Randomized complete blocks were set up with three replications at both sites and in both years to minimize experimental error. Plots were equally irrigated across the trials at the same stages: tillering and right after Ans. Plots were agronomically managed according to local practices at each site.

A natural population of 207 genotypes was used to validate the KASP markers results using chlorophyll content data. These genotypes were grown under two water treatments (i. e. full and limited irrigations) with two replications at Xinxiang (35°18′0″N, 113° 52′ 0″ E), Henan province, China during 2018-2019. The plot dimension was consistent (3. 6 m² area [3 m×1. 2 m]) under both treatments.

UAV platform, multispectral imaging, and pixel extraction

A DJI Phantom drone (SZ DJI Technology Co., Shenzhen, China) carrying a Micasense RedEdge multispectral sensor (Micasense, Seattle, WA, USA) (https: //support. micasense. com/ hc/en-us/ articles/235402807-Getting-Started-With-RedEdge) was used for all multispectral imagery (Supplemental Figure 1B). User manual of RedEdge sensor can be downloaded using link given in online links section. The DJI Phantom can fly slowly and at low altitude for 16 min. The red-edge sensor consists of five sensors, one for blue, green, red, red-edge, and NIR spectral bands. A sunshine sensor and GPS device were connected to the multispectral sensor on the top of the Phantom to calibrate environmental irradiation during light reflectance measurement and geo-referencing. Band values before and after flight were standardized using a calibration board with known reflectance provided by Micasense. DJI pro version 3. 6. 0 (https: // www. DJI. com) was used to design the flight mission. All flights were conducted at 30 m altitude at 2 m s⁻¹, maintaining 85% forward and side overlaps among images to generate a dense point cloud for good quality orthomosaic. Average GSD of sensor was between 2. 0 and 2. 5 cm.

Pi×4 Dmapper software (version 1. 4, PI×4d, Lausanne, Switzerland) was used for orthomosaic generation (Figure 1C). Key steps for the orthomosaic generation using Pi×4D mapper comprised camera alignment, georeferencing, point cloud creation, and orthomosaic generation as previously reported (Hassan et al., 2018, 2019a, 2019b; Figure 1, D and E). QGIS version 3.2 (an open-source geographic information software that support viewing and editing of data) was used for image segmentation of each plot. For this, polygon shapes were generated with the specific plot ID for a particular DH line. Spectral values in the form of pixels were extracted from the segmented parts of orthomosaic TIFF images of five different bands using polygon shapes as mask through a computer vision approach in the program IDL (version 8.6, Harris Geospatial Solutions, Inc., Broomfield, CO, USA) (Figure 1C). The plant density of each plot was well enough to ignore the data noise due to background soil. Multispectral data were collected by the UAV at five TPsthat is, H, Ans, 10 d (TP90), 17 d (TP50), and 24 d (TP10) after Ans (Figure 1f).

Five UAV-based VIs were quantified from multispectral image pixels captured from the reflectance of four spectral bands at five important GTPs (Figure 1G): red NDVI (RNDVI), chlorophyll VI (CVI) for green biomass, CCCI (GCI, and RECI for chlorophyll content estimation. The equations for the estimation of all VIs are given in Figure 1H and Supplemental Table S1. VIs were visualized through the combinations of particular multispectral pixels of light reflectance bands using Pi×4D calculator (Figure 1H). The assessment of GY through VIs was done by correlation analysis (Figure 1I).

Estimation of ground truth measurements

Ground NDVI was measured using a handheld GreenSeeker at five TPs that is, H, Ans, at 10 (TP90; 10% senescence), 17 (TP50; 50% senescence), and 24 (TP10; 90% senescence) days after Ans for DH lines. Agronomic and yield-related traits such as number of days to reach to H, flowering, physiological maturity, and GY were also recorded using standard procedures detailed previously (Gao et

al., 2017). Chlorophyll level of natural population was assessed using SPAD-502 Plus (Konica Minolta, Japan) at mid GF (MGF) stage to validate the KASP markers. Chlorophyll was averaged from flag leaves of 10 plants of each genotype.

Estimation of CS

Estimated valuesfor the UAV and ground-based VIs were highest at H stage compared with all other TPs; thus, the value at this stage was considered to be the maximum VI value (VI_{max}). The degree of senescence was estimated by subtracting the decreasing values of VIs measured at 10 (TP90; 10% senescence), 17 (TP50; 50% senescence), and 24 (TP10; 90% senescence) days after Ans. The mean senescence was estimated after each 100 degree-days since Ans using the following equation:

$$CS = VImax - VI_{(TP90;TP50\ and\ TP10)},\quad (1)$$

where CS is mean CS, VI_{max} indicates the maximum VI value (at H), GTP TP90, TP50, and TP10 indicate the VI values at 10%, 50%, and 90% senescence stages, respectively. These TPs were calculated according to thermal TPs after Ans (Lopes and Reynolds, 2012; Christopher et al., 2018). Genotypes with slow senescence were selected by plotting the UAV-based VIs derived senescence at TP90 to TP10 against the GY using a PCA-biplot analysis.

SNP genotyping and QTL mapping

The 198 DH lines and parents were genotyped using the commercially available Wheat 660K SNP array that was de-veloped by Affymetrix and Prof. Jizeng Jia at the Institute of Crop Sciences, CAAS (Capital Bio Corporation, Beijing, China). Previously, our research group constructed a genetic map for mapping QTL for key yield-related traits. Markers with distorted segregation, no polymorphisms among parents, and missing at a rate >20% were removed in a subsequent linkage analysis. Around 10, 242 markers, each representing a bin site, were selected to construct the linkage map for the DH population. The map comprises 25 link- age groups for all 21 chromosomes of the A, B, and D genomes. Inclusive composite-interval mapping was used for QTL analysis in IciMapping version 4. 1 software (Meng et al. , 2015). The averaged data of each trait and GTPs from 2 years at Xinxiang (E1) and Luohe (E2) separately, and best linear unbiased estimates (BLUEs) values of across environments (E1 and E2) were used for QTL detection. BLUEs values were calculated using a model explained by Alvarado et al. (2020) for a randomized complete block design (RCBD) experiment in META-R software (a multi-environment analysis tool developed at CIMMYT, Mexico City, Mexico). The genotype of Yangmai 16 was defined as A, that of Zhongmai 895 as B. Hence, alleles from Yangmai 16 reduced trait values when the additive effects were negative. Recombination frequencies were converted into map distance using the Kosambi mapping function. Locations of QTL for the traits were determined by ICIM-ADD using the same software as for the linkage analysis. The threshold for declaring the presence of a significant QTL for each trait was defined by 1, 000 permutations at $P < 0.05$, and the minimum LOD score of 2. 5 was chosen; the walking speed was set at 1. 0 cM.

KASP marker development

KASP markers for locus QTL-$caas. 5DL$ were developed based on two corresponding SNPs to validate the UAV-based Vis for genetic dissection of senescence (Supplemental Table S5). The flanking sequences of the SNPs were used as queries in a blast search against the reference genome of wheat using IWGSC (CS Refseq version 1. 0; IWGSC 2018). Chromosome-specific KASP primers were developed by alignment of homologous sequence. Allele-specific primers carrying FAM and HEX were designed with targeted SNPs at the 3'-end, and common reverse primers were designed with less than 200 bp of amplified sequences for chromosome-specific amplification. The detailed information about the two KASP markers is provided in Supplemental Table S7. The KASP assay mixture consisted of 40 mL of common primer (100 mL), 16 mL of each tailed primer (100 μL), and 60 μL ddH$_2$O. Each reaction mixture comprised 2. 5 μL of 2 × KASP master mixture (LGC Genomics, Hoddesdon, UK), 0. 056 μL of KASP assay mixture, and 2. 5 μL of DNA (30-50 ng/μL). PCR was performed in a 384-well plate with denaturation at 95℃ for 15 min; 9 touchdown cycles (95℃ for 20 s touchdown at 65℃ initially then deceasing by 1℃ per cycle for 1min); and 32 cycles of denaturing, annealing, and extension (95℃ for 10 s, 57℃ for 1min). KASP genotyping results were compared among 207 accessions of a natural population for chlorophyll level at mid to late GF (LGF) stage and original chip-based results using a t test in XLSTATE software to confirm the association of the KASP markers with the trait.

Analysis of putative candidate genes

The geneslocated in the physical intervals of the genomic region of identified QTL were screened based on annota- tions in the wheat reference genome (CS RefSeq version 1. 0; IWGSC 2018), and those related to growth, development, stress resistance, and nutrient mobilization were considered as candidate genes. Gene annotations were retrieved using EnsemblPlant and EMBL—EBI (http: //www. ebi. ac. uk/ inter pro) databases. Gene annotations for putative proteins were done using BLAST2GO (https: // www. blast2go. com/).

Statistical analyses

Phenotypic and genetic correlation matrices were calculated to evaluate the relationship between all observed parameters using META-R software (a multi-environment analysis tool developed at CIMMYT, Mexico City, Mexico; Alvarado et al. , 2020). Logistic regression was done to check the relevance of UAV-based VIs for GY using a generalized linear model in XLSTAT software by Addinsoft. A mixed linear model was used to test the significance of variation among DH lines, and G×E interaction for UAV-based VIs, ground truth NDVI and GY explained by Alvarado et al. (2020). The results were considered as significant at $P < 0.05$.

$$Y_{ijk} = \mu + \text{Loc}_i + \text{Rep}j \ (\text{Loc}_i) \ + \text{Gen}_k + \text{Gen}_k \times \text{Loc}_i + \varepsilon_{ijk} \qquad (2)$$

Where Y_{ijk} is the trait of interest, μ is the overall mean effect, Loc_i and Rep_j is the random effect of the i th location and j th replicate, Gen_k is the fixed effect of the k th genotype, Loc_i and Loc_i Gen_k are the random effects of the i th environment and the $\text{G} \times \text{E}$ interaction, respectively, and ε_{ijk} is the effect of the error associated with the i th location, j th replication and k th genotype, which is assumed to be independently and identically distributed (iid) normal with mean zero and variance σ_ε^2.

To confirm that phenotypic variations of traits were due to genetic diversity, we calculated broad-sense heritabilities for all traits in each and across the environments using entries as a random effect using the following equation (Sehgal et al., 2015):

$$H^2 = \sigma_g^2 / (\sigma_g^2 + \frac{\sigma_{ge}^2}{e} + \frac{\sigma_\varepsilon^2}{re}) \qquad (3)$$

where σ_g^2 and σ_ε^2 represent the genotypic and error variances, respectively. The term σ_{ge}^2 is the $\text{G} \times \text{E}$ interaction variance component, e is number of environments (locations and years) and r is the number of replications for each genotype in each environment. To calculate the H^2 for each of two locations e is considered number of years. H^2 of each component at all growth stages provides an indication of the consistency of the trait in a particular environment.

PCA was used to assess the diversity among DH lines for senescence. The basic equation used for the PCA in matrix notation was

$$Y = W'X, \qquad (4)$$

where W is a matrix of coefficients that is determined by PCA and X is an adjusted data matrix consisting of n observations (rows) on p variables (columns).

❖ References

Albacete AA, Martínez-Andújar C, Pérez-Alfocea F (2014) Hormonal and metabolic regulation of source-sink relations under salinity and drought: from plant survival to crop yield stability. Biotechnol Adv 32: 12-30.

Alvarado G, Rodríguez FM, Pacheco A, Burgueño J, Crossa J, Vargas M, Pérez-Rodríguez P, Lopez-Cruz MA (2020) META-R: a software to analyze data from multi-environment plant breeding trials. Crop J 8: 745-756.

Araus JL, Cairns JE (2014) Field high-throughput phenotyping: the new crop breeding frontier. Trends Plant Sci 19: 52-61.

Araus JL, Kefauver SC (2018) Breeding to adapt agriculture to climate change: affordable phenotyping solutions. Curr Opinion Plant Biol 45: 237-247.

Araus JL, Kefauver SC, Zaman-Allah M, Olsen MS, Cairns JE (2018) Translating high-throughput phenotyping into genetic gain. Trends Plant Sci 23: 451-466.

Borrell AK, Mullet JE, George-Jaeggli B, van Oosterom EJ, Hammer GL, Klein PE, Jordan DR (2014) Drought adaptation of stay-green sorghum is associated with canopy development, leaf anatomy, root growth, and water uptake. J Exp Bot 65: 6251-6263 Campbell MT, Du Q, Liu K, Brien CJ, Berger B, Zhang C, Walia H (2017) A comprehensive image-based phenomic analysis reveals the complex genetic architecture of shoot growth dynamics in rice (Oryza sativa). Plant Genome 10: 1-14.

Christopher JT, Christopher MJ, Borrell AK, Fletcher S, Chenu K (2016) Stay-green traits to improve wheat adaptation in well-watered and water-limited environments. J Exp Bot 67: 5159-5172.

Christopher M, Chenu K, Jennings R, Fletcher S, Butler D, Borrell A, Christopher J (2018) QTL for stay-green traits in wheat in well-watered and water-limited environments. Field Crops Res 217: 32-44.

Duan T, Chapman SC, Guo Y, Zheng B (2017) Dynamic monitoring of NDVI in wheat agronomy and breeding trials using an un-manned aerial vehicle. Field Crops Res 210: 71-80.

Eagles HA, Cane K, Vallance N (2009) The flow of alleles of important photoperiod and vernalisation genes through Australian wheat. Crop Pasture Sci 60：646-657.

Furbank RT，Tester M（2011）Phenomics-technologies to relieve the phenotyping bottleneck. Trend Plant Sci 16：635-644.

Gao F，Ma D，Yin G，Rasheed A，Dong Y，Xiao Y，Xia X，Wu X，He Z（2017）Genetic progress in grain yield and physiological traits in Chinese wheat cultivars of southern yellow and Huai Valley since 1950. Crop Sci 57：760-773.

Gregersen PL，Holm PB（2007）Transcriptome analysis of senescence in the flag leaf of wheat（*Triticum aestivum* L.）. Plant Biotechnol J 5：192-206.

Guo Z，Yang W，Chang Y，Ma X，Tu H，Xiong F，Jiang N，Feng H，Huang C，Yang P，et al.（2018）Genome-wide association studies of image traits reveal genetic architecture of drought resistance in rice. Mol Plant 11：789-805.

Hassan MA，Yang M，Fu L，Rasheed A，Zheng B，Xia X，Xiao Y，He Z（2019a）Accuracy assessment of plant height using an unmanned aerial vehicle for quantitative genomic analysis in bread wheat. Plant Methods 15：37.

Hassan MA，Yang M，Rasheed A，Jin X，Xia X，Xiao Y，He Z（2018）Time-series multispectral indices from unmanned aerial vehicle imagery reveal senescence rate in bread wheat. Remote Sens 10：809.

Hassan MA，Yang M，Rasheed A，Yang G，Reynolds M，Xia X，Xiao Y，He Z（2019b）A rapid monitoring of NDVI across the wheat growth cycle for grain yield prediction using a multi-spectral UAV platform. Plant Sci 282：95-103.

Hatfield JL，Gitelson AA，Schepers JS，Walthall CL（2008）Application of spectral remote sensing for agronomic decisions. Agron J 100：S-117-S-131.

Jin X，Zarco-Tejada P，Schmidhalter U，Reynolds MP，Hawkesford MJ，Varshney RK，Yang T，Nie C，Li Z，Ming B，et al.（2020）High-throughput estimation of crop traits：a review of ground and aerial phenotyping platforms. IEEE Geosci Remote Sens Magaz 9：200-231.

Lim PO，Kim Y，Breeze E，Koo JC，Woo HR，Ryu JS，Park DH，Beynon J，Tabrett A，Buchanan-Wollaston V（2007）Overexpression of a chromatin architecture-controlling AT-hook protein extends leaf longevity and increases the post-harvest storage life of plants. Plant J 52：1140-1153.

Liu L，Zhou Y，Zhou G，Ye R，Zhao L，Li X，LinY（2008）Identification of early senescence-associated genes in rice flag leaves. Plant Mol Biol 67：37-55.

Lopes MS，Reynolds MP（2012）Stay-green in spring wheat can be determined by spectral reflectance measurements（normalized difference vegetation index）independently from phenology. J Exp Bot 63：3789-3798.

Lyra DH，Virlet N，Sadeghi-Tehran P，Hassall KL，Wingen LU，Orford S，Griffiths S，Hawkesford MJ，Slavov GT（2020）Functional QTL mapping and genomic prediction of canopy height in wheat measured using a robotic field phenotyping platform. J Exp Bot 71：1885-1898.

Meng L，Li H，Zhang L，Wang J（2015）QTL IciMapping：integrated software for genetic linkage map construction and quantitative trait locus mapping in biparental populations. Crop J 3：269-283.

Mochida K，Koda S，Inoue K，Hirayama T，Tanaka S，Nishii R，Melgani F（2018）Computer vision-based phenotyping for im- provement of plant productivity：a machine learning perspective. GigaScience 8：giy153.

Pinto RS，Lopes MS，Collins NC，Reynolds MP（2016）Modelling and genetic dissection of staygreen under heat stress. Theor Appl Genet 129：2055-2074.

Pleban JR，Guadagno CR，Mackay DS，Weinig C，EwersBE（2020）Rapid chlorophyll fluorescence light response curves mechanisti- cally inform photosynthesis modeling. Plant Physiol 183：602-619.

Rasheed A，Takumi S，Hassan MA，Imtiaz M，Ali M，Morgunov AI，Mahmood T，He Z（2020）Appraisal of wheat genomics for gene discovery and breeding applications：a special emphasis on advances in Asia. Theor Appl Genet 133：1503-1520.

Rehman HM，Nawaz MA，Shah ZH，Ludwig-Müller J，Chung G，Ahmad MQ，Yang SH，Lee SI（2018）Comparative genomic and transcriptomic analyses of family-1 UDP glycosyltransferase in three Brassica species and Arabidopsis indicates stress-responsive regulation. Sci Rep 8：1875.

Ruckelshausen A，Busemeyer L（2015）Toward digital and image-based phenotyping. *In* J Kumar，A Pratap，S Kumar，eds，Phenomics in Crop Plants：Trends，Options and Limitations. Springer India，New Delhi，India，pp 41-60.

Sade N，del Mar Rubio-Wilhelmi M，Umnajkitikorn K，Blumwald E（2017）Stress-induced senescence and plant tolerance to abioticstress. J Exp Bot 69：845-853.

Salas Fernandez MG，Bao Y，Tang L，Schnable PS（2017）A high-throughput，field-based phenotyping technology for tall bio-mass crops. Plant Physiol 174：

2008-2022.

Schippers JH，Schmidt R，Wagstaff C，Jing HC（2015）Living to die and dying to live：the survival strategy behind leaf senescence. Plant Physiol 169：914-930.

Sehgal D，Skot L，Singh R，Srivastava RK，Das SP，Taunk J，Sharma PC，Pal R，Raj B，Hash CT，Yadav RS（2015）Exploring potential of pearl millet germplasm association panel for association mappping of drought tolerance traits. PLoS One 10：e0122165.

Senapati N，Stratonovitch P，Paul MJ，Semenov MA（2018）Drought tolerance during reproductive development is important for increasing wheat yield potential under climate change in Europe. J Exp Bot 70：2549-2560.

Shi Y，Phan H，Liu Y，Cao S，Zhang Z，Chu C，Schläppi MR（2020）Glycosyltransferase OsUGT90A1 helps protect the plasma membrane during chilling stress in rice. J Exp Bot 71：2723-2739.

Shi S，Azam FI，Li H，Chang X，Li B，Jing R（2017）Mapping QTL for stay-green and agronomic traits in wheat under diverse water regimes. Euphytica 213：246.

Su J，Liu C，Hu X，Xu X，Guo L，Chen WH（2019）Spatio-temporal monitoring of wheat yellow rust using UAV multispectral imagery. Comput Electron Agric 167：105035.

Tardieu F，Cabrera-Bosquet L，Pridmore T，Bennett M（2017）Plant phenomics，from sensors to knowledge. Curr Biol 27：R770-R783.

Tester M，Langridge P（2010）Breeding technologies to increase crop production in a changing world. Science 327：818-822.

Vijayalakshmi K，Fritz AK，Paulsen GM，Bai G，Pandravada S，Gill BS（2010）Modeling and mapping QTL for senescence-related traits in winter wheat under high temperature. Mol Breed 26：163-175.

Woo HR，Kim HJ，Nam HG，Lim PO（2013）Plant leaf senescence and death-regulation by multiple layers of control and implications for aging in general. J Cell Sci 126：4823-4833.

Yang M，Hassan MA，Xu K，Zheng C，Rasheed A，Zhang Y，Jin X，Xia X，Xiao Y，He Z（2020）Assessment of water and nitrogen use efficiencies through UAV-based multispectral phenotyping in winter wheat. Front Plant Sci 11：927.

Yang W，Guo Z，Huang C，Duan L，Chen G，Jiang N，Fang W，Feng H，Xie W，Lian X（2014）Combining high-throughput phenotyping and genome-wide association studies to reveal natural genetic variation in rice. Nat Commun 5：5087.

Yolcu S，Li X，Li S，Kim YJ（2017）Beyond the genetic code in leaf senescence. J Exp Bot 69：801-810.

York LM（2018）Functional phenomics：an emerging field integrating high-throughput phenotyping，physiology，and bioinformatics. J Exp Bot 70：379-386.

Zhou Y，Srinivasan S，Mirnezami SV，Kusmec A，Fu Q，Attigala L，Salas Fernandez MG，Ganapathysubramanian B，Schnable PS（2019）Semiautomated feature extraction from RGB images for sorghum panicle architecture GWAS. Plant Physiol 179：24-37.

Zhu JK（2016）Abiotic stress signaling and responses in plants. Cell 167：313-324.

Identification and validation of genetic loci for tiller angle in bread wheat

Dehui Zhao[1,2] · Li Yang[1] · Kaijie Xu[3] · Shuanghe Cao[1] · Yubing Tian[1] ·
Jun Yan[3] · Zhonghu He[1,4] · Xianchun Xia[1] · Xiyue Song[2,*] · Yong Zhang[1,*]

[1] Institute of Crop Sciences, National Wheat Improvement Centre, Chinese Academy of Agricultural Sciences (CAAS), 12 Zhongguancun South Street, Beijing 100081, China

[2] College of Agronomy, Northwest A & F University, Yangling 712100, Shaanxi province, China

[3] Institute of Cotton Research, CAAS, 38 Huanghe Dadao, Anyang 455000, Henan province, China

[4] CIMMYT-China Office, c/o CAAS, 12 Zhongguancun South Street, Beijing 100081, China

* Correspondence: Xiyue Song (songxiyue@nwafu.edu.cn, +86-29-87082845), Yong Zhang (zhangyong05@caas.cn, +86-10-82108745)

Key message: Two major QTL for tiller angle were identified on chromosomes 1AL and 5DL, and *TaTAC-D1* might be the candidate gene for *QTA. caas-5DL*.

Abstract: An ideal plant architecture is important for achieving high grain yield in crops. Tiller angle (TA) is an important factor influencing yield. In the present study, 266 recombinant inbred lines (RILs) derived from a cross between Zhongmai 871 (ZM871) and its sister line Zhongmai 895 (ZM895) was used to map TA by extreme pool-genotyping and inclusive composite interval mapping (ICIM). Two quantitative trait loci (QTL) on chromosomes 1AL and 5DL were identified with reduced tiller angle alleles contributed by ZM895. *QTA. caas-1AL* was detected in six environments, explaining 5.4-11.2% of the phenotypic variances. The major stable QTL, *QTA. caas-5DL*, was identified in all eight environments, accounting for 13.8-24.8% of the phenotypic variances. The two QTL were further validated using BC_1F_4 populations derived from backcrosses ZM871/ZM895//ZM871 (121 lines) and ZM871/ZM895//ZM895 (175 lines). Gene *TraesCS5D02G322600*, located in the 5DL QTL and designated *TaTAC-D1*, had a SNP in the third exon with 'A' and 'G' in ZM871 and ZM895, respectively, resulting in a *Thr169Ala* amino acid change. A KASP marker based on this SNP was validated in two sets of germplasm, providing further evidence for the significant effects of *TaTAC-D1* on TA. Thus extreme pool-genotyping can be employed to detect QTL for plant architecture traits and KASP markers tightly linked with the QTL can be used in wheat breeding programs targeting improved plant architecture.

Introduction

Plant architecture determined mainly by plant height, tiller number and angle, and panicle morphology affects photosynthetic efficiency, stress response, grain yield, and grain quality in rice (Wang and Li 2008). Wheat plants with desirable architecture are

Published in Theoretical and Applied Genetics, 2020, 133: 3037-3047.

able to produce higher grain yields, as was the case in the "Green Revolution" (GR) where grain yields were significantly increased by growing lodging resistant semi-dwarf varieties that could respond to higher nutrient inputs (Peng et al. 1999). A large number of QTL for plant height have been identified and 25 of them have been categorized as *Rht1-Rht25* (Griffiths et al. 2012; Würschum et al. 2015; Tian et al. 2017; Ford et al. 2018; Mo et al. 2018; Sun et al. 2019). Several QTL for tiller number have been identified, including *tin1*, *tin2*, *tin3* and *ftin* (Peng et al. 1998; Kato et al. 2000; Spielmeyer and Richards, 2004, Kuraparthy et al. 2007; Li et al. 2010; Deng et al. 2011; Naruoka et al. 2011; Zhang et al. 2013; Nasseer et al. 2016; Wang et al. 2016; Hu et al. 2017; Xu et al. 2017; Ren et al. 2018). TEOSINTE BRANCHED1 (*TB1*) affects tiller number (Lewis et al. 2008; Dixon et al. 2018). Quite a few QTL for flag leaf angle have been mapped on chromosomes 1B, 2A, 2B, 3A, 3B, 3D, 4B, 5A, 5B, 6B, 7B, 7A and 7D (Isidro et al. 2012; Wu et al. 2016; Liu et al. 2018). Leaf angle gene, *TaSPL8*, encoding a SQUAMOSA PROMOTER BINDING-LIKE (SPL) protein, was positionally cloned and a mutant conferred a larger leaf angle phenotype due to changes in the auxin signaling and brassinosteroid biosynthesis pathways (Liu et al. 2019). However, no QTL for tiller angle has been reported in wheat.

Many QTL underpinning tiller angle have been identified in rice (Xu et al. 1995; Li et al. 1999; Yan et al. 1999; Qian et al. 2001; Thomson et al. 2003; Li et al. 2006; Li et al. 2007; Yu et al. 2007; Dong et al. 2016; He et al. 2017) and a few rice genes controlling tiller angle have been isolated. Rice *D2* encoded a novel cytochrome brassinosteroid (BR) biosynthesis enzyme P450; a mutant with allele *d2* conferred erect leaves (Hong et al. 2003). *LAZY1* regulated shoot gravitropism and tiller angle through a negative role in polar auxin transport (Li et al. 2007). TILLER ANGLE CONTROL1 (*TAC1*) is a key regulator responsible for different tiller angles (Yu et al. 2007). A

candidate gene in peach named *PepTAC1*, obtained by using a next generation sequence-based mapping approach, was associated with vertically oriented growth of branches (Dardick et al. 2013). Dong et al. (2016) identified a tiller angle gene in rice, *TAC3*, contributing most of the natural variation of tiller angle together with *TAC1* and *D2*. These findings provided important reference information to mine genes for tiller angle in wheat through comparative genomic approaches. By homology cloning based on *OsTAC1* and *ZmTAC1* genes, Cao et al. (2017) identified two types of cDNA sequences on the A-genome homeolog (*TaTAC-A1*), Type I (with a premature stop codon) and Type II (with a CGCGCG insertion) in four wheat cultivars, and the expression profiling indicated that *TaTAC1* was mostly expressed in the leaf sheath and stem at the tillering stage. However, the chromosomal locations of the two types of *TaTAC1* were not determined.

Takagi et al. (2013) reported QTL-seq as a method for identification of QTL for partial resistance to blast and seedling vigor in recombinant inbred lines and F_2 populations of rice. The combination of bulked segregant analysis and RNA-sequencing (BSR-Seq) provided a fast and economic way for QTL mapping taking advantage of the recently developed reference genome sequence of Chinese Spring (IWGSC, https://urgi.versailles.inra.fr/blast iwgsc/blast.php) and high-throughput sequencing (Ramirez-Gonzalez et al. 2015a, b; Wang et al. 2017; Wu et al. 2018a, b; Mu et al. 2019).

Zhongmai 895 is an elite cultivar in the Yellow and Huai River Valleys Winter Wheat Region, with an annual production area around 0.70 million ha, and Zhongmai 871 is its near-isogenic line with contrasting plant architecture. The objectives of the present study were to: (1) identify loci controlling TA in a recombinant inbred line (RIL) population developed from a cross of sister lines, Zhongmai 871 and Zhongmai 895, (2) develop and validate breeder-friendly markers for marker-assisted selection (MAS), and

(3) investigate the effect of the *TaTAC-D1* allele in bread wheat.

Materials and methods

Plant materials

Zhongmai 895 (ZM895) has an erect stem, whereas Zhongmai 871 (ZM871) is more prostate (Fig. 1). Detailed information regarding ZM895 and ZM871 was provided in a previous study (Yang et al., 2020). Two hundred and sixty-six F_6 RILs from cross ZM871/ZM895 and two BC_1F_4 populations from backcrosses ZM871/ZM895//ZM871 and ZM871/ZM895//ZM895 with 121 and 175 lines, respectively, were used for QTL validation. Two germplasm panels with 156 and 93 cultivars/lines were used to validate the association between the *TaTAC-D1* alleles and the differences in tiller angle.

The ZM871/ZM895 mapping population was evaluated in eight environments including Anyang (Henan) in 2015-2016, 2016-2017 and 2017-2018, Shangqiu (Henan) in 2016-2017 and 2017-2018, Zhoukou (Henan) in 2015-2016 and 2016-2017, and Xianyang (Shaanxi) in 2015-2016. The BC_1F_4 populations for QTL validation were planted in six environments, Anyang in 2015-2016 and 2016-2017, Shangqiu in 2016-2017; Zhoukou in 2015-2016 and 2016-2017, and Xianyang in 2015-2016. The experimental design was randomized complete blocks with two replications in 2015-2016 season and three replications in 2016-2017 and 2017-2018 seasons. Each plot comprised two 1 m rows spaced 25 cm apart, with 30 seeds in each row. The validation panel (VP) with 156 cultivars/lines was grown at Anyang in 2017-2018 and 2018-2019 and the panel with 93 cultivars/lines was grown at Beijing in 2017-2018 and 2018-2019 (Table S1). Each line was grown in a two-row 2 m plot with row spacing of 25 cm and 60 seeds were sown in each row. A parental check was sown every 15th plot. All the field trials were managed according to local practices.

(a)　　　　　　　　　　　　(b)

Fig. 1　Plant architecture of two parent lines at 15 days after heading. (a) A single plant in pots (*left*, Zhongmai 871; *right*, Zhongmai 895); (b) a row in field (*left*, Zhongmai 871; *right*, Zhongmai 895)

The maximum distances (D in cm) between stems at 30 cm above ground level in three randomly selected plants each row were measured by the straightedge tool at 15 days post heading. Tiller angle (TA) was calculated by the arctan- gent function of ratios in a half-average D value to 30 cm, i. e. TA = 2 * arctan (D/2 * 30).

Phenotyping of tiller angles

Tiller angle in wheat differs from rice in that it can be measured by protractor. The morphological characteristics of wheat can be more accurately and reasonably quantified based on the distribution patterns of stem tillers at 30 cm above ground level using a straightedge tool.

Bulked segregant RNA-sequencing and molecular genotyping

Based on TA data obtained from the 2015-2016 season, two pools involving 30 lines from each tail (high and low) of the distribution of the ZM895/ZM871 mapping population were genotyped by BSR-Seq. Total RNA was isolated using the TRIzol protocol (Invitrogen, Carlsbad, CA, USA). After quality testing, a single RNA library was constructed for each pool and the library preparations were sequenced on an Illumina Hiseq platform at Novogene Bioinformatics Technology (Beijing, http://www.novog ene.com/). SNP and Indel genotype calling and allele clustering were processed with GATK2 (v3.2.1) software. SNP and Indel filtering criteria were as follows: monomorphic and poor quality SNPs with more than 10% missing values, ambiguous calling, and minor allele frequencies < 5% were excluded from subsequent analysis. The differences of the SNPs/InDels between two pools were calculated by QTL-seq as the delta SNP/InDel index (Takagi et al. 2013). Chromosomes with potential TA QTL were deter-mined based on the frequency distribution of polymorphic SNPs/Indels index. Genomic DNA was extracted from young leaves using the CTAB method for mapping and validation populations (Porebski et al. 1997).

Map construction and QTL analysis

SNP markers from BSR-Seq and the wheat 660 K SNP chip (Yang et al. 2020) were converted into KASP markers. Fourteen KASP markers on chromosomes 1AL and 5DL were used to genotype the whole mapping population, and to construct genetic maps using JoinMap 4.0 (Table S2). QTL analysis was performed by inclusive composite interval mapping (ICIM) using IciMapping 4.1. Original phenotypic data and adjusted mean phenotypic data from each environment in the 2015-16, 2016-17, and 2017-2018 seasons and the best linear unbiased estimators (BLUEs) across eight environments were used for QTL detection. A LOD threshold of 2.0 was set for declaring significant QTL based on 1000 permutations at $P = 0.05$. Physical positions of mapped SNPs in the QTL regions were obtained by blasting SNP flanking sequences against the reference genome sequence of Chinese Spring (IWGSC, https://urgi.versailles.inra.fr/blast_iwgsc/).

DNA sequencing of *TaTAC-D1*

The genomic fragments were amplified from ZM871 and ZM895 using primer pairs TAC-D1-2F: 5'-TTCAGGCGATCCCATTCC-3'/TAC-D1-2R: 5'-CAGATTCACCACTTTACTCTAC -3' and TAC-D1-3F: 5'-GCTACCATCAGT-TTCAAGGAAT-3'/TAC-D1-3R: 5'-CAAGCCCCTAAC-TACTGCC -3' to identify the 5'-UTR, coding and 3'-UTR sequences of *TaTAC-D1*. The polymerase chain reaction (PCR) products were purified and directly sequenced.

Quantitative real-time PCR

Total RNA was extracted from fresh roots, stems, tiller nodes, leaves, and leaf sheaths at the tillering, jointing and heading stages with three biological replications using a RNAprep pure Plant Kit (Tiangen Biotech Co., Ltd., http://www.tiangen.com/). Reverse transcription was performed with reverse transcriptase M-MLV (RNase H⁻) (Takara Biomedical Technology Co., Ltd., http://www.takar a.com). Quantitative real-time PCR (qPCR) was performed with three technology replications by the iTaq™ Universal SYBR Green Supermix on CFX Real-Time System (Bio-Rad). Gene-specific qPCR primer pairs for *TaTAC-D1* were designed according to gene annotation from the reference genome sequence of Chinese Spring (IWGSC, https://urgi.versailles.inra.fr/blast_iwgsc/). The expression of *TaTAC-D1* was using primer pairs qTAC-D1F: GAA GTGGAGGATGTTGCTAAAG and qTAC-D1R: TAATTTTCTCGTAGCCTTCAGC . The qPCR profile was: 95°C for 2 min, followed by 40 cycles of 95°C for 5 s and 58°C for 20 s. A final dissociation stage was run to generate a melting curve for judgment of amplification specificity. Quantification of qPCR was based on the $2^{-\Delta\Delta Ct}$ method (Schmittgen and Livak 2008).

Statistical analysis

Analysis of variance (ANOVA), phenotypic correlation coefficients, and Student's t-tests were conducted with SAS 9.2 software (SAS Institute Inc, Cary, NC). PROC MIXED was used in ANOVA to evaluate the contributions of lines (RILs) and environments, where lines, environments, line $\times$ environment interactions, and replicates nested in environments were all considered as random effects. In parallel, a model considering lines as fixed effect was fitted for estimating BLUEs of lines across environments. Adjusted means of each line for TA in individual and across environments were separately computed with PROC MIXED. Broad-sense heritabilities (H_b^2) on an across-environment genotype mean basis and standard error were calculated following Holland et al. (2003). Homozygous lines from backcrosses ZM871/ZM895//ZM871 and ZM871/ZM895//ZM895 were used to verify QTL effects. The differences of TA between two classes of homozygous genotypes (homozygous for ZM871 and homozygous for ZM895) were calculated by PROC MIXED, treating genotypes as fixed effect, with lines nested in genotype, environments, and their related interactions, and replicates nested in environments as random. The two validation panels (VP) were used to validate the association between the $TaTAC$-$D1$ alleles and the differences in tiller angle by Student's t-tests.

Results

Phenotypic evaluation

The difference in tiller angle started at the late jointing stage between parental lines, ZM871 and ZM895, which have average tiller numbers of 216 and 209, respectively, in a 1 m row, without any significant differences. Analysis of variance (ANOVA) showed that line and line $\times$ environment interaction effects were significant for TA. A moderate broad-sense heritability ($H_b^2 = 0.75$) was obtained across the eight environments (Table S3). Pearson's correlation co-

efficients for TA of the mapping population ranged from 0.46 to 0.75 among eight environments ($P < 0.001$) (Table S4). The mean value on an across-environment TA was 24° and 36° for ZM895 and ZM871, respectively, confirming that ZM895 was more erect than ZM871 (Fig. 1). Continuous distribution and transgressive segregation were observed for TA, indicative of polygenic inheritance (Fig. S1). Although semi-dwarf genes $Rht1$ and $Rht2$ usually have significant effects on plant architecture (Peng et al. 1999; Pearce et al. 2011; Wu et al. 2011), ZM895 and ZM871 had the same Rht-$B1a$ and Rht-$D1b$ alleles, with average plant height of 71 and 67 cm, respectively.

QTL mapping of TA

Two TA (erect vs. prostate stems) pools of 30 lines were genotyped using BSR-Seq, based on the data obtained from 2015 to 2016 season. Five chromosomes, 1A, 2A, 3A, 5B and 5D, were associated with TA based on the chromosomal distributions of clustered polymorphic SNPs and Indels above 0.5 or under -0.5 (Fig. S2). KASP markers from these chromosomes were selected to construct genetic maps. Thirty-eight KASP markers displaying polymorphisms between the parents were used to construct the linkage maps for QTL analysis, while only 14 KASP markers on chromosomes 1A (7) and 5D (7) showed consistent polymorphisms between lines in the two pools. The polymorphic KASP markers were that used to genotype all 266 F_6 RILs, and linkage groups were constructed for chromosomes 1A and 5D (Table S2, Fig. 2). Two QTL for TA were mapped on chromosomes 1AL and 5DL by inclusive composite interval mapping (ICIM) (Table 1, Fig. 2). $QTA.caas$-$1AL$ detected in six environments explaining 5.4-11.2% of the phenotypic variances was located in a 0.5 cM region between $Kasp\_1A18$ and $Kasp\_1A90$. $QTA.caas$-$5DL$ in a 3.0 cM region between $Kasp\_5D16$ and $Kasp\_5D17$ was identified in all eight environments, explaining 13.8-24.8% of the phenotypic variances. For both QTL, ZM895 contributed the alleles for a reduced tiller angle.

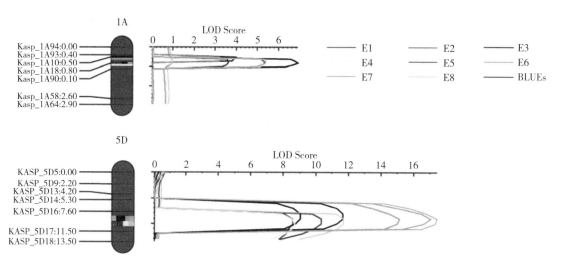

Fig. 2　LOD contours obtained by inclusive composite interval map- ping of QTL for tiller angle in the Zhongmai 871/Zhong-mai 895 RIL population. E1 to E8，Anyang 2015-2016，Xianyang 2015-2016，Zhoukou 2015-2016，Anyang 2016-2017，Shangqiu 2016-2017，Zhoukou 2016-2017，Anyang 2017-2018，and Shangqiu 2017-2018

Table 1　QTL for tiller angle detected by inclusive composite interval mapping（ICIM）in the Zhongmai 871/Zhongmai 895 RIL population

QTL	Environment	Marker interval	Confidence interval（cM）	Physical interval（Mb）[a]	LOD	PVE（%）[b]	Add[c]
QTA. caas-1AL	E1[d]	Kasp _ 1A18-Kasp _ 1A90	0. 6-1. 1	308. 8-356. 7	3. 6	6. 1	1. 8
	E2	Kasp _ 1A18-Kasp _ 1A90	0. 6-1. 1	308. 8-356. 7	5. 3	8. 9	2. 2
	E4	Kasp _ 1A18-Kasp _ 1A90	0. 6-1. 1	308. 8-356. 7	6. 8	11. 2	2. 8
	E5	Kasp _ 1A10-Kasp _ 1A18	0. 5-0. 8	307. 8-308. 8	4. 0	6. 6	1. 5
	E6	Kasp _ 1A10-Kasp _ 1A18	0. 5-0. 8	307. 8-308. 8	3. 2	5. 4	1. 6
	E7	Kasp _ 1A18-Kasp _ 1A90	0. 9-1. 1	308. 8-356. 7	5. 3	8. 6	2. 0
	BLUE	Kasp _ 1A18-Kasp _ 1A90	0. 6-1. 1	308. 8-356. 7	5. 4	9. 0	1. 7
QTA. caas-5DL	E1	Kasp _ 5D16-Kasp _ 5D17	7. 5-11. 5	408. 6-418. 4	10. 3	16. 3	3. 1
	E2	Kasp _ 5D16-Kasp _ 5D17	7. 5-13. 0	408. 6-418. 4	8. 6	13. 8	2. 8
	E3	Kasp _ 5D16-Kasp _ 5D17	7. 5-11. 5	408. 6-418. 4	15. 1	22. 9	3. 6
	E4	Kasp _ 5D16-Kasp _ 5D17	6. 5-11. 5	408. 6-418. 4	9. 0	14. 6	3. 2
	E5	Kasp _ 5D16-Kasp _ 5D17	6. 5-10. 5	408. 6-418. 4	11. 7	18. 5	2. 5
	E6	Kasp _ 5D16-Kasp _ 5D17	7. 5-10. 5	408. 6-418. 4	17. 0	24. 8	3. 6
	E7	Kasp _ 5D16-Kasp _ 5D17	7. 5-13. 0	408. 6-418. 4	8. 7	14. 0	2. 6
	E8	Kasp _ 5D16-Kasp _ 5D17	7. 5-11. 5	408. 6-418. 4	11. 7	18. 7	2. 4
	BLUE	Kasp _ 5D16-Kasp _ 5D17	7. 5-11. 5	408. 6-418. 4	17. 4	26. 2	2. 9

[a]Physical intervals（Mb）were obtained by blasting SNP flanking sequences against the reference genome sequence of Chinese Spring（IWGSC，https：//urgi. versailles. inra. fr/blastiwgsc/blast. php）

[b]Phenotypic variation explained by the QTL

[c]Estimated additive effect of the QTL；positive values indicated that favorable alleles came from Zhongmai 895

[d]E1 to E8，Anyang 2015-2016，Xianyang 2015-2016，Zhoukou 2015-2016，Anyang 2016-2017，Shangqiu 2016-2017，Zhoukou 2016-2017，Anyang 2017-2018，andShangqiu 2017-2018

Validation of QTL

Four KASP markers closely associated with QTA. caas-1AL and QTA. caas-5DL were used to test the two BC_1F_4 populations（Table 2，Fig. 3）. The ZM895 genotype had a significantly（$P<0.05$）smaller TA than the ZM871 genotype for QTA. caas-5DL in all six environments in both BC_1F_4 populations. Differences in TA between the two genotypes ranged from 2. 1° to 3. 8° (6. 9-13. 0%) in ZM871/ZM895//ZM871 and 2. 0° to 3. 9° (6. 8-13. 4%) in ZM871/ ZM895//ZM895 across environments. Likewise, the ZM895 genotype had a significantly ($P <$ 0. 05) smaller TA than the ZM871 genotype for *QTA. caas-1AL* in three environments, with TA differences between the two genotypic groups ranging from 1. 3° to 2. 5° (4. 2-8. 1%) in the ZM871/ZM895//ZM871 and 1. 3° to 1. 6° (4. 7-5. 9%) in the ZM871/ZM895//ZM895 populations.

Table 2　Mean tiller angles of the two genotypic classes in $BC_1 F_4$ populations across six environments

QTL region	Genotype	No. [a]	E1[b]	E2	E3	E4	E5	E6	BLUEs
Zhongmai 871/Zhongmai 895//Zhongmai 871									
1AL	ZM871	74	30. 8±3. 9[c]	29. 5±3. 6	32. 1±3. 6	33. 3±4. 5	32. 3±2. 7	33. 1±3. 3	33. 3±3. 3
	ZM895	24	29. 9±4. 1	27. 8±2. 5	32. 0±3. 8	30. 8±3. 9	31. 0±2. 7	32. 2±3. 5	32. 1±3. 5
			0. 9	1. 7 *[d]	0. 1	2. 5 *	1. 3 *	0. 7	1. 2
5DL	ZM871	80	31. 7±3. 9	29. 6±3. 6	33. 0±3. 4	33. 5±4. 4	32. 5±2. 6	33. 6±3. 3	32. 4±2. 8
	ZM895	22	28. 1±4. 1	27. 2±2. 8	29. 2±4. 4	29. 7±4. 3	30. 3±3. 1	29. 8±3. 2	29. 1±3. 0
			3. 4***	2. 4**	3. 8***	3. 8***	2. 1**	3. 8***	3. 3***
Zhongmai 871/Zhongmai 895//Zhongmai 895									
1AL	ZM871	32	29. 0±3. 8	28. 7±3. 3	30. 1±3. 1	30. 6±4. 4	29. 9±2. 7	29. 8±3. 5	30. 1±2. 9
	ZM895	102	27. 7±3. 5	27. 1±3. 4	29. 4±3. 2	29. 3±4. 3	29. 4±2. 9	29. 2±3. 3	29. 1±2. 8
			1. 3*	1. 6**	0. 7	1. 4*	0. 5	0. 6	1. 0*
5DL	ZM871	36	30. 8±2. 4	30. 5±2. 4	31. 9±2. 0	33. 0±3. 2	31. 3±1. 7	31. 8±2. 7	31. 9±1. 9
	ZM895	105	27. 6±3. 2	27. 2±3. 3	29. 0±2. 9	29. 1±3. 9	29. 3±2. 0	29. 0±2. 7	28. 9±2. 9
			3. 2***	3. 3***	2. 9***	3. 9***	2. 0***	2. 7***	3. 0***

[a]Number of lines within the corresponding genotypes

[b]E1 to E6, Anyang 2015-2016, Xianyang 2015-2016, Zhoukou 2015-2016, Anyang 2016-2017, Shangqiu 2016-2017 and Zhoukou 2016-2017

[c]Data was shown as mean±SD

[d]Phenotypic difference between Zhongmai 895 (ZM895) genotype and Zhongmai 871 (ZM871) genotype. *, * * and * * *, significant at $P <$ 0. 05, 0. 01 and 0. 001, respectively

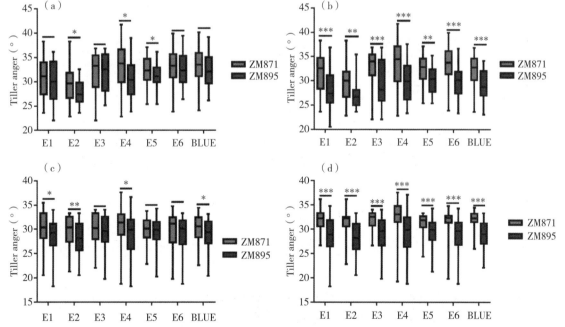

Fig. 3　Phenotypic differences in tiller angles between the Zhongmai 895 (ZM895) and Zhongmai 871 (ZM871) genotypes in two $BC_1 F_4$ populations. E1 to E6 represent Anyang 2015-2016, Xianyang 2015- 2016, Zhoukou 2015-2016, Anyang 2016-2017, Shangqiu 2016- 2017 and Zhoukou 2016-2017. (a) *QTA. caas-1AL* in Zhongmai 871/Zhongmai 895//Zhongmai 871; (b) *QTA. caas-5DL* in Zhongmai 871/ Zhongmai 895//Zhongmai 871; (c) *QTA. caas-1AL* in Zhongmai 871/ Zhongmai 895//Zhongmai 895; (d) *QTA. caas-5DL* in Zhongmai 871/ Zhongmai 895//Zhongmai 895. *, ** and ***, significant at $P <$ 0. 05, 0. 01 and 0. 001, respectively

A candidate gene and its validation in the two germplasm panels

QTA. caas-5DL, located in a 3.0 cM region between *Kasp _ 5D16* and *Kasp _ 5D17*, from 408.6 to 418.4 Mb according to Chinese Spring reference genome (IWGSC, https://urgi.versailles.inra.fr/blast _ iwgsc/), contained 127 annotated genes. *TaTAC-D1* (*TraesCS5D02G322600*), located at 414.1 Mb, showed a high similarity to *OsTAC1* and *ZmTAC1*, and a SNP is present in the third exon of *TaTAC-D1*, with 'A' in the ZM871 and 'G' in ZM895 alleles, respectively. The SNP, causing a change from hydrophilic threonine to hydrophobic alanine at the 169 amino acid position, was a neutral variant that the PROVEAN score was 5.0 of amino acids for *TaTAC-D1* (Fig. 4). However, the amino acid in homoeologous proteins was both threonine in *TaTAC-A1* and *TaTAC-B1*, the same as that in *OsTAC1* and *ZmTAC1*, according to the Chinese Spring reference genome (IWGSC, https://urgi.versailles.inra.fr/blast _ iwgsc/) (Fig. 4). Therefore, the KASP marker *KASP _ TAC-D1* based on the SNP was used to genotype the two validation panels, among which 42 and 28 cultivars had the *TaTAC-D1-G* allele, whereas the other 114 and 65 cultivars had the *TaTAC-D1-A* allele (Table 3, Fig. 5). The *TaTAC-D1-G* genotype had a significantly ($P < 0.05$) smaller TA than the *TaTAC-D1-A* in the two locations, with differences in TA between two genotypes ranging from 2.8° to 3.0° (9.8-10.3%) at Anyang and 3.5° to 3.6° (12.1-12.2%) at Beijing (Table 3, Fig. 5). These results provided further evidence for a significant effect of *TaTAC-D1* on TA and indicated that *TaTAC-D1* could be a candidate gene for *QTA. caas-5DL*.

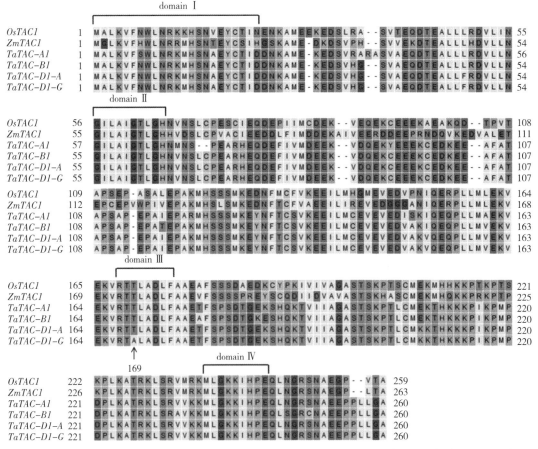

Fig. 4　Alignment of the gene *TAC1* among rice (*OsTAC1*), maize (*ZmTAC1*) and wheat (*TaTAC-A1/B1/D1*). Conserved domains are indicated and numbered with roman numerals (above each) by the reference Dardick et al. (2013)

Table 3   Mean tiller angles of two genotypic classes of *TaTAC-D1* in two validation panels

Genotype[b]	VP1[a]			VP2		
	No. [c]	2016 –2017TA[d]	2017 –2018TA	No.	2016 –2017TA	2017 –2018TA
TaTAC1-A	114	32. 2±3. 9[e]	31. 3±3. 3	65	32. 2±4. 0	33. 3±4. 2
TaTAC-D1-G	42	29. 2±2. 8	28. 5±3. 3	28	28. 6±3. 0	29. 7±3. 3
		3. 0***[f]	2. 8***		3. 5***	3. 6***

[a] VP，Validation panel

[b] Genotype of *TaTAC-D1* was identified by marker *KASP _ TAC-D1*

[c] Number of lines with corresponding genotypes

[d] TA，tiller angle

[e] Data was shown as means±SD

[f] Phenotypic difference between Zhongmai 895 (*TaTAC-D1-G*) and Zhongmai 871 (*TaTAC-D1-A*) genotypes. ***，significant at $P < 0.001$

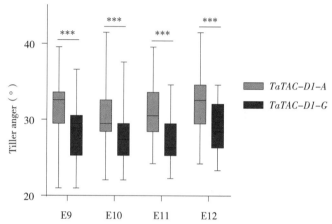

Fig. 5   Phenotypic differences in tiller angles between the Zhongmai 895 (*TaTAC-D1-G*) and Zhongmai 871 (*TaTAC-D1-A*) genotypes for *TaTAC-D1* in two valida-tion panels. E9 to E12 represent Anyang 2017 –2018，Anyang 2018 –2019，Bei-jing 2017 –2018，Beijing 2018 –2019. ***，significant at $P < 0.001$.

The expression levels of *TaTAC-D1* were measured in roots，stems，tiller nodes，leaves and leaf sheaths at the tillering，jointing and heading stages. The expres-sion levels of *TaTAC-D1* were not significantly different between ZM895 and ZM871，except that in stems and tiller nodes were significantly lower ($P < 0.05$) in ZM895 than those in ZM871 at the heading stage (Table S5，Fig S3).

The number of favorable alleles affecting TA

Factorial ANOVA indicated that TA was significantly ($P < 0.001$) influenced by *QTA. caas-1AL* and *QTA. caas-5DL*，but the effect of their interactions was not significant (Table S6). We also detected highly significant interactions between the environment and each of the two QTL，which indicates that the effect of *QTA. caas-1AL* and *QTA. caas-5DL* is modu-lated by the environment (Table S6). *QTA. caas-1AL* and *QTA. caas-5DL* explained 3. 1% and 32. 0% of the variances，respectively (Table S6). No epistatic in-teraction between the two QTL was identified using IciMapping 4. 1，indicating the importance of additive effects. There was significant difference between phenotypes and numbers of favorable alleles (ZM895 genotype)，with the combination of favorable alleles con-tributing to reduced TA (Fig. S4a). RILs carrying unfa-vorable alleles of both QTL exhibited 26. 7% higher TA (9. 4°) than those possessing both positive alleles.

Discussion

Comparisons with previous reports

QTA. caas-1AL

QTA. caas-1AL was located in a 0. 5 cM region，

flanked by markers *Kasp_1A18* and *Kasp_1A90*, from 308. 8 to 356. 7 Mb according to Chinese Spring reference genome（IWGSC，https：//urgi. versailles. inra. fr/blast_iwgsc/），explaining 5. 4-11. 2% of the phenotypic variances in six environments. Yang et al. （2020）reported a stable *Qtgw. caas-1AL* flanked by markers *Kasp_1A10 and Kasp_1A90*，from 307. 8 to 356. 7 Mb（IWGSC，https：//urgi. versailles. inra. fr/blast_iwgsc/），explaining 5. 2-15. 0% of the phenotypic variances of thousand grain weight （TGW）. The overlap between the TGW QTL from Yang et al. （2020）and our *QTA. caas-1AL* for tiller angle suggests that this 48. 9 Mb region either includes a single gene with pleiotropic effects on both traits or two linked genes with separate effects. Moreover，the haplotype for an erect tiller angle is associated with higher TGW in the parental lines，and the two favorable alleles were in phase. Therefore，these KASP markers could be used in breeding for improvement of both plant architecture and grain-related traits.

QTA. caas-5DL

Markers *Kasp_5D16* and *Kasp_5D17* flanking *QTA. caas-5DL* were located at 408. 6 and 418. 4 Mb. Based on gene annotations of the Chinese Spring reference genome（IWGSC，https：//urgi. versailles. inra. fr/blast_iwgsc/）*TraesC-S5D02G322600*，located at 414. 1 Mb，was a candidate gene for the QTL. The gene was designated *TaTAC-D1*.

Yu et al. （2007）identified a SNP in the 3′-splicing site of the fourth 1. 5 kb intron that resulted in failure to splice the fourth intron and a premature stop codon，which was associated with an erect plant architecture in rice. Dardick et al. （2013）named the *TAC1* gene family as *IGT* with four domains that were members of a previously unknown gene family. The *Thr169Ala* mutation described in wheat *TaTAC-D1* in this study is located in the third domain，and both rice，maize and the wheat homoeologous *TaTAC-A1* and *TaTAC-B1* carry a threonine at this position，suggesting that this is the ancestral allele （Fig. 4）.

In wheat，Cao et al. （2017）described two types of *TAC1*：one with a premature stop codon，which they designated as Type I and the other one with a CGCGCG insertion designated as Type II. However，they did not indicate the specific genome of these variants in their paper. Here，we found that these two variants correspond to the A-genome homology （*TaTAC-A1*）and that the *TaTAC-B1* and *TaTAC-D1* sequences do not have the CGCGCG insertion. Our study described a natural *Thr169Ala* differences in the D-genome copy *TaTAC-D1*.

If *TaTAC-D1* is demonstrated to be the causal gene for the difference in tiller angle，it would be interesting to investigate if the *Thr169Ala* mutation or the differences in expression contribute to the observed differences. The expression levels of *TaTAC-D1* measured in stems and tiller nodes were significantly （$P<$ 0. 05）lower in ZM895 than those in ZM871 at the heading stage. It seemed that TA was determined by lower nodes at or near crown，which was the relevant tissue for expression analysis of *TaTAC-D1*. Yu et al. （2007）reported that *tac1* with a lower expression level had a smaller tiller angle. Cao et al. （2017） showed that the expression abundance of *TaTAC1* from 15 to 45 days after sowing in tiller nodes was significantly and positively correlated with tiller angles at maturity stage.

A significant difference in TA was recorded between the ZM871 and ZM895 genotypes in germplasm panels，with the latter genotype having a smaller TA than the former. Though an amino acid difference in *TaTAC-D1* between the parental lines was identified，we cannot determine if the differences in TA are associated with the differences in protein sequence. We hypothesize that the *Thr169Ala* amino acid change may reduce *TAC1* activity in ZM895 in stems and tiller nodes at the heading stage，thus result in smaller tiller angles. Therefore，*TraesCS5D02G322600* was proposed as a candidate gene for *QTA. caas-5DL*，but further genetic studies are required for confirmation. Transgenic

Fielder plants overexpressing *TaTAC-D1* have been generated to determine the gene effect on TA.

Effect of number of favorable TA alleles on TGW

Wheat plant architecture is considered a key driver of yield and has long attracted the attention of breeders and physiologists in defining ideal plant architectures (ideotypes) for maximum yield potential. Plant architecture is mainly genetically determined but is significantly influenced by plant density, photosynthetic efficiency, lodging resistance and disease presence. Tiller angles determine the planting density per unit area and contribute greatly to grain yield in rice (Dong et al. 2016).

One of our interests was to investigate the effects of TA alleles on TGW. The stable QTL on chromosome 1AL also had a significant effect on TGW (Yang et al. 2020). To further understand the combined effects of TA alleles on TGW, the relationship between TGW and the number of favorable alleles of TA was analyzed (Fig. S4b). RILs carrying positive alleles of both QTL exhibited a 5.5% higher TGW (2.5 g) than those possessing contrasting pair of alleles, i. e. , genotypes with smaller TA had a higher TGW. Therefore, both QTL were not only favorable for plant architecture, but also favorable for TGW improvement. The tightly linked KASP markers, *Kasp _ 1A18* and *Kasp _ 1A90* for *QTA. caas-1AL* and *KASP _ TAC-D1* for *TaTAC-D1* should be useful in developing breeding lines with improved plant architecture and high TGW.

❈ Acknowledgements

The authors are grateful to Prof. R. A. McIntosh, Plant Breeding Institute, University of Sydney, for critical review of this manuscript. This work was funded by the CAAS-ZDRW202002, CAAS Science and Technology Innovation Program, National Key Research and Development Programs of China (2016YFD0101802, 2016YFD0100502, 2016YFE0108600).

❈ References

Cao X, Deng M, Zhang Z-L, Liu Y-J, Yang X-L, Zhou H, Zhang Z-J (2017) Molecular characterization and expression analysis of *TaTAC1* gene in *Triticum aestivum* L. J Plant Genet Resour 18: 125-132.

Dardick C, Callahan A, Horn R, Ruiz KB, Zhebentyayeva T, Hollender C, Whitaker M, Abbott A, Scorza R (2013) PpeTAC1 promotes the horizontal growth of branches in peach trees and is a member of a functionally conserved gene family found in diverse plants species. Plant J 75: 618-630.

Deng S, Wu X, Wu Y, Zhou R, Wang H, Jia J, Liu S (2011) Characterization and precise mapping of a QTL increasing spike number with pleiotropic effects in wheat. Theor Appl Genet 122: 281-289.

Dixon LE, Greenwood JR, Benvivegna S, Zhang P, Cockram J, Mellers G, Ramm K, Cavanagh C, Swain SM, Boden SA (2018) *TEOSINTE BRANCHED1* regulates inflorescence architecture and development in bread wheat (*Triticum aestivum*). Plant Cell 30: 563-581.

Dong H, Zhao H, Xie W, Han Z, Li G, Yao W, Bai X, Hu Y, Guo Z, Lu K, Yang L, Xing Y (2016) A novel tiller angle gene, *TAC3*, together with *TAC1* and *D2* largely determine the natural variation of tiller angle in rice cultivars. PLoS Genet 12: e1006412.

Ford BA, Foo E, Sharwood R, Karafiatova M, Vrána J, MacMillan C, Nichols DS, Steuernagel B, Uauy C, Doležel J, Chandler PM, Spielmeyer W (2018) *Rht18* semidwarfism in wheat is due to increased GA 2-oxidaseA9 expression and reduced GA content. Plant Physiol 177: 168-180.

Griffiths S, Simmonds J, Leverington M, Wang Y, Fish LJ, Sayers L, Alibert L, Orford S, Wingen L, Snape J (2012) Meta-QTL analysis of the genetic control of crop height in elite. European winter wheat germplasm. Mol Breed 29: 159-171.

He J, Shao G, Wei X, Huang F, Sheng Z, Tang S, Hu P (2017) Fine mapping and candidate gene analysis of *qTAC8*, a major quantitative trait locus controlling tiller angle in rice (*Oryza sativa* L.). PLoS One 12: e0178177.

Holland JB, Nyquist WE, Cervantes-Martínez CT (2003) Estimating and interpreting heritability for plant breeding: an update. PlantBreed Rev 22: 9-112.

Hong Z, Ueguchi-Tanaka M, Umemura K, Uozu S, Fujioka

S, Takat- suto S, Yoshida S, Ashikari M, Kitano H, Matsuoka M (2003) A rice brassinosteroid-deficient mutant, ebisu dwarf (*d2*), is caused by a loss of function of a new member of cytochrome P450. Plant Cell 15: 2900-2910.

Hu Y, Ren T, Li Z, Tang Y, Ren Z, Yan B (2017) Molecular mapping and genetic analysis of a QTL controlling spike formation rateand tiller number in wheat. Gene 634: 15-21.

Isidro J, Knox R, Clarke F, Singh A, Depauw R, Clarke J, Somers D (2012) Quantitative genetic analysis and mapping of leaf angle in durum wheat. Planta 236: 1713-1723.

Kato K, Miura H, Sawada S (2000) Mapping QTLs controlling grain yield and its components on chromosome 5A of wheat. Theor Appl Genet 101: 1114-1121.

Kuraparthy V, Sood S, Dhaliwal HS, Chhuneja P, Gill BS (2007) Identification and mapping of a tiller inhibition gene (*tin3*) in wheat. Theor Appl Genet 114: 285-294.

Lewis JM, Mackintosh CA, Shin S, Gilding E, Kravchenko S, Baldridge G, Zeyen R, Muehlbauer GJ (2008) Overexpression of the maize *Teosinte Branched1* gene in wheat suppresses tiller development. Plant Cell Rep 27: 1217-1225.

Li Z, Paterson AH, Pinson SRM, Stansel JW (1999) RFLP facilitated analysis of tiller and leaf angles in rice (*Oryza sativa* L.). Euphytica 109: 79-84.

Li C, Zhou A, Sang T (2006) Genetic analysis of rice domestication syndrome with the wild annual species, *Oryza nivara*. New Phytol 170: 185-194.

Li P, Wang Y, Qian Q, Fu Z, Wang M, Zeng D, Li B, Wang X, Li J (2007) *LAZY1* controls rice shoot gravitropism through regulating polar auxin transport. Cell Res 17: 402-410.

Li Z, Peng T, Xie Q, Han S, Tian J (2010) Mapping of QTL for tiller number at different stages of growth in wheat using doublehaploid and immortalized F₂ populations. J Genet 89: 409-415.

Liu K, Xu H, Liu G, Guan P, Zhou X, Peng H, Yao Y, Ni Z, Sun Q, Du J (2018) QTL mapping of flag leaf-related traits in wheat (*Triticum aestivum* L.). Theor Appl Genet 131: 839-849.

Liu K, Cao J, Yu K, Liu X Gao Y, Chen Q, Zhang W, Peng H, Du J, Xin M, Hu Z, Guo W, Rossi V, Ni Z, Sun Q, Yao Y (2019) Wheat *TaSPL8* modulates leaf angle through auxin and brassinosteroid signaling. Plant Physiol 181: 179-194.

Mo Y, Vanzetti LS, Hale I, Spagnolo EJ, Guidobaldi F, Al-Oboudi J, Odle N, Pearce S, Helguera M, Dubcovsky J (2018) Identification and characterization of

Rht25, a locus on chromosome arm 6AS affecting wheat plant height, heading time, and spike development. Theor Appl Genet 131: 2021-2035.

Mu J, Huang S, Liu S, Zeng Q, Dai M, Wang Q, Wu J, Yu S, Kang Z, Han D (2019) Genetic architecture of wheat stripe rust resistance revealed by combining QTL mapping using SNP-based genetic maps and bulked segregant analysis. Theor Appl Genet 132: 443-455.

Naruoka Y, Talbert LE, Lanning SP, Blake NK, Martin JM, Sherman JD (2011) Identification of quantitative trait loci for productive tiller number and its relationship to agronomic traits in spring wheat. Theor Appl Genet 123: 1043-1053.

Nasseer AM, Martin JM, Heo HY, Blake NK, Sherman JD, Pumphrey M, Kephart KD, Lanning SP, Naruoka Y, Talbert LE (2016) Impact of a quantitative trait locus for tiller number on plasticity of agronomic traits in spring wheat. Crop Sci 56: 595-602.

Pearce S, Saville R, VaughanSP, Chandler PM, Wilhelm EP, Al-Kaff N, Korolev A, Boulton MI, Phillips AL, Hedden P, Nicholson P, Thomas SG (2011) Molecular characterization of *Rht-1* dwarfing genes in hexaploid wheat. Plant Physiol 157: 1820-1831.

Peng Z-S, Yen C, Yang J-L (1998) Genetic control of oligo-culmscharacter in common wheat. Wheat Inf Serv 86: 19-24.

Peng J, Richards DE, Hartley NM, Murphy GP, Devos KM, Flintham JE, Beales J, Fish LJ, Worland AJ, Pelica F, Sudhakar D, Christou P, Snape JW, Gale MD, Harberd NP (1999) 'Green revolution' genes encode mutant gibberellin response modulators. Nature 400: 256-261.

Porebski S, Bailey LG, Baum BR (1997) Modification of a CTAB DNA extraction protocol for plants containing high polysaccharide and polyphenol components. Plant Mol Biol Rep 15: 8-15.

Qian Q, He P, Teng S, Zeng D-L, Zhu L-H (2001) QTLs analysis of tiller angle in rice (*Oryza sativa* L.). Acta Genetica Sin 28: 29-32.

Ramirez-Gonzalez RH, SegoviaV, Bird N, Fenwick P, Holdgate S, Berry S, Jack P, Caccamo M, Uauy C (2015a) RNA-Seq bulked segregant analysis enables the identification of high-resolution genetic markers for breeding in hexaploid wheat. Plant Biotechnol J 13: 613-624.

Ramirez-Gonzalez RH, Uauy C, Caccamo M (2015b) PolyMarker: a fast polyploid primer design pipeline. Bioinformatics 31: 2038-2039.

Ren T, HuY, Tang Y, Li C, Yan B, Ren Z, Tan F, Tang Z, Fu S, Li Z (2018) Utilization of a wheat 55 K SNP array for mapping of major QTL for temporal expression of the tiller number. Front Plant Sci 9: 333.

Schmittgen TD, Livak KJ (2008) Analyzing real-time PCR data by the comparative C_T method. Nat Protoc 3: 1101-1108.

Spielmeyer W, Richards RA (2004) Comparative mapping of wheat chromosome 1AS which contains the tiller inhibition gene (tin) with rice chromosome 5S. Theor Appl Genet 109: 1303-1310.

Sun L, Yang W, Li Y, Shan Q, Ye X, Wang D, Yu K, Lu W, Xin P, Pei Z, Guo X, Liu D, Sun J, Zhan K, Chu J, Zhang A (2019) A wheat dominant dwarfing line with Rht12, which reduces stem cell length and affects gibberellic acid synthesis, is a 5AL terminal deletion line. Plant J 97: 887-900.

Takagi H, Abe A, Yoshida K, Kosugi S, Natsume S, Mitsuoka C, Uemura A, Utsushi H, Tamiru M, Takuno S, Hideki I, Cano LM, Kamoun S, Terauchi R (2013) QTL-seq: rapid mapping of quantitative trait loci in rice by whole genome resequencing of DNA from two bulked populations. Plant J 74: 174-183.

Thomson MJ, Tai TH, McClung AM, Lai XH, Hinga ME, Lobos KB, Xu Y, Martinez CP, McCouch SR (2003) Mapping quantitative trait loci for yield, yield components and morphological traits in an advanced backcross population between Oryza rufipogon and the Oryza sativa cultivar Jefferson. Theor Appl Genet 107: 479-493.

Tian X, Wen W, Xie L, Fu L, Xu D, Fu C, Wang D, Chen X, Xia X, Chen Q, He Z, Cao S (2017) Molecular mapping of reduced plant height gene Rht24 in bread wheat. Front Plant Sci 8: 1379.

Wang Y, Li J (2008) Molecular basis of plant architecture. Annu RevPlant Biol 59: 253-279.

Wang Z, Liu Y, Shi H, Mo H, Wu F, Lin Y, Gao S, Wang J, Wei Y, Liu C, Zheng Y (2016) Identification and validation of novel low-tiller number QTL in common wheat. Theor Appl Genet 129: 603-612

Wang Y, Xie J, Zhang H, Guo B, Ning S, Chen Y, Lu P, Wu Q, Li M, Zhang D, Guo G, Zhang Y, Liu D, Zou S, Tang J, Zhao H, Wang X, Li J, Yang W, Cao T, Yin G, Liu Z (2017) Mapping stripe rust resistance gene YrZH22 in Chinese wheat cultivar Zhoumai 22 by bulked segregant RNA-Seq (BSR-Seq) and comparative genomics analyses. Theor Appl Genet 130: 2191-2201.

Wu J, Kong X, Wan J, Liu X, Zhang X, Guo X, Zhou

R, Zhao G, Jing R, Fu X, Jia J (2011) Dominant and pleiotropic effects of a GAI gene in wheat results from a lack of interaction between DELLA and GID1. Plant Physiol 157: 2120-2130.

Wu Q, ChenY, Fu L, Zhou S, Chen J, Zhao X, Zhang D, Ouyang S, Wang Z, Li D, Wang G, Zhang D, Yuan C, Wang L, You M, Han J, Liu Z (2016) QTL mapping of flag leaf traits in common wheat using an integrated high-density SSR and SNP genetic linkage map. Euphytica 208: 337-351.

Wu J, Liu S, Wang Q, Zeng Q, Mu J, Huang S, Yu S, Han D, Kang Z (2018a) Rapid identification of an adult plant stripe rust resistance gene in hexaploid wheat by high-throughput SNP array genotyping of pooled extremes. Theor Appl Genet 131: 43-58.

Wu J, Huang S, Zeng Q, Liu S, Wang Q, Mu J, Yu S, Han D, Kang Z (2018b) Comparative genome-wide mapping versus extreme pool- genotyping and development of diagnostic SNP markers linked to QTL for adult plant resistance to stripe rust in common wheat. Theor Appl Genet 131: 1777-1792.

Würschum T, Langer SM, Longin CF (2015) Genetic control of plant height in European winter wheat cultivars. Theor Appl Genet 128: 865-874.

Xu Y, Shen Z, Xu J, Zhu H, Chen Y, Zhu L (1995) Interval mapping of quantitative trait loci by molecular markers in rice (Oryza sativa L.). Sci China Ser B-Chem 38: 422-428.

Xu T, Bian N, Wen M, Xiao J, Yuan C, Cao A, Zhang S, Wang X, Wang H (2017) Characterization of a common wheat (Triticum aestivum L.) high-tillering dwarf mutant. Theor Appl Genet 130: 483-494.

Yan J, Zhu J, He C, Benmoussa M, Wu P (1999) Molecular marker- assisted dissection of genotype × environment interaction for plant type traits in rice (Oryza sativa L.). Crop Sci 39: 538-544.

Yang L, Zhao D, Meng Z, Xu K, Yan J, Xia X, Cao S, Tian Y, He Z, Zhang Y (2020) QTL mapping for grain yield-related traits in bread wheat via SNP-based selective genotyping. Theor Appl Genet 133: 857-872.

Yu B, Lin Z, Li H, Li X, Li J, Wang Y, Zhang X, Zhu Z, Zhai W, Wang X, Xie D, Sun C (2007) TAC1, a major quantitative trait locus controlling tiller angle in rice. Plant J 52: 891-898.

Zhang J, Wu J, Liu W, Lu X, Yang X, Gao A, Li X, Lu Y, Li L (2013) Genetic mapping of a fertile tiller inhibition gene, ftin, in wheat. Mol Breeding 31: 441-449.

Fine mapping and validation of a major QTL for grain weight on chromosome 5B in bread wheat

Dehui Zhao[1,2] · Li Yang[1] · Dan Liu[1] · Jianqi Zeng[1] · Shuanghe Cao[1] · Xianchun Xia[1] · Jun Yan[3] · Xiyue Song[2] · Zhonghu He[1,4] · Yong Zhang[1]

[1] Institute of Crop Sciences, National Wheat Improvement Centre, Chinese Academy of Agricultural Sciences (CAAS), 12 Zhongguancun South Street, Beijing 100081, China

[2] College of Agronomy, Northwest A & F University, Yangling 712100, Shaanxi, China

[3] Institute of Cotton Research, CAAS, 38 Huanghe Dadao, Anyang 455000, Henan, China

[4] CIMMYT-China Office, C/O CAAS, Beijing 100081, China

Abstract: Thousand grain weight (TGW), determined by grain length and width, and is an important yield component in wheat; understanding of the underlying genes and molecular mechanisms remains limited. A stable QTL *QTgw.caas-5B* for TGW was identified previously in a RIL population developed from a cross between Zhongmai 871 (ZM871) and a sister line Zhongmai 895 (ZM895), and the aim of this study was to perform fine mapping and validate the genetic effect of the QTL. It was delimited to an interval of approximately 2.0 Mb flanked by markers *Kasp_5B29* and *Kasp_5B31* (49.6-51.6 Mb) using 12 heterozygous recombinant plants obtained by selfing a residual BC_1F_6 line selected from the ZM871/ZM895//ZM871 population. A candidate gene was predicted following sequencing and differential expression analyses. Marker *Kasp_5B_Tgw* based on a SNP in *TraesCS5B02G044800*, the *QTgw.caas-5B* candidate, was developed and validated in a diversity panel of 166 cultivars. The precise mapping of *QTgw.caas-5B* laid a foundation for cloning of a predicted causal gene and provides a molecular marker for improving grain yield in wheat.

Key message: A major QTL *QTgw.caas-5B* for thousand grain weight in wheat was fine mapped on chromosome 5B, and TraesCS5B02G044800 was predicted to be the candidate gene.

Introduction

Wheat is one of the most important food crops in the world, providing approximately 20% of the calories and 25% of the protein for humans (FAO 2017, http://www.fao.org/faostat/en/). Although significant progress has already been made on yield improvement during the last 60 years, with a genetic gain of 0.7-1.0%

annually (Fischer and Edmeades 2010; Gao et al. 2017), it is estimated that yield must still increase by more than 60% to meet predicted growth in the world population by 2050, despite the restricting effects of climate change and the declining area of available land due to urbanization and degradation (Langridge 2013). Improved yield potential is, therefore, still a major breeding objective. Identification and mining of genetic loci for grain yield will provide genetic resources and tools to improve yield potential.

Published in Theoretical and Applied Genetics, 2021, 134: 3731-3741. (doi.org/10.1007/s00122-021-03925-9)

Yield is a complex quantitative trait determined by thousand grain weight (TGW), grain number per spike and spike number per unit area, with TGW having the highest heritability among three components (Simmonds et al. 2014, 2016; Chen et al. 2020; Yang et al. 2020). TGW is determined by grain size and grain filling characteristics. Grain size can be divided into components grain length (GL), grain width (GW) and grain thickness (Kuchel et al. 2007; Simmons et al. 2014; Yang et al. 2020). Numerous genes for grain weight have been cloned in rice (Li and Li 2016; Li et al. 2019b), and many genes associated with grain weight in wheat were cloned by comparative genomics (Su et al. 2011; Zhang et al. 2012; Dong et al. 2014; Zhang et al. 2014; Jiang et al. 2015; Wang et al. 2015; Yue et al. 2015; Ma et al. 2016; Wang et al. 2016; Hanif et al. 2016; Hu et al. 2016; Simmonds et al. 2016; Sajjad et al. 2017; Yang et al. 2019; Cao et al. 2020). A large number of quantitative trait loci (QTL) for TGW, GL and GW have also been identified (Huang et al. 2003; Quarrie et al. 2005; Prashant et al. 2012; Cui et al. 2014; Wu et al. 2015; Sun et al. 2017; Cabral et al. 2018; Guan et al. 2018; Ma et al. 2018; Su et al. 2018; Zhai et al. 2018; Yang et al. 2020). However, few TGW genes have been isolated by map-based cloning due to the complexity of the wheat genome. With the current availability of genotyping arrays and release of wheat reference genome sequences (IWGSC 2018, https://urgi.versa illes.inra.fr/blast_iwgsc/) fine mapping of major QTL for TGW has been achieved using near-isogenic lines (NILs) and residual heterozygosity (Brinton et al. 2017; Guan et al. 2019; Xu et al. 2019; Chen et al. 2020).

In a previous study (Yang et al. 2020) four QTL for TGW-related traits were identified on chromosomes 1AL, 2BS, 3AL and 5B. $QTgw.caas$-5B, detected across all of ten environments, explaining 5.7-17.1% of the phenotypic variances, was mapped to a 11.6 cM region between markers $Kasp\_5B5$ and $Kasp\_5B12$, extending from 45.3 to 394.2 Mb on chromosome 5B based on the Chinese Spring reference genome (IWGSC

2018, https://urgi.versailles.inra.fr/blast_iwgsc/). The present study was to precisely map $QTgw.caas$-5B by analysis of a residual heterozygous plant, predict one or more candidate genes and develop kompetitive allele-specific PCR (KASP) markers for accurate marker-assisted selection in breeding and research.

Materials and methods

Plant materials

Zhongmai 895 (hereafter ZM895), with a current production area around 0.7 million ha annually is a leading cultivar in the Yellow and Huai River Valleys Winter Wheat Region of China. ZM895 and Zhongmai 871 (hereafter ZM871), developed by pedigree selection and bulked as fixed lines at F_5, are sister lines that can be traced back to a single F_2 plant of cross Zhoumai 16/Liken 4. The detailed information of ZM895 and ZM871 was described in a previous study (Yang et al. 2020). One residual heterozygous line (L2925) within the marker interval of $QTgw.caas$-5B was selected from $BC_1 F_6$ generation of the ZM871/ZM895//ZM871 population (Fig. 1a, b). A heterozygous recombinant plant (HRL2925) from L2925 was self-pollinated, generating 12 heterozygous recombinant plants (designated RL1 to RL12) and 119 homozygous plants with 57 having 5B+alleles (ZM895 genotype) and 62 having 5B−(ZM871 genotype), in which two groups of homozygous plants were used for a preliminary evaluation of the phenotypic effects of $QTgw.caas$-5B on TGW, GL and GW (Fig. 1c, d). Genetic backgrounds of HRL2925 were evaluated by genotyping two 5B+ and two 5B− homozygous lines generated from two kinds of homozygous plants, respectively, using the wheat 50 K SNP array developed in collaboration with the Capital-Bio, Beijing, China (https://www.capitalbiotech.com/). Twenty 5B+ and 20 5B− homozygous lines were used to measure TGW, GL and GW at different grain developmental stages and for RNA-sequencing. In addition, 52-77 NILs from each of RL1 to RL12 were identified using $QTgw.caas$-5B-flanking markers to narrow the region of candidate genes (Fig. 1d; Table 2). A diverse panel of 166 cultivars (Li et al. 2019a) was used to validate the effects of *QTgw. caas-5B*.

Field trials and trait measurement

The progeny from HRL2925 were sown in ten 3.0 m rows spaced 0.3 m apart with 30 seeds per row at Xinxiang (Henan province) during the 2017-2018 cropping season. Twenty homozygous lines with 5B+ and 20 with 5B− selected from 119 self-pollinated homozygous plants of HRL2925 were evaluated at Xinxiang and Anyang (Henan province) during the 2018-2019 cropping season. The lines were grown as plots in randomized complete blocks with three replications. Each plot comprised two 1.0 m rows spaced 0.3 m apart, with 30 seeds per row. Twelve segregating populations derived from the recombinants were sown in plots of seven 3.0 m rows spaced 0.3 m apart with 30 seeds per row at Xinxiang (Henan province) during the 2018-2019 cropping season. The panel of 166 cultivars was sown in three 1.5 m rows spaced 0.2 m apart in 50 seeds each row with three replications at Suixi (Anhui province) during the 2012-2013, 2013-2014 and 2014-2015 cropping seasons, at Anyang (Henan province) during the 2012-2013 and 2013-2014 cropping seasons, and at Shijiazhuang (Hebei province) during the 2014-2015 cropping season.

Wanshen SC-G seed detector (Hangzhou Wanshen Detection Technology Co., Ltd) was used to record TGW, GL and GW. Thirty random spikes were harvested from each plot of all 20 homozygous individuals in the contrasting 5B+ and 5B− groups to measure TGW, GL and GW. The same parameters were measured on the 119 homozygous plants from HRL2925 were measured on grains from 6-10 spikes of each plant and 52-77 homozygous progenies from each of RL1-RL12. For the diversity panel, TGW was determined by weighing 500 grains, and GL and GW were measured on 20 random grains from each plot to obtain mean length and width values, respectively (Li et al. 2019a).

Grain sampling

Twenty 5B+ and 20 5B− homozygous lines grown at Xinxiang in 2018-2019 were used for the study. Main stems at anthesis were marked with red tags at 09：00-11：00 am every day. Ten grains from outer florets of five spikelets in the middle regions of tagged spikes were sampled at 09：00-11：00 am at 4, 8, 12, 16, 20, 25 and 30 days post-anthesis (DPA). At each time point 10 spikes per plot were sampled, including 200 grains (20 spikes×10 grains each spike) for each group. GL and GW were measured using image analysis software (Image-Pro Plus 6.0, http：//www. mediacy. com/) after scanning the grain samples placed on a scanner panel with grain creases placed downwards. Following measurement, the grains were dried in an oven at 135℃ for 15 min and then at 65 ℃ until a constant weight. The TGW of the dried grain samples harvested at various DPA was deter-mined, three biological replications were performed for each time point.

RNA and DNA extraction and RNA-sequencing

Two grains sampled from the outer florets of spikelets in the middle of tagged spikes of main stems at 4, 8, 12, 16, 20 and 25 DPA were snap-frozen in liquid nitrogen and stored at −80 ℃. For 20 homozygous lines with 5B+ and 20 with 5B− selected from 119 self-pollinated homozygous plants of HRL2925, 40 grains were sampled (20 spikes×2 grains each spike) with three biological replications at each time point.

Total RNA was isolated using the TRIzol protocol (Invitrogen, Carlsbad, CA, USA). After quality testing, a single RNA library for each sample was constructed, and the library preparations were sequenced on an Illumina Hiseq platform with 250-300 bp paired-end reads at Novogene Bioinformatics Technology in Beijing (http：//www. novog ene. com/). FeatureCounts v1.5.0-p3 was used to count the number of reads mapped to each gene. Then FPKM of each gene was calculated based on the length of the gene and reads count mapped. Differential expression analysis of 5B+ and 5B− genotypes were performed with three biological replications at each stage using the DESeq2 R package (1.10.1), which provides the statistical routines for deter-mining differential expression in digital gene expression data using a model based on a negative binomial distribution. The resulting P-value was adjusted using the Benjamini and Hochberg's approach for controlling false discovery rate. Genes with an adjusted $P<0.05$ determined by DESeq2 were considered as differentially expressed. Genomic DNA was extracted from young leaves of experimental lines using the CTAB method.

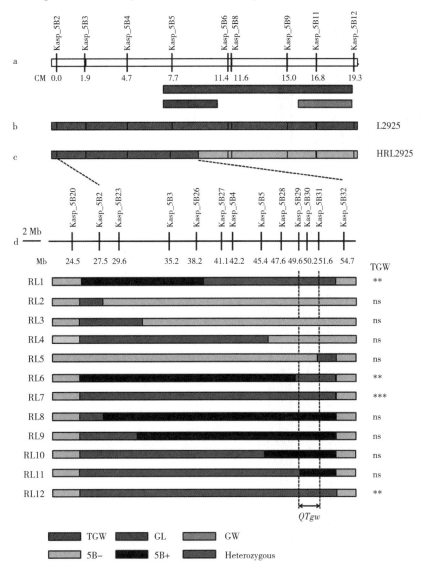

Fig. 1   Fine mapping of $QTgw. caas-5B$. **a** genetic map of 5B chromosome showing QTL for thousand grain weight (TGW), grain length (GL) and grain width (GW) in the Zhongmai 871/Zhongmai 895 RIL population from Yang et al. (2020), **b** residual heterozygous line (L2925), **c** heterozygous recombinant plant (HRL2925), **d** TGW map of 2. 0 Mb interval on chromosome 5B. Left upside of 13 markers used to screen recombinants are shown at the middle and graphical genotypes are shown for 12 recombinant lines (RL1-RL12). Statistical comparisons of TGW between 5B+and 5B-genotypes among self-pollinated progenies of each RL are shown at the right. *, **, *** and ns, significant at $P<$ 0. 05, $P<0.01$, $P<0.001$ and non-significant, respectively

Whole-genome resequencing

Quantified DNA samples of ZM871 and ZM895 were randomly fragmented by Covaris, and the fragments were collected by magnetic beads. Ligation products with end-repair and addition of 3' adenine DNA fragments were cycled and amplified by linear isothermal rolling-circle replication and DNA nanoball technology. Sequencing of these DNA libraries was performed by the BGISEQ sequencing platform at Shenzhen Huada Gene Technology

（Shenzhen，https：// www. genomics. cn/）. The re-maining high-quality paired-end reads following filtering a high proportion of adaptors and low-quality reads in the raw data were mapped to the Chinese Spring reference genome（IWGSC 2018，https：//urgi. versailles. inra. fr/ blast_iwgsc/）using the Burrows-Wheeler Aligner Tool （http：//bio-bwa. sourceforge. net/）with the com-mand "mem-t 4-k 32-M" . SNPs and small Insertion/Deletions （InDels）were detected by GATK（https：// www. broad institute. org/gatk/）with filtration parameters of "QD<2. 0 ‖ FS>60. 0 ‖ MQ<40. 0 ‖ MQRankSum<－12. 5 ‖ ReadPosRankSum<－8. 0" for SNP calling and "QD<2. 0 ‖ FS>200. 0 ‖ SOR>10. 0 ‖ MQRankSum<－12. 5 ‖ ReadPos-RankSum<－8. 0" for InDel calling in the same way as the whole-genome resequencing.

Array-based SNP markers or SNPs from RNA-sequencing and whole-genome resequencing upstream of the physical location of the *Kasp_5B2* locus and between *Kasp_5B2* and *Kasp_5B6* were converted to KASP markers for fine map-ping of *QTgw. caas-5B*. Allele-specific and common reverse primers for each KASP marker were de-signed using Poly-Marker（Ramirez-Gonzalez et al. 2015, http：//www. polym arker. info/）.

Statistical analysis

For the statistical analyses in progeny tests，grain de-velopment and diversity panel，phenotypic differences betweenthe allelic pairs were determined by Student's *t* tests with SAS 9. 2 software（SAS Institute Inc，Cary，NC，USA）. BLUE（best linear unbiased esti-mation）value of the phenotypic data for each line/cul-tivar in each environment was used for the analyses.

Results

Generation of fine mapping population usingresidual heterozygous lines

A heterozygous line L2925 screened from a BC_1F_6 popula-tion of the ZM871/ZM895//ZM871 had homozygous background across the mapping interval spanning *QTgw. caas-5B*（Fig. 1a，b；Yang et al. 2020）. One recombinant plant（HRL2925）from L2925 was self-pollinated and generated 12 heterozygous recombinant plants（Fig. 1c，d）. The heterozygous interval was flanked by SNP markers *AX-110372788* and *AX-95631395* from 24. 5 to 53. 7 Mb in the HRL2925 based on the wheat 50 K SNP array. The genetic simi-larity between homozygous progenies for 5B+ and 5B-was more than 99％ according to 50 K SNP array data，indicating that the segregating progenies from the HRL2925 were suitable for fine mapping（Table S2）. Within each family of selfed progenies from 12 recom-binant plants，homozygous non-recombinant plants，namely 5B+NILs and 5B-NILs，were genotyped with markers according to heterozygous interval and pheno-types evaluated for fine mapping.

Phenotypic validation of NILs for QTgw. caas-5B

After a progeny test，a significant difference in TGW was detected between the genotypes 5B＋ with 57 plants and 5B-with 62 plants from selfing progenies of HRL2925（Fig. 2）. In order to improve the accuracy of phenotypic evaluation，20 homozygous lines with 5B+and 20 with 5B-generated from two kinds of ho-mozygous plants were evaluated at Xinxiang and Anyang（Henan province）to verify the effects of *QT-gw. caas-5B*. Student's *t* tests indicated significantly （$P<0. 05$）higher TGW and GL in 5B+ lines than their contrasting 5B-lines（Fig. 3）. These demonstrated that the ZM895 allele at *QTgw. caas-5B* had a positive effect on TGW.

To further analyze the genetic effect of *QTgw. caas-5B*，the dynamic change of grain weight and size at different developmental stages between the above 5B+ and 5B-homozygous lines were investigated. Twenty lines with 5B+and 20 with 5B-were used to determine the differences on grain morphometric parameters. Student's *t* tests indicated significantly（$P<0. 05$）higher GL in 5B＋ lines than those in 5B-from the 12 DPA，while 5B+lines also had significantly（$P<0. 05$）higher TGW than those of 5B-from the 25 DPA（Fig. 4，Table 1）. No significant differences were observed in GW between the 5B＋ and 5B-genotypes at all the developmental stages. This suggests that the increased TGW is attributed to the increased grain length in the 5B+ genotypes.

Fine mapping of QTgw. caas-5B

For fine mapping of *QTgw. caas-5B*, eight new markers between *Kasp _ 5B2* and *Kasp _ 5B6* were developed from 660 K, 50 K SNP arrays and resequencing data (Fig. 1d, Table S1). Twelve heterozygous recombinant plants (RL1-RL12) identified from HRL2925 using 13 markers (Fig. 1d) were analyzed for fine mapping of *QTgw. caas-5B*. After progeny tests, significant differences in TGW were detected between 5B+ and 5B− genotypes within RL1, RL6, RL7 and RL12 (*P* < 0.05), whereas there were no significant differences within the other NILs from RL2 to RL5 and RL8 to RL11 (Fig. 1d, Table 2). *QTgw. caas-5B* was delimited to an interval of approximately 2.0 Mb flanked by markers *Kasp _ 5B29* and *Kasp _ 5B31* (49.6-51.6 Mb), with 17 high-confidence genes based on gene annotations for the Chinese Spring reference genome (IWGSC 2018, https://urgi. versa illes. inra. fr/blast _ iwgsc/).

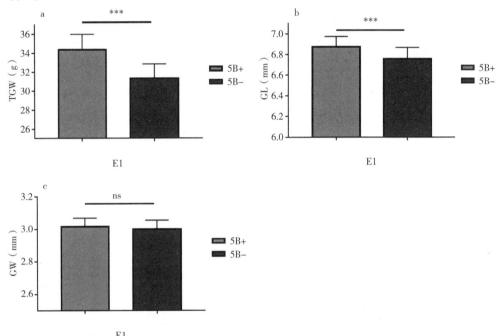

Fig. 2　Phenotypic differences in **a** thousand grain weight (TGW), **b** grain length (GL) and **c** grain width (GW) between 5B+ and 5B− genotypes among progenies of heterozygous recombinant plant

Prediction of candidate genes for QTgw. caas-5B

Based on resequencing data for the parents, SNPs or InDels were found in the coding or intron regions of *TraesCS5B02G044500*, *TraesCS5B02G044600*, *TraesCS5B02G044700*, *TraesCS5B02G044900*, *TraesCS5B02G045500*, *TraesCS5B02G045800*, *TraesCS5B02G045900* and *TraesCS5B02G046000*, whereas the other nine high-confidence genes lacked sequence polymorphisms between two parents. SNPs are synonymous mutation in *TraesCS5B02G044500*, *TraesCS5B02G044600*, *TraesCS5B02G044700*, *TraesCS5B02G045900* and *TraesCS5B02G046000*. They are not likely to cause delirious effects on the proteins. Whereas, SNPs are missense mutation in *TraesCS5B02G044900*, *TraesCS5B02G045500* and *TraesCS5B02G045800*.

The results of RNA-seq indicated that only *TraesCS5B02G044800* among the 17 high-confidence genes in the 49.6-51.6 Mb region on chromosome 5B showed higher expression level in the 5B-genotype, whereas the transcript was not detected in the genotype 5B+ (Fig. 5, Table 3). The other 16 high-confidence genes, including *TraesCS5B02G044900*, *TraesCS5B02G045500* and *TraesCS5B02G045800* which had missense mutations, showed no differential expression levels between homozygous 5B+ and 5B− genotypes (Table S3). A SNP was located at 824 bp upstream of the initiation codon ATG in the promoter region of *TraesCS5B02G044800* by resequencing the parents. Thus, *TraesCS5B02G044800* was considered a candidate gene for *QTgw. caas-5B*.

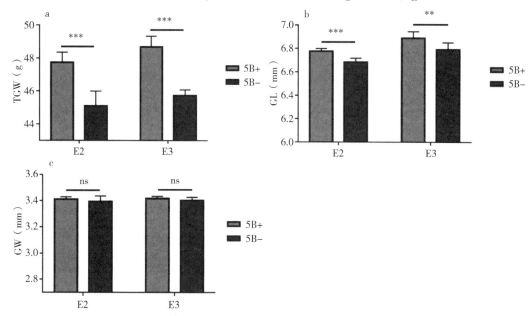

Fig. 3　Phenotypic differences in **a** thousand grain weight (TGW), **b** grain length (GL) and **c** grain width (GW) between 20 homozygous lines with the 5B+genotype and 20 with 5B−. E2 and E3, Anyang 2018-2019 and Xinxiang 2018-2019, respectively. ∗∗, ∗∗∗and ns, significant at $P<0.01$, $P<0.001$ and non-significant, respectively.

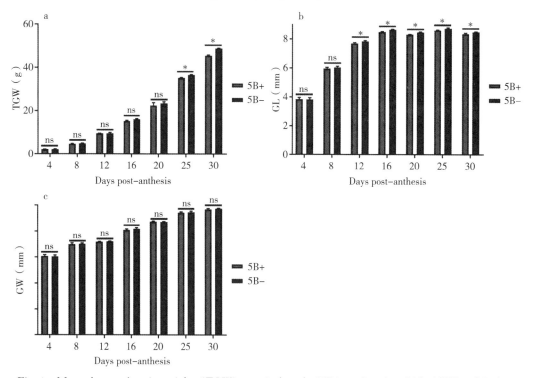

Fig. 4　Mean thousand grain weight (TGW), grain length (GL) and grain width (GW) of 20 homozygous lines with the 5B+ genotype and 20 with 5B− at different grain development stages. ∗ and ns, significant at $P<0.05$ and non-significant, respectively.

Table 1   Mean thousand grain weight (TGW), grain length (GL) and grain width (GW) of 20 homozygous lines with 5B+ and 20 with 5B− from progenies of L2925 at different grain development stages

Trait	Genotype	4 DPA[a]	8 DPA	12 DPA	16 DPA	20 DPA	25 DPA	30 DPA
TGW (g)	5B+ (ZM895)	2.1±0.1[b]A[c]	4.8±0.1A	9.6±0.1A	16.0±0.3A	23.4±1.1A	36.6±0.2A	48.7±0.2A
	5B− (ZM871)	2.1±0.2A	4.7±0.2A	9.5±0.3A	15.2±0.3A	22.4±0.6A	35.2±0.2B	45.5±0.4B
GL (mm)	5B+ (ZM895)	3.82±0.11A	6.04±0.07A	7.85±0.05A	8.66±0.03A	8.46±0.03A	8.76±0.04A	8.50±0.03A
	5B− (ZM871)	3.83±0.11A	5.97±0.07A	7.71±0.04B	8.51±0.02B	8.35±0.04B	8.63±0.02B	8.37±0.05B
GW (mm)	5B+ (ZM895)	3.02±0.05A	3.51±0.03A	3.61±0.01A	4.09±0.03A	4.36±0.01A	4.73±0.04A	4.87±0.02A
	5B− (ZM871)	3.04±0.05A	3.50±0.04A	3.59±0.01A	4.05±0.04A	4.37±0.01A	4.72±0.02A	4.84±0.03A

[a]DPA days post-anthesis

[b]Data are means±SD

[c]Phenotypic differences between contrasting genotypes followed by different letters are significant at $P<0.05$

Validation of KASP markers flanking QTgw. caas-5B in a germplasm panel

The *Kasp _ 5B _ Tgw* based on the SNP with 'C' in ZM871 and 'A' in ZM895 alleles in the promoter region of *TraesCS5B02G044800* for *QTgw. caas-5B* was used to genotype the diversity panel of 166 cultivars, among which 48 cultivars had the ZM895 genotype, and 118 had the ZM871 genotype (Tables 4 and S3). The ZM895 genotype showed significantly ($P<0.05$) higher TGW and GL than the ZM871 genotype in all six environments as well as BLUE value. Differences in TGW and GL between the two genotypes ranged from 2.1 to 2.9 g (5.5 to 5.9%) and 0.30 to 0.40 mm (2.0 to 3.0%), respectively (Table 4). Other three KASP markers based on SNPs in *TraesCS5B02G044900*, *TraesCS5B02G045500* and *TraesCS5B02G045800* were run on the same diversity panel, but no significant differences in TGW were observed between two genotypes. These results provided further evidence for a significant effect of *QTgw. caas-5B* on TGW, and *TraesCS5B02G044800* is probably a candidate gene for *QTgw. caas-5B*.

Table 2   Mean thousand grain weight (TGW), grain length (GL) and grain width (GW) of near-isogenic lines (NILs) 5B+ and 5B− progeny from each recombinant line (RL)

Line	Genotype	No.[a]	TGW (g)	GL (mm)	GW (mm)
RL1	5B+ (ZM895)	25	49.7±1.2[b]	6.88±0.05	3.29±0.05
	5B− (ZM871)	27	47.0±1.4	6.78±0.07	3.25±0.06
			2.7**[c]	0.10***	0.04*
RL2	5B+ (ZM895)	33	48.7±1.4	6.84±0.05	3.40±0.05
	5B− (ZM871)	42	48.0±1.4	6.77±0.05	3.38±0.05
			0.7	0.07***	0.02
RL3	5B+ (ZM895)	24	48.5±1.0	6.84±0.05	3.38±0.03
	5B− (ZM871)	32	48.2±2.1	6.78±0.07	3.39±0.06
			0.3	0.06**	−0.01
RL4	5B+ (ZM895)	28	47.9±2.7	6.89±0.10	3.37±0.09
	5B− (ZM871)	27	48.3±1.7	6.84±0.06	3.40±0.05
			−0.4	0.05*	−0.03
RL5	5B+ (ZM895)	30	48.8±1.7	6.82±0.11	3.42±0.07
	5B− (ZM871)	29	48.5±1.1	6.80±0.15	3.40±0.03
			0.03	0.02	0.02

（continued）

Line	Genotype	No. [a]	TGW (g)	GL (mm)	GW (mm)
RL6	5B+ (ZM895)	43	48.9±2.6	6.87±0.10	3.39±0.08
	5B— (ZM871)	33	47.0±1.2	6.80±0.09	3.36±0.04
			1.9**	0.07***	0.03
RL7	5B+ (ZM895)	29	49.7±0.9	6.98±0.06	3.44±0.02
	5B— (ZM871)	33	47.5±1.1	6.88±0.07	3.40±0.03
			2.2**	0.10***	0.04*
RL8	5B+ (ZM895)	33	48.6±2.1	6.96±0.09	3.40±0.07
	5B— (ZM871)	31	48.5±2.0	6.89±0.10	3.41±0.06
			0.1	0.07*	−0.01
RL9	5B+ (ZM895)	29	48.9±1.4	6.90±0.08	3.41±0.05
	5B— (ZM871)	26	48.4±1.4	6.84±0.09	3.41±0.05
			0.5	0.06**	0
RL10	5B+ (ZM895)	31	48.7±1.7	6.95±0.08	3.41±0.04
	5B— (ZM871)	29	48.0±1.1	6.86±0.09	3.39±0.04
			0.7	0.09***	0.02
RL11	5B+ (ZM895)	24	48.1±2.1	6.93±0.09	3.39±0.07
	5B— (ZM871)	31	47.8±1.4	6.86±0.07	3.39±0.05
			0.3	0.07**	0
RL12	5B+ (ZM895)	30	49.4±1.1	6.88±0.02	3.41±0.04
	5B— (ZM871)	33	47.1±1.2	6.80±0.07	3.38±0.04
			2.3**	0.08***	0.03

[a] Number of lines within the corresponding genotypic group

[b] Data are means±SD

[c] Phenotypic difference between the means of genotypes 5B+ and 5B—. Asterisks indicate significant differences determined by Student's t-tests. *, ** and ***, significant at $P<0.05$, $P<0.01$ and $P<0.001$, respectively

Table 3　Relative expression of *TraesCS5B02G044800* in genotypes 5B+ and 5B— at different grain development stages

Genotype	4 DPA[a]	8 DPA	12 DPA	16 DPA	20 DPA	25 DPA
5B+ (ZM895)	0B	0B	0B	0B	0B	0B
5B—(ZM871)	1.02±0.10[b]A[c]	1.11±0.06A	1.18±0.23A	0.56±0.18A	0.81±0.14A	0.82±0.34A

[a] *DPA*, days post-anthesis

[b] Data are means±SD.

[c] Differences of expression levels between two genotypes followed by different letters are significant at $P<0.001$.

Table 4　Mean thousand grain weight (TGW), grain length (GL) and grain width (GW) of genotypes 5B+ and 5B— in the germplasm panel of 166 wheat cultivars grown in six environments

Trait	Genotype[a]	No. [b]	E4[c]	E5	E6	E7	E8	E9	BLUE[d]
TGW (g)	5B+ (ZM895)	48	43.0±5.4[e]	43.9±5.0	51.9±5.3	47.7±6.2	41.6±5.0	41.3±5.0	45.1±4.8
	5B— (ZM871)	118	40.4±4.9	41.5±5.7	49.0±5.4	45.1±5.8	38.5±4.2	39.0±4.9	42.4±4.8
			2.6[f]**	2.4*	2.9**	2.6*	3.1***	2.3**	2.7**
GL(cm)[g]	5B+ (ZM895)	48	13.80±0.74	13.57±0.74	15.03±0.74	15.02±0.71	14.39±0.68	14.28±0.71	14.49±0.69

(continued)

Trait	Genotype[a]	No.[b]	E4[c]	E5	E6	E7	E8	E9	BLUE[d]
	5B− (ZM871)	118	13.40±0.67	13.17±0.63	14.73±0.68	14.71±0.69	14.05±0.67	13.93±0.71	14.15±0.64
			0.40**	0.40***	0.30*	0.31*	0.34**	0.35**	0.34**
GW(cm)[g]	5B+ (ZM895)	48	6.66±0.41	6.59±0.32	7.39±0.31	7.22±0.36	6.92±0.35	6.80±0.35	7.10±0.31
	5B− (ZM871)	118	6.59±0.34	6.49±0.43	7.30±0.36	7.14±0.40	6.77±0.34	6.63±0.39	7.00±0.34
			0.07	0.10	0.09	0.08	0.15**	0.17*	0.10

[a] Genotypes identified using marker *Kasp _ 5BTgw*

[b] Number of cultivars with corresponding genotype

[c] E4-E9, Anyang 2012−2013, Suixi 2012−2013, Anyang 2013−2014, Suixi 2013−2014, Suixi 2014−2015 and Shijiazhuang 2014−2015, respectively

[d] *BLUE*: Best linear unbiased estimation

[e] Data are shown as means±SD

[f] Phenotypic differences between two genotype 5B+ and 5B−. Asterisks indicate significant differences determined by Student's t-tests. *, ** and ***, significant at $P<0.05$, $P<0.01$ and $P<0.001$, respectively

[g] GL and GW are mean lengths and widths of 20 grains

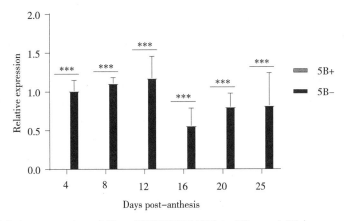

Fig. 5   Relative expression of *TraesCS5B02G044800* in 5B− and 5B+ genotypes at 4, 8, 12, 16, 20, 25 DPA (days post-anthesis). No transcript was detected in 5B+ genotypes. ***, significant at $P<0.001$

Discussion

Residual heterozygous recombinant lines are usefulstocks for fine mapping

Many QTL for grain-related traits have been identified in different genetic backgrounds (Huang et al. 2003; Quarrie et al. 2005; Prashant et al. 2012; Cui et al. 2014; Wu et al. 2015; Ma et al. 2018; Su et al. 2018; Zhai et al. 2018; Xu et al. 2019). Most were located in large chromosome intervals due to limited numbers of markers or lack of recombination events within the targeted QTL regions. Following release of the Chinese Spring reference genome sequence (IWGSC 2018, https：//urgi. versailles. inra. fr/blast

_ iwgsc/) and development of new sequencing technologies densely populated genetic maps can easily be constructed, and genetic information for a specific map interval can be searched. Consequently, many researchers have employed fine mapping approaches to validate QTL or narrow genomic intervals within targeted QTL regions (Brinton et al. 2017; Guan et al. 2019; Chen et al. 2020). In this study, progenies of 12 heterozygous recombinant plants were genotyped and phenotyped to fine map *QTgw. caas-5B* to an approximately 2.0 Mb physical interval containing 17 high-confidence annotated genes.

Candidate genes for QTL controlling TGW

Theapproximate 2.0 Mb interval of *QTgw. caas-5B*

was flanked by *Kasp _ 5B29* and *Kasp _ 5B31* （49.6-51.6 Mb）. To identify candidate genes for TGW expression levels of the annotated were determined. *TraesCS5B02G044800* showed as significantly lower expression level in the 5B+ genotype than in the 5B-genotype at all grain developmental stages. The ZM895 genotype had significantly higher TGW and GL than ZM871 genotype in the germplasm panel assessed with the *Kasp _ 5B _ Tgw* marker based on a SNP developed from *TraesCS5B02G044800*, thus indicating that *TraesCS5B02G044800* is a potential candidate gene for *QTgw. caas-5B*. According to the Chinese Spring reference genome sequence (IWGSC 2018, https：// urgi. versailles. inra. fr/blast _ iwgsc/), *TraesCS5-B02G044800* were predicted to encode a TIR-NBS-LRR disease resistance protein and another unknown functional protein. There was no significant difference in the disease resistance between 5B+ and 5B— lines based on the data for powdery mildew and leaf rust reactions. Now, it is necessary to undertake gene over-expression and knockout studies to confirm the role of this candidate gene on TGW and GL.

GL contributes to TGW at QTgw. caas-5B locus

Various studies suggest that the early stage of grain length development is important in determining final grain weight inwheat (Hasan et al. 2011; Guo et al. 2015; Simmonds et al. 2016; Brinton et al. 2017, 2018). This study initially detected a clear difference in GL between 5B+ and 5B— genotypes at 12 DPA, whereas a corresponding difference in TGW between 5B+ and 5B— genotypes was first observed at 25 DPA. This supported the contention that GL is a main factor contributing to grain weight (Brinton et al. 2017).

Applications in wheat breeding

Major stable QTL for yield-related traits with tightly linked markers is veryimportant for molecular breeding. In this study, a QTL for TGW on chromosome 5B showed stable effects on TGW and GL cross environments. Its presence in 48 accessions among a panel of 166 indicated that it had been a selected target for grain

weight（or even yield）in past breeding programs, and thus *Kasp _ 5B _ Tgw* represents a future target for marker-assisted selection to enhance grain size and weight. Moreover, the current results provide a basis for map-based cloning of the gene underlying the QTL.

❖ References

Brinton J, Simmonds J, Minter F, Leverington-Waite M, Snape J, Uauy C (2017) Increased pericarp cell length underlies a major quantitative trait locus for grain weight in hexaploid wheat. New Phytol 215：1026-1038.

Brinton J, Simmonds J, Uauy C (2018) Ubiquitin-related genes are differentially expressed in isogenic lines contrasting for pericarp cell size and grain weight in hexaploid wheat. BMC Plant Biol 18：22.

Cabral AL, Jordan MC, Larson G, Somers DJ, Humphreys DG, McCartney CA (2018) Relationship between QTL for grain shape, grain weight, test weight, milling yield, and plant height in the spring wheat cross RL4452/' AC domain '. PLoS One 13：e0190681.

Cao S, Xu D, Hanif M, Xia X, He Z (2020) Genetic architecture underpinning yield component traits in wheat. Theor Appl Genet 133：1811-1823.

Chen Z, Cheng X, Chai L, Wang Z, Bian R, Li J, Zhao A, Xin M, Guo W, Hu Z, Peng H, Yao Y, Sun Q, Ni Z (2020) Dissection of genetic factors underlying grain size and fine mapping of QTgw. cau-7D in common wheat (*Triticum aestivum* L.). Theor Appl Genet 133：149-162.

Cui F, Zhao C, Ding A, Li J, Wang L, Li X, Bao Y, Li J, Wang H (2014) Construction of an integrative linkage map and QTL mapping of grain yield-related traits using three related wheatRIL populations. Theor Appl Genet 127：659-675.

Dong L, Wang F, Liu T, Dong Z, Li A, Jing R, Mao L, Li Y, Liu X, Zhang K, Wang D (2014) Natural variation of TaGASR7-A1 affects grain length in common wheat under multiple cultivationconditions. Mol Breed 34：937-947.

FAO (2017) http：//www. fao. org/faostat/en/.

Fischer RA, Edmeades GO (2010) Breeding and cereal yield progress. Crop Sci 50：85-98.

Gao F, Ma D, Yin G, Rasheed A, Dong Y, Xiao Y, Xia X, Wu X, He Z (2017) Genetic progress in grain yield and physiological traits in Chinese wheat cultivars of southern yellow and Huai valleysince 1950. Crop Sci 57: 760-773.

Guan P, Lu L, Jia L, Kabir MR, Zhang J, Lan T, Zhao Y, Xin M, Hu Z, Yao Y, Ni Z, Sun Q, Peng H (2018) Global QTL analysis identifies genomic regions on chromosomes 4A and 4B harboring stable loci for yield-related traits across different environments in wheat (*Triticum aestivum* L.). Front Plant Sci 9: 529.

GuanP, Di N, Mu Q, Shen X, Wang Y, Wang X, Yu K, Song W, Chen Y, Xin M, Hu Z, Guo W, Yao Y, Ni Z, Sun Q, Peng H (2019) Use of near-isogenic lines to precisely map and validate a major QTL for grain weight on chromosome 4AL in bread wheat (*Triticum aestivum* L.). Theor Appl Genet 132: 2367-2379.

Guo Z, Chen D, Schnurbusch T (2015) Variance components, heritability and correlation analysis of anther and ovary size duringthe floral development of bread wheat. J Exp Bot 66: 3099-3111.

HanifM, Gao F, Liu J, Wen W, Zhang Y, Rasheed A, Xia X, He Z, Cao S (2016) TaTGW6-A1, an ortholog of rice TGW6, is associated with grain weight and yield in bread wheat. Mol Breed 36: 1.

Hasan AK, Herrera J, Lizana C, Calderini DF (2011) Carpel weight, grain length and stabilized grain water content are physiological drivers of grain weight determination of wheat. Field Crops Res123: 241-247.

Hu M-J, Zhang H-P, Cao J-J, Zhu X-F, Wang S-X, Jiang H, Wu ZY, Lu J, Chang C, Sun G-L, Ma C-X (2016) Characterization of an IAA-glucose hydrolase gene TaTGW6 associated with grain weight in common wheat (*Triticum aestivum* L.). Mol Breed 36: 25.

Huang XQ, Cöster H, GanalMW, Röder MS (2003) Advanced back-cross QTL analysis for the identification of quantitative trait loci alleles from wild relatives of wheat (*Triticum aestivum* L.). Theor Appl Genet 106: 1379-1389.

International Wheat Genome Sequencing Consortium (IWGSC) (2018) Shifting the limits in wheat research and breeding using a fully annotated reference genome. Science 361: eaar7191.

JiangY, Jiang Q, Hao C, Hou J, Wang L, Zhang H, Zhang S, Chen X, Zhang X (2015) A yield-associated gene TaCWI, in wheat: its function, selection and evolution in global breeding revealed by haplotype analysis. Theor Appl Genet 128: 131-143.

Kuchel H, Williams KJ, Langridge P, Eagles HA,

Jefferies SP (2007) Genetic dissection of grain yield in bread wheat I QTL Analysis. Theor Appl Genet 115: 1029-1041.

Langridge P (2013) Wheat genomics and the ambitious targets forfuture wheat production. Genome 56: 545-547.

Li N, Li Y (2016) Signaling pathways of seed size control in plants. Curr Opin Plant Biol 33: 23-32.

Li F, Wen W, Liu J, Zhang Y, Cao S, He Z, Rasheed A, Jin H, Zhang C, Yan J, Zhang P, Wan Y, Xian X (2019a) Genetic architecture of grain yield in bread wheat based on genome-wide association studies. BMC Plant Biol 19: 168.

Li N, Xu R, Li Y (2019b) Molecular networks of seed size control in plants. Annu Rev Plant Biol 70: 435-463.

Ma L, Li T, Hao C, WangY, Chen X, Zhang X (2016) TaGS5-3A, a grain size gene selected during wheat improvement for larger kernel and yield. Plant Biotechnol J 14: 1269-1280.

Ma F, Xu Y, Ma Z, Li L, An D (2018) Genome-wide association and validation of key loci for yield-related traits in wheat founderparent Xiaoyan 6. Mol Breed 38: 91.

PrashantR, Kadoo N, Desale C, Kore P, Dhaliwal HS, Chhuneja P, Gupta V (2012) Kernel morphometric traits in hexaploid wheat (*Triticum aestivum* L.) are modulated by intricate QTL$\times$QTL and genotype$\times$environment interactions. J Cereal Sci56: 432-439.

Quarrie SA, Steed A, Calestani C, Semikhodskii A, Lebreton C, Chinoy C, Steele N, Pljevljakusic D, Waterman E, Weyen J, Schondelmaier J, Habash DZ, Farmer P, Saker L, Clarkson DT, Abugalieva A, Yessimbekova M, Turuspekov Y, Abugalieva S, Tuberosa R, Sanguineti MC, Hollington PA, Aragues R, Royo A, Dodig D (2005) A high-density genetic map of hexaploid wheat (*Triticum aestivum* L.) from the cross Chinese spring$\times$SQ1 and its use to compare QTLs for grain yield across a rangeof environments. Theor Appl Genet 110: 865-880.

Ramirez-GonzalezRH, Uauy C, Caccamo M (2015) PolyMarker: a fast polyploid primer design pipeline. Bioinformatics 31: 2038-2039.

Sajjad M, Ma X, Habibullah Khan S, Shoaib M, Song Y, Yang W, Zhang A, Liu D (2017) TaFlo2-A1, an ortholog of rice Flo2, is associated with thousand grain weight in bread wheat (*Triticum aestivum* L.). BMC Plant Biol 17: 164.

Simmonds J, Scott P, Leverington-Waite M, Turner AS, Brinton J, Korzun V, Snape J, Uauy C (2014) Identification and independ-ent validation of a stable yield and thousand grain weight QTL on chromosome 6A of hexaploid wheat (*Triticum aestivum* L.). BMC Plant Biol 14: 191.

Simmonds J, Scott P, Brinton J, Mestre TC, Bush M, Del Blanco A, Dubcovsky J, Uauy C (2016) A splice acceptor site mutation in TaGW2-A1 increases thousand grain weight in tetraploid and hexaploid wheat through wider and longer grains. Theor ApplGenet 129: 1099-1112.

Su Z, Hao C, Wang L, Dong Y, Zhang X (2011) Identification and development of a functional marker of TaGW2 associated with grain weight in bread wheat (*Triticum aestivum* L.). Theor ApplGenet 122: 211-223.

Su Q, Zhang X, ZhangW, Zhang N, Song L, Liu L, Xue X, Liu G, Liu J, Meng D, Zhi L, Ji J, Zhao X, Yang C, Tong Y, Liu Z, Li J (2018) QTL detection for kernel size and weight in bread wheat (*Triticum aestivum* L.) using a high-density SNP and SSR-based linkage map. Front Plant Sci 9: 1484.

Sun C, Zhang F, Yan X, Zhang X, Dong Z, Cui D, Chen F (2017) Genome-wide association study for 13 agronomic traits reveals distribution of superior alleles in bread wheat from the yellow and Huai valley of China. Plant Biotechnol J 15: 953-969.

Wang S, Zhang X, Chen F, Cui D (2015) A single-nucleotide polymorphism of TaGS5 gene revealed its association with kernel weight in Chinese bread wheat. Front Plant Sci 6: 1166.

Wang S, Yan X, Wang Y, Liu H, Cui D, Chen F (2016) Haplotypes of the TaGS5-A1 gene are associated with thousand-kernel weight in Chinese bread wheat. Front Plant Sci 7: 783.

Wu Q, Chen Y, Zhou S, Fu L, Chen J, Xiao Y, Zhang D, Ouyang S, Zhao X, Cui Y, Zhang D, Liang Y, Wang Z, Xie J, Qin J, Wang G, Li D, Huang Y, Yu M, Lu P, Wang L, Wang L, Wang H, Dang C, Li J,

Zhang Y, Peng H, Yuan C, You M, Sun Q, Wang J, Wang L, Luo M, Han J, Liu Z (2015) High-density genetic linkage map construction and QTL mapping of grain shape and size in the wheat population Yanda 1817 x Beinong 6. PLoS One 10: e0118144.

Xu D, Wen W, Fu L, Li F, Li J, Xie L, Xia X, Ni Z, He Z, Cao S (2019) Genetic dissection of a major QTL for kernel weight spanning the Rht-B1 locus in bread wheat. Theor Appl Genet 132: 3191-3200.

Yang J, Zhou YJ, Wu QH, Chen YX, Zhang PP, Zhang YE, Hu WG, Wang XC, Zhao H, Dong LL, Han J, Liu Z, Cao TJ (2019) Molecular characterization of a novel TaGL3-5A allele and its association with grain length in wheat (*Triticum aestivum* L.). Theor Appl Genet 132: 1799-1814.

Yang L, Zhao D, Meng Z, Xu K, Yan J, Xia X, Cao S, Tian Y, He Z, Zhang Y (2020) QTL mapping for grain yield-related traits in bread wheat via SNP-based selective genotyping. Theor Appl Genet 133: 857-872.

Yue A, Li A, Mao X, Chang X, Li R, Jing R (2015) Identification and development of a functional marker from 6-SFT-A2 associated with grain weight in wheat. Mol Breed 35: 63.

Zhai H, Feng Z, Du X, Song Y, Liu X, Qi Z, Song L, Li J, Li L, Peng H, Hu Z, Yao Y, Xin M, Xiao S, Sun Q, Ni Z (2018) A novel allele of TaGW2-A1 is located in a finely mapped QTL that increases grain weight but decreases grain number in wheat (*Triticum aestivum* L.). Theor Appl Genet 131: 539-553.

Zhang L, Zhao YL, Gao LF, ZhaoGY, Zhou RH, Zhang BS, Jia JZ (2012) TaCKX6-D1, the ortholog of rice OsCKX2, is associated with grain weight in hexaploid wheat. New Phytol 195: 574-584.

Zhang Y, Liu J, Xia X, He Z (2014) TaGS-D1, an ortholog of rice OsGS3, is associated with grain weight and grain length in common wheat. Mol Breed 34: 1097-1107.

QTL mapping for grain yield-related traits in bread wheat via SNP-based selective genotyping

Li Yang[1] · Dehui Zhao[1] · Zili Meng[2] · Kaijie Xu[3] · Jun Yan[3] · Xianchun Xia[1] · Shuanghe Cao[1] · Yubing Tian[1] · Zhonghu He[1,4] · Yong Zhang[1]

[1] Institute of Crop Sciences, National Wheat Improvement Center, Chinese Academy of Agricultural Sciences (CAAS), 12 Zhongguancun South Street, Beijing 100081, China

[2] Shangqiu Academy of Agricultural and Forestry Sciences, 10 Shengli Road, Shangqiu 476000, Henan Province, China

[3] Institute of Cotton Research, CAAS, 38 Huanghe Dadao, Anyang 455000, Henan Province, China

[4] International Maize and Wheat Improvement Center (CIMMYT), China Office, c/o CAAS, Beijing 100081, China

Abstract: Identification of major stable quantitative trait loci (QTL) for grain yield-related traits is important for yield potential improvement in wheat breeding. In the present study, 266 recombinant inbred lines (RILs) derived from a cross between Zhongmai 871 (ZM871) and its sister line Zhongmai 895 (ZM895) were evaluated for thousand grain weight (TGW), grain length (GL), grain width (GW), and grain number per spike (GNS) in 10 environments and for grain filling rate in six environments. Sixty RILs, with 30 higher and 30 lower TGW, respectively, were genotyped using the wheat 660 K SNP array for preliminary QTL mapping. Four genetic regions on chromosomes 1AL, 2BS, 3AL, and 5B were identified to have a significant effect on TGW-related traits. A set of Kompetitive Allele Specific PCR markers were converted from the SNP markers on the above target chromosomes and used to genotype all 266 RILs. The mapping results confirmed the QTL named *Qgw. caas-1AL*, *Qgl. caas-3AL*, *Qtgw. caas-5B*, and *Qgl. caas-5BS* on the targeted chromosomes, explaining 5.0-20.6%, 5.7-15.7%, 5.5-17.3%, and 12.5-20.5% of the phenotypic variation for GW, GL, TGW, and GL, respectively. A novel major QTL for GNS on chromosome 5BS, explaining 5.2-15.2% of the phenotypic variation, was identified across eight environments. These QTL were further validated using BC_1F_4 populations derived from backcrosses ZM871/ZM895//ZM871 (121 lines) and ZM871/ZM895//ZM895 (175 lines) and 186 advanced breeding lines. Collectively, selective genotyping is a simple, economic, and effective approach for rapid QTL mapping and can be generally applied to genetic mapping studies for important agronomic traits.

Key message: We identified four chromosome regions harboring QTL for grain yield-related traits, and breeder-friendly KASP markers were developed and validated for marker-assisted selection.

Published in Theoretical and Applied Genetics, 2020, 133: 857-872.

Introduction

Wheat is one of the most important food crops in the serving as the major source of carbohydrate and protein for 35% of the human population (Paux et al. 2008). It has been estimated that wheat yields must increase by over 60% in the next 30 years to meet the demands of growing populations (Langridge 2013). Although significant progress has been made in wheat yield improvement during the last 50 years, yield growth rates are generally at no more than 1% per annum (Fischer and Edmeades 2010; Gao et al. 2017). Therefore, genetic improvement in yield potential is required, along with better crop management, to achieve further increases in wheat yield in farmers' fields.

Thousand grain weight (TGW), grain number per spike (GNS), and spike number per unit area are three major yield components of wheat. They showed less sensitivity to environment than yield itself and are treated as indirect traits for yield improvement (Xu et al. 2017). Historically, genetic gains in grain yield potential have been driven mainly by increased grain number per unit area, but positive contributions of TGW were also observed in China and other countries (Fischer 2008; Gao et al. 2017; Sadras and Lawson 2011). TGW is a complex trait determined by grain size and grain filling (Simmonds et al. 2014). Grain size can be broken into three components: grain length (GL), grain width (GW), and grain thickness. From a developmental point of view, GL is determined in the early stage of grain development and is less influenced by environment conditions, whereas GW and grain thickness are established later and are more environmentally sensitive (Lizana et al. 2010; Prashant et al. 2012; Xie et al. 2015). Large grains are generally favored in wheat breeding because they contribute not only to TGW, but also to seedling emergence/vigor and consumer preference (Chastain et al. 1995; Gegas et al. 2010). Grain filling can be divided into two components, namely rate and duration (Xie et al. 2015). Grain filling rate (GFR) reflects the efficiency of dry matter accumulation (Shewry 2009), and grain filling duration (GFD) reflects the time that it takes (Xie et al. 2015). Both rate and duration contribute to grain weight, with the former showing a stronger correlation with grain weight than the latter (Wang et al. 2009; Xie et al. 2015). Selection of cultivars with high GFR appears to be a promising choice to increase grain yield in the Yellow and Huai River Valleys Winter Wheat Region in China where grain filling duration cannot be freely prolonged under the annual wheat-maize double-cropping system (Wang et al. 2009). Negative correlation among TGW and GNS is a typical selection trade-off problem in breeding, but the molecular mechanisms underlying the individual traits and their interactions are still largely unknown (Xu et al. 2017).

With the use of molecular markers, a large number of quantitative trait loci (QTL) for TGW have been identified on all 21 wheat chromosomes (Cabral et al. 2018; Campbell et al. 1999; Cheng et al. 2017; Cui et al. 2014; Gao et al. 2015; Guan et al. 2018; Huang et al. 2003; Jahani et al. 2019; Jia et al. 2013; Li et al. 2018; Liu et al. 2018; 2019; Prashant et al. 2012; Quarrie et al. 2005; Simmonds et al. 2014; Su et al. 2018; Wang et al. 2019; Wu et al. 2015; Xu et al. 2017; Yan et al. 2017; Yu et al. 2017; Zhai et al. 2018). Many studies were conducted to identify QTL associated with GL and GW with the availability of high-throughput phenotyping pipelines for grain morphology traits (Cabral et al. 2018; Gegas et al. 2010; Li et al. 2018; Maphosa et al. 2014; Su et al. 2018; Wang et al. 2019; Williams and Sorrells 2014; Xiao et al. 2011; Zhai et al. 2018). However, only a few studies focused on GFR (Bhusal et al. 2017; Charmet et al. 2005; Griffiths et al. 2015; Wang et al. 2009; Xie et al. 2015). Charmet et al. (2005) mapped a stable QTL for maximum GFR on chromosome 2B and an environment-specific one on chromosome 3A. Wang et al. (2009) documented six stable QTL associated with mean and/or maximum GFR on chromosomes 1B, 2A, 3B, 5B, 6D, and 7D and seven environment-specific QTL on chromosomes 1A, 2D, 3A, 3B, 3D, 4D, and 5B. An environment-specific QTL on chromosome 7B was reported by Griffiths et al. (2015). Xie et al. (2015) conducted a comprehensive study on GFR by investigating the initial, rapid, late, mean, and maximum GFR of grains from the first, second and third florets within a spikelet. Important QTL clusters on chromosomes 2A, 3B, 4A, 5DL, and 7B, as well as other genetic regions on chromosomes 1A, 1DS, 2D, 3DL, 4DL, 5A, and 7D, were identified (Xie et al. 2015). Bhusal et al. (2017) detected a stable QTL for GFR on chromosome 2A and an environment-specific QTL on chromosome 6D. Although hundreds of QTL for grain-related traits have been identified, subsequent fine mapping was only reported in a few publications, and no gene has been isolated through map-based cloning due to the lack of genome sequence until *WAPO1* was identified as a promising candidate gene for a 7AL locus affecting spikelet number per spike (Brinton and Uauy 2018; Kuzay et al. 2019). With the release of wheat genome sequences, more effort will most likely be devoted to map-based cloning. Therefore, rapid identification and validation of major stable QTL for grain-related traits and tightly linked markers are important.

Traditionally, QTL are identified by linkage analysis using progenies derived from crosses between parents showing contrasting phenotypes. Genetically distant parents are chosen in order to generate complete linkage maps. This can lead to unwanted variation that decreases the accuracy of evaluation of target traits and increases the complexity of subsequent fine mapping (Takagi et al. 2013). On the other hand, whenever populations derived from closely related parents are used the number of DNA markers becomes a limiting factor (Takagi et al. 2013). Genotyping an entire segregating population with markers covering the whole genome is laborious, time-consuming, and expensive (Zou et al. 2016). Recent progress on wheat genome sequencing and availability of high-throughput chip-based markers have accelerated QTL analysis and make the use of populations derived from genetically

closely related parents possible. Selective genotyping of individuals from the high and/or low tails of a phenotypic distribution provides a cost-effective alternative approach for genetic mapping with negligible practical disadvantage in terms of detection power (Sun et al. 2010). This methodology has been successfully used in rice, maize, and rye (Farkhari et al. 2013; Gimhani et al. 2016; Myskow and Stojalowski 2016), whereas it has not been reported in wheat for identification of QTL associated with complex traits such as grain weight and size.

The objectives of this study were to: (1) identify loci controlling TGW, GL, GW, and GFR in a wheat recombinant inbred line (RIL) population developed from two sister lines by a selective genotyping approach, (2) investigate their effects on GNS, and (3) develop and validate breeder-friendly markers for marker-assisted selection (MAS) in wheat breeding.

Materials and methods

Plant materials

The mapping population of 266 F_6 RILs was derived from a cross between Zhongmai 871 and Zhongmai 895 (hereafter ZM871 and ZM895, respectively) by single-seed descent. ZM895 was released in 2012 jointly by the Institute of Crop Sciences and the Institute of Cotton Research, Chinese Academy of Agricultural Sciences. Now it is a leading cultivar in the Yellow and Huai River Valleys Winter Wheat Region, with an annual production area around 0.45 million ha. ZM871 and ZM895, developed by pedigree selection and fixed at F_5, are two sister lines that could be traced back to a single F_2 plant of the Zhoumai 16/Liken 4 cross. As revealed by the genotyping results of the 90 K SNP array, 3% of the markers were polymorphic between ZM871 and ZM895 compared with around 10-15% among genetically distant varieties (Dong et al. 2016; Wang et al. 2014). They exhibited similar agronomic traits such as plant height, heading, and maturity dates. However, ZM895 had higher TGW, larger grain, faster GFR, and lower GNS than ZM871

(Fig. S1). Two sets of materials were used for QTL validation. The first comprised backcross (BC$_1$) F$_4$ populations ZM871/ZM895//ZM871 and ZM871/ZM895//ZM895 of 121 and 175 recombinant inbred lines, respectively. The second consisted of 186 advanced breeding lines from a joint wheat breeding program conducted by the Institute of Crop Sciences and the Institute of Cotton Research, Chinese Academy of Agricultural Sciences. Pedigrees and other relevant information are listed in Table S1.

Two hundred and forty-six $F_{2:8}$ RILs from the Zhou 8425B/Chinese Spring cross and 275 $F_{2:6}$ RILs from the Doumai/Shi 4185 cross were used to examine the relation-ships between the 5B QTL for TGW and GL found in the present work and those reported in our previous studies (Gao et al. 2015; Li et al. 2018). Relevant information about these populations and their parents can be found in previous reports (Gao et al. 2015; Li et al. 2018).

Field trials and phenotyping

Details of the growing environments and traits evaluated in the RIL and BC$_1$F$_4$ populations are summarized in Table S2. Field trials were carried out over three cropping seasons at Anyang and Shangqiu in Henan province (2014-2015, 2015-2016 and 2016-2017) and two seasons at Zhoukou also in Henan (2015-2016 and 2016-2017) and at Xianyang in Shaannxi province (2014-2015 and 2015-2016). A randomized complete block design was used for all populations in all environments with no replication in 2014-2015, two replications in 2015-2016 and three replications in 2016-2017. Each plot comprised two 1-m rows spaced 25 cm apart. Thirty seeds were sown evenly in each row. TGW was evaluated in duplicate by weighing 200 grains after the grain had been dried to a constant moisture content at room temperature. GL and GW were calculated using image analysis software (Image-Pro Plus 6. 0, http: //www. mediacy. com/) after scanning 50 sound, fully developed grains placed on a scanner panel with grains crease-down. Flowering dates were visually assessed and recorded when 50% of

the spikes had extruded their anthers. Physiological maturity was recorded when 50% of the peduncles lacked green color. GFD was the number of days from flowering to maturity. Mean GFR (g/day) was calculated as GFR = TGW/GFD. GNS was calculated from the mean grain number of 30 representative spikes of each plot at physiological maturity. Collectively, TGW, GL, GW, and GNS of the ZM871/ZM895 RIL population were evaluated in 10 environments, and GFR of this population was evaluated in six environments, whereas TGW, GL, GW, and GNS of the two BC$_1$F$_4$ populations were evaluated in eight environments, and their GFR was evaluated in four environments.

Among the 186 advanced breeding lines, 62 were sown at Anyang, 26 at Xinxiang in Henan province and the remaining 98 lines were grown at both locations (Table S1). As a result, 160 lines were evaluated at Anyang and 124 lines were assessed at Xinxiang in the 2016-2017 cropping season in 4. 0×1. 6 m six row plots using a randomized design with no replication, and a check was added in every 12 plots. The planting density was 2. 4×10^6 plants/ha. TGW, GL, and GW were recorded using a scaled camera-assisted phenotyping system (Wanshen Detection Technology Co. , Ltd. , Hangzhou) and GNS was evaluated using the same method as the mapping population.

Field trials and phenotypic evaluation of the Zhou 8425B/Chinese Spring and Doumai/Shi 4185 populations were described in our previous studies (Gao et al. 2015; Li et al. 2018). In brief, the Zhou 8425B/Chinese Spring population was grown at Zhengzhou and Zhoukou in Henan province during the 2012-2013 and 2013-2014 cropping seasons, providing TGW data for four environments. The Doumai/Shi 4185 population was evaluated at Shunyi in Beijing and Shijiazhuang in Hebei province for three successive cropping seasons (2012-2013, 2013-2014 and 2014-2015), providing data of TGW, GL, and GW from six environments.

Genotyping

Genomic DNA was extracted from young leaves using the CTAB method (Doyle and Doyle1987). Based on mean values of TGW obtained from cropping season 2014-2015, 60 lines showing extreme phenotypes, including 30 lines exhibiting the highest TGW and 30 lines with the lowest TGW, were selected from the mapping population and genotyped using a Wheat 660 K SNP array (http://wheat.pw.usda.gov/ggpages/topics/Wheat660_SNP_array_devel oped_by_CAAS.pdf). The selected proportion was 11% at each tail for extreme phenotype, that is supposed to have a 95% probability of detecting QTL with large effects (that explain more than 10% of total phenotypic variation) and a 75% probability of detecting QTL with medium effects (that explain around 7% of total phenotypic variation) (Sun et al. 2010). Genotyping was performed by Capital Bio Corporation (http://www.capitalbio.com) according to the Affymetrix Axiom 2.0 Assay Manual Workflow protocol. Zhou 8425B/Chinese Spring and Doumai/Shi 4185 populations were genotyped using the 90 K SNP array as described by Gao et al. (2015) and Li et al. (2018).

Array-based SNP markers closely linked to the QTL for TGW-related traits were converted into Kompetitive Allele Specific PCR (KASP) markers for QTL confirmation. Allele-specific and common reverse primers for each KASP marker were designed using PolyMarker (http://polymarker.tgac.ac.uk/), a fast polyploid primer design pipeline. Newly designed KASP markers were evaluated for polymorphisms between the parents before genotyping the entire mapping population. Two BC_1F_4 populations and advanced breeding lines were genotyped with the flanking markers of the QTL regions on chromosomes 1AL, 2BS, and 3AL and three markers (Kasp_5B4, Kasp_5B8, and Kasp_5B11) representing the 5B QTL region. Zhou 8425B/Chinese Spring and Doumai/Shi 4185 populations were genotyped using KASP markers in the 5B QTL region that showed polymorphisms between two parents. KASP assays were performed in a 5 μl reaction volume containing 2.5 μl 2×KASP Master Mix, 0.056 μl KASP primer mix and 2.5 μl genomic DNA at 30 ng/μl. Fluorescence was detected in a Synergy H1 microplate reader (BioTek Instruments Inc., USA) and the data were analyzed using KlusterCaller 2.24 (KBioscience, UK).

Map construction and QTL analysis

ASNP-based genetic map was constructed using 60 selected RILs from the mapping population for preliminary QTL identification. Markers were discarded if they were monomorphic between parents or missing (treating heterozygous as missing) in either of two parents, contained > 20% missing data or showed minor allele frequencies < 0.2. The BIN function in IciMapping 4.1 (http://www.isbreeding.net/) was used to remove redun-dant markers (co-segregating markers) to reduce the complexity of calculation. Linkage analysis was performed with JoinMap 4.0 using the regression mapping algorithm. Linkage groups with less than five markers or markers with no linkage were discarded in the subsequent analysis. The remaining linkage groups were assigned to chromosomes based on the 660 K genetic map reported by Cui et al. (2017). Selective genotyping was subjected to marker-based analysis, in which trait means were compared between classes defined based on marker genotypes, or to 'trait-based' analysis, in which marker allele frequencies were compared between classes of progeny defined based on trait values (Navabi et al. 2009). Single marker analysis (SMA) and selective genotyping mapping (SGM) was performed with IciMapping 4.1 software to find potential chromosome regions responsible for TGW, GL, GW, and GFR. Phenotypic data for the mapping population obtained from three locations in cropping season 2014-2015 were used to declare significant associations between marker genotypes and traits, with default LOD thresholds of 2.5 in SMA and 5.0 in SGM. Proportions of the bottom and top tails used in SGM were set to 0.5. To minimize the probability of false positives in selective genotyping analysis and identify stable QTL, a QTL was declared and chosen for further confirmation only when at least two closely linked SNP markers simultaneously showed significant associations with the TGW-related traits and at least one marker was detected in two or more environments.

The genetic maps used for QTL confirmation were constructed with JoinMap 4.0 using 26 KASP markers (Table S3) that were converted from SNP markers closely linked to the preliminarily identified QTL. QTL analysis was performed by inclusive composite interval mapping (ICIM) using IciMapping 4.1. Phenotypic data obtained from individual environments and the best linear unbiased estimators (BLUEs) across 10 (TGW, GL, GW, and GNS) or six (GFR) environments were used for QTL detection. A LOD threshold of 2.0 was set based on 1000 permutation tests at $P < 0.01$.

Genotypes of KASP markers were merged with those of the 90 K SNP array, and new linkage maps of chromosome 5B were generated for Zhou 8425B/Chinese Spring and Doumai/Shi 4185 populations. Map construction and QTL analysis followed the procedures described in previous reports (Gao et al. 2015; Li et al. 2018).

Physical positions of mapped SNPs in the QTL regions were obtained by blasting SNP flanking sequences against the Chinese Spring reference genome sequence (IWGSC2018).

Statistical analysis

Phenotypic data analyses were conducted with SAS 9.2 soft-ware (SAS Institute Inc, Cary, NC, USA). PROC MIXED was used in the analysis of variance (ANOVA) to evaluate the contributions of lines (RILs) and environments, where environments, lines, line × environment interaction and replicates nested in environments were all considered as random effects. In parallel, a model considering lines as fixed factors was fitted for estimating BLUEs of lines across environments. Adjusted means of each line for each trait in individual environments were separately computed with PROC MIXED. Original phenotypic data

obtained from the 2014-2015 cropping season, adjusted mean phenotypic data of each environment obtained from cropping seasons 2015-2016 and 2016-2017 were used for broad-sense heritability (h_b^2) estimates and Pearson's correlation analyses. Broad-sense heritability on a genotype mean basis was estimated following Holland et al. (2003). Genotypic and phenotypic correlation coefficients among different traits were also estimated (Holland 2006). Homozygous lines in the ZM871/ZM895//ZM871 and ZM871/ZM895//ZM895 populations were used to verify QTL effects. The differences in TGW, GL, GW, GFR, and GNS between two classes of genotypes (homozygous for ZM871 and homozygous for ZM895) were calculated by PROC MIXED, treating genotypes as fixed effects, and lines nested in genotypes, environments, environment-related interactions, and replicates nested in environments as random effects. In the advanced breeding lines, QTL effects on TGW, GL, GW and GNS were evaluated by performing Student's t tests. The effects of the 2BS and 5B QTL were evaluated at individual marker level instead of interval level because of frequent recombination. Generally, QTL repeatedly detected in different environments and/or across multiple genetic backgrounds were considered to be stable. In the present study, a QTL was considered to be major and stable when it was detected in more than three environments and had significant effects in at least one set of the validation materials, accounting for more than 10% of the phenotypic variation. Associations between KASP markers and phenotypic values in the Zhou 8425B/Chinese Spring and Doumai/Shi 4185 populations were determined by Student's t tests.

Results

Phenotypic evaluations

ANOVA showed that line and line × environment interaction effects were significant for TGW, GL, GW, GFR, and GNS at $P < 0.001$, and environment effects were significant for TGW, GL, GW, and GNS at $P < 0.05$ (Table S4). All traits had broad-sense heritabilities exceeding 0. 85. Better among-environment correlations were observed for TGW-related traits than for GNS. Among TGW-related traits, TGW and GL had better among-environment correlations (Table S5). Pearson's correlation coefficients among environments ranged from 0. 30 to 0. 89 for TGW-related traits and from 0. 26 to 0. 75 for GNS.

Larger and heaviergrains, faster grain filling and more grains per spike are favorable for breeders. ZM895 had larger grain size, higher TGW, and GFR, but lower GNS than ZM871 under all the environments tested. Continuous distribution and transgressive segregation were observed for TGW, GL, GW, and GFR, indicating polygenic inheritance (Fig. S2). GNS showed a more-or-less bimodal distribution, suggesting the presence of potential major QTL for GNS in the ZM871/ZM895 RIL population (Fig. S2).

At both the phenotypic and genotypic levels, positive correlations among TGW-related traits and negative correlations between GNS and TGW-related traits were observed (Table S6). TGW, GW, and GFR were highly correlated with each other ($r = 0. 75$-0. 96) except GW with GFR ($r = 0. 64$) at the phenotypic level. GL was moderately correlated with GFR ($r = 0. 48$ and 0. 38) and weakly correlated with GW

($r = 0. 18$ and 0. 26). Weak to moderate negative correlations were observed between TGW and GNS ($r = -0. 37$ and $-0. 26$).

SNP-based genetic map construction and QTL identification

After filtering the genotypic data, 39, 189 high-quality polymorphic markers from the Wheat 660 K SNP chip were employed for subsequent analysis. By performing Bin function, 5745 non-redundant markers were identified and used for linkage analysis, of which 4231 were grouped into 65 linkage groups representing all chromosomes except 3D (Table S7; Fig. 1).

Eighty-one and 76 markers were significantly associated with TGW-related traits in SMA and SGM, of which 57 (70%) and 56 (74%) were mapped on four chromosomes 1AL, 2BS, 3AL and 5B, each containing 2-3 QTL (Tables 1, S8, S9). Thirty-two markers (56. 1 and 57. 1%) and five QTL (62. 5 and 71. 4%) in these four genetic regions were common between SMA and SGM. Collectively, the 1AL and 2BS QTL regions showed significant effects on TGW and GW, whereas the 3AL and 5B QTL regions had significant effects on TGW and GL. Two GFR QTL were identified on chromosomes 2BS and 3AL. All favorable alleles came from ZM895.

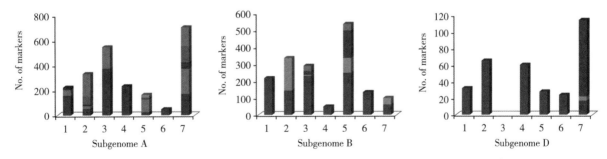

Fig. 1   Distribution of the 4231 loci in the 65 linkage groups belonging to 20 chromosomes. Linkage groups of the same chromosome are shown in different colors

Table 1   Four genomic regions harboring QTL for TGW, GL, GW and GFR identified by selective genotyping

Chromosome	Trait	QTL	SMA		SGM	
			Position (Mb)[a]	No. [b]	Position (Mb)	No.
1AL	TGW	*Qtgw. caas-1AL*	304. 2-461. 0	2 (1)		
	GW	*Qgw. caas-1AL*	203. 3-461. 0	8 (4)	202. 9-460. 6	16 (4)
2BS	TGW	*Qtgw. caas-2BS*	32. 9-44. 4	9 (5)	32. 9-46. 6	10 (4)

（continued）

Chromosome	Trait	QTL	SMA		SGM	
			Position（Mb）[a]	No.[b]	Position（Mb）	No.
	GW	*Qgw. caas-2BS*	32. 9-46. 6	5（1）		
	GFR	*Qgfr. caas-2BS*	32. 9-46. 6	13（6）	32. 9-44. 4	10（4）
3AL	TGW	*Qtgw. caas-3AL*			502. 5-503. 0	4（2）
	GL	*Qgl. caas-3AL*	502. 5-503. 5	5（2）	502. 5	1（0）
	GFR	*Qgfr. caas-3AL*			502. 4-503. 0	5（0）
5B	TGW	*Qtgw. caas-5B*	327. 1-466. 3	2（0）	35. 2-289. 1	2（0）
	GL	*Qgl. caas-5B*	18. 4-466. 6	13（3）	18. 4-378. 9	8（3）

TGW，thousand grain weight；*GL*，grain length；*GW*，grain width；*GFR*，mean grain filling rate

SMA，single marker analysis；*SGM*，selective genotyping mapping

[a] Physical positions （Mb） were obtained by blasting SNP flanking sequences against the Chinese Spring RefSeq v1. 0 sequence （IWGSC 2018）

[b] Number of markers significantly associated with corresponding traits and repeatedly detected in≥2 environments （shown in brackets）

QTL confirmation

To confirm the preliminarily identified QTL based on two tails of the mapping population，tightly linked SNP markers were converted into KASP markers for QTL analysis. In total，26 KASP markers were used for genetic map construction by genotyping 266 RILs of the ZM871/ZM895 population. The resulting linkage maps represented segments of chromosomes 1AL, 2BS，3AL and 5B which contained four，six，seven and nine markers，spanning 2. 2，6. 6，8. 2 and 19. 3 cM in length，respectively （Fig. 2）.

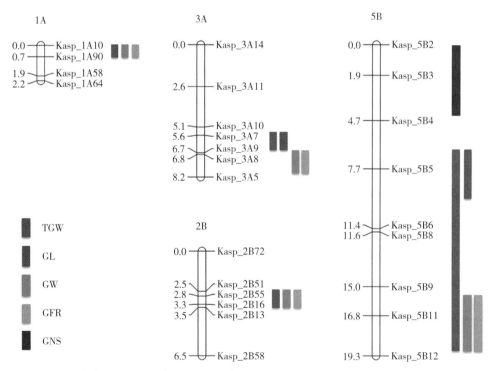

Fig. 2　Genetic maps of chromosomes showing QTL for TGW，GL，GW，GFR and GNS in the Zhongmai 871/ Zhongmai 895 RIL population.

　　TGW thousand grain weight，*GL* grain length，*GW* grain width，*GFR* mean grain filling rate，*GNS* grain number per spike

Inclusive composite interval mapping （ICIM） indicated that chromosomes 1AL，2BS，3AL，and 5B contained QTL for TGW，GL，GW and/or GFR with favorable alleles from ZM895 （Tables 2，S10；Fig. 2）. The QTL on chromosomes 1AL and 2BS showed significant effects on TGW, GW, and GFR, but no significant effect on GL. *Qgw. caas-1AL* was a major QTL associated with GW explaining 5.0-20.6% of the phenotypic variation. Chromosome 3AL possessed three QTL associated with TGW, GL, and GW, respectively. *Qgl. caas-3AL*, a GL QTL detected in all 10 environments, explained 5.7-15.7% of the phenotypic variation. Two major QTL for TGW and GL, respectively, and a minor QTL for GFR were identified on chromosome 5B. *Qtgw. caas-5B* and *Qgl. caas-5BS*, observed in all 10 environments, explained 5.7-17.1% and 12.0-19.3% of the phenotypic variation of TGW and GL, respectively.

No epistatic interaction among different QTL was identified using IciMapping 4.1, indicating all the QTL had additive effects. There was a linear relationship between phenotype and the number of favorable alleles (Fig. 3); with the addition of each favorable allele additively contributing to enhanced phenotype values.

The RILs carrying positive alleles at all four QTL regions exhibited 17.7% higher TGW (7.6 g), 4.1% higher GL (0.29 mm), 5.8% higher GW (0.20 mm), and 23.8% higher GFR (0.3 g/day) than those possessing contrasting alleles.

Pleiotropic effects on GNS

QTL mapping for GNS was conducted using the new KASP linkage maps for four chromosomes to determine whether these QTL regions had significant effects on GNS. A major GNS QTL, explaining 5.2-15.2% of the phenotypic variation, was identified on chromosome 5BS with ZM871 con-tributing the favorable allele (Tables 2, S10; Fig. 2). RILs carrying ZM871 alleles had 5.7% higher GNS (2.5 grains) than those having ZM895 alleles. Interestingly, this locus did not share a common marker interval with *Qtgw. caas-5B* in seven of the eight environments (Table S10), suggesting that variation of TGW and GNS was probably controlled by different genes.

Table 2 QTL for TGW, GL, GW, GFR, and GNS identified on chromosomes 1AL, 2BS, 3AL and 5B using the entire mapping population

Chromosome	Trait	QTL[a]	Maker Interval	Physical Interval (Mb)[b]	No. [c]	LOD	PVE (%)[d]	Add[e]
1AL	TGW	*Qtgw. caas-1AL*	*Kasp _ 1A10-Kasp _ 1A90*	307.8-356.7	9/10	3.1-9.2	5.2-15.0	0.7-1.2
	GW	*Qgw. caas-1AL*	*Kasp _ 1A10-Kasp _ 1A90*	307.8-356.7	9/10	2.8-13.0	5.0-20.6	0.03-0.05
	GFR	*Qgfr. caas-1AL*	*Kasp _ 1A10-Kasp _ 1A90*	307.8-356.7	5/6	2.8-7.5	4.9-12.2	0.02-0.03
2BS	TGW	*Qtgw. caas-2BS*	*Kasp _ 2B55-Kasp _ 2B16*	41.4-44.3	8/10	2.4-5.6	4.4-9.3	0.7-0.9
	GW	*Qgw. caas-2BS*	*Kasp _ 2B55-Kasp _ 2B16*	41.4-44.3	6/10	2.4-4.8	4.2-8.1	0.02-0.03
	GFR	*Qgfr. caas-2BS*	*Kasp _ 2B55-Kasp _ 2B16*	41.4-44.3	4/6	2.6-6.8	4.6-11.2	0.02-0.03
3AL	TGW	*Qtgw. caas-3AL*	*Kasp _ 3A7-Kasp _ 3A9*	497.7-511.0	8/10	2.8-7.1	7.2-13.3	0.7-1.1
	GL	*Qgl. caas-3AL*	*Kasp _ 3A7-Kasp _ 3A9*	497.7-511.0	10/10	3.1-9.4	5.7-15.7	0.05-0.08
	GW	*Qgw. caas-3AL*	*Kasp _ 3A8-Kasp _ 3A5*	516.1-533.0	6/10	2.6-7.3	4.8-12.3	0.02-0.03
	GFR	*Qgfr. caas-3AL*	*Kasp _ 3A8-Kasp _ 3A5*	516.1-533.0	2/6	3.7-4.0	6.2-6.9	0.02-0.03
5B	TGW	*Qtgw. caas-5B*	*Kasp _ 5B5-Kasp _ 5B12*	45.3-394.2	10/10	3.2-10.6	5.5-17.3	0.7-1.4
	GL	*Qgl. caas-5BS*	*Kasp _ 5B5-Kasp _ 5B6*	45.3-68.8	10/10	7.2-12.1	12.5-20.5	0.07-0.10
	GW	*Qgw. caas-5B*	*Kasp _ 5B11-Kasp _ 5B12*	383.0-394.2	3/10	3.5-3.9	6.0-6.5	0.02-0.03
	GFR	*Qgfr. caas-5B*	*Kasp _ 5B11-Kasp _ 5B12*	383.0-394.2	4/6	2.8-7.8	4.7-13.0	0.02-0.04
	GNS	*Qgns. caas-5BS*	*Kasp _ 5B3-Kasp _ 5B4*	35.1-42.1	8/10	2.7-9.5	5.2-15.2	-1.6 to- 1.0

TGW thousand grain weight, *GL* grain length, *GW* grain width, *GFR* mean grain filling rate, *GNS* grain number per spike

[a] QTL identified in more than three environments are shown in bold and QTL detected in the analysis of across-environment BLUEs (best linear unbiased estimators) are underlined

[b] Physical intervals (Mb) were obtained by blasting SNP flanking sequences against the Chinese Spring RefSeq v1.0 sequence (IWGSC 2018)

[c] Number of detected environments among total environments

[d] Phenotypic variation explained by the QTL

[e] Estimated additive effect of the QTL. Positive and negative values indicate favorable allele coming from Zhongmai 895 and Zhongmai 871, respectively

QTL validation

ANOVA of data from the BC_1F_4 populations indicated a significant influence of genotypes on TGW, GL, GW, GFR, and GNS (Table 3). Significant differences between the ZM871 and ZM895 genotypes in TGW, GW and GFR at the 1AL locus, in TGW, GL, GW, and GFR at the 2BS locus, and in TGW, GL and GW at the 3AL locus were present in both populations. Lines with homozygous ZM895 alleles exhibited significantly higher phenotypic values than those with ZM871 alleles irrespective of QTL region, with the differences ranging from 1.3 to 2.3 g for TGW, from 0.11 to 0.14 mm for GL, from 0.04 to 0.10 mm for GW, and from 0.06 to 0.11 g/day for GFR. Unexpectedly, a significant negative effect on GNS contributed by the ZM895 allele was observed for the 3AL QTL in both populations and for the 1AL and 2BS QTL in the ZM871/ZM895//ZM895 population, with differences ranging from 1.3 to 1.9 grains per spike. Differences in GL (0.15 and 0.16 mm) and GNS (2.1 and 2.4 grains per spike) associated with the 5B QTL were significant in both populations, whereas differences in TGW (2.0 g), GW (0.04 mm) and GFR (0.04 g/day) were significant only in the ZM871/ZM895//ZM895 population. Lines homozygous for the ZM895 5B allele had larger grain size, higher rate of grain filling and grain weight, but lower GNS than those possessing the ZM871 allele.

Experiments on the advanced breeding lines provided further evidence for significant effects of all four QTL (Table 4). At the 1AL locus, ZM895 allele was significantly associated with higher TGW (4.0 and 5.1 g), GL (0.32 and 0.42 mm) and GW (0.10 and 0.12) at both locations. Eighty-one percent of advanced breeding lines grown at Anyang and Xinxiang had the ZM895 genotype, indicating a strong past, positive field selection on the ZM895 allele. The ZM895 genotype for the 3AL QTL was also present in high frequencies at both locations (72 and 80%, respectively). Significant differences in GW (0.07 and 0.21 mm) were observed at both locations, whereas differences in TGW (5.5 g), GL (0.33 mm) and GNS (3.0 grains) were identified only at Xinxiang, with the ZM895 allele contributing positive effects on TGW-related traits and negative effect on GNS.

Effects of the 2BS and 5B QTL were evaluated by investigating the association between marker genotype and phenotype because frequent recombination was apparent among markers used to genotype the lines. As indicated by *Kasp _ 2B55* in the 2BS QTL region, lines carrying homozygous alleles from ZM895 had significantly higher TGW (2.0 and 2.4 g), GL (0.23 and 0.36 mm) and GW (0.05 and 0.06 mm) than those with ZM871 alleles. The ZM895 genotype was present at lower frequencies (29 and 37%, respectively). For the 5B QTL, *Kasp _ 5B4* and *Kasp _ 5B8* were significantly associated with GL and GNS at both locations, with differences between the two genotypes ranging from 0.13 to 0.28 mm for GL and from 1.9 to 3.4 grains per spike for GNS, respectively. In addition, significant differences in TGW (3.0 and 2.8 g) and GW (0.10 mm) were detected between the ZM895 and ZM871 genotypes at the *Kasp _ 5B8* locus. Seventy-nine and 84% of advanced breeding lines tested at the two locations, respectively, had the ZM895 genotype at the *Kasp _ 5B4* locus compared with 54 and 61% at the *Kasp _ 5B8* locus, indicating strong selection on the ZM895 allele at *Kasp _ 5B4* locus.

Comparison of the 5B QTL

ATGW QTL, flanked by *wsnp _ Ra _ c5634 _ 9952011* and *RAC875 _ c14882 _ 275*, was previously identified in the Zhou 8425B/Chinese Spring population (Gao et al. 2015). In the present study, *RAC875 _ c14882 _ 275* was 4.3 cM from *Kasp _ 5B11* and 5.7 cM from *Kasp _ 5B12* on the new linkage map of chromosome 5B (Fig. S3a). Moreover, the TGW QTL was mapped to a marker interval (*JD _ c20126 _ 516-Kukri _ rep _ c105540 _ 177*) next to the original one, with Zhou 8425B contributing the favorable allele (Fig. S3a). This QTL was detected in two environments, explaining 5. 2 and 8. 9% of the phenotypic variation, respectively, in agreement with previous results. Zhou 8425B had the ZM871 genotype, whereas Chinese Spring had the ZM895 genotype at *Kasp _ 5B11* and *Kasp _ 5B12* loci. RILs with the ZM871 genotype exhibited significantly higher TGW than those with the ZM895 genotype (Fig. S3b, c). It is possible that the QTL for TGW in the Zhou 8425B/Chinese Spring and ZM871/ZM895 populations are controlled by the same gene.

Previously, a QTL for GL was mapped near *Excalibur _ c4232 _ 2834* in the Doumai/Shi 4185 population (Li et al. 2018). This locus was confirmed using the new linkage map comprising *Kasp _ 5B4* and *Kasp _ 5B8* (Fig. S4a), with the favorable allele from Doumai. *Kasp _ 5B8*, 0. 8 cM from *Excalibur _ c4232 _ 2834*, showed significant effects on GL in all six environments (Fig. S4c). *Kasp _ 5B4* was 8. 2 cM from *Excalibur _ c4232 _ 2834*, significantly associated with GL in four out of the six environments (Fig. S4b). Shi 4185 had the ZM871 genotype, whereas Doumai had the ZM895 genotype at *Kasp _ 5B4* and *Kasp _ 5B8* loci. Lines with the ZM895 genotype exhibited significantly higher GL than those with the ZM871 genotype (Fig. S4b, c), suggesting that the QTL for GL in the Dou-mai/Shi 4185 and ZM871/ZM895 populations are likely the same.

Table 3   Comparison of TGW, GL, GW, GFR and GNS between Zhongmai 895 (ZM895) and Zhongmai 871 (ZM871) genotypes in the two BC_1F_4 populations

QTL	Genotype	No. [a]	TGW (g)	GL (mm)	GW (mm)	GFR (g/day)	GNS
ZM871/ZM895//ZM871							
1AL	ZM871	74	44. 4±2. 8[b]	7. 09±0. 19	3. 48±0. 09	1. 34±0. 08	48. 2±2. 7
	ZM895	24	46. 5±3. 6	7. 13±0. 25	3. 55±0. 10	1. 41±0. 11	47. 7±3. 0
			2. 1** [c]	0. 04	0. 07***	0. 07***	−0. 6
2BS	ZM871	64	44. 3±2. 5	7. 05±0. 19	3. 49±0. 09	1. 35±0. 08	48. 4±2. 6
	ZM895	22	46. 6±3. 8	7. 16±0. 23	3. 54±0. 11	1. 40±0. 11	47. 2±3. 2
			2. 3**	0. 11*	0. 06*	0. 06**	−1. 2
3AL	ZM871	67	44. 6±2. 6	7. 06±0. 19	3. 49±0. 09	1. 36±0. 08	48. 2±2. 4
	ZM895	28	46. 7±3. 5	7. 19±0. 19	3. 56±0. 10	1. 40±0. 11	46. 8±3. 0
			2. 1**	0. 13**	0. 06**	0. 04	−1. 4*
5B	ZM871	78	45. 0±2. 9	7. 07±0. 20	3. 51±0. 09	1. 37±0. 09	48. 2±2. 8
	ZM895	18	46. 4±4. 1	7. 22±0. 20	3. 52±0. 12	1. 39±0. 12	45. 8±1. 8
			1. 4	0. 15**	0. 02	0. 02	−2. 4***
ZM871/ZM895//ZM895							
1AL	ZM871	32	47. 8±2. 7	7. 17±0. 18	3. 58±0. 09	1. 41±0. 06	45. 0±3. 6
	ZM895	102	51. 2±2. 8	7. 25±0. 18	3. 68±0. 07	1. 51±0. 08	43. 1±2. 6
			1. 5***	0. 07	0. 10***	0. 11***	−1. 9**
2BS	ZM871	33	49. 0±2. 8	7. 15±0. 17	3. 62±0. 10	1. 45±0. 07	44. 3±3. 1
	ZM895	98	51. 2±2. 7	7. 29±0. 18	3. 67±0. 07	1. 51±0. 08	42. 9±2. 4
			2. 3***	0. 14***	0. 05**	0. 06***	−1. 4**
3AL	ZM871	36	49. 5±3. 7	7. 15±0. 17	3. 62±0. 12	1. 48±0. 09	44. 6±3. 5
	ZM895	105	50. 8±2. 9	7. 27±0. 19	3. 66±0. 07	1. 49±0. 08	43. 2±2. 6
			1. 3*	0. 12***	0. 04*	0. 02	−1. 3*
5B	ZM871	27	49. 1±3. 5	7. 15±0. 21	3. 63±0. 10	1. 47±0. 09	44. 8±3. 5
	ZM895	92	51. 1±2. 3	7. 30±0. 17	3. 66±0. 06	1. 50±0. 07	42. 7±2. 3
			2. 0***	0. 16***	0. 04*	0. 04*	−2. 1***

TGW, thousand grain weight; *GL*, grain length; *GW*, grain width; *GFR*, mean grain filling rate; *GNS*, grain number per spike

[a] Number of lines with corresponding genotypes

[b] Data are shown as means ± SD

[c] Phenotypic difference between ZM895 and ZM871 genotypes. Asterisks indicate significance determined by ANOVA. Significant at *, $P<0.05$; **, $P<0.01$; ***, $P<0.001$

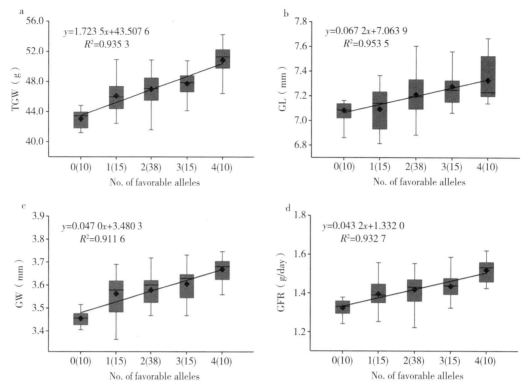

Fig. 3　Linear regressions between number of favorable alleles and across-environment BLUEs of TGW，GL，GW and GFR in the Zhongmai 871/Zhongmai 895 RIL population. *BLUE*，best linear unbiased estimator；*TGW*，thousand grain weight；*GL*，grain length；*GW*，grain width；*GFR*，mean grain filling rate. Numbers of lines carrying the corresponding number of favorable alleles are shown in brackets. x and Y in the equations represent number of favorable alleles and across-environment BLUEs of TGW （a），GL （b），GW （c） and GFR （d），respectively

Discussion

Selective genotyping is an economical and effective approach for QTL mapping

Using a mapping population with 266 RILs，a 11% selection proportion at each tail for extreme phenotype and a high-density genetic map，we identified QTL associated with TGW-related traits that explain 5-19% of the phenotypic variation，indicating the effectiveness of selective genotyping in genetic analysis of complex traits in wheat. Confirmation of QTL using the entire mapping population is not required，although it could provide a better estimation of QTL effects that are less accurately assessed in selective genotyping. The probability of finding false positives decreases with increased numbers of markers that simultaneously show significant associations （Sun et al. 2010）. In the present study，all QTL were repre-

sented by more than one marker and were confirmed by ICIM. No other stable QTL was identified by rerunning the selective genotyping analysis using phenotypic data from six or 10 environments （data not shown），in agreement with the similar among-environment correlations （Table S5） of the phenotypic data for 60 extreme lines observed in the cropping season 2014-2015 and those in the following seasons.

Compared with conventional QTL mapping，selective genotyping is cost-effective when the ratio of genotyping to phenotyping costs is higher than one （Gallais et al. 2007）. The overall expenses could be further reduced by excluding an appropriate proportion of individuals with intermediate phenotypes after each round of evaluation （Myskow and Stojalowski 2016），or replacing complex and expensive techniques with quicker，easier and cheaper ones that are accurate e-nough for identifying extreme phenotypes from the intermediate phenotypes. However, we have to notice that selective genotyping is limited to only one or a few correlated traits in a study, while an entire population may need to be genotyped if many traits are considered.

Generally, a lot of crosses and selections are involved in breeding programs every year, resulting in many progenies or lines in which multiple favorable alleles from different genetic resources are present. Selective genotyping provides an excellent choice for breeders to explore these materials for QTL detection underlying the variation of targeted traits, making QTL identification a co-product of breeding programs. This is particularly attractive to breeders who are mainly interested in identification of QTL for marker-assisted selection of traits of interests (Gallais et al. 2007). Effectiveness and allele frequency changes of the QTL regions on chromosomes 1AL, 3AL and 5B in the advanced breeding lines indicate that combination of selective genotyping and breeding practice is feasible. Though marker-based and trait-based analyses are equally powerful in biparental populations (Tables 1, S8, S9; Navabi et al. 2009), the former may be more appropriate for breeding populations because not all loci respond to selection.

Comparison with previous reports

In the present study, QTL for TGW-related traits were mapped on chromosomes 1AL, 2BS, 3AL and 5B, and a QTL for GNS was mapped on chromosome 5BS. Previously identified QTL and cloned genes on the chromosomes mentioned above are summarized in Tables S11 and S12, respectively. In addition to consensus maps, the IWGSC (2018) Chinese Spring reference sequence was used as a common coordinating system for comparisons of QTL identified in different studies.

1AL QTL

The linkage map we generated for mapping the 1AL QTL contained only four KASP markers, spanning 2.2cM and corresponding to an interval of 307.8-439.0

Mb in the IWGSC reference sequences. This low recombination rate (0.017 cM/Mb) informed us that the 1AL QTL region located in the pericentromeric region. It is difficult to compare its position with previously reported QTL due to strong suppression of recombination and the poor relationship between physical and genetic distance of pericentromeric region (Campbell et al. 1999; Su et al. 2018; Wang et al. 2009; Xiao et al. 2011). Using the 1A consensus map of Somers et al. (2004) as an example, the marker order in intervals *Glu-A3-Xwmc24* and *Xwmc312-Xgwm99* are in accordance with their physical positions whereas those in *Xgwm357-Xcfd22*, a 13 cM interval correspond-ing to about 50-500 Mb, were not (Fig. S5a). In another high-density consensus map, a pericentromeric region of chromosome 1A corresponds to an about 1 cM interval covering 100-300 Mb (Fig. S5b; Maccaferri et al. 2015). We could not distinguish QTL when they co-located in 50-500 Mb, especially those with large confidence intervals identified using low-density maps (Wang et al. 2009; Xiao et al. 2011). *TaSnRK2.3-1A*, a homoeologue of plant-specific protein kinase gene *SnRK*, was reported to have significant effects on TGW and plant height (Miao et al. 2017). This gene corresponds to *TraesCS1A02G215900*, located at 381.8 Mb in the IWGSC reference sequences. No variation was detected by sequencing *TaSnRK2.3-1A* from ZM871 and ZM895, indicating that *TaSnRK2.3-1A* was not the gene underlying the 1AL QTL.

2BS QTL

The 2BS QTL located in the 41.4 to 44.3 Mb region over-lapped with several QTL identified for TGW and yield components (Cabral et al. 2018; Kumar et al. 2006; Prashant et al. 2012; Xu et al. 2017). A copy of the photoperiod response gene *Ppd-B1* was found at 56.2 Mb by blasting its sequence (Beales et al. 2007) against the IWGSC reference sequences. It is not likely that this is the causal gene underlying the 2BS QTL, because markers distributed from 50 to 100 Mb on chromosome 2B were not polymorphic in the mapping population and there was no significant effect on heading date. Another gene associated with TGW on the short arm of chromosome 2B is *TaSus2-2B* (Jiang et al. 2011). It was annotated as *TraesCS2B02G194200* and located at 171.0 Mb, clearly different from the present QTL. *Qgfr.caas-2B* probably represents a new locus because only one GFR QTL has been identified on chromosome 2B in *Xgwm148-Xgwm388* (100.8-555.7 Mb, Charmet et al. 2005).

3AL QTL

The 3AL QTL waslocated in a 2.6 cM interval that corresponds to 497.7-533.0 Mb in the IWGSC reference sequences (Fig. S5c). Although many QTL controlling TGW, GNS and other agronomic traits have been reported on chromosome 3A (Ali et al. 2011; Bennett et al. 2012; Gao et al. 2015; Huang et al. 2004; Jia et al. 2013; Li et al. 2018; Wu et al. 2012; Zhai et al. 2018; Zhang et al. 2014), most of them were either in the pericentromeric region (a-bout 100-450 Mb; Fig. S5c) or located in the distal regions of 3AS (<25 Mb) and 3AL (>625 Mb). In addition, Ma et al. (2018) mapped a QTL for plant height, spike length and TGW at 53.6-57.7 Mb and Ali et al. (2011) identified a QTL for GNS in 597.5-624.0 Mb. Few QTL for grain size or GFR have been documented on chromosome 3A compared to agronomic traits. Gegas et al. (2010) detected two QTL for grain size and shape close to *Xgwm2* at 60.2 Mb and *Xbarc19* at 310.7 Mb. Wang et al. (2009) identified a genetic region affecting GFR and GNS but not TGW in the interval *Xwmc505-Xwmc264* corresponding to 90.0-625.7 Mb. Two genes on 3AL, *Tackx4* and *TaTGW6-A1*, were reported to have significant effects on grain weight at 712.1 and 722.4 Mb, respectively (Chang et al. 2015; Hanif et al. 2015). The present QTL is likely to be a new locus associated with TGW, GL and GW.

Table 4　Comparison of TGW, GL, GW and GNS between Zhongmai 895 (ZM895) and Zhongmai 871 (ZM871) genotypes in advanced breeding lines

QTL[a]	Genotype	Anyang					Xinxiang				
		No.[b]	TGW (g)	GL (mm)	GW (mm)	GNS	No.	TGW (g)	GL (mm)	GW (mm)	GNS
1AL	ZM871	25	45.1±3.9[c]	6.61±0.23	3.47±0.15	31.7±4.1	20	46.4±4.5	6.50±0.21	3.50±0.17	36.4±3.7
	ZM895	103	49.2±3.9	6.93±0.30	3.57±0.12	31.1±4.2	79	51.5±3.7	6.92±0.29	3.62±0.11	37.0±3.9
			4.0***[d]	0.32***	0.10***	−0.6		5.1***	0.42***	0.12**	−0.6
2BS	ZM871	104	47.7±4.4	6.81±0.30	3.53±0.14	31.0±4.1	83	49.5±4.8	6.71±0.28	3.58±0.16	37.2±4.2
(Kasp_2B55)	ZM895	43	49.6±3.5	7.04±0.28	3.58±0.11	32.4±4.3	33	51.9±3.3	7.07±0.31	3.64±0.07	37.0±3.2
			2.0*	0.23***	0.05*	1.4		2.4**	0.36***	0.06**	−0.2
3AL	ZM871	37	47.6±5.0	6.81±0.31	3.50±0.15	30.7±4.4	15	45.0±3.9	6.52±0.17	3.41±0.14	40.2±4.4
	ZM895	96	48.6±3.8	6.88±0.31	3.58±0.13	32.0±3.9	77	50.6±4.2	6.86±0.34	3.62±0.13	37.2±3.5
			1.0	0.07	0.07**	1.3		5.5***	0.33***	0.21***	−3.0**
5B	ZM871	32	47.5±4.6	6.74±0.22	3.56±0.18	33.5±2.7	20	50.3±4.9	6.73±0.18	3.65±0.21	39.5±3.6
(Kasp_5B4)	ZM895	118	48.6±4.3	6.92±0.33	3.54±0.12	30.7±4.1	95	50.2±4.6	6.86±0.35	3.58±0.12	36.5±3.7
			1.2	0.19**	−0.01	−2.8***		−0.09	0.13*	−0.07	−3.0**
5B	ZM871	61	47.2±3.8	6.76±0.28	3.52±0.14	32.4±4.3	37	49.1±3.7	6.69±0.26	3.59±0.15	38.9±3.4
(Kasp_5B8)	ZM895	71	50.1±3.8	7.03±0.30	3.60±0.11	30.4±3.9	62	51.5±3.9	6.97±0.32	3.61±0.11	35.5±3.7
			3.0***	0.27***	0.10***	−1.9**		2.4**	0.28***	0.02	−3.4***

TGW, thousand grain weight; *GL*, grain length; *GW*, grain width; *GNS*, grain number per spike

[a] Only significant loci are shown

[b] Number of lines with corresponding genotypes

[c] Data are shown as means ± SD

[d] Phenotypic difference between ZM895 and ZM871 genotypes. Asterisks indicate significance determined by *t*-test. Significant at **, $P < 0.01$; ***, $P < 0.001$

5B QTL

We constructed a 19.3 cM linkage map spanning 27.5 to 394.2 Mb in chromosome 5B on which QTL for TGW, GL, GW, GFR, and GNS were identified (Table 2; Fig. S5d). Many stable QTL for TGW have been identified in the 5B QTL region (Cui et al. 2014; Huang et al. 2003; Ma et al. 2018; Prashant et al. 2012; Quarrie et al. 2005; Su et al. 2018; Wu et al. 2015; Zhai et al. 2018) and some coincide with QTL for GL and/or GW (Cui et al. 2014; Prashant et al. 2012; Su et al. 2018; Wu et al. 2015; Zhai et al. 2018). This genetic region co-located with a TGW QTL in the Zhou 8425B/Chinese Spring population and a GL QTL in the Doumai/Shi 4185 population (Gao et al. 2015; Li et al. 2018). Zhou 8425B contributed the favorable allele, but its genotype was different from ZM895. There are several possible explanations for this phenomenon. The peaks of LOD contours presented near *Kasp _ 5B12* in the ZM871/ZM895 population and between *JD _ c2012 _ 516* and *RAC875 _ 14882 _ 275* in the Zhou 8425B/Chinese Spring population are 5-20 cM apart. We cannot rule out the possibility for the presence of two different genes responsible for the QTL, which is supported by many reported QTL in the 5B QTL region. Co-location at the present mapping level cannot guarantee the same gene. Relative effects caused by different alleles of the same gene could also lead to this result. *TaSAP7-B*, corresponding to *TraesCS5B02G200000* at 360.9 Mb on chromosome 5B, was significantly associated with TGW and plant height (Wang et al. 2018). However, no difference was observed in the complete coding sequence and partial promoter sequence (about 700 bp upstream from ATG) between ZM871 and ZM895. Both carried the supe-rior allele. Wang et al. (2009) identified two QTL for GFR in marker intervals *Xcfd7-Tx37-38* and *Xbarc232-Xbarc275*, respectively. *Xbarc74* (402.7 Mb) was 14.1 cM from *Xcfd7* and 80.4 cM from *Xbarc232* at 619.8 Mb, suggesting that *Qgfr. caas-5B* may be a new QTL for GFR. As for GNS, several environment-specific QTL have been detected in the pericentromeric region of chromosome 5B (Cui et al. 2014; Li et al. 2015; Tang et al. 2011); none of them located in the 35.1-42.1 Mb (*Kasp _ 5B3-Kasp _ 5B4*) interval. *DEP1* gene plays important roles in regulating grain number per panicle and grain yield in rice (Huang et al. 2009). One of its homologs is located at 378.5 Mb (between *Kasp _ 5B9* and *Kasp _ 5B11*) on chromosome 5B, and at least 10 cM from *Kasp _ 5B4* (42.1 Mb). There is no direct evidence that *DEP1* plays the same role in wheat, indicating that *DEP1* is unlikely to be the gene underlying *Qgns. cass-5BS*. These results suggest that *Qgns. caas-5BS* is a new GNS QTL.

Applications in wheat breeding

Major stable QTL for yield-related traits and their tightly linked markers are of high importance in molecular breeding. In this study, the QTL for TGW, GW, and GFR on chromosome 1AL showed constant effects on TGW and GW and negligible TGW-GNS tradeoffs in different genetic backgrounds, and were strongly selected in breeding, represents a valuable target for MAS to enhance grain size and weight. The availability of time-saving and cost-effective KASP markers could facilitate its use in wheat breeding.

As shown in the present and previous studies, the 5B QTL was located in an important but complex region with more than one gene responsible for TGW, GL, and GNS (Table S11). Results from the Zhou 8425B/Chinese Spring population indicated that associations between the favorable allele and marker genotype depended upon the genetic background. This is a major limitation in the application of MAS in breeding (Liu et al. 2012). In the present study, *Qtgw. caas-5B*, *Qgl. caas-5BS* and *Qgns. caas-5BS* were major stable loci located in different marker intervals. A more precise delimitation of these QTL is needed to determine if they are caused by the same or closely linked genes before using them to improve grain weight and size or GNS. Therefore, to initiate fine mapping of these QTL, two $BC_1 F_4$ lines from the ZM871/ZM895//ZM871 population with residual heterozygosity at the 5B QTL were identified and self-pollinated to generate heterogeneous inbred lines（HILs）. The a-vailability of genome sequences and high-throughput KASP genotyping system will make fine mapping easier and faster. Exploring high-resolution genotyping data in large diversity collections showed great potential in candidate gene elucidation（Voss-Fels et al. 2019）. Application of this approach in our future fine mapping work may save years of self-pollinating, genotyping and phenotyping.

❖ Acknowledgements

The authors are grateful to Prof. R. A. McIntosh, Plant Breeding Institute, University of Sydney, for critical review of this manuscript. This work was funded by the National Basic Research Program of China(2014CB138105), National Natural Science Foundation of China（31461143021）, National Key Research and Development Programs of China（2016YFD0101802, 2016YFD0100502, 2016YFE0108600）, and CAAS Science and Technology Innovation Program.

❖ References

Ali ML, Baenziger PS, Ajlouni ZA, Campbell BT, Gill KS, Eskridge KM, Mujeeb-Kazi A, Dweikat I（2011）Mapping QTL for agronomic traits on wheat chromosome 3A and a comparison of recombinant inbred chromosome line populations. Crop Sci 51: 553-566. https: //doi. org/10. 2135/cropsci2010. 06. 0359

Beales J, Turner A, Griffiths S, Snape JW, Laurie DA（2007）A pseudo-response regulator ismisexpressed in the photoperiod insensitive *Ppd-D1a* mutant of wheat（*Triticum aestivum* L.）. Theor Appl Genet 115: 721-733. https: //doi. org/10. 1007/ s0012 2-007-0603-4

Bennett D, Izanloo A, Reynolds M, Kuchel H, Langridge P, Schnur-busch T（2012）Genetic dissection of grain yield and physical grain quality in bread wheat（*Triticum aestivum* L.）under water-limited environments. Theor Appl Genet 125: 255-271. https: //doi. org/10. 1007/s00122-012-1831-9

Bhusal N, Sarial AK, Sharma P, Sareen S（2017）Mapping QTLs for grain yield components in wheat under heat stress. PLoS ONE 12: e0189594. https: //doi. org/10. 1371/journal. pone. 0189594

Brinton J, Uauy C（2018）A reductionist approach to dissecting grain weight and yield in wheat. J Integr Plant Biol 61: 337-358. https: //doi. org/10. 1111/ jipb. 12741

Cabral AL, Jordan MC, Larson G, Somers DJ, Humphreys DG, McCartney CA（2018）Relationship between QTL for grain shape, grain weight, test weight, milling yield, and plant height in the spring wheat cross RL4452/ 'AC Domain'. PLoS ONE 13（1）: e0190681. https: //doi. org/10. 1371/journal. pone. 0190681

Campbell KG, Bergman CJ, Gualberto DG, Anderson JA, Giroux MJ, Hareland G, Fulcher RG, Sorrells ME, Finney PL（1999）Quanti-tative trait loci associated with kernel traits in a soft×hard wheat cross. Crop Sci 39: 1184-1195. https: //doi. org/10. 2135/cropsci199 9. 0011183X003900040039x

Chang C, Lu J, Zhang HP, Ma CX, Sun GL（2015）Copy number variation of cytokinin oxidase gene *Tackx4* associated with grain weight and chlorophyll content of flag leaf in common wheat. PLoS ONE 10: e0145970. https: //doi. org/10. 1371/ journ al. pone. 0145970

Charmet G, Robert N, Branlard G, Linossier L, Martre P, Triboï E（2005）Genetic analysis of dry matter and nitrogen accumulation and protein composition in wheat kernels. Theor Appl Genet 111: 540-550. https: //doi. org/10. 1007/s00122-005-2045-1

Chastain TG, Ward KJ, Wysocki DJ（1995）Standestblishment response of soft white winter wheat to seedbed residue and seed size. Crop Sci 35: 213-218. https: //doi. org/10. 2135/cropsci199 5. 0 011183X003500010040x

Cheng RR, Kong ZX, ZhangLW, Xie Q, Jia HY, Yu D, Huang YL, Ma ZQ（2017）Mapping QTLs controlling kernel dimensions in a wheat intervarietal RIL mapping population. Theor Appl Genet 130: 1405-1414. https: //doi. org/10. 1007/s00122-017-

2896-2

Cui F, Zhao CH, Ding AM, Li J, Wang L, Li XF, Bao YG, Li JM, Wang HG (2014) Construction of an integrative linkage map and QTL mapping of grain yield-related traits using three related wheat RIL populations. Theor Appl Genet 127: 659-675. https: //doi. org/10. 1007/s00122-013-2249-8

Cui F, Zhang N, Fan XL, Zhang W, Zhao CH, Yang LJ, Pan RQ, Chen M, Han J, Zhao XQ, Ji J, Tong YP, Zhang HX, Jia JZ, Zhao GY, Li JM (2017) Utilization of a Wheat660K SNP array-derived high-density genetic map for high-resolution mapping of a major QTL for kernel number. Sci Rep 7: 3788. https: //doi. org/10. 1038/s4159 8-017-04028-6

Dong Y, Liu JD, Zhang Y, Geng HW, Rasheed A, Xiao YG, Cao SH, Fu LP, Yang J, Wen WE, Zhang Y, Jing RL, Xia XC, He ZH (2016) Genome-wide association of stem water soluble carbo-hydrates in bread wheat. PLoS ONE 11 (11): e-0164293. https: //doi. org/10. 1371/journal. pone. 0164293

Doyle JJ, Doyle JL (1987) A rapid DNA isolation procedure for smallquantities of fresh leaf tissue. Phytochem Bull 19: 11-15

Farkhari M, Krivanek A, Xu YB, Rong Th, Naghavi MR, Samadi BY, Yl Lu (2013) Root-lodging resistance in maize as an example for high-throughput genetic mapping via single nucleotide polymorphism-based selective genotyping. Plant Breed 132: 90-98. https: //doi. org/10. 1111/pbr. 12010

Fischer RA (2008) The importance of grain or kernel number in wheat: a reply to Sinclair and Jamie-son. Field Crop Res 105: 15-21. https: //doi. org/10. 1016/j. fcr. 2007. 04. 002

Fischer RA, Edmeades GO (2010) Breeding and cereal yield progress. Crop Sci 50: 85-98. https: //doi. org/10. 2135/cropsci2009. 10. 0564

Gallais A, Moreau L, Charcosset A (2007) Detection of marker-QTL associations by studying change in marker frequencies with selection. Theor Appl Genet 114: 669-681. https: //doi. org/

10. 1007/s00122-006-0467-z

Gao FM, Wen WE, Liu JD, Rasheed A, Yin GH, Xia XC, Wu XX, He ZH (2015) Genome-wide linkage mapping of QTL for yield components, plant height and yield-related physiological traits in the Chinese wheat cross Zhou 8425B/Chinese Spring. Front PlantSci 6: 1099. https: //doi. org/10. 3389/fpls. 2015. 01099

Gao FM, Ma DY, Yin GH, Rasheed A, Dong Y, Xiao YG, Wu XX, Xia XC, He ZH (2017) Genetic progress in grain yield and physiological traits in Chinese wheat cultivars of Southern Yellow andHuai valley winter wheat zone since 1950. Crop Sci 57: 760-773. https: //doi. org/10. 2135/crops-ci2016. 05. 0362

Gegas VC, Nazari A, Griffiths S, Simmonds J, Fish L, Orford S, Sayers L, Doonan JH, Snape JW (2010) A genetic framework for grain size and shape variation in wheat. Plant Cell 22: 1046-1056. https: //doi. org/10. 1105/tpc. 110. 074153

Gimhani DR, Gregorio GB, Kottearachchi NS, Samarasinghe WLG (2016) SNP-based discovery of salinity-tolerant QTLs in a biparental population of rice (Oryza sativa). Mol Genet Genomics 291: 2081-2099. https: //doi. org/10. 1007/s00438-016-1241-9

Griffiths S, Wingen L, Pietragalla J, Garcia G, Hasan A, Miralles D, Calderini DF, Ankleshwaria JB, Waite ML, Simmonds J, Snape J, Reynolds M (2015) Genetic dissection of grain size and grain number trade-offs in CIMMYT wheat germplasm. PLoS ONE 10: e0118847. https: //doi. org/10. 1371/journal. pone. 0118847

Guan PF, Lu LH, Jia LJ, Kabir MR, Zhang JB, Lan TY, Zhao Y, Xin MM, Hu ZR, Yao YY, Ni ZF, Sun QX, Peng HR (2018) Global QTL analysis identifies genomic regions on chromosomes 4A and 4B harboring stable loci for yield-related traits across different environments in wheat (Triticum aestivum L.). Front Plant Sci9: 529. https: //doi. org/10. 3389/fpls. 2018. 00529

Hanif M, Gao FM, Liu JD, Wen WE, Zhang YJ, Rasheed A, Xia XC, He ZH, Cao SH (2015)

TaTGW6-A1，an ortholog of rice *TGW6*，is associated with grain weight and yield in bread wheat. Mol Breed36：1. https：//doi. org/10. 1007/s11032-015-0425-z

Holland JB（2006）Estimating genotypic correlations and their standard errors using Multivariate restricted maximum likelihood estimation with SAS Proc MIXED. Crop Sci 46：642-654. https：//doi. org/10. 2135/cropsci2005. 0191

Holland JB，Nyquist WE，Cervantes-Martínez CT（2003）Estimating and interpreting heritability for plant breeding：an update. PlantBreed Rev 22：9-112

Huang XQ，Coster H，Ganal MW，Roder MS（2003）Advanced back-cross QTL analysis for the identification of quantitative trait loci alleles from wild relatives of wheat（*Triticum aestivum* L.）. Theor Appl Genet 106：1379-1389. https：//doi. org/10. 1007/s0012 2-002-1179-7

Huang XQ，Kempf H，Ganal MW，Roder MS（2004）Advanced back-cross QTL analysis in progenies derived from a cross between a German elite winter wheat variety and a synthetic wheat（*Triticum aestivum* L.）. Theor Appl Genet 109：933-943. https：//doi. org/10. 1007/s00122-004-1708-7

Huang XZ，Qian Q，Liu ZB，Sun HY，He SY，Luo D，Xia GM，Chu CC，Li JY，Fu XD（2009）Natural variation at the *DEP1* locus enhances grain yield in rice. Nat Genet 41：494-497. https：//doi. org/10. 1038/ng. 352

IWGSC（2018）Shifting the limits in wheat research and breeding using a fully annotated reference genome. Science 361：eaar7191. https：//doi. org/10. 1126/science. aar7191

Jahani M，Mohammadi-Nejad G，Nakhoda B，Rieseberg LH（2019）Genetic dissection of epistatic and QTL by environment interaction effects in three bread wheat genetic backgrounds for yield-related traits under saline conditions. Euphytica 215：103. https：//doi. org/10. 1007/s10681-019-2426-1

Jia HY，Wan HS，Yang SH，Zhang ZZ，Kong ZX，Xue SL，Zhang LX，Ma ZQ（2013）Genetic dis-section of yield-related traits in a recombinant inbred line population created using a key breeding parent in China's wheat breeding. Theor Appl Genet 126：2123-2139. https：//doi. org/10. 1007/s00122-013-2123-8

Jiang QY，Hou J，Hao CY，Wang LF，Ge HM，Dong YS，Zhang XY（2011）The wheat（*T. aestivum*）sucrose synthase 2 gene（*TaSus2*）active in endosperm development is associated with yield traits. Funct Integr Genomics 11：49-61. https：//doi. org/10. 1007/s1014 2-010-0188-x

Kumar N，Kulwal PL，Gaur A，Tyagi AK，Khurana JP，Khurana P，Balyan HS，Gupta PK（2006）QTL analysis for grain weight in common wheat. Euphytica 151：135-144. https：//doi. org/10. 1007/s12298-018-0552-1

Kuzay S，Xu YF，Zhang JL，Katz A，Pearce S，Su ZQ，Fraser M，Anderson JA，Brown-Guedira G，DeWitt N，Haugrud AP，Faris JD，Akhunov E，Bai GH，Dubcovsky J（2019）Identifcation of a candidate gene for a QTL for spikelet number per spike on wheat chromosome arm 7AL by high-resolution genetic mapping. Theor Appl Genet 132：2689-2705. https：//doi. org/10. 1007/s00122-019-03382-5

Langridge P（2013）Wheat genomics and the ambitious targets for future wheat production. Genome 56：545-547. https：//doi. org/10. 1139/gen-2013-0149

Li XM，Xia XC，Xiao YG，He ZH，Wang DS，Trethowan R，Wang HJ，Chen XM（2015）QTL mapping for plant height and yield components in common wheat under water-limited and full irrigation environments. Crop Pasture Sci 66：660-670. https：//doi. org/10. 1071/cp14236

Li FJ，Wen WE，He ZH，Liu JD，Jin H，Cao SH，Geng HW，Yan J，Zhang PZ，Wan YG，Xia XC（2018）Genome-wide linkage mapping of yield-related traits in three Chinese bread wheat populations using high-density SNP markers. Theor Appl Genet 131：1903-1924. https：//doi. org/10. 1007/s00122-018-3122-6

Liu Y，He Z，Appels R，Xia X（2012）Functional markers in wheat：current status and future prospects. Theor Appl Genet 125: 1-10. https://doi.org/10.1007/s00122-012-1829-3

Liu K, Sun XX, Ning TY, Duan XX, Wang QL, Liu TT, An YL, Guan X, Tian JC, Chen JS (2018) Genetic dissection of wheat panicle traits using linkage analysis and a genome-wide association study. Theor Appl Genet 131: 1073-1090. https://doi.org/10.1007/s0012 2-018-3059-9

Liu J, Wu BH, Singh RP, Velu G (2019) QTL mapping for micronutrients concentration and yield component traits in a hexaploid wheat mapping population. J Cereal Sci 88: 57-64. https://doi.org/10.1016/j.jcs.2019.05.008

Lizana XC, Riegel R, Gomez LD, Herrera J, Isla A, McQueen-Mason SJ, Calderini DF (2010) Expansins expression is associated with grain size dynamics in wheat (Triticum aestivum L.). J Exp Bot 61: 1147-1157. https://doi.org/10.1093/jxb/erp380

Ma FF, Xu YF, Ma ZQ, Li LH, An DG (2018) Genome-wide association and validation of key loci for yield-related traits in wheat founder parentXiaoyan 6. Mol Breed 38: 91. https://doi.org/10.1007/s11032-018-0837-7

Maccaferri M, Ricci A, Salvi S, Milner SG, Noli E, Martelli PL, Casa-dio R, Akhunov E, Scalabrin S, Vendramin V (2015) A high-density, SNP-based consensus map of tetraploid wheat as a bridge to integrate durum and bread wheat genomics and breeding. Plant Biotechnol J 13: 648-663. https://doi.org/10.1111/pbi.12288

Maphosa L, Langridge P, Taylor H, Parent B, Emebiri LC, Kuchel H, Reynolds MP, Chalmers KJ, Okada A, Edwards J, Mather DE (2014) Genetic control of grain yield and grain physical char-acteristics in a bread wheat population grown under a range of environmental conditions. Theor Appl Genet 127: 1607-1624. https://doi.org/10.1007/s00122-014-2322-y

Miao LL, Mao XG, Wang JY, Liu ZC, Zhang B, Li WY, Chang XP, Reynolds M, Wang ZH, Jing RL (2017) Elite haplotypes of a protein kinase gene TaSnRK2.3 associated with important agro-nomic traits in common wheat. Front Plant Sci 8: 368. https://doi.org/10.3389/fpls.2017.00368

Myskow B, Stojalowski S (2016) Bidirectional selective genotyping approach for the identification of quantitative trait loci controlling earliness per se in winter rye (Secale cereale L.). J Appl Genet 57: 45-50. https://doi.org/10.1007/s13353-015-0294-5

Navabi A, Mather DE, Bernier J, Spaner DM, Atlin GN (2009) QTL detection with bidirectional and unidirectional selective geno-typing: marker-based and trait-based analyses. J Appl Genet 118: 347-358. https://doi.org/10.1007/s00122-008-0904-2

Paux E, Sourdille P, Salse J, Saintenac C, Choulet F, Leroy P, Korol A, Michalak M, Kianian S, Spielmeyer W, Lagudah E, Somers D, Kilian A, Alaux M, Vautrin S, Bergès H, Eversole K, Appels R, Safar J, Simkova H, Dolezel J, Bernard M, Feuillet C (2008) A physical map of the 1-gigabase bread wheat chromosome 3B. Science 322: 101-104. https://doi.org/10.1126/science.1161847

Prashant R, Kadoo N, Desale C, Kore P, Dhaliwal HS, Chhuneja P, Gupta V (2012) Kernel morphometric traits in hexaploid wheat (Triticum aestivum L.) are modulated by intricate QTL×QTL and genotype×environment interactions. J Cereal Sci 56: 432-439. https://doi.org/10.1016/j.jcs.2012.05.010

Quarrie SA, Steed A, Calestani C, Semikhodskii A, Lebreton C, Chinoy C, Steele N, Pljevljakusic D, Waterman E, Weyen J, Schondelmaier J, Habash DZ, Farmer P, Saker L, Clarkson DT, Abugalieva A, Yessimbekova M, Turuspekov Y, Abugalieva S, Tuberosa R, Sanguineti MC, Hollington PA, Aragues R, Royo A, Dodig D (2005) A high-density genetic map of hexaploid wheat (Triticum aestivum L.) from the cross Chinese Spring×SQ1 and its use to compare QTLs for grain yield across a range of environments. Theor Appl Genet 110: 865-880. https://doi.org/10.1007/s00122-004-1902-7

Sadras VO, Lawson C (2011) Genetic gain in yield and associated changes in phenotype, trait plasticity and competitive ability of South Australian wheat varieties released between 1958 and 2007. Crop Pasture Sci 62：533-549. https：//doi. org/10. 1007/s00122-004-1902-7

Shewry PR（2009）Wheat. J Exp Bot 60：1537-1553. https：//doi. org/10. 1093/jxb/erp058

Simmonds J，ScottP，Leverington-Waite M，Turner AS，Brinton J，Korzun V，Snape J，Uauy C（2014）Identification and independent validation of a stable yield and thousand grain weight QTL on chromosome 6A of hexaploid wheat（*Triticum aestivum* L.）. BMC Plant Biol 14：1-13. https：//doi. org/10. 1186/s12870-014-0191-9

Somers DJ，Isaac P，Edwards K（2004）A high-density microsatellite consensus map for bread wheat（*Triticum aestivum* L.）. Theor Appl Genet 109：1105-1114. https：//doi. org/10. 1007/s0012-004-1740-7

Su QN，Zhang XL，Zhang W，Zhang N，Song LQ，Liu L，Xue X，Liu GT，Liu JJ，Meng DY，Zhi LY，Ji J，Zhao XQ，Yang CL，Tong YP，Liu ZY，Li JM（2018）QTL detection for kernel size and weight in bread wheat（*Triticum aestivum* L.）using a high-density SNP and SSR-based linkage map. Front Plant Sci 9：1484. https：//doi. org/10. 3389/fpls. 2018. 01484

Sun YP，Wang JK，Crouch JH，Xu YB（2010）Efficiency of selective genotyping for genetic analysis of complex traits and potential applications in crop improvement. Mol Breed 26：493-511. https：//doi. org/10. 1007/s11032-010-9390-8

Takagi H，Abe A，Yoshida K，Kosugi S，Natsume S，Mitsuoka C，Uemura A，Utsushi H，Tamiru M，Takuno S，Innan H，Cano LM，Kamoun S，Terauchi R（2013）QTL-seq：rapid mapping of quantitative trait loci in rice by whole genome resequencing of DNA from two bulked populations. Plant J 74：174-183. https：//doi. org/10. 1111/tpj. 12105

Tang YL，Li J，Wu YQ，Wei HT，Li CS，Yang WY，Chen F（2011）Identification of QTLs for yield-related traits in theecombi-nant inbred line population derived from the cross between a synthetic hexaploid wheat-derived variety Chuanmai 42 and a Chinese elite variety Chuannong 16. Agric

Sci China 10：1665-1680. https：//doi. org/10. 1016/s1671-2927（11）60165-x

Voss-FelsKP，Keeble-Gagnère G，Hickey LT，Tibbits J，Nagornyy S，Hayden MJ，Pasam R，Kant S，Friedt W，Snowdon RJ，Appels R，Wittkop B（2019）High-resolution mapping of rachis nodes per rachis，a critical determinant of grain yield components in wheat. Theor Appl Genet 132：2707-2719. https：//doi. org/10. 1007/s00122-019-03383-4

Wang RX，Hai L，ZhangXY，You GX，Yan CS，Xiao SH（2009）QTL mapping for grain filling rate and yield-related traits in RILs of the Chinese winter wheat population Heshangmai × Yu8679. Theor Appl Genet 118：313-325. https：//doi. org/10. 1007/s0012-008-0901-5

Wang SC，Wong D，Forrest K，Allen A，Chao S，Huang BE，Mac-caferri M，Salvi S，Milner SG，Cattivelli L，Mastrangelo AM，Whan A，Stephen S，Barker G，Wieseke R，Plieske J et al（2014）Characterization of polyploid wheat genomic diversity using a high-density 90，000 single nucleotide polymorphism array. Plant Biotechnol J 12：787-796. https：//doi. org/10. 1111/pbi. 12183

Wang YX，Xu QF，Chang XP，Hao CY，Li RZ，Jing RL（2018）A dCAPS marker developed from a stress associated protein gene *TaSAP7-B* governing grain size and plant height in wheat. J Integr Agric 17：276-284. https：//doi. org/10. 1016/s2095-3119（17）61685-x

Wang XQ，Dong LH，Hu JM，Pang YL，Hu LQ，Xiao GL，Ma X，Kong XY，Jia JZ，Wang HW，Kong LR（2019）Dissecting genetic loci afecting grain morphological traits to improve grain weight via nested association mapping. Theor Appl Genet 132：3115-3128. https：//doi. org/10. 1007/s00122-019-03410-4

WilliamsK，Sorrells ME（2014）Three-dimensional deed size and shape QTL in hexaploid wheat（*Triticum aestivum* L.）populations. Crop Sci 54：98-110. https：//doi. org/10. 2135/cropsci2012. 10. 0609

Wu XS，ChangXP，Jing RL（2012）Genetic insight into yield-associated traits of wheat grown in multiple rain-fed environments. PLoS ONE 7：

e31249. https：//doi. org/10. 1371/journal. pone. 00312 49

WuQH, Chen YX, Zhou SH, Fu L, Chen JJ, Xiao Y, Zhang D, Ouyang SH, Zhao XJ, Cui Y, Zhang DY, Liang Y, Wang ZZ, Xie JZ, Qin JX, Wang GX, Li DL, Huang YL, Yu MH, Lu P, Wang LL, Wang L, Wang H, Dang C, Li J, Zhang Y, Peng HR, Yuan CG, You MS, Sun QX, Wang JR, Wang LX, Luo MC, Han J, Liu ZY (2015) High-density genetic linkage map construction and QTL mapping of grain shape and size in the wheat population Yanda 1817 × Beinong6. PLoS ONE 10： e0118144. https：//doi. org/10. 1371/journal. pone. 0118144

Xiao YG, He SM, Yan J, Zhang Y, Zhang YL, Wu YP, Xia XC, Tian JC, Ji WQ, He ZH (2011) Molecular mapping of quantitative trait loci for kernel morphology traits in a non-1BL. 1RS × 1BL. 1RS wheat cross. Crop Pasture Sci 62： 625-638. https：//doi. org/10. 1071/CP11037

Xie Q, Mayes S, Sparkes DL (2015) Carpel size, grain filling, and morphology determine individual grain weight in wheat. J Exp Bot 66： 6715-6730. https：//doi. org/10. 1093/jxb/erv378

Xu YF, Li SS, Li LH, Ma FF, Fu XY, Shi ZL, Xu HX, Ma PT, An DG (2017) QTL mapping for yield and photosynthetic related traits under different water regimes in wheat. Mol Breed 37： 34. https：//doi. org/10. 1007/s11032-016-0583-7

YanL, Liang F, Xu HW, Zhang XP, Zhai HJ, Sun QX, Ni ZF (2017) Identification of QTL for grain size and shape on the D genome of natural and synthetic allohexaploid wheats with near-identical AABB genomes. Front Plant Sci 8： 1705. https：//doi. org/10. 3389/fpls. 2017. 01705

Yu M, Zhang H, Zhou XL, Hou DB, Chen GY (2017) Quantitative trait loci associated with agronomic traits and stripe rust in winter wheat mapping population using single nucleotide polymorphic markers. Mol Breed 37： 105. https：//doi. org/10. 1007/s1103 2-017-0704-y

Zhai HJ, Feng ZY, Du XF, Song YE, Liu XY, Qi ZQ, Song L, Li J, Li LH, Peng HR, Hu ZR, Yao YY, Xin MM, Xiao SH, Sun QX, Ni ZF (2018) A novel allele of *TaGW2-A1* is located in a finely mapped QTL that increases grain weight but decreases grain number in wheat (*Triticum aestivum* L.). Theor Appl Genet 131： 539-553. https：//doi. org/10. 1007/s00122-017-3017-y

Zhang XY, Deng ZY, Wang YR, Li JF, Tian JC (2014) Unconditional and conditional QTL analysis of kernel weight related traits in wheat (*Triticum aestivum* L.) in multiple genetic backgrounds. Genetica 142： 371-379. https：//doi. org/10. 1007/s1070 9-014-9781-6

Zou C, Wang PX, Xu YB (2016) Bulked sample analysis in genetics, genomics and crop improvement. PlantBiotechnol J 14： 1941-1955. https：//doi. org/10. 1111/pbi. 12559

Genome-wide association analysis of stem water-soluble carbohydrate content in bread wheat

Luping Fu[1] · Jingchun Wu[1] · Shurong Yang[2] · Yirong Jin[3] · Jindong Liu[4] · Mengjiao Yang[1] · Awais Rasheed[1,5] · Yong Zhang[1] · Xianchun Xia[1] · Ruilian Jing[1] · Zhonghu He[1,6] · Yonggui Xiao[1]

[1] Institute of Crop Sciences, National Wheat Improvement Center, Chinese Academy of Agricultural Sciences (CAAS), Beijing 100081, China

[2] College of Agronomy, Gansu Agricultural University, Lanzhou 730000, Gansu, China

[3] Dezhou Institute of Agricultural Sciences, Dezhou 253000, Shandong, China

[4] Agricultural Genomics Institute at Shenzhen, Chinese Academy of Agricultural Sciences (CAAS), Shenzhen 518000, China

[5] Department of Plant Sciences, Quaid-i-Azam University, Islamabad 45320, Pakistan

[6] International Maize and Wheat Improvement Center (CIMMYT) China Office, Beijing 100081, China

Key message: GWAS identified 36 potentially new loci for wheat stem water-soluble carbohydrate (WSC) contents and 13 pleiotropic loci affecting WSC and thousand-kernel weight. Five KASP markers were developed and validated.

Abstract: Water-soluble carbohydrates (WSC) reserved in stems contribute significantly to grain yield (GY) in wheat. However, knowledge of the genetic architecture underlying stem WSC content (SWSCC) is limited. In the present study, 166 diverse wheat accessions from the Yellow and Huai Valleys Winter Wheat Zone of China and five other countries were grown in four well-watered environments. SWSCC at 10 days post-anthesis (10DPA), 20DPA and 30DPA, referred as WSC10, WSC20 and WSC30, respectively, and thousand-kernel weight (TKW) were assessed. Correlation analysis showed that TKW was significantly and positively correlated with WSC10 and WSC20. Genome-wide association study was performed on SWSCC and TKW with 373, 106 markers from the wheat 660 K and 90 K SNP arrays. Totally, 62 stable loci were detected for SWSCC, with 36, 24 and 19 loci for WSC10, WSC20 and WSC30, respectively; among these, 36 are potentially new, 16 affected SWSCC at two or three time-points, and 13 showed pleiotropic effects on both SWSCC and TKW. Linear regression showed clear cumulative effects of favorable alleles for increasing SWSCC and TKW. Genetic gain analyses indicated that pyramiding favorable alleles of SWSCC had simultaneously improved TKW. Kompetitive allele-specific PCR markers for five pleiotropic loci associated with both SWSCC and TKW were developed and validated. This study provided a genome-wide landscape of the genetic architecture of SWSCC, gave a perspective for understanding the relationship between WSC and GY and

Published in Theoretical and Applied Genetics, 2020, 133: 2897-2914.

explored the theoretical basis for co-improvement of WSC and GY. It also provided valuable loci and markers for future breeding.

Abbreviations:

BLUE	Best linear unbiased estimation
FarmCPU	Fixed and random model circulating prob-ability unification
GWAS	Genome-wide association study
GY	Grain yield
H^2	Broad-sense heritability
KASP	Kompetitive allele-specific PCR
LD	Linkage disequilibrium
MAF	Minor allele frequency
MAS	Marker-assisted selection
MTA	Marker-trait association
NIRS	Near-infrared spectroscopy
PIC	Polymorphism information content
QTL	Quantitative trait locus/loci
SWSCC	Stem water-soluble carbohydrate content
TKW	Thousand-kernel weight
WSC	Water-soluble carbohydrate
WSC10	Stem WSC content at 10 days post-anthesis
WSC20	Stem WSC content at 20 days post-anthesis
WSC30	Stem WSC content at 30 days post-anthesis
YHVWWZ	Yellow and Huai Valleys Winter Wheat Zone

Introduction

Bread wheat(*Triticum aestivum* L.) is among the most important food crops worldwide. It was estimated that a genetic gain of 50% in yield, or an annual gain of ~2%, is essential to meet predicted global requirements over the next 20 years (Lopes et al. 2012). However, annual gains in yield were only 0.6-0.7% in past decades (Sharma et al. 2012; Gao et al. 2017), mainly achieved by conventional breeding. Therefore, it is urgent to improve grain yield (GY) potential with better dissecting the genetic basis of yield and related traits in wheat.

Grain filling in wheat relies on two major carbon sources, namely direct photosynthetic assimilation from green leaves and reserved carbohydrates in stems and leaf sheaths (Ehdaie et al. 2008). When the photosynthetic source is depressed either by leaf senescence or by drought/heat stresses, grain filling becomes more dependent on mobilized resources (Bidinger et al. 1977; Kobata et al. 1992; Blum et al. 1994). Water-soluble carbohydrates (WSC) are stored in stems and leaf sheaths during vegetative and early reproductive stages, and they are remobilized and transported to grains at the later grain filling stages (Pheloung and Siddique 1991; Wardlaw and Willenbrink 2000). Stem-reserved WSC could account for 10-20% and 30-50% of the wheat GY under well-watered and terminal drought conditions, respectively (Aggarwal and Sinha 1984; Wardlaw and Willenbrink 2000; Foulkes et al. 2010; Ovenden et al. 2017). There-fore, improvement in SWSCC can be a valuable approach to improve GY (Shearman et al. 2005; Ruuska et al. 2008; Sadras and Lawson 2011; Xiao et al. 2012; Gao et al. 2017).

In addition to improving GY, WSC also play important roles in coping with abiotic stresses caused by water deficiency and/or high temperature (Livingston et al. 2009). SWSCC are higher in drought-tolerant cultivars than in sensitive ones (Foulkes et al. 2002; Goggin and Setter 2004), and increasing the genetic capacity for WSC accumulation was considered an approach to improve drought tolerance in wheat (Ovenden et al. 2017). In addition, the components of WSC, e. g., fructose, glucose and sucrose, are involved in plant immunity as signaling molecules for regulation of defense genes on biotic stress (Bolouri-Moghaddam and van den Ende 2013; Trouvelot et al. 2014). It is obvious that WSC are involved in a complex system of plant growth, development and diverse biotic and abiotic stress responses (Rolland et al. 2006; Trouvelot et al. 2014). And researchers indicated selection for higher SWSCC has potential in breed-ing for improved adaptation across a range of environmental stresses (Rebetzke et al. 2008).

Many studies have shown that genotypic differences in SWSCC are repeatable across diverse environments with high broad-sense heritability (H^2) of 0. 7-0. 9 (Zhang et al. 2014; Dong et al. 2016a, b). This indicates that variation in SWSCC is largely genetically determined; however, SWSCC is also significantly affected by environmental factors like drought and heat stresses (Ovenden et al. 2017). Quantitative trait loci (QTL) mapping researches of SWSCC have been reported in barley (Teulat et al. 2001), rice (Nagata et al. 2002; Wang et al. 2017; Phung et al. 2019), maize (Thévenot et al. 2005; Bian et al. 2015), perennial ryegrass (Turner et al. 2006), and sorghum (Brenton et al. 2016). In wheat, diverse bi-parental populations were used in identification of QTL or genomic regions associated with SWSCC (Snape et al. 2007; Yang et al. 2007; Rebetzke et al. 2008; Dong et al. 2016b). Nevertheless, these studies didn't provide a genome-wide landscape of the complex genetic architecture, as family-based genetic populations have limited diversity. In addition, the amount of recombination places a limit on mapping resolution in family-based

QTL mapping (Korte and Farlow 2013). It is also possible that QTL with moderate or small effects may be missed in QTL mapping, and the Beavis effect can cause a biased estimation of QTL effects especially when the population size is small (Xu 2003). Moreover, the relatively long genetic distances between linked markers and causal genes limit their use in marker-assisted selection (MAS, Platten et al. 2019).

Genome-wide association studies (GWAS) on SWSCC were performed at flowering, mid-grain filling and maturity stages (Zhang et al. 2014; Li et al. 2015), but the detailed genetic architecture was not revealed due to use of only 209 SSR markers. Dong et al. (2016a) conducted GWAS on SWSCC at 14 days postanthesis (DPA) using 18, 207 markers from the wheat 90 K SNP array (Wang et al. 2014), but there were still large gaps in the genetic map, and marker coverage for the D genome was particularly sparse (Liu et al. 2017). Therefore, it is necessary for a more precise dissection of the genetic architecture underlying the complexity of stem carbohydrate metabolism. In addition, genes controlling SWSCC express dynamically at different growth stages (Veenstra et al. 2017; Yáñez et al. 2017; Hou et al. 2018), and knowledge of basis of SWSCC over time would provide more valuable information for breeding. In maize, QTL analyses of stalk sugar contents at different growth stages were conducted by Bian et al. (2015). Although QTL mapping works on wheat SWSCC have been performed at different stages (Snape et al. 2007; Yang et al. 2007; Rebetzke et al. 2008; Zhang et al. 2014; Li et al. 2015; Dong et al. 2016a, b) knowledge of the dynamic expression patterns of genes associated with WSC remains limited. It is therefore important to gain deep insights into the genetic expression patterns related to wheat SWSCC at different stages.

In the present study, a diversity panel of 166 winter wheat accessions was planted in four well-watered environments, and SWSCC at three time-points at the grain filling stage and thousand-kernel weight (TKW)

were investigated. GWAS was performed on SWSCC and TKW using a high-density physical map constructed with the markers from wheat 660 K and 90 K SNP arrays. The objectives were to (1) evaluate the relationship between SWSCC and TKW, (2) identify loci and candidate genes associated with SWSCC and TKW, providing insights into the genetic basis of SWSCC in wheat and (3) develop high-throughput kompetitive allele-specific PCR (KASP) markers for MAS targeting WSC and TKW improvement.

Materials and methods

Plant materials and field trials

Adiversity panel of 166 representative wheat accessions chosen from more than 400 cultivars was used for GWAS on SWSCC and TKW, including 144 accessions collected from the Yellow and Huai Valleys Winter Wheat Zone (YHVWWZ) of China, and 22 from five other countries (Liu et al. 2017; Zhai et al. 2018; Li et al. 2019; Table S1). Among them, 130 Chinese wheat cultivars released from 1947 to 2016 (Table S1) were used to investigate the genetic progress in improvement in SWSCC and TKW. These cultivars were divided into five groups, i. e. , 9 cultivars released during 1947-1979, 13 in the 1980s, 36 in the 1990s, 59 in the 2000s and 13 released in the 2010s. Furthermore, cultivars released after 1990 from five main wheat producing provinces in China, i. e. , Anhui (9 cultivars), Hebei (13), Henan (44), Shandong (21) and Shaanxi (17), were used to investigate inter-province differences of SWSCC and TKW (Table Sl) .

All accessions were grown in four environments including Dezhou (37° 27′ N, 116° 18′ E; Shandong Province) and Gaoyi (37° 37′ N, 114° 34′ E; Hebei Province) during the 2016-2017 cropping season, and Luohe (33° 36′ N, 113°58′ E; Henan Province) and Xinxiang (35° 18′ N, 113° 51′E; Henan Province) during 2017-2018. These environments were designated as 17DZ, 17GY, 18LH and 18XX, respectively. The field trials at each location were carried out under well-watered conditions with two flood irrigations during jointing and flowering stages. All locations experienced warm temperatures during the later grain filling stage (Fig. S1). The cultivars were planted in randomized complete blocks with three replications. Each plot contained two 2-m rows spaced 20 cm apart, with about 50 seeds sown per row. Field managements followed local practices and fungicide applications were made to control diseases (powdery mildew, stripe rust and leaf rust). All accessions are available from the National Gene Bank of China, Chinese Academy of Agricultural Sciences.

Phenotypic evaluation

SWSCC was assayed at 10DPA, 20DPA and 30DPA using a near-infrared spectroscopy (NIRS) method following Wang et al. (2011), and the corresponding phenotypic data are referred to as WSC10, WSC20 and WSC30, respectively. About ten main culms were randomly taken from each plot, the leaf blades were removed, and the spikes were cut off at the spike collars. Fresh samples from each plot were put into a labeled paper bag and exposed to 105℃ for 30 min and then oven-dried at 80℃ for 24 h. The dried samples were cut into 3-5 mm lengths. Before NIRS assays, the cut samples were re-dried at 80℃ until a constant weight, and brought to room temperature in vacuum bags. The detailed procedure for NIRS assay was reported in Dong et al. (2016b), and SWSCC was reported as a percentage on a dry weight basis. Three technically independent assays were performed for each sample at 10DPA and 20DPA, and five assays were made for samples at 30DPA. The mean values for each sample with extreme outliers discarded were used in subsequent statistical analyses. TKW of wheat accessions were obtained using an automatic seed character analyzer with the SC-G V2. 1. 2. 3 software (Wanshen Detection Technology Co. , Ltd. , Hangzhou, China, http: // www. wseen. com/).

Statistical analysis

Analysis of variance (ANOVA) was performed using SAS 9. 2 software (SAS Institute Inc. , Cary, NC,

USA). Mean squares of each source of variation were used to estimate the variance components for genotypes (σ_G^2), genotype×environment interaction (σ_{GE}^2) and residual error (σ_ϵ^2), respectively, and heritabilities were estimated using the formula $H^2 = \sigma_G^2 / (\sigma_G^2 + \dfrac{\sigma_{GE}^2}{e} + \dfrac{\sigma_\epsilon^2}{re})$, in which e and r were the numbers of environments and replicates per environment, respectively (Holland et al. 2003; Yin et al. 2015). Best linear unbiased estimations (BLUE) for phenotypic data across environments were extracted using the linear model described in Yin et al. (2015) which had been implemented in the ANOVA function in QTL IciMapping v4.1 software (Li et al. 2007). Correlation analyses and t-tests were performed using SAS 9.2.

Genotyping and physical map construction

Genomic DNA was extracted from young leaves using a modified CTAB method (Murray and Thompson 1980). All accessions were genotyped using the Affymetrix 660 K wheat SNP array (containing 630, 517 SNPs, Cui et al. 2017) and the Illumina 90 K wheat SNP array (containing 81, 587 SNPs, Wang et al. 2014) by CapitalBio Technology Co., Ltd. (http://www.capitalbiotech.com/). Minor allele frequency (MAF), polymorphism information content (PIC) and genetic diversity were computed by PowerMarker v3.25 (Liu and Muse 2005, http://statgen.ncsu.edu/powermarker/). The heterozygous genotypes were considered as missing data; markers with MAF<5% and missing data > 20% were excluded to avoid spurious marker-trait associations (MTAs) in subsequent association mapping. Flanking sequences of SNP markers were used to blast against the Chinese Spring (CS) reference genome in IWGSC (RefSeq v1.0, http://www.wheatgenome.org/; IWGSC 2018), and corresponding physical positions were determined according to the best blast hit results. The positions of 1212 markers with multiple hit positions on different chromosomes were assigned according to the consensus 660 K-SNP (Cui et al. 2017) and 90 K-SNP (Wang et al. 2014) genetic linkage maps. High-quality markers from the two SNP arrays were integrated into a common physical map for association study.

Linkage disequilibrium and population structure

Linkage disequilibrium (LD) and population structure of the 166 accessions were analyzed in a previous study using the same population (Liu et al. 2017). Briefly, 12, 324 SNPs evenly distributing on 21 wheat chromosomes were used to calculate LD using a full matrix and sliding window method implemented in Tassel v5.0 (Bradbury et al. 2007). It showed that the average LD decay distance for the whole genome was about 8 Mb; and LD decays were 6, 4 and 11 Mb for the A, B and D genomes, respectively (Fig. S2; Liu et al. 2017). A total of 2000 evenly distributed polymorphic SNPs were chosen to analyze the population structure and estimate the Q matrix by the software Structure v2.3.4 (Pritchard et al. 2000). A neighbor-joining tree was constructed and principal components analysis was performed by Tassel v5.0 to verify the population stratification. Obvious population stratification was observed and the entire panel comprised three subgroups (Fig. S3, Table S1; Liu et al. 2017).

Genome-wide association study

The mean values of three replicates in each environment and the BLUE values across environments for each trait were used for GWAS. To control background variation and eliminate spurious MTAs, associations between markers and traits were estimated using a $Q + K$ mixed linear model (MLM, Yu et al. 2006; Zhang et al. 2010) which was implemented in software TASSEL v5.0 (Bradbury et al. 2007). The Q matrix estimated by Structure v2.3.4 (Pritchard et al. 2000), defining the population structure was considered a fixed-effect factor. The kinship matrix (K matrix) computed by TASSEL v5.0, reflecting relation-ships among individuals, was incorporated as the variance-covariance structure of the random effect for individuals (Zhang et al. 2010). As the $Q + K$ MLM may also compromise true positives in some cases when it controls false positives, the fixed and random model circulating probability unifica-tion (FarmCPU) method, which was demonstrated to have a more improved statistical power than MLM (Liu et al. 2016), was also used to perform GWAS on SWSCC by R software (R Version 3.5.1, https: // www. r-project. org/; FarmCPU package at http: // zzlab. net/FarmCPU/FarmCPU _ functions. txt).

Different methodsfor multiple testing corrections were tried, including the Bonferroni-Holm correction method (Holm 1979) and the Benjamini and Hoch-berg' s false discovery rate (FDR) procedure (Ben-jamini and Hoch-berg 1995), but few SNPs could be declared significant in some of the trait-by-environment conditions in the present study which may due to the higher extent of LD in wheat (Chao et al. 2010; Hao et al. 2011; Chen et al. 2012; Liu et al. 2017) and/or the complex underlying genetic architecture for WSC. Finally, a threshold of $P=1.0\times10^{-3}$ ($-\log_{10}$ (P) = 3.0) was adopted for calling significant MTAs. This threshold was also used in some other as-sociation studies on complex traits in hexaploid wheat (Liu et al. 2017; Muqaddasi et al. 2019; Rahimi et al. 2019).

The adjacent associated markers were grouped togeth-eras one locus if the inter-marker distance is smaller than the average LD decay for specific chromosome, which was reported in Liu et al. (2017). The most significant marker across environments for each locus was considered the representative, and the corre-sponding effect and R^2 (phenotypic variance explained) were estimated and outputted by TASSEL v5.0 (Bradbury et al. 2007). To further control the FDR of WSC-associated loci, those detected in at least two environments by either MLM or FarmCPU were considered to be stable. The genome-wide MTAs were visualized by Manhattan plots with $-\log_{10}$ (P) for each SNP displayed on the Y-axis and the corresponding genomic coordinates displayed along the X-axis. The quantile-quantile ($Q - Q$) plots (observed *versus* expected $-\log_{10}$ (P) values) were used to assess the association mapping mod-els. Manhattan plots and $Q-Q$ plots of the GWAS re-sults were drawn using the CMplot code (https: // githu b. com/YinLiLin/R-CMplot) in R software (Version 3.5.1).

Analyses of allele frequencies and effectsof identified QTL in subgroups

Allele frequencies and effects ofthe identified QTL were analyzed in three subgroups (Table S1) of the population based on representative markers of each lo-cus, and the averaged BLUE values for two genotypes at each locus were used to compare the effects among subgroups by t-test.

Comparison of two models for GWAS

To compare the GWAS performance of MLM and Farm-CPU, the linear model (LM) fitting analysis was performed using representative markers for the identified QTL, where markers were fitted as inde-pendent variables and observed phenotypes as dependent variables. The coefficients of determination from the LM were then calculated using the LM func-tion in R 3.5.1 (https: //stat. ethz. ch/R-manual/R-patch ed/library/stats/html/lm. html). Dot plots with the observed phenotypes on the X-axis and predicted values on the Y-axis were drawn to show the LM fitting results.

Comparison of identified WSC loci with reported QTL or genes

The WSC loci detected in this study were compared with WSC- and GY-related QTL or genes that were searched from the literature based on the physical posi-tions (CS RefSeq v1.0; IWGSC 2018) of their flanking or associated markers. If the physical distances between two QTL were smaller than the average LD decay for a specific chromosome (Liu et al. 2017), they were considered to be at the same locus.

Putative candidate gene analyses

The genes located inthe physical intervals of WSC-as-sociated loci were screened based on the annotations in the wheat reference genome (CS RefSeq v1.0;

IWGSC 2018），and those related to sugar metabolism or transportation were regarded as candidate genes. In addition，the sequences and corresponding physical positions of some known genes involved in WSC synthesis，degradation，and remobilization were obtained from the NCBI（https：//www. ncbi. nlm. nih. gov/）and IWGSC（http：//www. wheatgenome. org/）. The positions of the known genes and their homoeologs were compared with the WSC loci，and some candidates were identified.

KASP marker development

High-throughput KASP markers for potentially important loci were developed based on corresponding representative SNPs. Flanking sequences of SNPs were used as queries to blast against the wheat reference genome in IWGSC（CS RefSeq v1. 0；IWGSC 2018），and chromosome-specific KASP primers were developed based on alignment of homologous sequences. Allele-specific primers carrying FAM（5′ GAAGGTGAC-CAAGTTCATGCT 3′）and HEX（5′ GAAGGTCG-GAGTCAACGGATT 3′）tails were designed with the targeted SNP at the 3′ end，and common reverse primer was designed for a chromosome-specific amplification with less than 200 bp of amplified sequence. The KASP assay mixture was prepared with 40 μL of common primer（100 μM），16 μL of each tailed primer（100 μM），and 60 μL of ddH$_2$O. Each reaction mixture comprised 2. 5 μL of 2 × KASP master mixture（ LGC Genomics，https：// www. biosearchtech. com/），0. 056 μL of the assay mixture，and 2. 5 μL of DNA（30-50 ng/μL）. PCR were performed in a 384-well plate as follows：denaturation at 95℃ for 15 min，followed by 9 touchdown cycles（95℃ for 20 s；touchdown at 65℃ initially then decreasing by 1℃ per cycle for 1 min），and 32 additional cycles of denaturing，annealing and extension（ 95℃ for 10 s；57℃ for 1 min）. Consistency between KASP genotyping results and the original chip-based genotypes was investigated，and t-tests were conducted to confirm the effectiveness of the KASP markers.

Results

Phenotypic variation

Continuous variations among the 166 wheat accessions were observed for WSC10，WSC20，WSC30 and TKW（Table S1，Fig. S4）. The resulting BLUE values for WSC10，WSC20，WSC30 and TKW across the four environments were 9. 41-18. 81%（mean，14. 71%），8. 49-17. 87%（mean，13. 62%），1. 04-9. 71%（mean，4. 14%）and 26. 9-56. 6 g（mean，43. 1 g），respectively（Tables S1，S2）. Correlations among environments for WSC10，WSC20，WSC30 and TKW showed ranges of 0. 56-0. 75，0. 51-0. 71，0. 50-0. 60 and 0. 66-0. 87 with $P<0. 0001$，respectively（Table S3）. ANOVA revealed that genotypes，environments and genotype × environment interactions had significant effects on SWSCC at all three time-points（Table 1）. The H^2 of WSC10，WSC20，WSC30 and TKW across four environments were 0. 90，0. 87，0. 85 and 0. 93，respectively（Table 1），suggesting that most of the phenotypic variation was determined by genetic factors.

Correlation analyses showed that WSC10 was significantly correlated with WSC20 with correlation coefficients（r）ranging from 0. 27（$P<0. 001$）to 0. 61（$P<0. 0001$）in different environments and in BLUE；WSC20 was significantly correlated with WSC30，with r = 0. 43-0. 61（ $P < 0. 0001$ ），whereas the correlations between WSC10 and WSC30 were much lower，ranging from 0. 04（not significant）to 0. 31（$P<0. 001$）（Table 2）. Correlations between WSC10 and TKW（ r = 0. 33-0. 63，$P < 0. 0001$ ）were similar to those between WSC20 and TKW（r =0. 32-0. 59，$P<0. 0001$），whereas r between WSC30 and TKW ranged from 0. 15（not significant）to 0. 42（$P<0. 0001$）（Table 2）.

Table 1   Analysis of variance for stem WSC contents and thousand-kernel weight in the 166 wheat accessions

| Trait[a] | Analysis of variance[b] | | | | | Broad-sense heritability |
	Genotype (G)	Environment (E)	G×E interaction	Replicate	Error	
WSC10	42. 09***	341. 64***	4. 59***	6. 78***	1. 17	0. 90
WSC20	41. 35***	340. 88***	6. 02***	3. 90***	1. 51	0. 87
WSC30	33. 84***	408. 22***	5. 88***	6. 29***	1. 40	0. 85
TKW	268. 55***	6701. 84***	18. 48***	94. 10***	3. 64	0. 93

[a] WSC10, WSC20 and WSC30 indicate stem WSC contents at 10 days post-anthesis (10DPA), 20DPA and 30DPA, respectively. TKW, thousand-kernel weight

[b] Mean square values from the analysis of variance are reported. ***$P<0.0001$

Table 2   Correlation analyses for stem WSC contents and thousand-kernel weight of the 166 wheat accessions in different environments

Environment[a]	Trait[b]	WSC10	WSC20	WSC30
17DZ	WSC20	0. 27**		
	WSC30	0. 04	0. 57***	
	TKW	0. 45***	0. 32***	0. 15
17GY	WSC20	0. 44***		
	WSC30	0. 16*	0. 61***	
	TKW	0. 33***	0. 53***	0. 42***
18LH	WSC20	0. 55***		
	WSC30	0. 19*	0. 43***	
	TKW	0. 51***	0. 50***	0. 20*
18XX	WSC20	0. 61***		
	WSC30	0. 31***	0. 52***	
	TKW	0. 63***	0. 59***	0. 34***
BLUE	WSC20	0. 57***		
	WSC30	0. 20**	0. 52***	
	TKW	0. 56***	0. 51***	0. 34***

[a] 17DZ and 17GY, Dezhou and Gaoyi locations, respectively, 2016-2017; 18LH and 18XX, Luohe and Xinxiang locations, respectively, 2017-2018; BLUE, best linear unbiased estimation across environments for each trait

[b] WSC10, WSC20 and WSC30 indicate stem WSC contents at 10 days post-anthesis (10DPA), 20DPA and 30DPA, respectively. TKW, thousand-kernel weight

***$P<0.0001$; **$P<0.001$; *$P<0.01$

Marker coverage and genetic diversity

A total of 373, 106 high-quality SNPs with 359, 760 (96. 42%) and 13, 346 (3. 58%) from the 660 K and 90 K SNP arrays, respectively, were used in GWAS (Table S4). The A, B and D genomes were represented by 39. 8, 49. 3 and 10. 8% of the markers, respectively. Chromosome 3B possessed the most markers (46, 708), whereas chromosome 4D had the least (2375). The markers covered the whole genome (14, 061. 15 Mb) with an average density of 0. 038 Mb per marker, and for respective chromosomes, the marker densities ranged from 0. 018 (3B) to 0. 214 (4D). The B genome showed the highest marker density (0. 028 Mb per marker), genetic diversity (0. 367) and PIC (0. 293) compared with the A (0. 033, 0. 357 and 0. 286) and D (0. 098, 0. 334 and 0. 270) genomes. The detailed information of marker number, marker density, genetic diversity and PIC is provided in Table S4, in addition to a density map showing marker distribution along chromosomes (Fig. S5).

Marker-trait associations and stable WSC-associated loci

Significant MTAs for WSC10, WSC20 and WSC30 analyzed using the MLM and FarmCPU methods are listed in Tables S5-S10. In total, 1095, 652 and 597 significant markers corresponding to 168, 93 and 14 loci were detected for WSC10, WSC20 and WSC30, respectively, using MLM in TASSEL; and 813, 1344 and 1415 significant markers corresponding to

134，112 and 23 loci were identified for the three traits using FarmCPU. Manhattan plots for WSC contents analyzed by MLM and FarmCPU using BLUE values are shown in Fig. 1, and Manhattan and $Q-Q$ plots for each trait in each environment analyzed by both methods are shown in Figs. S6 and S7.

The WSC loci detected in at least three out of the four environments by either MLM or FarmCPU are summarized in Table 3. Furthermore, all stable WSC loci that detected in at least two environments by either

MLM or FarmCPU are reported in Table S11. The numbers of stable loci for WSC10，WSC20 and WSC30 were 36，24 and 19，respectively（Table S11）. Overall，62 stable loci for SWSCC across the three developmental time-points were detected on all 21 chromosomes except 5D，and 16 of these loci were associated with SWSCC at two or more time-points （Table S11，Fig. 2）. In terms of the stable loci，6 （17%），7 （29%）and 8 （42%）for WSC10, WSC20 and WSC30，respectively，were detected by both MLM and FarmCPU.

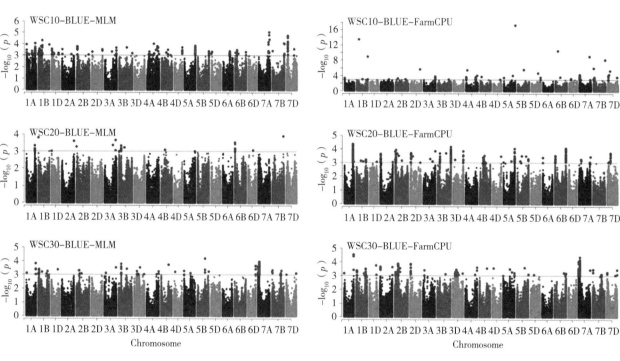

Fig. 1　Manhattan plots for stem WSC contents analyzed by $Q+K$ mixed linear model （MLM）and FarmCPU. The threshold of $P=1.0\times10^{-3}$ （$-\log_{10}$ （P）$=3.0$）was used for calling significant marker-trait associations （MTAs）. WSC10，WSC20 and WSC30 indicate stem WSC contents at 10 days post-anthesis （10DPA），20DPA and 30DPA, respectively. BLUE indicates the best linear unbiased estimations for each trait across four environments in this study

Allele frequencies and QTL effects in subgroups

For each WSC QTL，the allele frequencies and effects in three subgroups of the population were investigated （Table S12）. For most of the loci，similar allele frequencies and effects were observed among subgroups. But some large inter-subgroup variations of frequencies and effects were detected in a few loci. For instance，the allele frequency of a WSC10 locus on chromosome 5B （692. 45 Mb，*AX-111100349*）were

0. 37 for the entire panel，while the frequencies were 0. 11，0. 57 and 0. 49 in subgroups 1，2 and 3，respectively（Table S12）. The effect of a WSC10 locus on 1B （681. 04 Mb，*AX-110000934*）was 0. 57% for the entire population，whereas the effects were 2. 48, 0. 28 and 0. 05% in subgroups 1，2 and 3, respectively（Table S12）. These indicated that the allele frequencies and effects of a few loci might be significantly affected by the population stratification，and we need to further confirm their effects in future stud-

ies.

Pleiotropic loci for SWSCC and TKW

By comparing the physical positions of significant markers, 13 stable loci associated with TKW were co-localized with the WSC loci on chromosomes 1A, 1B (2), 1D, 2A, 2B, 3A (2), 4A, 4B, 5B and 6B (2), explaining 7.69-16.38% and 8.17-15.53% of the phenotypic variations for TKW and SWSCC, respectively (Tables 4, S11). Because high TKW is a main objective in wheat breeding, the alleles associated with increased TKW and SWSCC were considered favorable. The frequencies of favorable alleles at five loci on 1A, 1B, 2B and 3A (2) were 0.86-0.93 (Table 4), indicating that the favorable alleles at these loci had already been widely selected in traditional breeding programs, whereas the other eight loci exhibited favorable allele frequencies (FAF) of 0.45-0.70 (Table 4), indicating more potential possibilities for selection in future breeding.

Cumulative effect of increasing-effect alleles on SWSCC and TKW

To further investigate the effects of combined alleles on SWSCC, the number of increasing-effect alleles in each accession was investigated. The number of WSC-increasing alleles possessed by each accession had ranges of 6-32, 5-22 and 2-16 for WSC10, WSC20 and WSC30, respectively (Table S1, Fig. S8). Significant correlations were observed between SWSCC and number of WSC-increasing alleles with $r = 0.84$, 0.68 and 0.63 ($P < 0.001$) for WSC10, WSC20 and WSC30, respectively. Linear regressions using the BLUE values were determined to further investigate the relationships between SWSCC and the number of WSC-increasing alleles, and it showed significant linear associations between SWSCC and number of alleles with regression slopes of 0.31 (coefficient of determination (r^2) = 0.95), 0.44 ($r^2 = 0.95$) and 0.40 ($r^2 = 0.95$) for WSC10, WSC20 and WSC30,

respectively (Fig. 3).

The accessions contained 0-13 increasing-effect alleles at the 13 pleiotropic loci for both SWSCC and TKW (Table S1, Fig. S8). There were clear cumulative effects on WSC10, WSC20, WSC30 and TKW, with cumulative number of increasing-effect pleiotropic alleles (Fig. 4). Linear regression slopes for WSC10, WSC20, WSC30 and TKW versus corresponding numbers of increasing-effect alleles were 0.51 ($r^2 = 0.88$), 0.49 ($r^2 = 0.91$), 0.30 ($r^2 = 0.82$) and 1.56 ($r^2 = 0.96$), respectively.

Pyramiding WSC-increasing alleles improved SWSCC and TKW over past decades

To explore the roles of WSC-associated loci in improving GY, the genetic progresses of SWSCC and TKW have been investigated. The results showed that WSC10, WSC20 and WSC30 had increased on average from 13.02%, 12.55% and 3.45% before 1980 (9 cultivars) to 16.27%, 15.41% and 4.63% after 2010 (13 cultivars), respectively (Fig. 5a-c); accompanied by increased TKW from 41.1 to 46.4 g (Fig. 5d) and increased numbers of increasing-effect alleles from 15, 14, 7 and 7 to 26, 18, 10 and 11 for WSC10, WSC20, WSC30 and TKW, respectively (Fig. 5e-h).

In addition, we investigated the frequency-changes in favorable alleles for the 13 pleiotropic loci (Table4). It showed that FAFs for all these loci increased in recent decades except the one on 4AL with a representative marker *AX-109832317* (Fig. 6). Furthermore, differences of SWSCC and TKW among five main wheat producing provinces in China have been investigated. The results showed that cultivars in Henan, which is the top wheat producing province in China, had the most WSC10 and TKW, as well as the most numbers of increasing-effect WSC alleles (Fig. S9).

Table 3　WSC-associated loci detected in at least three environments by either MLM or FarmCPU

Trait[a]	Environment-MLM[b]	Environment-FarmCPU[b]	Chr[c]	Marker interval (Mb)[d]	Representative SNP[e]	Major/Minor allele[f]	MAF[f,g]	P value[f]	R² (%)[f,h]	Effect[f]
WSC10	2	1, 2, 4, B	1A	445.51-461.52	AX-109989656	C/T	0.22	2.72E-04	8.86	1.80
	1, 2, 4	1	1A	513.74-522.15	AX-109293426	T/C	0.12	2.29E-05	12.98	2.68
	2, 3	2, 3, 4, B	1B	39.60-41.64	AX-111648440	C/T	0.24	1.09E-05	13.68	1.89
	2, 3, 4, B	—	1D	4.25-8.62	AX-94811887	C/A	0.06	2.09E-04	10.38	2.99
	—	1, 3, 4	2A	701.48-709.03	BS0009423_51	C/A	0.14	3.38E-06		1.21
	1, 3, 4, B	—	4A	735.30-737.58	AX-110010963	C/T	0.08	2.52E-04	10.16	2.42
	2, 3, 4	2, 4	6B	0.49-14.26	AX-108959247	A/T	0.37	8.86E-05	12.33	1.58
	1, 2, B	1, 2, 4, B	7D	60.58-78.26	AX-110514480	T/G	0.24	5.50E-06	15.43	2.19
WSC20	4	1, 3, 4	1B	4.35-9.56	AX-111668935	G/A	0.27	1.21E-04	12.07	2.22
	3	1, 3, 4, B	1D	414.59-420.74	AX-111640220	C/T	0.16	4.07E-04	9.85	2.12
	1, 3, 4	1, 2, 3, B	2A	27.51-32.40	AX-111552571	T/C	0.16	2.06E-05	14.58	2.70
	1, 3, 4	1, 3	2B	747.81-752.49	AX-108729610	T/C	0.45	1.37E-04	11.24	1.67
	—	1, 2, 3, B	3A	737.44-744.29	AX-94829137	A/G	0.2	7.37E-04		1.61
	2, 3, 4, B	2, 3, 4, B	5A	688.14-690.41	AX-111031397	A/G	0.06	1.55E-05	13.8	3.56
	2	1, 2, 3, 4, B	6B	674.84-677.48	AX-111274903	A/G	0.42	8.97E-04	8.17	1.31
WSC30	2, 3, 4, B	2, 3, 4, B	1A	572.25-581.75	BobWhite_c39668_143	G/A	0.21	1.57E-04	11.77	1.89
	1, 2	1, 2, 3, B	2A	763.02-769.51	AX-110028549	C/G	0.49	2.52E-05	17.44	1.78
	—	1, 2, 3, B	7A	38.33-46.90	AX-108784564	G/A	0.07	9.48E-04		2.49
	2, B	1, 2, 4, B	7A	81.88-90.63	AX-108972618	T/C	0.36	1.32E-04	11.59	1.23

[a] WSC10, WSC20 and WSC30 indicate stem WSC contents at 10 days post-anthesis (10DPA), 20DPA and 30DPA, respectively

[b] Environments of the corresponding locus have been detected by mixed linear model (MLM) or FarmCPU; 1, 2, 3 and 4 indicate Dezhou, Gaoyi, Luohe and Xinxiang locations, and B indicates the best linear unbiased estimation across four environments; "—" indicates the locus has not been detected by the corresponding GWAS model

[c] The chromosome of the corresponding locus located on

[d] Physical positions of SNP markers were based on the Chinese Spring reference genome in IWGSC (RefSeq v1.0, http://www.wheatgenome.org/)

[e] The most significant SNP across environments for the corresponding locus was reported as a representative

[f] The information in corresponding columns are based on the representative SNP

[g] MAF indicates the minor allele frequency

[h] R^2 indicates the percentage of phenotypic variance explained by the SNP marker; the data were not provided if a locus was detected only by FarmCPU

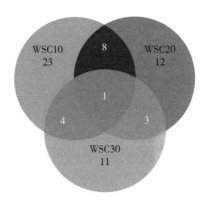

Fig. 2　The Venn diagram of stable loci associated with stem WSC contents. Loci detected in at least two environments by eithermixed linear model （MLM） or FarmCPU were considered stable （Table S11）. WSC10，WSC20 and WSC30 indicate the stem WSC contents at 10 days post-anthesis （10DPA），20DPA and 30DPA，respectively

Table 4　Thirteen TKW-associated loci simultaneously associated with stem WSC contents

Representative marker[a]	Chromosome	Physical interval （Mb）[b]	P value	R^2 （%）[c]	Effect （g）	Allele[d]	Frequency of favorable allele	Environment[e]
AX-110587308	1AL	587.83-587.84	3.22E-04	8.50	5.29	G̲/A	0.86	E3，E4
BS00023084_51	1BS	7.48	1.49E-04	9.44	3.41	A̲/G	0.68	E1，E2，BLUE
AX-109463291	1BL	580.63-581.21	1.72E-04	9.56	5.90	T̲/C	0.91	E1，E2，E3，E4，BLUE
AX-111496323	1DS	8.01-11.49	2.87E-04	8.51	3.86	C̲/G	0.70	E1，E3，BLUE
AX-109293110[†]	2AS	27.33-31.96	2.44E-04	8.89	3.89	T̲/C	0.51	E2，E4，BLUE
wsnp_Ex_c163_320858	2BS	41.96-46.88	3.92E-05	11.55	5.31	G̲/A	0.89	E1，E2，E3，E4，BLUE
Kukri_c7087_896	3AL	594.00-595.44	3.06E-04	8.45	5.81	A̲/G	0.93	E2，E4
AX-109483618	3AL	714.36-716.25	2.56E-04	8.76	5.72	G̲/C	0.93	E2，E4，BLUE
AX-109832317[†]	4AL	738.89-742.84	3.49E-04	9.55	3.63	C̲/T	0.45	E1，E2
AX-110974144[†]	4B	298.11-302.17	1.63E-04	11.37	4.60	A̲/G	0.50	E3，E4，BLUE
AX-111044647[†]	5BL	694.03-697.61	3.00E-04	8.33	3.56	A̲/G	0.67	E1，E3，BLUE
AX-94910312	6BS	30.01-32.33	5.57E-04	7.69	3.62	G̲/T	0.69	E1，E3
AX-111494281[†]	6BL	670.21-678.06	1.35E-06	16.38	4.51	T̲/C	0.50	E1，E2，E3，E4，BLUE

[a]SNP markers indicated with † were converted to KASP markers （Tables S12，S13；Figs. 7，S10）

[b]Physical positions of SNP markers were based on the Chinese Spring reference genome in IWGSC （RefSeq v1.0，http：// www.wheatgenom e.org/）

[c]R^2，percentage of the phenotypic variance explained by the QTL

[d]The underlined allele indicates the favorable allele that showed increasing-effects on thousand-kernel weight and stem WSC content

[e]E1，Dezhou 2016-2017；E2，Gaoyi 2016-2017；E3，Luohe 2017-2018；E4，Xinxiang 2017-2018；BLUE，best linear unbiased estimation

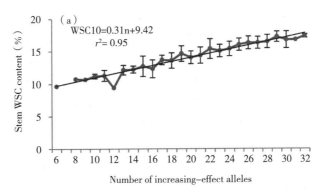

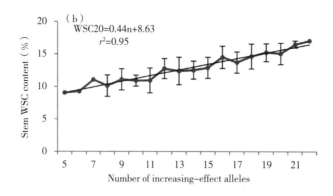

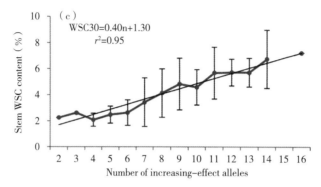

Fig. 3　Cumulative effects of increasing-effect alleles on stem WSC contents. WSC10, WSC20 and WSC30 indicate stem WSC contents at 10, 20 and 30 days post-anthesis, respectively. "n", number of increasing-effect alleles. Dots and bars represent the mean values and standard deviations, respectively. Linear regressions were per-formed to investigate the relationships between stem WSC contents and number of increasing-effect alleles in the 166 wheat accessions. Calculations were based on representative markers of the 36, 24 and 19 stable loci associated with WSC10, WSC20 and WSC30, respectively (Table S11). Best linear unbiased estimations (BLUE) for each trait across four environments were used

KASP marker development

Among the 13 pleiotropic loci showing effects on both WSC and TKW, the alleles at loci on 1AL, 1BL, 2BS and 3AL (2) were almost fixed in this panel of wheat accessions with FAF of 0.86-0.93, whereas the FAF of 0.45-0.70 were found in the other eight loci leaving more values for MAS, and these have been focused to develop KASP markers (Table 4). Finally, five KASP markers were successfully developed for representative SNP markers of pleiotropic loci on chromosomes 2A (physical interval 27.51-34.67 Mb), 4A (728.51-739.58 Mb), 4B (298.15-299.53 Mb), 5B (689.91-696.19 Mb) and 6B (674.84-677.48 Mb) (Table 4; Fig. 7). The primers and information for the corresponding loci are listed in Table S13. Comparison between the KASP genotyping results

and the chip-based genotypes showed consistencies of 0.96-0.98. T-tests in different environments confirmed the effectiveness of the KASP markers (Fig. S10).

Discussion

Comparison of GWAS results by MLM and FarmCPU

In the present study, 19 and 20 stable WSC loci identified by MLM and FarmCPU, respectively, were located at similar positions with reported WSC-related QTL or genes (Table S11); 6, 7 and 8 loci for WSC10, WSC20 and WSC30, respectively, were detected by both MLM and FarmCPU (Table S11). All these indicated the reliability of GWAS results from both models. Some different results were also observed in two models. The WSC20 loci on chromosomes 1D (414.59-420.74 Mb), 3A (737.44-744.29 Mb) and

6B（674.84-677.48 Mb）were detected by FarmCPU in three or more environments but not stably identified by MLM，whereas the WSC10 loci on 1B（236.96-237.38 Mb）and 4A（735.30-737.58 Mb）were specifically detected by MLM（Table S11）.

Liuet al.（2016）indicated that the $Q + K$ MLM may lead to false negatives in some cases，although it generally performs well in controlling false positives，while FarmCPU could control false positives and simultaneously avoid model over-fitting by using fixed and random effect models iteratively. In the present study，the Q-Q plots showed that the population structure and kinship were over-corrected by MLM in some environments，especially for WSC20 and WSC30（Figs. S6，S7），indicating that there might be some false negatives for MLM. In addition，when we focused on the WSC loci detected in at least three environments，more loci were observed by FarmCPU（Table 3）.

To further compare the results from MLM and Farm-CPU，LM fitting analysis was conducted based on representative markers identified by each model. For WSC10，the GWAS performance of MLM and Farm-CPU was similar with coefficients of determination（r^2）of 0.76 and 0.72，respectively（Fig. S11）. While for WSC20 and WSC30，FarmCPU performs better with $r^2 = 0.56$ and 0.58，respectively，than MLM with r^2 values of 0.43 and 0.34，respectively（Fig. S11）. In addition，LM fitting analyses using all the identified representative markers from MLM and FarmCPU（Table S11）gave the best results with $r^2 = 0.86$，0.62 and 0.58 or WSC10，WSC20 and WSC30，respectively（Fig. S11）. All these indicated that the detection powers of MLM and FarmCPU may vary according to different traits，and the two models are complementary in detecting QTL.

Additionally，although many loci were detected by both FarmCPU and MLM（Table S11），it seems that the Manhattan plots corresponding to the two models are quite different（Figs. S6，S7）. Some single significant SNPs hang on the plots for FarmCPU results，while strings of SNPs clustered as peaks in the plots for MLM. This may be attributed to different methodologies for the two methods（Liu et al. 2016）.

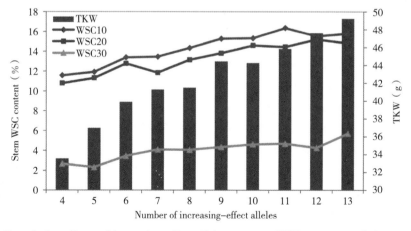

Fig. 4   Cumulative effects of increasing-effect alleles on stem WSC contents and thousand-kernel weight based on the 13 pleiotropic loci. Calculations were based on representative markers for the 13 pleiotropic loci simultaneously affecting stem WSC contents and TKW（Table 4）. Best linear unbiased estimations（BLUE）for phenotypic data across four environments were used. WSC10，WSC20 and WSC30 indicate stem WSC contents at 10，20 and 30 days post-anthesis，respectively. TKW，thousand-kernel weight

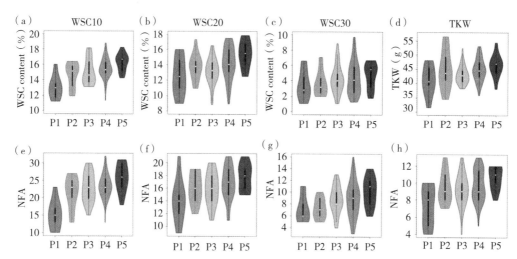

Fig. 5 Genetic progresses of stem WSC contents and thousand-kernel weight over past decades. Genetic progress was based on five groups from a panel of 130 Chinese wheat cultivars released from 1947 to 2016 (Table S1). WSC10, WSC20 and WSC30 indicate the stem WSC contents at 10 days post-anthesis (10DPA), 20DPA and 30DPA, respectively. TKW, thousand-kernel weight. P1, P2, P3, P4 and P5 indicate 1947-1979, the 1980s, the 1990s, the 2000s and 2010-2016, respectively. Violin plots (a) to (d), phenotypic changes in WSC10, WSC20, WSC30 and TKW, respectively; (e) to (h), changes in numbers of increasing-effect alleles for WSC10, WSC20, WSC30 and TKW, respectively. Each violin plot shows the median (indicated by the small, white dot), first through third interquartile range (black, thick, solid vertical band), and estimator of the density (thin vertical curves) of the corresponding observations. NFA indicates the number of increasing-effect alleles contained in a cultivar

Comparison of identified WSC loci with reported QTL or genes

Based on the physical positions (CSRefSeq v1.0; IWGSC 2018) of markers or genes, the WSC loci identified in this study were compared with reported QTL or genes. Of the 62 WSC loci, 26 were located at similar positions to previously reported WSC-related QTL or genes (Table S11), including 15 (41.7%), 9 (37.5%) and 8 (42.1%) stable loci for WSC10, WSC20 and WSC30, respectively. This indicated the importance of these loci, reflecting the reliability of our findings. The remaining 36 new WSC loci, comprising 21, 15 and 11 loci for WSC10, WSC20 and WSC30, respectively (Table S11), provided us a basis to more comprehensively understand the complex genetic architecture underlying SWSCC.

It should be mentioned that some loci might be related to WSC content in different parts of stem, and some were associated with SWSCC under diverse condi-

tions. For instance, a locus on 2DS ($Xcfd$53, 23.02 Mb) was associated with WSC in the uppermost internode at 14DPA under drought stress conditions (Zhang et al. 2014), whereas in the present study, a similar locus (29.17 Mb) was associated with WSC content of the whole stem (Table S11). A locus on 3AL ($Xbarc$314, 712.49 Mb) was associated with WSC in the lower internodes at the grain filling stage under simulated terminal drought stress conditions (Zhang et al. 2014), in agreement with the WSC10- and WSC20-associated locus (711.30-726.13 Mb) identified in the present study (Table S11).

Although diverse bi-parental and natural populations were used across the world, many WSC-associated loci identified in this study were located at similar positions to the previously reported ones based on the physical positions (CS RefSeq v1.0; IWGSC 2018) of QTL-flanking or associated markers (Table S11). This indicated that (a) linkage mappingand GWAS are complementary in identifying genes, and (b) many major

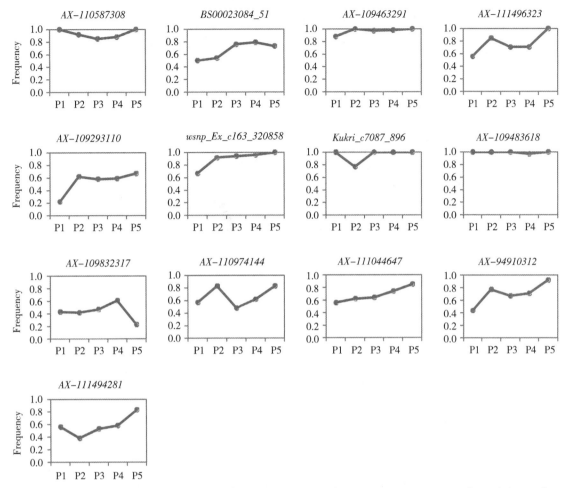

Fig. 6   Frequency changes in increasing-effect alleles in past decades based on representative markers of the 13 pleiotropic loci. The 13 pleiotropic loci were associated with both WSC and thousand-kernel weight (Table 4). The analysis was based on 130 wheat cultivars in five groups released from 1947 to 2016 (Table S1). P1, P2, P3, P4 and P5 indicate 1947-1979, the 1980s, the 1990s, the 2000s and 2010-2016, respectively. The name of each representative marker is indicated in the corresponding chart

genes controlling carbohydrate metabolism might be common in different wheat production regions. Considering the high marker density in the present study and consistency of many loci between the present and past researches, the loci and associated SNP markers detected here appear to be reliable and hence valuable for further genetic research on WSC metabolism and for MAS in breeding.

Putative candidate genes related to stem WSC content

Fructans are major components of WSC (Ruuska et al. 2006), and genes involved in fructan metabolism play important roles in controlling SWSCC. In regions adjacent to the WSC loci on chromosomes 2A (763. 02-769. 51 Mb), 4A (735. 30-737. 58 Mb) and

7D (2. 23-17. 34 Mb), some known genes involved in fructan synthesis and hydrolysis were identified, including *1-SST*, *6-SFT*, *1-FFT* and *6-FEH* (Table S11; McIntyre et al. 2011). SWSCC was significantly associated with *1-FFT-A1* (Yue et al. 2017) and *1-SST-D1* (Dong et al. 2016b), and the corresponding gene-specific markers had been reported. In addition to the genes directly involved in carbohydrate metabolism, transcription factors also play important roles in modulating metabolic pathways. *TaMYB13* was identified as a transcriptional activator of fructosyltransferase genes, and expression levels of *TaMYB13* were positively correlated with the mRNA levels of *1-SST* and *6-SFT* in wheat stems (Xue et al. 2011; Kooiker et al. 2013). Based on the genomic

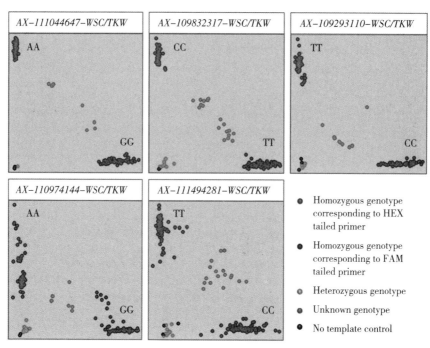

Fig. 7 Genotype calling screen-shots of five KASP markers. The KASP markers were developed based on representative SNP markers for five pleiotropic loci on both stem WSC contents (SWSCC) and thousand-kernel weight (TKW) (Tables 4, S11). Corresponding names for the KASP markers are presented on screenshots. Two homozygous genotypes are indicated beside the corresponding dots, and favorable genotypes showed increasing-effects on SWSCC and TKW (Fig. S10) are shown in red font (color figure online)

DNA sequences, we found that *TaMYB13-1* (ID: TraesCS3A02G535100, 746.63 Mb) was located near the WSC20 locus on 3A (737.44-744.29 Mb).

Stem reserved WSC can be remobilized and transported to sink organs such as seeds and roots, and sugar transporters are critical in these processes (Halford et al. 2011). Sucrose transporters (SUT) and hexose and sucrose transporters are major sugar transporter families (Eom et al. 2015). *TaSUT1* genes on homoeologous group 4 chromosomes were the predominant sucrose transporter group in wheat stems and leaf sheaths, and significantly positive correlations were observed between TKW and expression level of *TaSUT1* genes (Ahmed et al. 2018). A *TaSUT* gene (TraesCS2A02G505000, 733.56 Mb) was identified in the vicinity of the WSC30 locus on chromosome 2A (734.16-734.26 Mb) (Table S11). In addition, two potentially new sugar transporter genes (TraesCS6B 01G421500, 692.23 Mb; TraesCS7D01G521400, 618.68 Mb) belonging to the hexose and sucrose trans-

porters family were identified in the intervals of WSC loci on 6B (692.03-702.39 Mb) and 7D (616.95-621.14 Mb) (Table S11), indicating that these genes might have important roles in WSC remobilization.

Sucrose synthase (SUS) controls carbon flow in starchbiosynthesis, and dry matter accumulation of plants is positively correlated with SUS activity (Kato 1995), and *TaSus2-2B* was significantly associated with TKW (Jiang et al. 2011). Based on gene sequences, *TaSus2* (171.03 Mb) was located in a WSC30 locus on 2B (168.78-184.77 Mb) (Table S11). A diverse family of glycosyltransferases participates in glycan and glycoside biosynthesis during plant development and growth (Lao et al. 2014), and glycoside hydrolases are involved in hydrolysis of complex sugars (Bourne and Henrissat 2001). In the present study, six potentially new glycosyltransferase candidate genes were located in the intervals of WSC loci on chromosomes 4B, 5B, 6D, 7A, 7B and 7D, and two glycoside hydrolase genes were in the intervals of WSC loci on 6B and 7A (Table S11). These candidate genes provide a basis for further exploration of the genetic mechanism of WSC metabolism, and might be useful in breeding programs targeting increased TKW.

Relationship between WSC and GY

In the present study, TKW was significantly and positively correlated with SWSCC, especially with WSC10 and WSC20 (Table 2). This was consistent with previous findings about relationship of WSC and TKW (Zhang et al. 2015; del Pozo et al. 2016). In addition, there were evidences that cultivars carrying favorable WSC alleles had higher TKW than those without favorable alleles in both well-watered and drought and heat stress conditions (Zhang et al. 2014). Similar findings were also observed in the present study that 13 pleiotropic loci simultaneously affecting SWSCC and TKW (Table 4) showed clear cumulative effects on both SWSCC and TKW (Fig. 4). Besides the 13 pleiotropic loci, other 20 WSC loci also located at similar positions to reported GY-related QTL, including 12, 5, 2 and 1 related to TKW, spike number, GY and grain filling rate, respectively (Table S11). All these indicated the positive contributions of WSC to TKW.

In many studies, significant and positive correlations ($r = 0.47-0.80$) were observed between SWSCC and GY under diverse conditions (Foulkes et al. 2007; Snape et al. 2007; Xue et al. 2008; Gao et al. 2017), whereas in some other researches, low or non-significant correlations between SWSCC and GY were observed (Ruuska et al. 2006; Zhang et al. 2015; del Pozo et al. 2016). Using 384 wheat cultivars, del Pozo et al. (2016) found that WSC content at anthesis was negatively correlated with spikes per square meter, but positively correlated with kernels per spike and TKW under water stress and full irrigation conditions; consequently, the correlation between WSC with GY were low or not significant.

By comparing the WSC QTL and reported GY-related QTL, we found some clues that might genetically explain the complicated relationship between WSC and GY. A pleiotropic locus simultaneously affecting WSC20, WSC30 and TKW was identified on chromosome 2AS (27.33-34.67 Mb) in this study (Table S11). At a similar position, a QTL for TKW and grain filling rate (Xgwm359, 28.20 Mb; Wang et al. 2009) and a QTL for GY (30.3-31.9 Mb; Li et al. 2019) were reported. These indicate that the underlying gene (or genes) might control grain filling rate by affecting WSC con-tent in stems, and finally affect the TKW and GY.

Plant density and flowering date could also affect SWSCC (Rebetzke et al. 2008; del Pozo et al. 2016). In the present study, five WSC loci were located at similar positions with reported plant density-related QTL (Table S11), indicating that SWSCC might be affected by plant density, and this also explains the complex relationship between SWSCC and GY. At the WSC10-associated locus on 5A (697.77-699.48 Mb), a vernalization gene Vrn2-5A (698.2 Mb) had been reported (Yan et al. 2004). To further investigate the potential impact of vernalization on stem WSC, we conducted GWAS by con-trolling the Vrn2-5A (using its genotypic data) as a fixed-effect in the $Q + K$ MLM. The results showed that 66.7% (24 out of 36) of the loci could still be detected (Table S14), including the WSC locus that collocated with Vrn2-5A. It seems that the effect of this WSC locus is independent from Vrn2-5A.

Pyramiding WSC-increasing alleles by MAS could improve GY

Significant and positive correlations between SWSCC and TKW (Table 2) were consistent with previous reports (Ehdaie et al. 2008; Li et al. 2015; Gao et al. 2017). Increased SWSCC has contributed to genetic gains in GY in the UK, Australia and China (Shearman et al. 2005; Sadras and Lawson 2011; Xiao et al. 2012; Gao et al. 2017). Similar results were evident in the present study that improvements in TKW from pre-1980 to post-2010 were accompanied by increased WSC10, WSC20 and WSC30 (Fig. 5a-d), demonstrated by increased numbers of favorable alleles (Fig. 5e-h). Linear regressions also showed that pyramiding favorable alleles was effective for improving both SWSCC and TKW (Fig. 4). Furthermore, cultivars from Henan and Shandong, two high-yield wheat producing provinces in China, also had higher SWSCC and more favorable WSC alleles (Fig. S9). All these indicated that improvement in stem WSC content is a promising way to improve GY.

In addition, representative markers for pleiotropic loci on chromosomes 1BS (*BS00023084_51*), 2AS (*AX-109293110*), 4BS (*AX-110974144*), 5BL (*AX-111044647*), 6BS (*AX-94910312*) and 6BL (*AX-111494281*) exhibited FAF of 0.67-0.92 for cultivars in the 2010s (Fig. 6). These indicate that there are still opportunities for further improve-ment of SWSCC and TKW by pyramiding more increasing-effect alleles. High-throughput KASP markers for five pleiotropic loci were developed and validated in different environments (Fig. S10). These markers should be useful for MAS targeting improved SWSCC and TKW. Addition-ally, considering the important role of WSC in coping with diverse stresses (Rebetzke et al. 2008; Livingston et al. 2009; Trouvelot et al. 2014; Ovenden et al. 2017), these markers may also have potential values in developing stress-resistant cultivars by MAS.

The present study was conducted in well-watered field conditions with the same field management as used in YHVWWZ of China, and the findings would be useful for wheat breeding in this area and similar environ-ments. Cultivars Luyuan 502, Huaimai 20, Jinmai 61, Luomai 21, Yumai 13, Zhongmai 875, Zhongmai 895, Zhou 8425B, Zhoumai 22, Zhoumai 30, Zhoumai 31, Zhoumai 32 and Zimai 12 possess high SWSCC (WSC10 and WSC20) and TKW with high numbers of increasing-effect alleles (Tables 5, S11). These cultivars could be used as elite germplasms for improving WSC and TKW.

Advantages and disadvantages of this study

In the present study, GWAS was conducted to identify

QTL associated with SWSCC based on a high density of SNP markers. The high marker density has significant impact on the number of haplotypes within trait-associated loci by change in LD pattern and would result in discovering marker-trait associations in low-recombination regions on chromosomes (Kim and Yoo 2016; Andrade et al. 2019). Previously, Dong et al. (2016a) performed GWAS on SWSCC with 18,207 markers from the 90 K SNP array using the same panel of wheat accessions as that used in the present study, but only 11 significant MTAs were identified at two or more environments due to the presence of large gaps, particularly in the D genome (Liu et al. 2017). To resolve this problem, the present GWAS for WSC content was performed using 373,106 SNPs from the 660 K and 90 K SNP arrays. The integrated physical map showed a significantly improved marker density of 0.038 Mb per marker for the whole genome compared with 0.772 Mb per marker (14,061 Mb/18,207 markers) of the 90 K genetic map in Dong et al. (2016a). Consequently, 36, 24 and 19 stable loci were detected for WSC10, WSC20 and WSC30, respectively (Table S11). This indicated that the high-density physical map constructed here gave a significant advantage for GWAS on complex traits like SWSCC.

A relatively small population in this study might be the reason why a stringent threshold like Bonferroni correction (Holm 1979) could not be used for GWAS. Nevertheless, the 166 wheat accessions employed in this study were chosen from more than 400 cultivars, and they are good representatives of wheat germplasms from YHVWWZ of China. This population was previously used for GWAS on yield and quality-related traits, and many important loci were identified and validated (Liu et al. 2017; Zhai et al. 2018; Li et al. 2019). In the present study, 26 out of 62 WSC loci were identified at similar positions to previously reported QTL or genes (Table S11), also indicating the reliability of the results. Furthermore, based on the identified WSC loci, we observed a clear pyramiding effect of favorable WSC alleles in wheat accessions released in past decades, which was consistent with the improvement in SWSCC and TKW (Fig. 5). Therefore, the population is suitable for GWAS of complex traits. In addition, the present study was conducted under well-watered conditions following the local practices in YHVWWZ, so the findings would be useful for breeding in this zone and similar environments.

Conclusion

TKW was significantly correlated with WSC10 and WSC20, indicating the important contribution of WSC to GY. We identified 62 stable loci for WSC at three grain filling time-points, with 36, 24 and 19 loci for WSC10, WSC20 and WSC30, respectively. Of these loci, 36 are potentially new, 16 affected SWSCC at two or more time-points, and 13 showed pleiotropic effects on both WSC and TKW. Linear regression analyses showed clear cumulative effects of increasing-effect alleles on SWSCC and TKW. In past decades, SWSCC and TKW of wheat cultivars in the YHVWWZ of China were significantly improved due to the pyramiding of WSC-increasing alleles. The present study showed a genome-wide genetic landscape of WSC, providing a perspective for understanding the relationship between WSC and GY. The identified WSC loci, especially the 13 pleiotropic loci for both WSC and TKW, are valuable targets for further dissection of the genetic basis underlying SWSCC and TKW. The five WSC-and TKW-associated KASP markers would be valuable tools for improving WSC and GY by MAS.

Table 5   Elite germplasms for improving stem WSC contents and grain yield

Cultivar	Stem WSC content and TKW[a]			Number of favorable alleles[a,b]		
	WSC10 (%)	WSC20 (%)	TKW (g)	WSC10	WSC20	TKW
Huaimai 20	17.59	16.91	45.6	32	18	10
Jinmai 61	17.83	16.06	47.6	29	19	10
Luomai 21	16.70	15.18	49.9	29	19	12
Luyuan 502	16.73	15.83	50.5	26	18	11
Yumai 13	17.72	16.53	46.8	30	18	10
Zhongmai 875	16.69	16.74	54.5	30	21	12
Zhongmai 895	17.04	15.49	46.8	26	19	10
Zhoumai 22	17.20	16.31	51.3	32	20	12
Zhoumai 30	16.69	17.38	48.2	26	20	12
Zhoumai 31	16.47	15.91	47.2	26	20	10
Zhoumai 32	18.28	17.87	47.4	30	19	12
Zimai 12	17.92	17.03	45.6	27	21	10

[a]WSC10 and WSC20 indicate stem WSC contents at 10 and 20 days post-anthesis, respectively. TKW, thousand-kernel weight. The best linear unbiased estimations across four environments for WSC10, WSC20 and TKW are reported here. The ranges for WSC10, WSC20 and TKW were 9.41-18.81%, 8.49-17.87% and 26.9-56.6 g, respectively, in the 166 accessions (Table S1)

[b]Ranges for number of favorable alleles of WSC10, WSC20 and TKW were 6-32, 5-22 and 0-13, respectively, in the 166 accessions (Table S1, Figs. S5, S8). And the potential numbers of favorable alleles in an accession were 36 and 24 for WSC10 and WSC20, respectively (Table S11), and 13 for TKW based on the 13 pleiotropic loci in Table 4

❖ Acknowledgements

The authors are grateful to Prof. R. A. McIntosh, Plant Breeding Institute, University of Sydney, for review of this manuscript. This work was funded by the National Key Research and Development Programs of China (2016YFD0101804, 2016YFD0101802), National Natural Science Foundation of China (31461143021, 31671691), and CAAS Science and Technology Innovation Program.

❖ References

Aggarwal PK, Sinha SK (1984) Effect of water stress on grain growth and assimilate partitioning in two cultivars of wheat contrasting in their yield stability in a drought-environment. Ann Bot 53: 329-340.

Ahmed SAS, Zhang J, Ma W, Dell B (2018) Contributions of *TaS-UTs* to grain weight in wheat under drought. Plant Mol Biol 98: 333-347.

Andrade ACB, Viana JMS, Pereira HD, Pinto VB, Fonseca e Silva F (2019) Linkage disequilibrium and haplotype block patterns inpopcorn populations. PLoS ONE 14: e0219417.

Benjamini Y, Hochberg Y (1995) Controlling the false discovery rate: a practical and powerful approach to multiple testing. J R Stat Soc B 57: 289-300.

Bian Y, Gu X, Sun D, Wang Y, Yin Z, Deng D, Wang Y, Li G (2015) Mapping dynamic QTL of stalk sugar content at different growth stages in maize. Euphytica 205: 85-94.

Bidinger F, Musgrave RB, Fischer RA (1977) Contribution of stored pre-anthesis assimilate to grain yield in wheat and barley. Nature 270: 431-433.

Blum A, Sinmena B, Mayer J, Golan G, Shpiler L (1994) Stem reserve mobilisation supports wheat-grain filling under heat stress. Aust J Plant Physiol 21: 771-781.

Bolouri-Moghaddam MR, van den Ende W (2013) Sweet immunity in the plant circadian regulatory network. J Exp Bot 64: 1439-1449.

Bourne Y, Henrissat B (2001) Glycoside hydrolases and glycosyltransferases: families and functional modules. Curr Opin Struct Biol 11: 593-600.

Bradbury PJ, Zhang Z, Kroon DE, Casstevens TM, Ramdoss Y, Buckler ES (2007) TASSEL: software for association mapping of complex traits in diverse samples. Bioinformatics 23: 2633-2635.

Brenton ZW, Cooper EA, Myers MT, Boyles RE, Shakoor N, Zielinski KJ et al (2016) A genomic resource for the development, improvement, and exploitation of sorghum for bioenergy. Genetics 204: 21-33.

Chao S, Dubcovsky J, Dvorak J, Luo MC, Baenziger SP, Matnyazov R et al (2010) Population-and genome-specific patterns of linkage disequilibrium and SNP variation in spring and winter wheat (Triticum aestivum L.). BMC Genom 11: 727.

Chen X, Min D, Yasir TA, Hu YG (2012) Genetic diversity, population structure and linkage disequilibrium in elite Chinese winter wheat investigated with SSR markers. PLoS ONE 7: e44510.

Cui F, Zhang N, Fan XL, Zhang W, Zhao CH, Yang LJ et al (2017) Utilization of a Wheat660K SNP array-derived high-density genetic map for high-resolution mapping of a major QTL for kernel number. Sci Rep 7: 3788.

del Pozo A, Yáñez A, Matus IA, Tapia G, Castillo D, Sanchez-Jardón L, Araus JL (2016) Physiological traits associated with wheat yield potential and performance under water-stress in a Mediter-ranean environment. Front Plant Sci 7: 987.

Dong Y, Liu J, Zhang Y, Geng H, Rasheed A, Xiao Y et al (2016a) Genome-wide association of stem water soluble carbohydrates in bread wheat. PLoS ONE 11: e0164293.

Dong Y, Zhang Y, Xiao Y, Yan J, Liu J, Wen W et al (2016b) Cloning of *TaSST* genes associated with water soluble carbohydrate content in bread wheat stems and development of a functional marker. Theor Appl Genet 129: 1061-1070.

Ehdaie B, Alloush GA, Waines JG (2008) Genotypic variation in linear rate of grain growth and contribution of stem reserves to grain yield in wheat. Field Crops Res 106: 34-43.

Eom JS, Chen LQ, Sosso D, Julius BT, Lin IW, Qu XQ et al (2015) SWEETs, transporters for intracellular and intercellular sugar translocation. Curr Opin Plant Biol 25: 53-62.

Foulkes MJ, Scott RK, Sylvester-Bradley R (2002) The ability of wheat cultivars to withstand drought in UK conditions: formation of grain yield. J Agr Sci 138: 153-169.

Foulkes MJ, Sylvester-Bradley R, Weightman R, Snape JW (2007) Identifying physiological traits associated with improved drought resistance in winter wheat. Field Crops Res 103: 11-24.

Foulkes MJ, Slafer GA, Davies WJ, Berry PM, Sylvester-Bradley R, Martre P et al (2010) Raising yield potential of wheat. III. Optimizing partitioning to grain while maintaining lodging resistance. J Exp Bot 62: 469-486.

Gao F, Ma D, Yin G, Rasheed A, Dong Y, Xiao Y et al (2017) Genetic progress in grain yield and physiological traits in Chinese wheat cultivars of Southern Yellow and Huai Valley since 1950. Crop Sci 57: 760-773.

Goggin DE, Setter TL (2004) Fructosyltransferase activity

and fructan accumulation during development in wheat exposed to terminal drought. Funct Plant Biol 31: 11-21.

Halford NG, Curtis TY, Muttucumaru N, Postles J, Mottram DS (2011) Sugars in crop plants. Ann Appl Biol 158: 1-25.

Hao C, Wang L, Ge H, Dong Y, Zhang X (2011) Genetic diversity and linkage disequilibrium in Chinese bread wheat (*Triticum aestivum* L.) revealed by SSR markers. PLoS ONE 6: e17279.

Holland JB, Nyquist WE, Cervantes-Martínez CT (2003) Estimating and interpreting heritability for plant breeding: an update. PlantBreed Rev 22: 11-112.

Holm S (1979) A simple sequentially rejective multiple test procedure. Scand J Stat 6: 65-70.

Hou J, Huang X, Sun W, Du C, Wang C, Xie Y et al (2018) Accumulation of water-soluble carbohydrates and gene expression in wheat stems correlates with drought resistance. J PlantPhysiol 231: 182-191.

IWGSC (2018) Shifting the limits in wheat research and breeding using a fully annotated reference genome. Science 361: eaar7191.

Jiang Q, Hou J, Hao C, Wang L, Ge H, Dong Y, Zhang X (2011) The wheat (*T. aestivum*) sucrose synthase 2 gene (*TaSus2*) active in endosperm development is associated with yield traits. Funct Integr Genomics 11: 49-61.

Kato T (1995) Change of sucrose synthase activity in developingendosperm of rice cultivars. Crop Sci 35: 827-831.

Kim SA, Yoo YJ (2016) Effects of single nucleotide polymorphism marker density on haplotype block partition. Genomics Inform 14: 196-204.

Kobata T, Palta JA, Turner NC (1992) Rate of development of postanthesis water deficits and grain filling of spring wheat. Crop Sci 32: 1238-1242.

Kooiker M, Drenth J, Glassop D, McIntyre CL, Xue GP (2013) TaMYB13-1, a R2R3 MYB transcription factor, regulates the fructan synthetic pathway and contributes to enhanced fructan accumulation in bread wheat. J Exp Bot 64: 3681-3696.

Korte A, Farlow A (2013) The advantages and limitations of trait analysis with GWAS: a review. Plant Methods 9: 29.

Lao J, Oikawa A, Bromley JR, McInerney P, Suttangkakul A, Smith-Moritz AM et al (2014) The plant glycosyltransferase clone col-lection for functional genomics. Plant J 79: 517-529.

Li H, Ye G, Wang J (2007) A modified algorithm for the improvement of composite interval mapping. Genetics 175: 361-374.

Li W, Zhang B, Li R, Chang X, Jing R (2015) Favorable alleles for stem water-soluble carbohydrates identified by association analysis contribute to grain weight under drought stress conditions in wheat. PLoS ONE 10: e0119438.

Li F, Wen W, Liu J, Zhang Y, Cao S, He Z et al (2019) Genetic architecture of grain yield in bread wheat based on genome-wide association studies. BMC Plant Biol 19: 168.

Liu K, Muse SV (2005) PowerMarker: an integrated analysis environment for genetic marker analysis. Bioinformatics 21: 2128-2129.

Liu X, Huang M, Fan B, Buckler ES, Zhang Z (2016) Iterative usage of fixed and random effect models for powerful and efficient genome-wide association studies. PLoS Genet 12: e1005767.

Liu J, He Z, Rasheed A, Wen W, Yan J, Zhang P et al (2017) Genome-wide association mapping of black point reaction in common wheat (*Triticum aestivum* L.). BMC Plant Biol 17: 220.

Livingston DP, Hincha DK, Heyer AG (2009) Fructan and its relationship to abiotic stress tolerance in plants. Cell Mol Life Sci 66: 2007-2023 Lopes MS, Reynolds MP, Manes Y, Singh RP, Crossa J, Braun HJ (2012) Genetic yield gains and changes in associated traits of CIMMYT spring bread wheat in a "historic" set representing 30 years of breeding. Crop Sci 52: 1123-1131.

McIntyre CL, Casu RE, Rattey A, Dreccer MF, Kam JW, van Her-waarden AF et al (2011) Linked gene networks involved in nitro-gen and carbon metabolism and levels of water-soluble carbo-hydrate accumulation in wheat stems. Funct Integr Genomics 11: 585-597.

Muqaddasi QH, Zhao Y, Rodemann B, Plieske J, Ganal MW, Röder MS (2019) Genome-wide association mapping and prediction of adult stage *Septoria tritici* blotch infection in European winter wheat via high-density marker arrays. Plant Genome 12: 1.

Murray MG, Thompson WF (1980) Rapid isolation of high molecularweight plant DNA. Nucl Acids Res 8: 4321-4326.

Nagata K, Shimizu H, Terao T (2002) Quantitative trait loci for nonstructural carbohydrate accumulation in leaf sheaths and culms of rice (*Oryza sativa* L.) and their effects on grain filling. Breed Sci 52: 275-283.

Ovenden B, Milgate A, Lisle C, Wade LJ, Rebetzke GJ, Holland JB (2017) Selection for water-soluble carbohydrate accumulation and investigation of genetic × environment interactions in an elite wheat breeding population. Theor Appl Genet 130: 2445-2461.

Pheloung PC, Siddique KHM (1991) Contribution of stem dry matter to grain yield in wheat cultivars. Funct Plant Biol 18: 53-64.

Phung HD, Sugiura D, Sunohara H, Makihara D, Kondo M, Nishiuchi S, Doi K (2019) QTL analysis for carbon assimilate translocation-related traits during maturity in rice (Oryza sativa L.). Breeding Sci 69: 289-296.

Platten JD, Cobb JN, Zantua RE (2019) Criteria for evaluating molecular markers: comprehensive quality metrics to improve marker-assisted selection. PLoS ONE 14: e0210529.

Pritchard JK, Stephens M, Donnelly P (2000) Inference of populationstructure using multilocus genotype data. Genetics 155: 945-959.

Rahimi Y, Bihamta MR, Taleei A, Alipour H, Ingvarsson PK (2019) Genome-wide association study of agronomic traits in bread wheat reveals novel putative alleles for future breeding programs. BMC Plant Biol 19: 541.

Rebetzke GJ, van Herwaarden AF, Jenkins C, Weiss M, Lewis D, Ruuska S et al (2008) Quantitative trait loci for water-soluble carbohydrates and associations with agronomic traits in wheat. Aust J Agr Res 59: 891-905.

Rolland F, Baena-Gonzalez E, Sheen J (2006) Sugar sensing and sign-aling in plants: conserved and novel mechanisms. Annu Rev Plant Biol 57: 675-709.

Ruuska SA, Rebetzke GJ, van Herwaarden AF, Richards RA, Fettell NA, Tabe L, Jenkins CL (2006) Genotypic variation in water-soluble carbohydrate accumulation in wheat. Funct Plant Biol 33: 799-809.

Ruuska SA, Lewis DC, Kennedy G, Furbank RT, Jenkins CL, Tabe LM (2008) Large scale transcriptome analysis of the effects of nitrogen nutrition on accumulation of stem carbohydrate reserves in reproductive stage wheat. Plant Mol Biol 66: 15-32.

Sadras VO, Lawson C (2011) Genetic gain in yield and associated changes in phenotype, trait plasticity and competitive ability of South Australian wheat varieties released between 1958 and 2007. Crop Pasture Sci 62: 533-549.

Sharma RC, Crossa J, Velu G, Huerta-Espino J, Vargas M, Payne TS, Singh RP (2012) Genetic gains for grain yield in CIMMYT spring bread wheat across international environments. Crop Sci 52: 1522-1533.

Shearman VJ, Sylvester-Bradley R, Scott RK, Foulkes MJ (2005) Physiological processes associated with wheat yield progress inthe UK. Crop Sci 45: 175-185.

Snape JW, Foulkes MJ, Simmonds J, Leverington M, Fish LJ, Wang Y, Ciavarrella M (2007) Dissecting gene×environmental effects on wheat yields via QTL and physiological analysis. Euphytica 154: 401-408.

Teulat B, Borries C, This D (2001) New QTLs identified for plant water status, water-soluble carbohydrate and osmotic adjustment in a barley population grown in a growth-chamber under two water regimes. Theor Appl Genet 103: 161-170.

Thévenot C, Simond-Côte E, Reyss A, Manicacci D, Trouverie J, Le Guilloux M et al (2005) QTLs for enzyme activities and soluble carbohydrates involved in starch accumulation during grain filling in maize. J Exp Bot 56: 945-958.

Trouvelot S, Héloir MC, Poinssot B, Gauthier A, Paris F, Guillier C et al (2014) Carbohydrates in plant immunity and plant protection: roles and potential application as foliar sprays. Front Plant Sci 5: 592.

Turner LB, Cairns AJ, Armstead IP, Ashton J, Skøt K, Whittaker D, Humphreys MO (2006) Dissecting the regulation of fructan metabolism in perennial ryegrass (Lolium perenne) with quantitative trait locus mapping. New Phytol 169: 45-58.

Veenstra LD, Jannink JL, Sorrells ME (2017) Wheat fructans: a potential breeding target for nutritionally improved, climate-resilient varieties. Crop Sci 57: 1624-1640.

Wang RX, Hai L, Zhang XY, You GX, Yan CS, Xiao SH (2009) QTL mapping for grain filling rate and yield-related traits in RILs of the Chinese winter wheat population Heshangmai × Yu8679. Theor Appl Genet 118: 313-325.

Wang Z, Liu X, Li R, Chang X, Jing R (2011) Development of near-infrared reflectance spectroscopy models for quantitative determination of water-soluble carbohydrate content in wheat stem and glume. Anal Lett 44: 2478-2490.

Wang S, Wong D, Forrest K, Allen A, Chao S, Huang BE et al (2014) Characterization of polyploid wheat genomic diversity using a high-density 90000 single nucleotide polymorphism array. Plant Biotechnol J 12: 787-796.

Wang DR, Han R, Wolfrum EJ, McCouch SR (2017)

The buffering capacity of stems: genetic architecture of nonstructural carbohydrates in cultivated Asian rice, *Oryza sativa*. New Phytol 215: 658-671.

Wardlaw IF, Willenbrink J (2000) Mobilization of fructan reserves and changes in enzyme activities in wheat stems correlate with water stress during kernel filling. New Phytol 148: 413-422.

Xiao YG, Qian ZG, Wu K, Liu JJ, Xia XC, Ji WQ, He ZH (2012) Genetic gains in grain yield and physiological traits of winter wheat in Shandong Province, China, from 1969 to 2006. Crop Sci 52: 44-56 .

Xu S (2003) Theoretical basis of the Beavis effect. Genetics165: 2259-2268.

Xue GP, McIntyre CL, Jenkins CLD, Glassop D, Herwaarden AF, Shorter R (2008) Molecular dissection of variation in carbohydrate metabolism related to water soluble carbohydrate accumulation in stems of wheat (*Triticum aestivum* L.). Plant Physiol 146: 441-454.

Xue GP, Kooiker M, Drenth J, McIntyre CL (2011) TaMYB13 is a transcriptional activator of fructosyltransferase genes involved in β-2, 6-linked fructan synthesis in wheat. Plant J 68: 857-870.

Yan L, Loukoianov A, Blechl A, Tranquilli G, Ramakrishna W, San Miguel P et al (2004) The wheat VRN2 gene is a flowering repressor down-regulated by vernalization. Science 303: 1640-1644.

Yáñez A, Tapia G, Guerra F, del Pozo A (2017) Stem carbohydrate dynamics and expression of genes involved in fructan accumulation and remobilization during grain growth in wheat (*Triticum aestivum* L.) genotypes with contrasting tolerance to water stress. PLoS ONE 12: e0177667.

Yang DL, Jing RL, Chang XP, Li W (2007) Identification of quantitative trait loci and environmental interactions for accumulation and remobilization of water-soluble carbohydrates in wheat (*Triticum aestivum* L.) stems. Genetics 176: 571-584.

Yin C, Li H, Li S, Xu L, Zhao Z, Wang J (2015) Genetic dissection on rice grain shape by the two-dimensional image analysis in one *japonica×indica* population consisting of recombinant inbred lines. Theor Appl Genet 128: 1969-1986.

Yu J, Pressoir G, Briggs WH, Bi IV, Yamasaki M, Doebley JF et al (2006) A unified mixed-model method for association map-ping that accounts for multiple levels of relatedness. Nat Genet 38: 203-208.

Yue AQ, Ang LI, Mao XG, Chang XP, Li RZ, Jing RL (2017) Single-nucleotide polymorphisms, mapping and association analysis of 1-FFT-A1 gene in wheat. J Integr Agr 16: 789-799.

Zhai S, Liu J, Xu D, Wen W, Yan J, Zhang P et al (2018) A Genome-wide association study reveals a rich genetic architecture of flourcolor-related traits in bread wheat. Front Plant Sci 9: 1136.

Zhang Z, Ersoz E, Lai CQ, Todhunter RJ, Tiwari HK, Gore MA et al (2010) Mixed linear model approach adapted for genome-wide association studies. Nat Genet 42: 355-360.

Zhang B, Li W, Chang X, Li R, Jing R (2014) Effects of favorable alleles for water-soluble carbohydrates at grain filling on grain weight under drought and heat stresses in wheat. PLoS ONE 9: e102917.

Zhang J, Xu Y, Chen W, Dell B, Vergauwen R, Biddulph B et al (2015) A wheat *1-FEH w3* variant underlies enzyme activity for stem WSC remobilization to grain under drought. New Phytol 205: 293-305.

Genetic architecture of grain yield in bread wheat based on genome-wide association studies

Faji Li[1,2], Weie Wen[1,2], Jindong Liu[2], Yong Zhang[2], Shuanghe Cao[2], Zhonghu He[2,3], Awais Rasheed[2,3], Hui Jin[2,4], Chi Zhang[5], Jun Yan[6], Pingzhi Zhang[7], Yingxiu Wan[7] and Xianchun Xia[2]

[1] College of Agronomy, Xinjiang Agricultural University, Urumqi 830052, Xinjiang, China.

[2] Institute of Crop Sciences, National Wheat Improvement Center, Chinese Academy of Agricultural Sciences (CAAS), 12 Zhongguancun South Street, Beijing 100081, China.

[3] International Maize and Wheat Improvement Center (CIMMYT) China Office, c/o CAAS, 12 Zhongguancun South Street, Beijing 100081, China.

[4] Sino-Russia Agricultural Scientific and Technological Cooperation Center, Heilongjiang Academy of Agricultural Sciences, 368 Xuefu Street, Harbin 150086, Heilongjiang, China.

[5] School of Chemical Science and Engineering, Royal Institute of Technology, Teknikringen 42, SE-100 44 Stockholm, Sweden.

[6] Institute of Cotton Research, Chinese Academy of Agricultural Sciences (CAAS), 38 Huanghe Street, Anyang 455000, Henan, China.

[7] Crop Research Institute, Anhui Academy of Agricultural Sciences, 40 Nongke South Street, Hefei 230001, Anhui, China

Background: Identification of loci for grain yield (GY) and related traits, and dissection of the genetic architecture are important for yield improvement through marker-assisted selection (MAS). Two genome-wide association study (GWAS) methods were used on a diverse panel of 166 elite wheat varieties from the Yellow and Huai River Valleys Wheat Zone (YHRVWD) of China to detect stable loci and analyze relationships among GY and related traits.

Results: A total of 326, 570 single nucleotide polymorphism (SNP) markers from the wheat 90 K and 660 K SNP arrays were chosen for GWAS of GY and related traits, generating a physical distance of 14, 064. 8 Mb. One hundred and twenty common loci were detected using SNP-GWAS and Haplotype-GWAS, among which two were potentially functional genes underpinning kernel weight and plant height (PH), eight were at similar locations to the quantitative trait loci (QTL) identified in recombinant inbred line (RIL) populations in a previous study, and 78 were potentially new. Twelve pleiotropic loci were detected on eight chromosomes; among these the interval 714. 4-725. 8 Mb on chromosome 3A was significantly associated with GY, kernel number per spike (KNS), kernel width (KW), spike dry weight (SDW), PH, uppermost internode length (UIL), and flag leaf length (FLL). GY shared five loci with thousand kernel weight (TKW) and PH, indicating significantly affected by two

Published in BMC Plant Biology, 2019, 19 (1): 1-19. (doi. org/10. 1186/s12870-019-1781-3)

traits. Compared with the total number of loci for each trait in the diverse panel, the average number of alleles for increasing phenotypic values of GY, TKW, kernel length (KL), KW, and flag leaf width (FLW) were higher, whereas the numbers for PH, UIL and FLL were lower. There were significant additive effects for each trait when favorable alleles were combined. UIL and FLL can be directly used for selecting high-yielding varieties, whereas FLW can be used to select spike number per unit area (SN) and KNS.

Conclusions: The loci and significant SNP markers identified in the present study can be used for pyramiding favorable alleles in developing high-yielding varieties. Our study proved that both GWAS methods and high-density genetic markers are reliable means of identifying loci for GY and related traits, and provided new insight to the genetic architecture of GY.

Keywords: GWAS, Marker-assisted selection, Single nucleotide polymorphism, *Triticum aestivum*

Background

Bread wheat is an important crop cultivated on ~200 million hectares worldwide, and provides one fifth of the total needs of the global population [1-3]. Grain yield (GY) improvement is one of the most challenging objectives in wheat breeding due to the complex genetic architecture and low heritability. The Yellow and Huai River Valleys Wheat Zone (YHRVWZ) is the major wheat-producing region in China, and yield potential in this region has been improved over recent decades [4-6]. However, wheat production in the region is facing problems of decreasing groundwater and hence reduced irrigation frequency and decreasing growing area in the northern part, and frequent occurrence of Fusarium head blight in the southern part. Moreover, there is a decline in the rate of increase of yield potential in conventional breeding.

GY is a complex trait, significantly associated with spike number per unit area (SN), kernel numberper spike (KNS) and thousand-kernel weight (TKW). However, grain shape, spike architecture, plant height (PH), and flag leaf related traits can also affect GY through effects on photosynthetic intensity, grain filling and dry matter translocation[5-8]. These traits have higher heritabilities (h^2) than GY and are easier to select in small plots at the early stages of breeding programs. Previous studies showed that increased yield potential in the YHRVWZ was largely

associated with increased kernels per square meter, biomass and harvest index, and reduced PH[5,6]. Those improvements were mainly at- tributed to the use of dwarfing genes (*Rht1*, *Rht2*, *Rht8* and *Rht24*) and the 1BL.1RS translocation lines[8-13]. However, with the current widespread near-fixation of these genes new variation must be sought. It is now believed that further improvement in yield potential can be achieved only by a detailed understanding of its genetic architecture combined with marker-assisted selection (MAS).

MAS is considered to be a key technique to break through yield bottleneck of conventional breeding for further improvement of yield potential of wheat. The application potential of MAS depends on the number of available genes and tightly linked molecular markers. To date, about 65 genes have been cloned in wheat, among which 40 are associated with GY and related traits[14-17]. For all cloned genes, around 150 functional markers have been converted to kompetitive allele-specific PCR (KASP) formats convenient for high-throughput genotyping[15]. Although there are many reports on quantitative trait loci (QTL) mapping and genome-wide association study (GWAS) of yield and related trait loci[18-23], relatively few outcomes have been applied in selection of wheat lines in actual breeding programs. To enhance the application of MAS, more detailed studies on genetic architecture and identification of related loci for GY should be taken.

Single nucleotide polymorphism (SNP) arrays developed from the transcriptomes of plants and animals[24] providing the most advanced approach in searching for candidate genes for economic traits by QTL mapping or GWAS. The wheat 90 K and 660 K SNP arrays are gradually replacing simple sequence repeat (SSR) markers in genetic studies of yield, quality, disease resistance and stress tolerance[25-28]. In our previous study, 23 new stable QTL and 11 QTL clusters were identified for 12 yield related traits using high-density linkage maps constructed with the wheat 90 K SNP array in three RIL populations[17]. Wheat 50 K and 15 K SNP arrays now available for selecting important traits in wheat programs, include SNP markers derived from the wheat 35 K, 90 K and 660 K arrays, functional markers of cloned genes, and closely linked markers identified by QTL mapping and GWAS. SNP markers are becoming the main tool for genetic studies and breeding of crop species.

Analysis of GWAS data is based on linkage disequilibrium (LD) and provides a much higher resolutioncapacity to capture insights into the genetic architecture of complex traits than traditional QTL mapping[29]. Un- like QTL mapping, GWAS uses available germplasm as materials and bypasses the time of developing segregating populations. Moreover, QTL mapping by bi-parental populations focuses on specific traits, whereas a wider range of germplasm can be used in GWAS to phenotype many traits with one cycle of genotyping. Genetic variance of traits in crop species may be caused by a single SNP, but is more often attributed to several SNPs within a haplotype block[30]. Therefore, SNP-GWAS and Haplotype-GWAS can be complementary and verifiable in identifying genes controlling complex traits. SNP-GWAS is commonly applied in genetic studies of crop species, whereas Haplotype-GWAS has been mostly used in detecting heterozygous chromosome segments in cross-pollinated crops[31,32].

The aims of the present study were to: 1) identify stable loci for GY and related traits using both SNP-GWAS and Haplotype-GWAS based on high-density SNP markers, 2) investigate genetic relationships among yield and related traits, and 3) detect available loci for MAS of traits in breeding for high yield.

Results

Phenotypic evaluation

There was significant and continuous variation in GY and related traits across the diverse panel (Additional file1: Table S1; Additional file 2: Figure S1). ANOVA showed highly significant effects ($P < 0.01$) of lines, environments and line × environment interactions on all traits (Additional file 3: Table S2). GY in the panel was moderately heritable ($h^2 = 0.72$), whereas the other 12 traits showed high h^2 ($>$ 0.89), indicating that most of the traits were stable and largely determined by genetic factors (Additional file 3: Table S2).

GY showed significant ($P < 0.01$) and positive correlations with TKW and kernel width (KW); but significant and negative correlations with PH, uppermost internode length (UIL) and flag leaf length (FLL) (Additional file 4: Table S3); SN was significantly ($P < 0.01$) and negatively correlated with KNS, TKW, KW, spike dry weight (SDW) and flag leaf width (FLW) ($r = 0.38$ to 0.70); KNS exhibited significant ($P < 0.01$) and positive correlations with spike length (SL), SDW and FLW ($r = 0.36$ to 0.64); TKW was significantly ($P < 0.01$) and positively correlated with kernel length (KL), KW and SDW ($r = 0.50$ to 0.83).

Marker coverage and genetic diversity

After filtering, 326,570 polymorphic SNPs were employed for GWAS analysis; 10,780 were from the wheat 90 K SNP array and 315,790 came from the wheat 660 K SNP array (Additional file 5: Table S4; Additional file 6: Figure S2a). Among polymorphic SNP markers, 39.7, 49.4 and 10.9% were from the

A, B and D genomes, respectively. Chromo- some 3B had the most SNP markers (41, 439), whereas chromosome 4D possessed the least (2061). The total markers spanned a physical distance of 14, 064. 8 Mb, with an average marker density of 0. 043 Mb per marker. The average genetic diversity and polymorphism information content (PIC) for the whole genome were 0. 34 and 0. 27, respectively, and the average genetic diversities for A, B and D genomes were 0. 34, 0. 35 and 0. 32, and average PIC were 0. 28, 0. 28 and 0. 26, respectively.

Haplotype composition and coverage

Among all polymorphic SNP markers, 275, 000 were assigned to 31, 748 haplotype blocks, and 116, 555 haplotypes were generated based on 4-gamete LD analyses (Additional file 6: Figure S2b; Additional file 7: Table S5). The D genome had the least haplotype blocks and haplotypes (3384 and 10, 579), followed by the A (12, 574 and 46, 891) and B (15, 790 and 59, 085) genomes. Like the SNP marker coverage, chromosomes 3B and 4D had the most and least haplotype blocks, respectively. The average number of SNP markers for one haplotype block was 7. 9, and the average length was 74. 7 kb. For A, B and D genomes, the average numbers of SNP markers were 8. 8, 8. 7 and 6. 1, and the average lengths were 85. 9, 93. 6 and 44. 7 kb, respectively. Haplotype blocks on chromosomes 3B and 5B harbored the most SNP markers (10. 8 and 11. 0) with maximum length of 112. 7 and 97. 9 kb, whereas the haplotype blocks on chromosome 4D had the least SNP markers (4. 3) and the minimum length (24. 8). The range of haplotype block length was 0. 001-200. 0 kb.

Population structure and linkage disequilibrium

As shown in Liu et al. [26], the germplasm consisted of three subgroups: Subgroup I contained 62 varieties mainly from Shandong province and foreign countries; Subgroup II had 54 varieties mainly from Henan, Anhui and Shaanxi provinces; and Subgroup III comprised 50 varieties mainly from Henan. Average LD for the whole genome was 8 Mb, and for A, B and D genomes, 6, 4 and 11 Mb, respectively.

Genome-wide association studies

Totals of 239 and 248 loci for GY and related traits were identified on all 21 chromosomes using Tassel v5. 0 and PLINK, respectively (Additional file 8: Table S6; Additional file 9: Figure S3; Additional file 10: Figure S4). In SNP-GWAS 18, 13, 20, 22, 14, 23, 19, 10, 21, 28, 20, 16 and 15 loci were detected for GY, SN, KNS, TKW, KL, KW, SL, SDW, heading date (HD), PH, UIL, FLL and FLW, respectively; in Haplotype-GWAS the corresponding numbers were 20, 13, 9, 27, 13, 32, 11, 13, 21, 27, 27, 24 and 11. In both methods, the D genome possessed the lowest number of loci, consistent with its lowest diversity. One hundred and twenty loci were common in SNP-GWAS and Haplotype-GWAS, 49, 56 and 15 in the A, B and D genomes, respectively (Table 1).

GY and yield components

Twelve common loci for GY were identified on chromosomes 1A (2), 1B (2), 2A, 2B, 2D, 3A, 3B, 3D, 5A and 5B, with single loci explaining 6. 9-17. 7% and 9. 1-22. 6% of the phenotypic variances in SNP-GWAS and Haplotype-GWAS, respectively. Seven loci, on 1A (AX _ 110387060 and AX _ 110418502), 1B (AX _ 110508372 and AX _ 109820171), 2D (AX _ 109941480), 3B (AX _ 109881378), and 3D (AX _ 95257733) were detected in four environments and best linear unbiased estimation (BLUE) values by both methods. The 1A (AX _ 110418502) and 1B (H4268) loci explained the largest of phenotypic variances in SNP-GWAS and Haplotype-GWAS, respectively.

Six common loci for SN were detected on chromosomes 5B (2), 6B (2), 6D and 7D, explaining 7. 1-17. 1% and 9. 1- 23. 3% of the phenotypic variances in SNP-GWAS and Haplotype-GWAS, respectively. The 5B locus (IWB56499) was significant in three environments and

Table 1　Loci for grain yield and related traits in the diverse panel identified by both SNP-GWAS and Haplotype-GWAS and comparisons with previous studies

Trait	Chr	Physical position (Mb)[a]	Marker[b]	SNP-GWAS Environment	R^2 (%)[c]	Haplotype-GWAS Haplotype[d]	Environment	R^2 (%)[c]	QTL/marker/gene[e]
GY	1A	402.9-407.4	AX_110387060	E3/E4/E5/E6/B	6.9-15.5	H1290	E4/E5/E6/B	10.1-19.8	
	1A	434.0-439.9	AX_110418502	E4/E5/E6/B	6.9-17.7	H1313	E4/E5/E6/B	9.3-19.2	QYld.abrii-1A₁.2 [23]
	1B	539.6-542.6	AX_110508372	E4/E5/E6/B	7.1-12.8	H3650	E4/E5/E6/B	9.3-21.5	
	1B	673.6-675.7	AX_109820171	E3/E5/E6/B	7.6-14.3	H4268	E4/E5/E6/B	10.7-22.6	
	2A	30.3-31.9	AX_94546135	E4/E6/B	7.1-11.6	H5187	E4/E6/B	9.3-17.6	
	2B	106.1-108.9	AX_111210290	E2/E4/B	7.0-12.5	H7877	E1/E4/E5/E6/B	9.1-20.8	
	2D	643.1-650.8	AX_109941480	E4/E5/E6/B	7.0-12.3	H10738	E4/E5/E6/B	9.3-14.6	
	3A	721.3-724.0	AX_111492146	E5/E6/B	7.1-11.7	H11976	E3/E5/E6/B	9.1-18.2	
	3B	20.5-22.0	AX_109881378	E4/E5/E6/B	7.4-10.7	H12363	E4/E5/E6/B	9.3-17.6	
	3D	570.8-572.8	AX_95257733	E3/E4/E6/B	7.1-9.9	H15950	E1/E4/E4/E6/B	9.2-16.6	
	5A	570.2-570.3	AX_110523824	E3/E4/B	7.8-14.8	H19931	E1/E3/E4/E5/E6/B	10.0-20.1	barc151 [35]
	5B	587.1-588.2	AX_110995303	E3/E6/B	7.1-10.2	H22191	E1/E2/E3/E5/E6/B	9.4-15.7	
SN	5B	530.8-534.0	IWB56499	E1/E3/E4/B	7.1-11.9	H21909	E2/E3/E4/B	9.1-12.2	
	5B	696.3-700.9	AX_109936345	E2/E4/B	7.1-12.1	H22717	E1/E2/E3/E4/B	9.1-15.3	
	6B	605.8-607.1	IWB51629	E1/E3/B	7.5-8.1	H26344	E1/E2/E3/E4/B	9.5-16.6	
	6B	696.6-697.6	AX_95260682	E1/E3/B	7.4-9.3	H26799	E1/E2/E3/B	10.5-13.6	
	6D	468.8-471.0	AX_110652999	E2/E3/B	7.2-17.1	H27181	E1/E2/E3/E4/B	9.5-18.8	
	7D	75.9-76.9	AX_111490489	E2/E3/B	7.7-9.0	H31325	E1/E2/E3/E4/B	9.8-23.3	
KNS	1A	36.8-42.5	AX_111579941	E1/E2/E3/E5/E6/B	7.1-11.9	H322	E1/E2/E3/E4/E5/E6/B	9.2-15.8	QGns.uca-1AS.e1 [43]
	1A	497.5-498.1	AX_108737858	E1/E2/E3/E4/E5/E6/B	7.1-12.7	H1479	E1/E5/E6/B	9.3-12.5	QGnu.abrii-1A₁.1 [23]
	2A	117.2-117.3	IWB45503	E3/E4/E5/E6	7.2-8.6	H5556	E3/E4/E5/B	9.5-11.5	QGps.ccsu-2A.3 [34]; umc63 [41]; gwm122 [42]
	2D	633.2-634.5	IWB57054	E1/E2/E3/E4/E5/E6/B	7.4-17.7	H10675	E1/E3/E6/B	9.3-15.4	
	3A	36.8-37.2	AX_110657474	E1/E2/E3/E5/E6/B	7.4-11.1	H11017	E1/E2/E3/E5/E6/B	9.7-14.5	

(continued)

Trait	Chr	Physical position (Mb)[a]	SNP-GWAS			Haplotype-GWAS			QTL/marker/gene[e]
			Marker[b]	Environment	R^2 (%)[c]	Haplotype[d]	Environment	R^2 (%)[c]	
	3A	719.3–725.8	AX_108992368	E2/E3/E4/E5/E6/B	7.1–12.4	H11953	E3/E4/E5/E6/B	9.1–15.1	QGnu.abrii-3A [23]; QKNS.caas-3AL [40]
	5B	416.4–418.0	AX_109538915	E1/E4/E5/E6/B	7.1–10.5	H21421	E1/E4/E6/B	9.3–14.6	gwm213 [41]
	5B	489.3–492.9	AX_109537496	E1/E2/E3/E4/E6/B	7.4–10.9	H21713	E1/E2/E3/E4/E5/E6/B	9.5–14.8	
	7A	49.1–50.0	AX_89571435	E2/E3/E4/E5/E6/B	7.2–12.6	H27643	E1/E3/E4/E5/E6/B	9.3–13.9	
TKW	1B	658.7–662.5	AX_111147652	E1/E2/E4/E6/B	7.1–10.2	H4181	E1/E2/E3/E4/E5/E6/B	9.2–15.4	QTgw.cau-1B, QGl.cau-1B.1 [51]
	2A	760.6–760.7	AX_111579921	E2/E4/E5/B	7.0–8.5	H7248	E1/E2/E3/E4/E5/E6/B	9.3–18.1	
	2B	106.0–107.1	IWB50438	E1/E2/E4	6.7–9.1	H7872	E1/E2/E3/E4/E5/E6/B	9.3–27.0	
	4B	40.4–44.9	AX_110713957	E1/E2/E3/E4/E5/E6/B	7.1–12.7	H17589	E1/E2/E3/E4/E5/E6/B	9.4–22.0	QTgw-4B1 [48]
	4B	159.2–163.0	AX_109427900	E1/E2/E6/B	7.1–9.2	H17841	E1/E2/E3/E4/E5/E6/B	9.4–25.1	
	4B	670.4–670.5	IWB47765	E1/E5/E6/B	6.9–9.2	H18552	E1/E2/E3/E4/E5/E6/B	10.2–17.9	
	5A	706.2–708.0	AX_110958315	E1/E3/E4/E6/B	8.2–13.0	H20367	E1/E2/E3/E4/E5/E6/B	9.2–19.8	QTKW.caas-5AL.1 [40] Excalibur_c23801_115 [27]; wsnp_Ex_c16045_2471413 [45]; Ku_c7546_861 [46]
	5B	692.7–696.4	AX_108769612	E3/E4/E5/E6/B	7.2–9.9	H22684	E1/E2/E3/E4/E5/E6/B	9.3–15.8	
	6B	450.6–454.1	AX_110368497	E2/E3/E5/E6/B	6.8–8.3	H25900	E1/E2/E3/E5/E6/B	12.2–14.1	
	6B	675.4–677.5	AX_109917592	E1/E3/E6/B	7.2–9.7	H26662	E1/E3/E4/E5/E6/B	15.8–20.6	QTKW.caas-6BL [17]
	6D	461.4–468.0	AX_109481324	E2/E3/E6	6.8–8.3	H27181	E1/E2/E3/E4/E5/E6/B	9.3–16.5	
	7D	63.0–69.7	AX_109922697	E2/E3/E4/E5/E6/B	7.1–12.3	H31315	E1/E2/E3/E4/E5/E6/B	10.4–17.8	QKL.caas-7DS [17] QGw.ccsu-7D.1 [44]
KL	1B	26.9–30.8	AX_110032293	E3/E4/E5/E6/B	7.1–12.7	H2345	E2/E3/E4/E5/E6/B	9.2–13.5	
	1B	642.6–642.7	AX_108849700	E1/E2/E3/E5/E6/B	7.0–9.6	H4081	E2/E5/E6/B	9.2–12.0	QKL.caas-1BL [17]; tplb0043a07_1411 [45]
	2A	740.4–745.0	IWB32119	E1/E2/E3/E4/E5/E6/B	7.2–14.2	H7042	E1/E2/E3/E4/E5/E6/B	9.2–13.8	TaFlo2-A1 [49]
	3B	782.9–783.0	AX_111116403	E1/E2/E5/E6/B	7.2–9.9	H15574	E2/E5/E6/B	9.3–11.9	

(continued)

Trait	Chr	Physical position (Mb)[a]	SNP-GWAS Marker[b]	SNP-GWAS Environment	SNP-GWAS R² (%)[c]	Haplotype-GWAS Haplotype[d]	Haplotype-GWAS Environment	Haplotype-GWAS R² (%)[c]	QTL/marker/gene[e]
	5B	52.9–55.2	IWB50649	E1/E2/E3/E4/E5/E6/B	7.2–11.7	H20632	E3/E5/E6/B	10.5–12.6	QTgw.abrii-5B₁.1 [23]; QTgw.cau-5B. QGl.cau-5B.2 [47]
	5D	475.8–476.3	AX_111122970	E1/E2/E3/E4/E6/B	7.2–12.4	H22993	E1/E2/E3/E4/E5/E6/B	9.4–15.4	
KW	1A	9.6–12.0	IWB6999	E2/E3/E4/B	7.1–9.3	H46	E1/E2/E3/E4/E5/E6/B	9.1–17.5	QGw.ccsu-1A.1 [44]
	1A	532.6–533.4	IWB7676	E2/E3/E4/E6/B	7.0–10.4	H1679	E1/E2/E3/E4/E5/E6/B	9.2–16.3	QTgw.cau-1A [47]
	2A	27.3–29.9	AX_111037158	E1/E3/E4/E5/E6	7.2–9.6	H5176	E1/E2/E3/E4/E5/E6/B	9.4–20.2	wmc177 [42]
	2A	758.6–760.7	AX_111579921	E2/E4/E5/E6	7.1–10.9	H7248	E2/E3/E4/E5/E6/B	9.2–22.4	
	2B	105.8–108.7	AX_111634754	E1/E2/E3/E4/E5/E6/B	7.0–15.0	H7877	E1/E2/E3/E4/E5/E6/B	9.1–36.7	
	2B	415.0–418.9	AX_111819405	E2/E3/E4/E5/E6/B	7.0–10.6	H8418	E1/E2/E3/E4/E5/E6/B	9.4–16.3	
	3A	714.4–716.3	AX_111047166	E2/E3/E4/E5/B	7.0–10.9	H11918	E1/E2/E3/E4/E5/E6/B	9.1–27.2	QTgw-3A1 [48]
	3D	570.2–575.5	IWB17930	E2/E3/E4/E5/E6/B	6.8–10.3	H15950	E1/E2/E3/E4/E5/E6/B	9.2–21.4	
	4A	670.0–676.0	AX_110046841	E2/E3/E4/E6/B	7.0–10.3	H17106	E1/E2/E3/E4/E5/E6/B	12.1–19.1	QTKW.caas-4AL [40]; QTgw.cau-4A.2. QGl.cau-4A.2 [47]
	4B	670.4–672.9	IWB47765	E1/E3/E6/B	7.0–12.4	H18552	E2/E3/E4/E5/E6/B	11.5–18.8	
	5A	569.2–573.0	AX_110523824	E3/E4/E6/B	7.0–11.7	H19921	E2/E3/E4/E6/B	9.2–20.9	QGl.cau-5A.1 [47]
	5A	702.1–708.1	AX_110958315	E3/E4/E5/E6/B	7.4–13.3	H20333	E2/E3/E4/E5/E6/B	9.4–22.2	QTKW.caas-5AL.1 [40]
	5B	520.8–520.9	IWB20926	E1/E2/E3/E6/B	7.2–10.1	H21840	E1/E2/E3/E4/E5/E6/B	9.2–19.4	QTgw.abrii-5B₁.2 [23]
	6B	706.9–709.2	AX_109820966	E3/E4/E5/E6/B	6.9–10.3	H26817	E1/E2/E3/E4/E5/E6/B	9.4–17.9	
	7D	58.5–66.5	AX_109396082	E1/E2/E3/E4/E5/E6/B	7.0–12.5	H31312	E3/E4/E6/B	9.5–20.0	QKL.caas-7DS [17]; QGw.ccsu-7D.1 [44]
SL	2B	44.6–47.3	AX_109985540	E1/E2/E3/E4/E5/B	7.0–14.3	H7643	E1/E3/E4/E5/B	9.2–14.0	BS00022060_51 [27]
	5A	510.1–510.3	AX_109367907	E1/E3/E4/E5/E6/B	7.4–12.8	H19589	E4/E5/B	9.0–11.8	QSL.caas-5AL.2 [17]
	5A	534.9–540.1	AX_110919697	E1/E2/E3/E4/B	6.7–11.8	H19737	E1/E2/E3/E4/B	9.3–14.9	
	5A	568.3–574.3	AX_110523824	E1/E2/E3/E4/E5/E6/B	7.3–13.0	H19954	E2/E5/E6/B	9.8–10.3	QSl-5A1 [48]

(continued)

Trait	Chr	Physical position (Mb)[a]	Marker[b]	SNP-GWAS Environment	SNP-GWAS R^2 (%)[c]	Haplotype[d]	Haplotype-GWAS Environment	Haplotype-GWAS R^2 (%)[c]	QTL/marker/gene[e]
	5B	398.2-398.3	AX_94562344	E3/E4/E5/E6/B	8.0-11.0	H21291	E2/E3/E5/E6/B	9.2-15.1	
	5B	590.5-591.1	AX_110971192	E1/E3/E4/E6/B	7.4-10.3	H22208	E1/E3/E4/E6/B	9.1-12.6	
	7B	723.9-727.6	IWB71567	E1/E2/E3/E4/E5/E6/B	7.1-13.8	H30972	E1/E2/E3/E4/E6/B	9.1-19.3	
	7D	621.9-630.4	AX_110645784	E1/E2/E3/E4/E5/E6/B	7.6-15.0	H31712	E1/E2/B	10.7-20.0	
SDW	1A	568.5-573.5	AX_95255804	E3/E4/E5/E6/B	7.2-9.7	H2018	E3/E4/E6/B	9.1-12.6	
	3A	721.3-725.8	IWA94	E3/E4/E5/E6/B	7.1-19.0	H11992	E3/E4/E5/E6/B	9.5-15.2	
	4B	89.5-95.1	IWB35533	E3/E4/E6/B	7.1-10.7	H17675	E4/E5/B	9.3-11.4	
	5B	473.9-477.8	AX_111183518	E3/E5/E6/B	7.0-16.3	H21628	E3/E4/E5/E6/B	9.7-18.5	
	5B	698.0-698.5	AX_111051286	E3/E4/E6/B	7.1-10.6	H22718	E3/E4/B	9.9-11.2	
HD	2A	27.3-27.4	AX_111037158	E1/E2/E3/E4/E5/B	7.5-12.2	H5170	E1/E2/E3/E4/E5/E6/B	11.3-15.7	
	2A	704.8-710.1	AX_89674107	E1/E2/E3/E4/E6/B	6.8-8.6	H6769	E1/E2/E6	9.3-13.9	
	2A	755.8-757.0	AX_109964711	E1/E2/E3/E4/E6/B	6.8-9.4	H7248	E1/E2/E4/E6	9.2-13.8	wPt-1499 [51]
	2B	106.0-108.7	AX_110624209	E1/E2/E3/E4/E5/E6/B	6.6-10.7	H7877	E1/E2/E4/E6/B	9.4-22.0	
	5B	520.1-524.4	IWB20926	E1/E2/E4/E6/B	6.6-10.7	H21844	E1/E2/E4/E6	9.1-15.9	wPt-1409 [51]
	7A	511.5-517.4	AX_109921812	E1/E2/E3/E4/E5/B	6.9-9.4	H28667	E1/E3/E5/B	9.2-14.2	
	7A	556.2-561.0	AX_111160137	E1/E2/E3/E4/E5/B	6.7-12.3	H28779	E1/E2/E4/E5/E6	9.4-12.9	wPt-4796 [51]
	7B	701.0-703.7	IWB75191	E1/E2/E3/E4/E5/E6/B	6.6-13.1	H30856	E1/E2/E3/E4/E5/E6/B	9.3-16.3	
PH	1A	434.1-440.2	AX_109449226	E1/E2/E3/E4/E5/E6/B	6.9-13.1	H1306	E1/E2/E3/E4/E5/E6/B	9.3-16.1	
	1B	539.6-542.6	AX_94564150	E1/E2/E4/E5/E6/B	7.0-21.1	H3651	E1/E2/E3/E4/E5/E6/B	10.4-22.8	
	1B	673.9-675.7	AX_109820171	E1/E2/E4/E5/E6/B	6.8-30.8	H4268	E1/E2/E3/E4/E5/E6/B	9.5-38.0	
	2A	30.9-32.0	AX_110988136	E1/E2/E4/E5/B	7.0-11.3	H5187	E1/E2/E3/E4/E5/B	11.4-19.8	QUIL.caas-2AS.1 [17]
	2A	715.3-721.6	AX_94494373	E1/E2/E4/E6/B	6.7-10.5	H6911	E1/E2/E3/E4/E5/E6/B	10.9-16.5	QPH.caas-2AL, QUIL.caas-2AL [17]
	3A	716.5-721.3	AX_111577195	E1/E2/E4/E5/E6/B	6.9-14.8	H11943	E1/E2/E3/E4/E5/E6/B	9.2-18.4	barc1113 [19]
	3B	116.1-120.4	AX_109413472	E1/E2/E3/E4/E5/E6/B	6.8-12.8	H12783	E1/E2/E3/E4/E5/E6/B	9.0-17.4	gwm566 [19]

(continued)

Trait	Chr	Physical position (Mb)[a]	SNP-GWAS Marker[b]	Environment	R^2 (%)[c]	Haplotype-GWAS Haplotype[d]	Environment	R^2 (%)[c]	QTL/marker/gene[e]
	4D	16.6-19.7	AX_108916749	E1/E2/E4/E5/E6/B	6.7-14.5	H18598	E1/E2/E3/E4/E5/E6/B	10.2-16.1	QPH.caas-4DS, QUIL.caas-4DS [17]; Rht-D1b [52]
	5A	669.2-671.2	AX_110446653	E1/E2/E3/E4/E5/E6/B	6.9-14.7	H20207	E1/E2/E3/E4/E5/E6/B	10.8-17.5	
	5A	708.4-708.8	IWA2646	E2/E3/E5/B	6.7-8.0	H20356	E1/E2/E3/E4/E5/E6/B	9.4-18.0	QPh.hwwgr5AL [53]
	5B	576.3-580.9	AX_108921249	E1/E2/E4/E5/B	7.0-13.3	H22142	E1/E2/E3/E4/E5/E6/B	12.2-20.4	
	6B	1.4-4.9	AX_110482029	E1/E2/E4/E5/B	7.0-18.1	H24673	E1/E2/E3/E4/E5/E6/B	9.4-25.9	
	6B	42.7-42.8	AX_110671479	E1/E2/E4/B	7.2-10.8	H24862	E1/E2/E3/E4/E5/E6/B	9.4-19.0	
	7A	556.2-560.9	AX_109384874	E1/E2/E3/E5/E6/B	6.9-15.1	H28784	E1/E2/E3/E4/E5/E6/B	10.5-18.6	
UIL	1A	142.6-148.6	AX_109901254	E1/E5/E6/B	7.5-11.1	H608	E1/E2/E5/E6/B	9.2-17.4	
	1A	434.4-440.2	AX_109449226	E1/E2/E5/E6/B	6.9-12.9	H1306	E1/E2/E5/E6/B	9.1-16.3	
	1B	539.6-542.6	AX_94564150	E1/E2/E5/B	6.8-16.1	H3650	E1/E2/E5/B	9.6-17.9	
	1B	674.8-675.7	AX_109820171	E1/E2/E5/B	8.4-16.4	H4239	E1/E2/E5/E6/B	9.2-24.0	
	3A	721.3-723.7	AX_111610555	E1/E2/E5/B	7.3-9.7	H11922	E1/E2/E5/E6/B	9.1-17.3	barc1113 [19]
	5A	665.6-671.2	IWA5929	E1/E2/E5/E6/B	7.1-11.3	H20205	E1/E2/E5/E6/B	9.6-18.5	
	6B	1.4-4.9	AX_109526332	E1/E2/E5/B	7.0-11.4	H24673	E1/E2/E5/E6/B	9.3-24.8	
	6B	42.7-46.9	IWB12568	E1/E2/E5/E6/B	7.1-9.5	H24865	E1/E2/E5/E6/B	9.8-20.4	
	6B	563.3-567.8	AX_86165710	E1/E2/E5/E6/B	6.7-10.5	H26273	E1/E2/E5/E6/B	9.3-15.6	
	6D	396.5-403.7	AX_109331000	E1/E2/E5/E6/B	7.0-12.6	H27113	E1/E2/E5/E6/B	15.3-19.0	barc96 [19]
	6D	460.7-465.0	IWB2743	E1/E2/E5	7.2-10.5	H27272	E1/E2/E5/E6/B	9.4-14.1	
	7B	630.6-632.7	AX_109492373	E1/E2/E5/B	7.1-12.3	H30631	E1/E2/E5/E6/B	9.9-22.1	
FLL	1A	8.3-11.0	AX_109621606	E3/E4/E5/E6/B	7.1-16.9	H99	E3/E4/E5/E6/B	9.0-20.8	
	2A	2.5-7.8	AX_109880304	E2/E3/E5/E6/B	7.4-19.6	H5121	E2/E3/E4/E5/E6/B	10.9-29.1	
	2A	87.8-87.9	AX_95085564	E4/E5/E6/B	8.6-12.7	H5461	E2/E3/E5/E6/B	9.6-17.5	
	2B	46.4-50.4	AX_111027654	E2/E6/B	7.0-19.6	H7632	E2/E3/E4/E5/E6/B	9.4-25.5	QFLL-2B [57]

(continued)

Trait	Chr	Physical position (Mb)[a]	SNP-GWAS			Haplotype-GWAS			QTL/marker/gene[e]
			Marker[b]	Environment	R^2 (%)[c]	Haplotype[d]	Environment	R^2 (%)[c]	
	3A	716.9-722.5	AX_108908243	E3/E4/E6	7.2-9.7	H11976	E3/E4/E5/E6/B	9.4-22.7	
	5A	556.6-559.7	IWB4576	E3/E4/E6/B	6.9-9.6	H19845	E2/E3/E4/E5/E6/B	9.6-16.2	QFlw.cau-5A.2, QFlan.cau-5A.3 [56]
	6B	704.9-708.7	AX_109459603	E5/E6/B	8.8-9.6	H26801	E2/E3/E4/E5/E6/B	9.5-22.0	
	6D	455.5-462.2	AX_110087641	E3/E6/B	7.6-8.8	H27138	E2/E3/E4/E5/E6/B	9.8-18.0	QFll.cau-6D [56]
FLW	1A	440.4-445.5	AX_111540798	E2/E3/E4/E6/B	7.0-11.4	H1318	E2/E3/E4/B	9.1-17.3	
	3B	23.9-24.4	AX_111655083	E4/E6/B	7.3-9.7	H12389	E2/E3/E4/E5/E6/B	9.2-19.7	
	5B	531.5-533.5	IWB41225	E2/E6/B	6.9-7.9	H21876	E2/E3/E4/E5/E6/B	9.1-14.7	
	5B	557.1-559.4	AX_109519234	E2/E3/E4/E5/B	7.0-9.0	H22030	E2/E3/E4/E5/E6/B	9.3-16.7	
	6B	461.4-465.5	AX_108771909	E2/E3/E5/B	7.0-9.9	H25955	E2/E3/E4/E5/E6/B	9.6-12.2	QFlw.cau-6B [56]

[a] The physical positions of SNP markers based on wheat (Chinese Spring) genome sequences from the International Wheat Genome Sequencing Consortium

[b] Representative markers

[c] Percentage of phenotypic variance explained

[d] Representative haplotypes

[e] The previously reported QTL, markers or genes near the loci identified in the present study

GY grain yield, SN spike number per square meter, KNS kernel number per spike, TKW thousand-kernel weight, KL kernel length, KW kernel width, SL spike length, SDW spike dry weight, HD heading date, PH plant height, UIL uppermost internode length, FLL flag leaf length, FLW flag leaf width; E1: 2012-2013 Anyang; E2: 2012-2013 Suixi; E3: 2013-2014 Anyang; E4: 2013-2014 Suixi; E5: 2014-2015 Anyang; E6: 2014-2015 Shijiazhuang; B: Best linear unbiased estimation

BLUE value, whereas the 6D locus (AX _ 110652999) explained the largest phenotypic variance (7.2-17.1%) in SNP-GWAS. Loci on chromosomes 5B (H22717), 6B (H26344), 6D (H27181) and 7D (H31325) were identified in four environments and BLUE values, among which the 7D locus (H31325) accounted for the largest of phenotypic variance (9.8-23.3%) in Haplotype-GWAS.

Nine common loci for KNS were found on chromosomes 1A (2), 2A, 2D, 3A (2), 5B (2) and 7A, accounting for 7.1-17.1% and 9.1-15.8% of the phenotypic variances in SNP-GWAS and Haplotype-GWAS, respectively. In SNP-GWAS, the loci on chromosomes 1A (AX _ 108737858) and 2D (IWB57054) were identified in all six environments and BLUE values; the 2D locus (IWB57054) accounted for the largest phenotypic variance (7.4-17.7%). In Haplotype- GWAS, the loci on chromosomes 1A (H322) and 5B (H21713) were significant in all six environments and BLUE values; the 1A locus (H322) explained the largest phenotypic variance (9.2-15.8%). Four loci, including 1A (AX _ 111579941), 3A (AX _ 110657474), 5B (AX _ 109537496) and 7A (AX _ 89571435), were identified in five or more environments and BLUE values in both methods and were therefore stable.

Twelve common loci for TKW were identified on chromosomes 1B, 2A, 2B, 4B (3), 5A, 5B, 6B (2), 6D and 7D, explaining 6.7-13.0% and 9.2-27.0% of the phenotypic variances in SNP-GWAS and Haplotype-GWAS, respectively. In SNP-GWAS, the 4B locus (AX _ 110713957) and 7D locus (AX _ 109927697) were significant in at least five environments and BLUE values; in Haplotype-GWAS, all the loci were significant in at least five environments and BLUE values. The 5A (AX _ 110958315) and 2B (H7872) loci explained the largest phenotypic variances in SNP- GWAS and Haplotype-GWAS, respectively.

Kernel shape related traits

Six common loci for KL on chromosomes 1B (2), 2A, 3B, 5B and 5D explained 7.0-14.2% and 9.2-15.4% of the phenotypic variances in SNP-GWAS and Haplotype-GWAS, respectively. The 2A locus (IWB32119) was significant in all six environments and BLUE value, whereas the 5D locus (AX _ 111122970) was identified in five or six environments and BLUE value by both methods.

Fifteen common loci for KW on chromosomes 1A (2), 2A (2), 2B (2), 3A, 3D, 4A, 4B, 5A (2), 5B, 6B and 7D accounted for 6.8-15.0% and 9.1-36.7% of the phenotypic variances in SNP-GWAS and Haplotype-GWAS, respectively. The 2B locus (AX _ 111634754) was significant in all six environments and BLUE value with the largest contribution to phenotypic variance in both methods. The 2B (AX _ 111819405) and 3D (IWB17930) loci were significant in five environments and BLUE values in SNP-GWAS, and significant in all six environments and BLUE values in Haplotype-GWAS.

Spike related traits

Eight common loci for SL were identified on chromosomes 2B, 5A (3), 5B (2), 7B and 7D, explaining 6.7- 15.0% and 9.0-20.0% of the phenotypic variances in SNP-GWAS and Haplotype-GWAS, respectively. Locus (IWB71567) on chromosome 7B was significant in five or more environments and BLUE value in both methods; the 7D locus (AX _ 110645784) explained 7.6-15.0% and 10.7-20.0% of the phenotypic variances in SNP-GWAS and Haplotype-GWAS, respectively.

Five common loci for SDW detected on chromosomes 1A, 3A, 4B, and 5B (2) explained 7.0-19.0% and 9.1- 18.5% of the phenotypic variances in SNP- GWAS and Haplotype-GWAS, respectively. The 3A locus (IWA94) was significant in all four environments and BLUE value in both methods; the 5B locus (AX _ 111183518) was stable across three or four environments and BLUE value in both methods.

Heading date

Eight common loci for HD on chromosomes 2A (3), 2B, 5B, 7A (2) and 7B accounted for 6. 6-13. 1% and 9. 1- 22. 0% of the phenotypic variances in SNP-GWAS and Haplotype-GWAS, respectively. The locus (IWB75191) on chromosome 7B was significant in all six environments and BLUE value in both methods, whereas locus (AX _ 111037158) on chromosome 2A was stably detected in five and six environments and BLUE value in SNP-GWAS and Haplotype-GWAS, respectively.

Plant height related traits

Fourteen common locifor PH were identified on chromosomes 1A, 1B (2), 2A (2), 3A, 3B, 4D, 5A (2), 5B, 6B (2) and 7A, explaining 6. 7-30. 8% and 9. 0-38. 0% of the phenotypic variances in SNP-GWAS and Haplotype-GWAS, respectively. The loci on chromosomes 1A (AX _ 109449226), 3B (AX _ 109413472) and 5A (AX _ 110446653) were significant in all six environments and BLUE values, whereas the other five loci on chromosomes 1B (AX _ 94564150 and AX _ 109820171), 3A (AX _ 111577195), 4D (AX _ 108916749) and 7A (AX _ 109384874) were stably identified in five or six environments and BLUE values in both methods; the 1B (AX _ 109820171) locus was the most significant, explaining 6. 8-30. 8% and 9. 5-38. 0% of the phenotypic variances in SNP-GWAS and Haplotype-GWAS, respectively.

Twelve common loci for UIL were detected on chromosomes 1A (2), 1B (2), 3A, 5A, 6B (3), 6D (2) and 7B, with single loci explaining 6. 7-16. 4% and 9. 1-24. 8% of the phenotypic variances in SNP-GWAS and Haplotype-GWAS, respectively. Five loci on chromosomes 1A (AX _ 109449226), 5A (IWA5929), 6B (IWB12568 and AX _ 86165710) and 6D (AX _ 109331000) were identified in all four investigated environments and BLUE values by the two methods. Locus (AX _ 109820171) on chromosome

1B had a large effect on phenotypic variance in both methods.

Flag leaf related traits

Eight common loci for FLL on chromosomes 1A, 2A (2), 2B, 3A, 5A, 6B and 6D explained 6. 9-19. 6% and 9. 0- 29. 1% of the phenotypic variances in SNP-GWAS and Haplotype-GWAS, respectively. The 2A locus (AX _ 109880304) was significant in four or five environments and BLUE value and presented the largest effect on phenotypic variance in both methods. The 1A locus (AX _ 109621606) was also detected in four environments and BLUE value in both methods.

Five common loci for FLW were identified on chromosomes 1A, 3B, 5B (2) and 6B, accounting for 6. 9- 11. 4% and 9. 1-19. 7% of the phenotypic variances in SNP-GWAS and Haplotype-GWAS, respectively. The locus on chromosome 5B (AX _ 109519234) was significant in four or five environments and BLUE value in both methods, whereas 1A (AX _ 111540798) and 3B (AX _ 111655083) loci explained the highest phenotypic variances in SNP-GWAS and Haplotype-GWAS, respectively.

Pleiotropic loci

Twelve pleiotropic loci were associated with three or more traits on chromosomes 1A, 1B (2), 2A (2), 2B, 3A, 5A (2), 5B (2) and 6D based on the common loci detected by both methods (Table 2). The interval 714. 4-725. 8 Mb on chromosome 3A was associated with GY, KNS, KW, SDW, PH, UIL and FLL, showing a significant effect on GY. Seven pleiotropic loci were associated with GY, among which four were related to KW and five to PH or UIL. Three SN loci on chromosomes 5B (IWB56499 and AX _ 109936345) and 6D (AX _ 110652999) were located in pleiotropic loci; four HD loci on chromosomes 2A (AX _ 111037158 and AX _ 111579921), 2B (AX _ 111634754) and 5B (IWB56499) were also located in pleiotropic loci; these loci were both accompanied with TKW or KW loci. Finally, nine pleiotropic loci for TKW or KW and seven loci for PH or UIL should be crucial in determining GY. Of all common loci identified by both methods, more than half were co-localized.

Table 2   Distribution of pleiotropic loci associated with three or more grain yield related traits on wheat chromosomes

Chr	Trait	Marker[a]	Interval (Mb)[b]
1A	GY/PH/UIL/FLW	AX _ 110418502	434. 0-445. 5
1B	GY/PH/UIL	AX _ 94564150	539. 6-542. 6
1B	GY/PH/UIL	AX _ 109820171	673. 6-675. 7
2A	GY/KW/HD/PH	AX _ 111037158	27. 3-32. 0
2A	TKW/KW/HD	AX _ 111579921	755. 8-760. 7
2B	GY/TKW/KW/HD	AX _ 111634754	105. 8-108. 9
3A	GY/KNS/KW/SDW/PH/UIL/FLL	IWA94	714. 4-725. 8
5A	GY/KW/SL	AX _ 110523824	568. 3-574. 8
5A	TKW/KW/PH	AX _ 110958315	702. 1-708. 8
5B	SN/KW/HD/FLW	IWB56499	520. 1-534. 0
5B	SN/TKW/SDW	AX _ 109936345	692. 7-700. 9
6D	SN/TKW/UIL/FLL	AX _ 110652999	455. 5-471. 0

[a] Representative markers

[b] The physical positions of SNP markers based on wheat (Chinese Spring) genome sequences from the International Wheat Genome Sequencing Consortium

GY grain yield, *SN* spike number per square meter, *KNS* kernel number per spike, *TKW* thousand-kernel weight, *KL* kernel length, *KW* kernel width, *SL* spike length, *SDW* spike dry weight, *HD* heading date, *PH* plant height, *UIL* uppermost internode length, *FLL* flag leaf length, *FLW* flag leaf width

Relationships between trait performances and number of alleles for increasing phenotypic values

For most traits, ranges in the number of alleles for increasing phenotypic values across the panel were large (Table 3). The average number of alleles for increasing GY was 10. 0. Compared with the higher numbers of alleles for increasing TKW, KL, KW and FLW, those for SN, KNS, SL, SDW, HD, PH, UIL and FLL were lower.

Favorable alleles at each locus for GY exhibited significant and positive effects on phenotypic values (Fig. 1). Effects of number of alleles for increasing phenotypic values for each trait were also estimated (Fig. 2), and the results showed that the phenotypic traits were dependent on the number of alleles for increasing phenotypic value.

Table 3   Number of alleles for increasing phenotypic values of grain yield and related traits in the diverse panel

Trait	Total number of favorable alleles	Average number of favorable alleles	Range
GY	12	10. 0	3-12
SN	6	1. 7	0-5
KNS	9	3. 7	0-7
TKW	12	8. 5	2-12
KL	6	3. 6	1-6
KW	15	11. 6	7-14
SL	8	1. 3	0-7
SDW	5	2. 0	0-5
HD	8	3. 5	0-8
PH	14	1. 3	2-7
UIL	12	3. 4	1-8
FLL	8	2. 3	0-6
FLW	5	3. 9	0-5

GY grain yield, *SN* spike number per square meter, *KNS* kernel number per spike, *TKW* thousand-kernel weight, *KL* kernel length, *KW* kernel width, *SL* spike length, *SDW* spike dry weight, *HD* heading date, *PH* plant height, *UIL* uppermost internode length, *FLL* flag leaf length, *FLW* flag leaf width

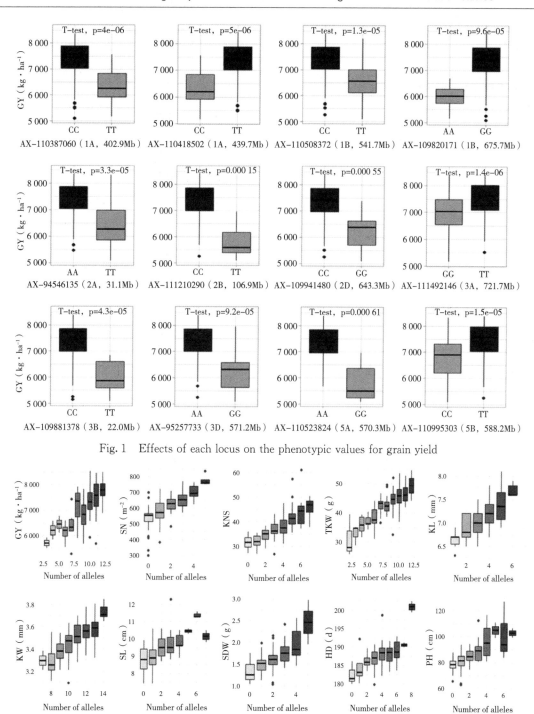

Fig. 1   Effects of each locus on the phenotypic values for grain yield

Fig. 2   Effects of the number of alleles for increasing phenotypic values of grain yield and related traits. GY; grain yield; SN; spike number per square meter; KNS; kernel number per spike; TKW; thousand-kernel weight; KL; kernel length; KW; kernel width; SL; spike length; SDW; spike dry weight; HD; heading date; PH; plant height; UIL; uppermost internode length; FLL; flag leaf length; FLW; flag leaf width

Discussion

Advantages of two methods of GWAS using high density SNP markers

SNP arrays based on Next Generation Sequencing Technology permit identification of many SNP markers, and represent very high throughput and multiple genotyping compared with traditional molecular markers[24]. In differing from QTL mapping, GWAS is performed by significance testing between phenotypic values and single markers or haplotype blocks comprised of contiguous SNP markers with similar genotype. The accuracy of GWAS results thus depends on the coverage of markers used for analysis. In the present study, 326, 570 SNP markers from the wheat 90 K and 660 K SNP arrays were used for GWAS of GY and related traits, with a physical distance of 0. 043 Mb per marker. The average LD for the whole genome was 8 Mb, and the high-density of SNP markers ensured multiple markers in each haplotype block and high efficiency in identifying significant loci.

SNPs are very common in the genomes of most crop species and result in a variety of genetic variances. However, genetic variance in crops can sometimes be caused by single SNP, but mostly there are numerous closely linked SNPs[30]. In order to avoid the disadvantage of SNP-GWAS in detecting genetic multiple variances caused by numerous SNP and false positives identified by Haplotype-GWAS, both methods were used in the present study to identify loci with significant effects. As already mentioned above, 275, 000 of a total 326, 570 SNP markers were sorted into 31, 748 haplotype blocks, remaining 51,570 single SNP markers. A total of 239 and 248 significant loci were detected and about half the loci were common in both methods. This indicated that the detection intensity of SNP-GWAS and Haplotype-GWAS differs between chromosome positions. Loci identified in three or more environments in both methods were regarded as the main loci affecting GY and related traits.

Comparison with the QTL identified in previous studies

GY and related traits are basic observable and measurable agronomic traits extensively reported in the literature. Being limited to low-density molecular markers, significant influence by environments, and likely presence of linkage drag, marker loci for GY and related traits identified by QTL mapping or GWAS are seldom used in wheat breeding programs. In the present study, associations of GY and related traits with single SNPs and haplotype blocks were conducted separately. Loci identified by both methods were compared with QTL previously reported on physical or linkage maps.

GY and its components

GY related QTL have been reported on all 21 wheat chromosomes[18,23,33-38]. Azadi et al. [23] reported a GY QTL on chromosome 1A tightly linked with SSR marker *gwm357*, which was also located between the two GY QTL by Cuthbert et al. [18] and Huang et al. [33]. The 1A locus (*AX _ 110418502*) for GY is about 0. 21 cM from *gwm357* on the consensus linkage map[39], indicating that these two loci are likely to be the same. Reif et al. [35] identified a GY QTL on chromo- some 5A linked with SSR marker *barc151*, at a similar position to the present GY locus (*AX _ 110523824*). The remaining loci are likely to be new.

Numerous reports indicate that SN is controlled by polygenes and significantly influenced by environment. Nine SN QTL were recently mapped using the wheat 90 K SNP array on three RIL populations[17]. *QSN. caa-s-3AL. 1* and *QSN. caas-6AL* were at similar positions to the QTL reported in Lee et al. [36] and Gao et al. [40], whereas the effect of *QSN. caas-4BS* was contributed by *Rht-B1b*. However, the six SN loci detected in this study are likely to be at different positions to the QTL reported previously.

Azadi et al. [23] detected KNS QTL on chromosomes 1A, 3A and 5B, linked with DArT markers *wPt-665*, *590*, *wPt-5133* and *wPt-3661*, respectively;

the 1A QTL is about 1.5 cM from the KNS locus (*AX _ 108737858*) identified in this study and they are likely to be the same; the 3A QTL is about 2 cM from the KNS locus *AX _ 108992368* and close to a QTL mapped by Gao et al. [40]; the 5B QTL is about 6.5 cM from the KNS locus *AX _ 109537496* and therefore might be different. Zhang et al. [41] identified SSR markers *wmc63* and *gwm213* significantly associated with KNS on chromosomes 2A and 5B, respectively; *wmc63* is about 2 cM from the present 2A locus *IWB45503* and close to a QTL re- ported by Kumar et al. [34] and Yao et al. [42]; *gwm213* is at the same position as the present KNS locus *AX _ 109538915*. In addition, the locus *AX _ 111579941* on chromosome 1A is about one LD from a QTL reported in Wang et al. [43]. The stable loci on chromosomes 3A (*AX _ 110657474*), 5B (*AX _ 109537496*) and 7A (*AX _ 89571435*) identified in five or more environments and BLUE values by both methods are likely to be new.

TKW locus *AX _ 109917592* on chromosome 6B is within the confidence interval of *QTKW. caas-6BL* detected in the D × S (Doumai × Shi 4185) population in Li et al. [17]. The 7D locus (*AX _ 109927697*) is at the similar position to *QKL. caas-7DS* located in the G × Z (Gaocheng 8901 × Zhoumai 16) population[17] and *QGw. ccsu-7D. 1* reported by Mir et al. [44]. TKW locus *AX _ 108769612* on chromosome 5B is at the same position as loci for KL, KW and TKW detected by Chen et al. [45], Mohler et al. [46] and Sun et al. [27], respectively, indicating that this should be an important locus in determining kernel weight. Wu et al. [47] reported a locus affecting both TKW and KL on chromosome 1B, located about one LD from the present TKW locus *AX _ 111147652*. Chromosomes 5A locus *AX _ 110958315* is about one LD from a TKW QTL reported by Gao et al. [40], whereas 4B locus *AX _ 110713957* is at a similar position to a QTL reported in Liu et al. [48]. The other six loci are likely to be new.

Kernel shape related traits

Sajjad et al. [49] cloned *TaFlo2-A1* for TKW on chromo- some 2A, at the same position as the present stable KL locus *IWB32119*. 1B locus *AX _ 108849700* is within the confidence interval of *QKL. caas-1BL* mapped in the L × Z (Linmai 2 × Zhong 892) popu- lation in Li et al. [17] and within one LD of the signifi- cantly associated SNP marker *tplb0043a07 _ 1411* for TKW[48]. KL locus *IWB50649* on chromosome 5B is very close to QTL reported by Azadi et al. [23] and Wu et al. [47], and within the interval of a TKW QTL mapped by Zhai et al. [50]. Locus *AX _ 111122970* on chromosome 5D stably detected in five or six environ- ments and BLUE value by both methods is probably new.

Theco-localized KW and TKW locus *AX _ 110958315* is at the same position as a TKW QTL on chromosome 5A reported by Gao et al. [40]. Another KW locus *AX _ 109396082* co-localized with TKW locus *AX _ 109927697* on chromo- some 7D is at a similar position to QTL reported in Mir et al. [44] and Li et al. [17]. Wu et al. [47] reported three TKW or KL QTL on chromo- somes 1A, 4A and 5A, which are within one LD from the KW loci *IWB7676*, *AX _ 110046841* and *AX _ 110523824*, respectively; 4A locus *AX _ 110046841* is also close to a TKW QTL mapped by Gao et al. [40]. The KW loci on chromosomes 1A (*IWB6999*), 2A (*AX _ 111037158*) and 3A (*AX _ 111047166*) are about one LD from TKW QTL reported by Mir et al. [44], Yao et al. [42] and Liu et al. [48], respective- ly. Locus *IWB20926* on chromosome 5B is about 2 cM from DArT marker *wPt-5851* linked to a TKW QTL in Azadi et al. [23]. The stable loci on chromosomes 2B (*AX _ 111634754* and *AX _ 111819405*) and 3D (*IWB17930*) identified in five or six environments and BLUE values by both methods are likely to be new.

Spike related traits

QSL. caas-5AL. 2 identified in the G × Z population[17] is at a similar position to the present SL locus *AX _ 109367907*. Sun et al. [27] reported SNP marker *BS00022060 _ 51* associated with SL on chromosome 2B. This gene is about one LD from the SL locus *AX _ 109985540*, indicating they are likely to be the same. Liu et al. [48] mapped a SL QTL on chromosome 5A about one LD from SL locus *AX _ 110523824* in this study. The stable locus *IWB71567* on chromosome 7B detected in five or six environments and BLUE value by both methods is likely to be new.

Compared with other traits there are few reports on QTL mapping of SDW. Li et al. [17] mapped 10 SDW QTL; among them *QSDW. caas-6BL* and *QSDW. caas-7BL* are at similar positions to SNPs *RAC875 _ c31299 _ 1302* and *BS00055584 _ 51* identified by Valluru et al. [28]. All five SDW loci identified in this study appear to be new.

Heading date

Le Gouis et al. [51] reported DArT markers *wPt-1499*, *wPt-1409* and *wPt-4796* associated with HD on chromosomes 2A, 5A and 7A, respectively; these three markers are close to the HD loci *AX _ 109964711*, *IWB20926* and *AX _ 111660137*, respectively, on the consensus linkage map[39]. As the majority of varieties in the present study were from the YHRVWZ with similar vernalization and photoperiod characteristics, and no variation associated with known *Vrn* and *Ppd* genes was detected. Stable loci on chromosomes 2A (*AX _ 111037158*) and 7B (*IWB75191*) detected in most environments and BLUE values by both methods are likely to be new.

Plant height related traits

Rht-D1b is widely present in wheat varieties in YHRVWZ[6]. The PH locus *AX _ 108916749* on chromosome 4D is at the same position as *Rht-D1*[52], indicating that the effect on PH is from *Rht-D1b*, and is the same as QTL or loci reported by Li et al. [17], Sun et al. [27] and Gao et al. [40]. Loci *AX _ 110988136* and *AX _ 94494373* on chromosome 2A are at similar positions to *QUIL. caas-2AS. 1* and *QPH. caas-2AL* (co-localized with *QUIL. caas-2AL*), respectively[17]. Cui et al. [19] identified QTL for PH or UIL on chromosomes 3A and 3B; these QTL are close to the present PH loci *AX _ 111577195* and *AX _ 109413472*, respectively. 3B locus *AX _ 109413472* is about 14 cM

from *Rht5* [7] and therefore should be different. 5A locus *IWA2646* is about one LD from a QTL in Li et al. [53], and about 25 Mb and 8. 9 cM from *Rht12* [54], respectively, on the physical and consensus linkage maps[39]. 5B locus *AX _ 108921249* is about 2 Mb from *Vrn-B1* [55], but there is no reported relationship between vernalization response and PH. The five loci identified in 1A (*AX _ 109449226*), 1B (*AX _ 94564150* and *AX _ 109820171*), 5A (*AX _ 110446653*) and 7A (*AX _ 109384874*) identified in five or more environments and BLUE values by both methods are likely to be new.

The UIL locus *AX _ 111610555*, co-localized with PH locus *AX _ 111577195*, is likely to be the same as a QTL on chromosome 3A for both PH and UIL reported by Cui et al. [19]. Another UIL locus (*AX _ 109331000*) on chromosome 6D is about one LD from a QTL associated with PH and third internode length reported in Cui et al. [19]; they are likely to be the same. Apart from 3A locus, the remaining six loci co-localized with PH loci are likely to be new.

Flag leaf related traits

Wu et al. [56] mapped a FLL QTL on chromosome 6D that overlapped with FLL locus *AX _ 110876641*. They also reported a pleiotropic locus for FLW and flag leaf angle at about one LD from the present 5A FLL locus *IWB4576*. Another FLL QTL linked with the SSR marker *barc318* identified on chromosome 2B[57] is about 1. 2 cM from the present FLL locus *AX _ 111027654* based on the consensus linkage map[39]. Loci on chromosomes 1A (*AX _ 109621606*) and 2A (*AX _ 109880304*) that were stable in four or more environments and BLUE values by both methods are probably new.

A FLW QTL mapped on chromosome 6B by Wu et al. [56] is at the same position as *AX _ 108771909*, and are probably the same gene. Two stable loci on chromosomes 1A (*AX _ 111540798*) and 5B (*AX _ 109519234*) identified in four or five environments and BLUE values in both methods are likely to be new.

Among the 120 loci for GY and related traits, 42 could be the same as QTL reported in previous studies, whereas the remaining are likely to be new. Stable loci identified in both GWAS and QTL mapping showed that they are widespread in varieties. Our results indicated that the methods of GWAS used in the present study were reliable and efficient in detecting loci for GY and related traits.

Genetic relationships among grain yield and related traits

High-yielding varieties should have good adaptability to prevailing environments, strong resistance to abiotic and biotic stresses, and highly coordinated agronomic traits. Previous studies have showed that improvemen-tsin agronomic traits made significant contributions to in- creased yield potential[4-6]. Many studies have reported interaction effects or genetic linkages among yield related traits, especially in regard to the reduced height loci *Rht-B1* and *Rht-D1*[17,18,40,41,58]. In the present study, 12 pleiotropic loci involving three or more traits were identified, and more than half of the common loci were co-localized. Previously, three QTL clusters associated with yield related traits were detected at different positions on chromosome 3A[40,41,58]; among these the QTL cluster detected by Xu et al.[58] over-lapped with the pleiotropic locus *IWA94* in the present study. Many studies have reported that chromosome 5A carries productivity and adaptability related genes[18,33,59,60]. Li et al.[17], Cuthbert et al.[18], Zhang et al.[41] and Liu et al.[48] all reported QTL clusters for yield related traits at different positions on chromosome 5A; however, they are likely to be different from two pleiotropic loci detected in this study. Another locus on chromosome 1B related to GY, PH and UIL is about 15 Mb from the QTL cluster for KNS, KL, PH and FLW identified in Li et al.[17].

Relationships between GY and yield components are discussed in several publications[18,58,61-63]. Many studies demonstrated that GY is significantly correlated with SN and KNS. For example, by unconditional and conditional QTL analysis, Xu et al.[58] found that spike number per plant and KNS have larger

effects on GY than TKW. Miralles and Slafer[63] reviewed reports on factors influencing GY and concluded that increased GY was associated with increased grain number, but associated with a negative relationship between grain number and grain weight. Huang et al.[61] and Li et al.[62] reported that GY was significantly correlated with kernel size. However, in the present study, co-localization of related loci and phenotypic correlations showed that TKW and KW were more highly correlated with GY than were SN and KNS. Recently, McIntyre et al.[64] detected six putative QTL that increased grain weight and co-located with QTL for SN, KNS and harvest index, and three putative QTL for increased KNS co-located with QTL for increased grain weight, fewer spikes and earlier flowering. In this study, three loci associated with SN and TKW showed opposite effects on these traits due to negative correlation.

Keyes et al.[65] reported that plants with the *Rht-B1b*, *Rht-B1e* and *Rht-D1b* alleles are GA-insensitive, and the reduced PH was induced by decreased sensitivity of their vegetative tissues to endogenous gibberellin (GA). Chebotar et al.[66] pointed out that both GA-sensitive (*Rht8*) and GA-insensitive (*Rht-B1* and *Rht-D1*) dwarfing alleles had effects on almost all investigated traits. Our earlier study on QTL mapping of yield related traits showed that the *Rht-B1* and *Rht-D1* loci, as well as other PH QTL, had significant influences on other traits[17]. In the present study, more than half of the PH and UIL loci were co-localized with other traits, indicating that genes underlying have multiple effects on other traits, including GY.

The growth of wheat is controlled by many genes expressed at different growth stages. Heading and flowering represent a node of spike development and grain-filling, and are affected by environmental conditions as well as the many genes associated with plant development[67]. As a result, HD is crucial in optimising agronomic traits like kernel and spike related phenotypes. However, in the present study, HD exhibited no significant correlations with traits other than

FLW. Through co-localization, early heading is likely to benefit kernel development at lower temperatures.

Flag leaves account for 45-58% of the total photo- synthetic activity of the plant and contributed 41- 43% of the carbohydrates required for grain-filling[68,69]. Previously, Li et al. [17] found that FLW was important in determining KNS. In the present study, FLL was negatively correlated with GY, whereas it was positively correlated with PH an UIL. However, only few FLL loci were co-localized with loci for GY, PH or UIL. FLW was negatively correlated with SN, but positively correlated with KNS and SDW.

Potential implications in wheat breeding

The YHRVWZ is the major wheat growing area in China, producing ~65% of national production[4]. Comparison of the 20 highest-yielding and other varieties in the germplasm panel showed that KNS, TKW, KW, SDW and FLW in the high-yield group were 2.0, 5.6, 2.6, 7.0 and 4.1%, respectively, higher than the other group, whereas PH, UIL and FLL were 3.5, 10.1 and 5.6% lower. The numbers of alleles for increasing phenotypic values for each trait assessed in the panel were in agreement with the results mentioned above and in favor of Xiao et al. [5] and Gao et al. [6]. However, with the anomaly change of climate and decreased use value of germplasm, yield potential of new varieties is increasing slowly in this area. As a result, new methods and technologies that assisted in se- lection are essential for further improvement of GY.

High-yielding lines are difficult to select in the early stages of breeding programs as significantly influenced by other traits and environments. Li et al. [17] showed that FLW can be used to select lines with large KNS. In the present study, UIL showed a significant, negative correlation with GY, indicating that larger UIL was associated with decreased carbohydrate transportation to grain. FLL, significantly and positively correlated with UIL, also showed a significant, negative association with GY. SN and KNS were significantly and negatively correlated with each other, as reflected by FLW. Larger FLW was significantly associated with larger KNS and smaller SN in the same variety. Therefore, selection for shorter UIL and FLL would be helpful in selection for higher GY of wheat lines, whereas FLW is convenient to reflect SN and KNS.

Favorable alleles at each locus affecting GY exhibited positive effects on phenotypic values. As a result, the GY loci are valuable for selecting high-yielding varieties in breeding programs. The alleles for increasing phenotypic values presented significant additive effects on each trait, indicating that pyramiding favorable alleles is feasible to improve trait performances using the loci listed inTable 1. Besides, the 12 pleiotropic loci are important in determining GY and related traits, especially the loci that related to GY; the eight loci for TKW (2), KL, KW, SL and PH (3) that at similar positions with the QTL identified in our previous study are also credible. As GY related traits are mostly controlled by polygenes with small effect each, a genome-wide selection would be more powerful in gene discovery and pyramiding breeding with high-density genetic markers or genotyping by sequencing in future. However, MAS may be more feasible as long as only a few QTL need to be tracked in wheat breeding.

Among the 11 varieties with GY potential higher than 8200 kg ha^{-1}, Luyuan 502, Luomai 21, Yannong 18, Shannong 20, Zhongmai 875 and Wanmai 52 possess all 12 favorable alleles for GY. They are good parents to develop new high-yielding varieties. Four varieties, Lumai 8, Zhou 8425B, Zhongmai 875 and 85 Zhong 33 have large TKW, with more than 10 favorable alleles for that trait. These varieties should be valuable germplasms to develop large kernel varieties and for cloning genes related to TKW. Lankao 906 has large spikes with an average KNS of 60.2 and possesses all the favorable alleles identified in the present study for KNS. As KNS in the YHRVWZ is currently not large, this variety can be used to improve

KNS. The superior germplasm and favorable alleles of markers identified or confirmed in this study can be used in breeding new high-yielding varieties.

Conclusions

In the present study, SNP-GWAS and Haplotype-GWAS for GY and related traits, were performed in a diverse panel of 166 varieties with the wheat 90 K and 660 K SNP arrays. One hundred and twenty loci were identified by two methods, and 78 of these are likely to be new. Varieties with higher yield potential identified in the study can be used as parents in breeding programs aimed to accumulate further favorable alleles by marker-assisted selection. Our study proved that two GWAS methods with high-density SNP markers were reliable in identifying genes for GY and related traits, and provided new insight into the genetic architecture of GY.

Materials and methods

Plant materials and field trials

The diverse panel used in the present study contained 166 varieties, comprising 144 accessions from the YHRVWZ of China, and 22 accessions from other countries[26].

The diverse panel was grown at Anyang in Henan province and Suixi in Anhui province during the 2012-2013 and 2013-2014 cropping seasons, and at Shijiazhuang in Hebei province and Suixi in Anhui province during the 2014-2015. A randomized complete block design with three replicates was employed in field trials. Each plot comprised three 1.5 m rows spaced 20 cm apart, with 50 plants in each row. Agronomic management was performed according to local practices at each location. All wheat accessions are deposited in the National Genebank of China, Chinese Academy of Agricultural Sciences, and available after approval. All wheat varieties were collected in accordance with national regulations, and the experiments comply with the ethical standards and legislations in China.

Phenotyping and statistical analysis

Thirteen phenotypic traits, GY, SN, KNS, TKW, KL, KW, SL, SDW, HD, PH, UIL, FLL, and FLW were assessed in the diverse panel (Additional file 1: Table S1).

All plants were harvested in each plot at physiological maturity and GY as kg ha^{-1} were measured when the moisture declined to 14%. Investigation of the other 12 traits and statistical analyses followed Li et al.[17]. The phenotypic traits GY, KNS, TKW, KL, KW, SL, HD, and PH were assessed in all six environments, whereas data for FLL and FLW and those for SN, SDW and UIL were obtained in five and four environments, respectively. The phenotypic values in each environment and BLUE values were used for GWAS.

Genotyping, quality control and construction of the physical map

The diverse panel was genotyped using both the wheat 90 K SNP and 660 K SNP arrays[26]. Minor allele frequency (MAF), genetic diversity and PIC were calculated using PowerMarker v3.25 (http://statgen. ncsu. edu/powermar- ker/). To avoid spurious alleles, SNP with missing data > 20% and MAF < 0.05 were removed. Flanking sequences of SNPs were used to blast against the CSS database (IWGSC RefSeq v1.0, https://urgi. versailles. inra. fr/blast_ iwgsc/blast. php) to identify their positions on the physical map. Markers from the two SNP arrays were ordered based on their positions on chromosomes and integrated into a common physical map for GWAS.

Haplotype analysis

Based on 4 gametes and default parameters as used by the Haploview 4.2 software package (http://www. broad- institute. org/haploview/haploview), genome-wide haplo- type blocks were constructed with PLINK. The number of haplotypes, genetic length (bp) for each block, and the number of tag SNPs based on the 'solid spine' of LD were also provided

(Extend spine if $D' > 0.8$). Haplotype frequency was calculated using a custom Perl script and haplotypes with low frequency ($F < 0.05$) were removed.

Population structure and linkage disequilibrium

The SNP markers and estimated methods for population structure and LD were the same as in Liu et al. [26] . For population structure, 2000 polymorphic SNP markers evenly distributed on all 21 chromosomes were analyzed in Structure v2. 3. 4[70] (http: //pritchardlab. stanford. edu/struc- ture. html). PCA and NJ trees were estimated using the soft- ware Tassel v5. 0[71] and PowerMarker v3. 25[72] (http: //www. maizegenetics. net), respectively, to verify the results.

A total of 12, 324 evenly distributed SNP markers were chosen to calculate LD for theA, B and D and entire genomes using the full matrix and sliding window options in Tassel v5. 0[73].

Genome-wide association studies

SNP-GWAS and Haplotype-GWAS were used to identify the associations between phenotypic and genotypic data. For SNP-GWAS, the mixed linear model (MLM) in Tassel v5. 0 was used including kinship matrix and population structure. The kinship matrix was treated as a random effect and calculated by the Tassel v5. 0 software, whereas the subpopulation data was considered a fixed effect and estimated by Structure v2. 3. 4 in MLM analysis. The P value indicated the degree of association between a SNP marker and a trait, and the R^2 was the variation explained by the significantly associated markers. As the Bonferroni-Holm correction for multiple testing ($\alpha = 0.05$) was too conserved for the traits in the present study, markers with an adjusted -log10 (P-value) $\geqslant 3.0$ were regarded as significant for all traits. For Haplotype-GWAS, PLINK was used in consideration of population structure. According to the results, markers with -log10 (P-value) $\geqslant 4.0$ were considered to be significant. Manhattan plots for both methods were drawn using the ggplot2 code in R Lan-

guage with the P value estimated between the marker and trait in Tassel v5. 0 and PLINK. In both cases loci identified in onehalf or more environments were taken as stable.

Loci position comparison

For each trait, significant SNP markers within one LD on the same chromosome and identified by the same method were considered to represent one locus. Overlapping loci identified by the two methods for same trait were regarded as common loci. For loci or QTL reported in previous studies, two steps were followed to decide whether currently identified loci were the same as previously found. Firstly, the sequences of the tightly linked or significant markers of the QTL or loci were used to blast against the CSS database (IWGSC RefSeq v1. 0, https: //urgi. versailles. inra. fr/blast _ iwgsc/blast. php). If the marker was less than one LD from the locus for the same trait detected in the present study, they were considered to be the same. Secondly, the consensus linkage map constructed by Maccaferri et al. [39] was used to compare different types of markers. Therefore, loci or QTL were considered to be the same if the tightly linked or significantly associated markers were less than 2. 1, 1. 2 and 3. 9 cM from each other on the A, B and D genomes, respectively.

Effects of alleles on grain yield and related traits

For eachcommon locus, the most significant SNP markers and haplotypes were chosen as representative markers and haplotypes. The effects of each locus on phenotypic values for GY and the effects of the number of alleles for increasing phenotypic values for each trait were estimated based on the representative markers using R Language.

❖ Acknowledgements

The authors are grateful to Prof. R. A. McIntosh, Plant Breeding Institute, University of Sydney, for critical review of this manuscript.

❖ References

[1] Schulte D, Close TJ, Graner A, Langridge P, Matsumoto T, Muehlbauer G, Sato K, Schulman AH, Waugh R, Wise RP, Stein N. The international barley sequencing consortium at the threshold of efficient access to the barley genome. Plant Physiol. 2009; 149: 142-7.

[2] Tester M, Langridge P. Breeding technologies to increase crop production in a changing world. Science. 2010; 327: 818-22.

[3] FAO. FAOSTAT . 2017. http: //www. fao. org/faostat/en/#data/QC/visualize.

[4] He ZH, Xia XC, Bonjean APA. Wheat improve-ment in China. In: He ZH, Bonjean APA, editors. Cereals in China. Mexico DF. : CIMMYT; 2010. p. 51-68.

[5] Xiao YG, Qian ZG, Wu K, Liu JJ, Xia XC, Ji WQ, He ZH. Genetic gains in grain yield and physiological traits of winter wheat in Shandong province, China, from 1969 to 2006. Crop Sci. 2012; 52: 44.

[6] Gao FM, Ma DY, Yin GH, Rasheed A, Dong Y, Xiao YG, Xia XC, Wu XX, He ZH. Genetic progress in grain yield and physiological traits in Chinese wheat cultivars of southern yellow and Huai Valley since 1950. Crop Sci. 2017; 57: 760-73.

[7] Ellis MH, Rebetzke GJ, Azanza F, Richards RA, Spielmeyer W. Molecularmapping of gibberellin-responsive dwarfing genes in bread wheat. Theor Appl Genet. 2005; 111: 423-30.

[8] Zhou Y, He ZH, Sui XX, Xia XC, Zhang XK, Zhang GS. Genetic improvement of grain yield and associated traits in the northern China winter wheat region from 1960 to 2000. Crop Sci. 2007; 47: 245-53.

[9] He ZH, Liu L, Xia XC, Liu JJ, Pena RJ. Composition of HMW and LMW glutenin subunits and their effects on dough properties, pan bread, and noodle quality of Chinese bread wheat. Cereal Chem. 2005; 82: 345-50.

[10] Zhang XK, Yang SJ, Zhou Y, Xia XC, He ZH. Distribution of *Rht-B1b*, *Rht-D1b* and *Rht8* genes in autumn-sown Chinese wheats detected by molecular markers. Euphytica. 2006; 152: 109-16.

[11] Tian XL, Wen WE, Xie L, Fu LP, Xu DA, Fu C, Wang DS, Chen XM, Xia XC, Chen QJ, He ZH, Cao SH. Molecular mapping of reduced plant height gene *Rht24* in bread wheat. Front Plant Sci. 2017; 8: 1379.

[12] Tian XL, Zhu ZW, Xie L, Xu DA, Li JH, Fu C, Chen XM, Wang DS, Xia XC, He ZH, Cao SH. Preliminary exploration of the source, spread and distribution of *Rht24* reducing height in bread wheat. Crop Sci. 2019; 59: 19-24.

[13] Würschum T, Langer SM, Longin CFH, Tucker MR, Leiser WL. A modern green revolution gene for reduced height in wheat. Plant J. 2017; 92: 892-903.

[14] Liu YN, He ZH, Appels R, Xia XC. Functional markers in wheat: current status and future prospects. Theor Appl Genet. 2012; 125: 1-10.

[15] Rasheed A, Wen WE, Gao FM, Zhai SN, Jin H, Liu JD, Guo Q, Zhang Y, Dreisigacker S, Xia XC, He ZH. Development and validation of KASP assays for genes underpinning key economic traits in bread wheat. Theor Appl Genet. 2016; 129: 1843-60.

[16] Nadolska-Orczyk A, Rajchel IK, Orczyk W, Gasparis S. Major genes determining yield-related traits in wheat and barley. Theor Appl Genet. 2017; 130: 1081-98.

[17] Li FJ, Wen WE, He ZH, Liu JD, Jin H, Geng HW, Yan J, Zhang PZ, Wan YX, Xia XC. Genome-wide linkage mapping of yield related traits in three Chinese bread wheat populations using high-density SNP markers. Theor Appl Genet. 2018; 131: 1903-24.

[18] Cuthbert JL, Somers DJ, Brûlé-Babel AL, Brown PD, Crow GH. Molecular mapping of quantitative trait loci for yield and yield components in spring wheat (*Triticum aestivum* L.). Theor Appl Genet. 2008; 117: 595-608.

[19] Cui F, Li J, Ding AM, Zhao CH, Wang L, Wang XQ, Li SS, Bao YG, Li XF, Feng DS, Kong LR, Wang HG. Conditional QTL mapping for plant height with respect to the length of the spike and internode in two mapping populations of wheat. Theor Appl Genet. 2011; 122: 1517-36.

[20] Cui F, Zhao CH, Ding AM, Li J, Wang L, Li XF, Bao YG, Li JM, Wang HG. Construction of an integrative linkage map and QTL mapping of grain yield-related traits using three related wheat RIL populations. Theor Appl Genet. 2014; 127: 659-75.

[21] Jia HY, Wan HS, Yang SH, Zhang ZZ, Kong ZX, Xue SL, Zhang LX, Ma ZQ. Genetic dissection of yield-related traits in a recombinant inbred line population created using a key breeding parent in China's

wheat breeding. Theor Appl Genet. 2013；126：2123-39.

[22] Edae EA，Byrne PF，Haley SD，Lopes MS，Reynolds MP. Genome-wide association mapping of yield and yield components of spring wheat under contrasting moisture regimes. Theor Appl Genet. 2014；127：791-807.

[23] Azadi A，Mardi M，Hervan EM，Mohammadi SA，Moradi F，Tabatabaee MT，Pirseyedi SM，Ebrahimi M，Fayaz F，Kazemi M，Ashkani S，Nakhoda B，Mohammadi-Nejad G. QTL mapping of yield and yield components under normal and salt-stress conditions in bread wheat (*Triticum aestivum* L.). Plant Mol Biol Rep. 2015；33：102-20.

[24] Wang SC，Wong D，Forrest K，Allen A，Chao S，Huang BE，Maccaferri M，Salvi S，Milner SG，Cattivelli L，Mastrangelo AM，Whan A，Stephen S，Barker G，Wieseke R，Plieske J，Lillemo M，Mather D，Appels R，Dolferus R，Guedira GB，Korol A，Akhunova AR，Feuillet C，Salse J，Morgante M，Pozniak C，Luo MC，Dvorak J，Morell M，Dubcovsky J，Ganal M，Tuberosa R，Lawley C，Mikoulitch I，Cavanagh C，Edwards KJ，Hayden M，Akhunov E. Characterization of polyploid wheat genomic diversity using a high-density 90000 single nucleotide polymorphism array. Plant Biotechnol J. 2014；12：787-96.

[25] Jin H，Wen WE，Liu JD，Zhai SN，Zhang Y，Yan J，Liu ZY，Xia XC，He ZH. Genome- wide QTL mapping for wheat processing quality parameters in a Gaocheng 8901/Zhoumai 16 recombinant inbred line population. Front Plant Sci. 2016；7：1032.

[26] Liu JD，He ZH，Rasheed A，Wen WE，Yan J，Zhang PZ，Wan YX，Zhang Y，Xie CJ，Xia XC. Genome-wide association mapping of black point reaction in common wheat (*Triticum aestivum* L.). BMC Plant Biol. 2017；17：220.

[27] Sun CW，Zhang FY，Yan XF，Zhang XF，Dong ZD，Cui DQ，Chen F. Genome- wide association study for 13 agronomic traits reveals distribution of superior alleles in bread wheat from the yellow and Huai Valley of China. Plant Biotechnol J. 2017；15：953-69.

[28] Valluru R，Reynolds MP，Davies WJ，Sukumaran S. Phenotypic and genome- wide association analysis of spike ethylene in diverse wheat genotypes under heat stress. New Phytol. 2017；214：271-83.

[29] Scherer A，Christensen GB. Concepts and relevance of genome-wide association studies. Sci Prog. 2016；99：59-67.

[30] Lorenz AJ，Hamblin MT，Jannink JL. Perfor-mance of single nucleotidepolymorphisms versus haplotypes for genome-wide association analysis in barley. PLoS One. 2010；5：e14079.

[31] Chia JM，Song C，Bradbury PJ，Costich D，de Leon N，Doebley J，Elshire RJ，Gaut B，Geller L，Glaubitz JC，Gore M，Guill KE，Holland J，Hufford MB，Lai J，Li M，Liu X，Lu Y，McCombie R，Nelson R，Poland J，Prasanna BM，Pyhajarvi T，Rong T，Sekhon RS，Sun Q，Tenaillon MI，Tian F，Wang J，Xu X，Zhang Z，Kaeppler SM，Ross-Ibarra J，McMullen MD，Buckler ES，Zhang G，Xu Y，Ware D. Maize HapMap2 identifies extant variation from a genome in flux. Nat Genet. 2012；44：803-7.

[32] Thomson MJ. High-throughput SNP genotyping to accelerate crop improvement. Plant Breed Biotechnol. 2014；2：195-212.

[33] Huang XQ，Kempf H，Ganal MW，Röder MS. Advanced backcross QTL analysis in progenies derived from a cross between a German elite winter wheat variety and a synthetic wheat (*Triticum aestivum* L.). Theor Appl Genet. 2004；109：933-43.

[34] Kumar N，Kulwal PL，Balyan HS，Gupta PK. QTL mapping for yield and yield contributing traits in two mapping populations of bread wheat. Mol Breeding. 2007；19：163-77.

[35] Reif JC，Maurer HP，Korzun V，Ebmeyer E，Miedaner T，Würschum T. Mapping QTLs with main and epistatic effects underlying grain yield and heading time in soft winter wheat. Theor Appl Genet. 2011；123：283-92.

[36] Lee HS，Jung JU，Kang CS，Heo HY，Park CS. Mapping of QTL for yield and its related traits in a doubled haploid population of Korean wheat. Plant Biotechnol Rep. 2014；8：443-54.

[37] Lopes MS，Dreisigacker S，Peña RJ，Sukumaran S，Reynolds MP. Geneticcharacterization of the wheat association mapping initiative (WAMI) panel for dissection of complex traits in spring wheat. Theor Appl Genet. 2015；128：453-64.

[38] Sukumaran S，Dreisigacker S，Lopes M，Chavez P，Reynolds MP. Genome- wide association study for grain yield and related traits in an elite spring wheat

population grown in temperate irrigated environments. Theor Appl Genet. 2015; 128: 353-63.

[39] Maccaferri M, Zhang JL, Bulli P, Abate Z, Chao S, Cantu D, Bossolini E, Chen XM, Pumphrey M, Dubcovsky J. A genome-wide association study of resistance to stripe rust (Puccinia striiformis f. sp. tritici) in a worldwide collection of hexaploid spring wheat (Triticum aestivum L.). Genetics. 2015; 5: 449-65.

[40] Gao FM, Wen WE, Liu JD, Rasheed A, Yin GH, Xia XC, Wu XX, He ZH. Genome-wide linkage mapping of QTL for yield components, plant height and yield-related physiological traits in the Chinese wheat cross Zhou 8425B/Chinese spring. Front Plant Sci. 2015; 6: 1099.

[41] Zhang HX, Zhang FN, Li GD, Zhang SN, Zhang ZG, Ma LJ. Genetic diversity and association mapping of agronomic yield traits in eighty six synthetic hexaploid wheat. Euphytica. 2017; 213: 111.

[42] Yao J, Wang LX, Liu LH, Zhao CP, Zheng YL. Association mapping ofagronomic traits on chromosome 2A of wheat. Genetica. 2009; 137: 67-75.

[43] Wang JS, Liu WH, Wang H, Li LH, Wu J, Yang XM, Li XQ, Gao AN. QTL mapping of yield-related traits in the wheat germplasm 3228. Euphytica. 2011; 177: 277-92.

[44] Mir RR, Kumar N, Jaiswal V, Girdharwal N, Prasad M, Balyan HS, Gupta PK. Genetic dissection of grain weight in bread wheat through quantitative trait locus interval and association mapping. Mol Breeding. 2012; 29: 963-72.

[45] Chen GF, Zhang H, Deng ZY, Wu RG, Li DM, Wang MY, Tian JC. Genome- wide association study for kernel weight-related traits using SNPs in a Chinese winter wheat population. Euphytica. 2016; 212: 173-85.

[46] Mohler V, Albrecht T, Castell A, Diethelm M, Schweizer G, Hartl L. Considering causal genes in the genetic dissection of kernel traits in common wheat. J Appl Genet. 2016; 57: 467-76.

[47] Wu QH, Chen YX, Zhou SH, Fu L, Chen JJ, Xiao Y, Zhang D, Ouyang SH, Zhao XJ, Cui Y, Zhang DY, Liang Y, Wang ZZ, Xie JZ, Qin JX, Wang GX, Li DL, Huang YL, Yu MH, Lu P, Wang LL, Wang L, Wang H, Dang C, Li J, Zhang Y, Peng HR, Yuan CG, You MS, Sun QX, Wang JR, Wang LX, Luo MC, Han J, Liu ZY. High-density genetic linkage map construction and QTL mapping of

grain shape and size in the wheat population Yanda1817   ×   Beinong6. PLoS   One. 2015; 10: e0118144.

[48] Liu G, Jia LJ, Lu LH, Qin DD, Zhang JP, Guan PF, Ni ZF, Yao YY, Sun QX, Peng HR. Mapping QTLs of yield-related traits using RIL population derived from common wheat and Tibetan semi-wild wheat. Theor Appl Genet. 2014; 127: 2415-32.

[49] Sajjad M, Ma XL, Habibullah Khan S, Shoaib M, Song YL, Yang WL, Zhang AM, Liu DC. TaFlo2-A1, an ortholog of rice Flo2, is associated with thousand grain weight in bread wheat (Triticum aestivum L.). BMC Plant Biol. 2017; 17: 164.

[50] Zhai HJ, Feng ZY, Du XF, Song YE, Liu XY, Qi ZQ, Song L, Li J, Li LH, Peng HR, Hu ZR, Yao YY, Xin MM, Xiao SH, Sun QX, Ni ZF. A novel allele of TaGW2-A1 is located in a finely mapped QTL that increases grain weight but decreases grain number in wheat (Triticum aestivum L.). Theor Appl Genet. 2018; 131: 539-53.

[51] Le Gouis J, Bordes J, Ravel C, Heumez E, Faure S, Praud S, Galic N, Remoue C, Balfourier F, Allard V, Rousset M. Genome-wide association analysis to identify chromosomal regions determining components of earliness in wheat. Theor Appl Genet. 2012; 124: 597-611.

[52] Peng J, Richards D-E, Hartley N-M, Murphy GP, Devos KM, Flintham JE, Beales J, Fish LJ, Worland AJ, Pelica F, Sudhakar D, Christou P, Snape JW, Gale MD, Harberd NP. 'Green revolution' genes encode mutant gibberellin response modulators. Nature. 1999; 400: 256-61.

[53] Li C, Bai G, Carver BF, Chao S, Wang Z. Mapping quantitative trait loci for plant adaptation and morphology traits in wheat using single nucleotide polymorphisms. Euphytica. 2016; 208: 299-312.

[54] Korzun V, Roder MS, Worland AJ, Borner A. Intrachromosomal mappingof genes for dwarfing (Rht12) and vernalization response (Vrn1) in wheat by using RFLP and microsatellite markers. Plant Breed. 1997; 116: 227-32.

[55] Fu DL, Szucs P, Yan LL, Helguera M, Skinner JS, Zitzewitz J, Hayes PM, Dubcovsky J. Large deletions within the first intron in VRN-1 are associated with spring growth habit in barley and wheat. Mol Gen Genomics. 2005; 273: 54-65.

[56] Wu QH, Chen YX, Fu L, Zhou SH, Chen JJ,

Zhao XJ, Zhang D, Ouyang SH, Wang ZH, Li D, Wang GX, Zhang DY, Yuan CG, Wang LX, You MS, Han J, Liu ZY. QTL mapping of flag leaf traits in common wheat using an integrated high-density SSR and SNP genetic linkage map. Euphytica. 2016; 208: 337-51.

[57] Liu KY, Xu H, Liu G, Guan PF, Zhou XY, Peng HR, Yao YY, Ni ZF, Sun QX, Du JK. QTL mapping of flag leaf-related traits in wheat (*Triticum aestivum* L.). Theor Appl Genet. 2018; 131: 839-49.

[58] Xu YF, Li SS, Li LH, Ma FF, Fu XY, Shi ZL, Xu HX, Ma PT, An DG. QTL mapping for yield and photosynthetic related traits under different water regimes in wheat. Mol Breeding. 2017; 37: 34.

[59] Marza F, Bai G-H, Carver BF, Zhou W-C. Quantitative trait loci for yield and related traits in the wheat population Ning7840 × Clark. Theor Appl Genet. 2005; 112: 688-98.

[60] Quarrie SA, Steed A, Calestani C, Semikhodskii A, Lebreton C, Chinoy C, Steele N, Pljevljakusic D, Waterman E, Weyen J, Schondelmaier J, Habash DZ, Farmer P, Saker L, Clarkson DT, Abugalieva A, Yessimbekova M, Tururuspekov Y, Abugalieva S, Tuberosa R, Sanguineti M-C, Hollington PA, Aragues R, Royo A, Dodig D. A high density genetic map of hexaploid wheat (*Triticum aestivum* L.) from the cross Chinese Spring × SQ1 and its use to compare QTLs for grain yield across a range of environments. Theor Appl Genet. 2005; 110: 865-80.

[61] Huang XQ, Cloutier S, Lycar L, Radovanovic N, Humphreys DG, Noll JS, Somers DJ, Brown PD. Molecular dissection of QTLs for agronomic and quality traits in a doubled haploid population derived from two Canadian wheats (*Triticum aestivum* L.). Theor Appl Genet. 2006; 113: 753-66.

[62] Li S, Jia J, Wei X, Zhang X, Li L, Chen H, Fan Y, Sun H, Zhao X, Lei T, Xu Y, Jiang F, Wang H, Li L. A intervarietal genetic map and QTL analysis for yield traits in wheat. Mol Breeding. 2007; 20: 167-78.

[63] Miralles DJ, Slafer GA. Sink limitations to yield in wheat, how could it be reduced? J Agric Sci. 2007; 145: 139-49.

[64] McIntyre CL, Mathews KL, Rattey A, Chapman SC, Drenth J, Ghaderi M, Reynolds M, Shorter R. Molecular detection of genomic regions associated with grain yield and yield-related components in an elite bread wheat cross evaluated under irrigated and rainfed conditions. Theor Appl Genet. 2010; 120: 527-41.

[65] Keyes GJ, Paolillo DJ, Sorrells ME. The effects of dwarfing genes *Rht1* and *Rht2* on cellular dimensions and rate of leaf elongation in wheat. Ann Bot. 1989; 64: 683-90.

[66] Chebotar GA, Chebotar SV, Motsnyy II. Pleiotropic effects of gibberellin- sensitive and gibberellin-insensitive dwarfing genes in bread wheat of the southern step region of the Black Sea. Cytol Genet. 2016; 50: 20-7.

[67] Kamran A, Iqbal M, Spaner D. Flowering time in wheat (*Triticum aestivum* L.): a key factor for global adaptability. Euphytica. 2014; 197: 1-26.

[68] Xu HY, Zhao JS. Canopy photosynthesis capacity and the contributionfrom different organs in high-yielding winter wheat. Acta Agron Sin. 1995; 21: 204-9.

[69] Sharma SN, Saini RS, Sharma PK. The genetic control of flag leaf length in normal and late sown durum wheat. J Agric Sci (Camb). 2003; 141: 323-31.

[70] Botstein D, White RL, Skolnick M, Davis RW. Construction of a genetic linkage map in man using restriction fragment length polymorphisms. Am J Hum Genet. 1980; 32: 314-9.

[71] Yu J, Buckler ES. Genetic association mapping and genome organization of maize. Curr Opin Biotechnol. 2006; 17: 155-60.

[72] Liu K, Muse SV. PowerMarker: an integrated analysis environmentfor genetic marker analysis. Bioinformatics. 2005; 21: 2128-9.

[73] Breseghello F, Sorrells ME. Association mapping of kernel size and milling quality in wheat (*Triticum aestivum* L.) cultivars. Genetics. 2006; 172: 1165-77.

根系与肥料利用效率

QTL mapping of seedling biomass and root traits under different nitrogen conditions in bread wheat (*Triticum aestivum* L.)

YANG Meng-jiao[1,2], WANG Cai-rong[2,3], Muhammad Adeel HASSAN[2],

WU Yu-ying[2], XIA Xian-chun[2], SHI Shu-bing[1], XIAO Yong-gui[2], HE Zhong-hu[2,4]

[1] College of Agriculture, Xinjiang Agricultural University, Urumqi 830052, P. R. China

[2] National Wheat Improvement Centre, Institute of Crop Sciences, Chinese Academy of Agricultural Sciences (CAAS), Beijing 100081, P. R. China

[3] Institute of Agricultural Science of Yili Prefecture, Yining 835000, P. R. China

[4] International Maize and Wheat Improvement Centre (CIMMYT) China Office, c/o CAAS, Beijing 100081, P. R. China

Abstract: Plant nitrogen assimilation and use efficiency in the seedling's root system are beneficial for adult plants in field condition for yield enhancement. Identification of the genetic basis between root traits and N uptake plays a crucial role in wheat breeding. In the present study, 198 doubled haploid lines from the cross of Yangmai 16/Zhongmai 895 were used to identify quantitative trait loci (QTLs) underpinning four seedling biomass traits and five root system architecture (RSA) related traits. The plants were grown under hydroponic conditions with control, low and high N treatments [Ca (NO$_3$)$_2$ • 4H$_2$O at 0, 0.05 and 2.0 mmol L^{-1}, respectively]. Significant variations among the treatments and genotypes, and positive correlations between seedling biomass and RSA traits ($r=0.20$ to 0.98) were observed. Inclusive composite interval mapping based on a high-density map from the Wheat 660K single nucleotide polymorphisms (SNP) array identified 51 QTLs from the three N treatments. Twelve new QTLs detected on chromosomes 1AL (1) in the control, 1DS (2) in high N treatment, 4BL (5) in low and high N treatments, and 7DS (3) and 7DL (1) in low N treatments, are first reported in influencing the root and biomass related traits for N uptake. The most stable QTLs (*RRS. caas-4DS*) on chromosome 4DS, which were related to ratio of root to shoot dry weight trait, was in close proximity of the *Rht-D1* gene, and it showed high phenotypic effects, explaining 13.1% of the phenotypic variance. Twenty-eight QTLs were clustered in 12 genetic regions. SNP markers tightly linked to two important QTLs clusters C10 and C11 on chromosomes 6BL and 7BL were converted to kompetitive allele-specific PCR (KASP) assays that underpin important traits in root development, including root dry weight, root surface area and shoot dry weight. These QTLs, clusters and KASP assays can greatly improve the efficiency of selection for root traits in wheat breeding programmes.

Keywords: KASP marker, QTL analysis, root traits, SNP array, *Triticum aestivum*

Published in Journal of Intergrative Agriculture, 2021, 20 (5): 1180-1192.

1 Introduction

Nitrogen (N) is the principal nutrient element in root development and photosynthate accumulation during the plant growth cycle (Hermans *et al.* 2006; Forde *et al.* 2014; Cormier *et al.* 2016). N fertilizers are essential for the maximum ecosystem productivity and meeting food demand (Moore and Lobell 2015). In recent years, N fertilizer use in terms of pure N increased from 92 to 108 Mt in the world (FAOSTAT 2020). Over-application and low nitrogen use efficiency (NUE) have caused major resource concerns and substantial greenhouse gas emissions (Diaz and Rosenberg 2008; Guo *et al.* 2010; Zhang and Wang 2015; Zhang W *et al.* 2016). In China, the agricultural system generally uses high to excessive N fertilizer, and the total average application of N for winter wheat (*Triticum aestivum* L.) has increased to more than 500 kg N ha^{-1}. Whereas the NUE in the wheat production system was lower than in maize (*Zea mays* L.) and rice (*Oryza sativa* L.), by approximately 25% (Zhang W *et al.* 2016; Cui *et al.* 2018), substantial regulation and breeding new varieties for high N acquisition would be an effective approach for improving NUE and yield potential (Rengel and Marschner 2005).

In wheat, root system architecture (RSA) defines the spatial configuration of root structure that includes the root's number, length, tip number, emergence angles, width, depth, convex hull area, and root mass center. An increased root biomass could help plants to maintain a balance between the shoot and root under N deficiency, and also has a significant impact on yield enhancement (Barraclough *et al.* 1989; Robinson *et al.* 2001; Reynolds *et al.* 2007; Cormier *et al.* 2016; Bettembourg *et al.* 2017). Development of RSA-related traits depend on genetic features of individual plants and the growing environment. It is difficult for genetic studies to investigate underground root traits for large sets of samples. Thus, there is a need for high-throughput methods for root phenotyping to determine the best-performing wheat

genotypes for precise selection.

Identifying multiple genes for root growth in response to N uptake provides a promising way to optimize RSA, and reduces N fertilizer requirements while maintaining yield stability and to overcome the deceasing resource problems in the future (Wasson *et al.* 2012). The *Rht* genes controlling shoot height have been reported to reduce root proliferation and promote increased N uptake from underground resources by manipulating root systems (Lynch *et al.* 2007; Hund *et al.* 2009; Bai *et al.* 2013; Cui *et al.* 2014; Narayanan and Prasad 2014; Ryan *et al.* 2015; Aziz *et al.* 2017). Quantitative trait loci (QTLs) mapping based on high-density maps has increased the understanding of the genetic basis for root traits, and marker-assisted selection (MAS) in multiple traits has contributed to genetically improved wheat varieties (Guo *et al.* 2012; Atkinson *et al.* 2015; Subira *et al.* 2016). An increased grain yield was achieved through introgression of QTLs into backcrossderived lines, and some studies have provided evidence for the feasibility of improving grain yield by manipulating root systems (Li P C *et al.* 2015). Moreover, a major locus on chromosome 2BS, *qTaLRO -B1*, was determined to affect root length and biomass accumulation of wheat seedlings, and the linked markers were developed for further breeding purposes (Cao *et al.* 2014).

The objectives of this study were to screen the doubledhaploid (DH) lines derived from the Yangmai 16/ Zhongmai 895 cross for seedling biomass and root traits using hydroponic culture, to identify QTLs at the seedling stage under three N conditions using 660K single nucleotide polymorphisms (SNP) array, and to develop kompetitive allele-specific PCR (KASP) markers associated with important loci to increase N uptake capacity in further wheat breeding programs.

2 Materials and methods

2.1 Plant materials

The mapping population used in the present study included 198 DH lines derived from a cross between two

Chinese wheat varieties, Yangmai 16 and Zhongmai 895. Zhongmai 895 is a facultative variety from the southern part of the Yellow and Huai River Valley Winter Wheat Zone, China with high yield potential and strong root vigor, whereas Yangmai 16 is a spring wheat variety widely grown in the Middle and Lower Yangtze River Valley Wheat Zone, China.

2.2 Hydroponic culture and experimental design

Root screening in hydroponics at the seedling stage was performed using Hoagland's nutrient solution (Hoagland and Arono 1950), with three N treatments, viz., control, low and high N based on Ca $(NO_3)_2$ · $4H_2O$ concentrations of 0, 0.05 and 2.0 mmol L^{-1}, respectively. To maintain the Ca nutrient in a common concentration, $CaCl_2$ · $2H_2O$ solution (2.0, 1.95 and 0 mmol L^{-1}, respectively) was applied corresponding to the Ca in the three N levels (Li F *et al*. 2015) (Table 1). A randomized complete block design was used to minimize the experimental errors,

with three replications in a temperature- controlled greenhouse from March 15 to April 28, 2016.

Thirty seedsof each line were surface-sterilized in a 10% H_2O_2 solution for 15-20 min, rinsed in sterilized water 5-6 times, and germinated on moist germination paper in Petri dishes. The germinated seeds were transferred to plastic trays filled with quartz sand (2 mm diameter) and held in darkness at 24℃ for 72 h. Nine seedlings for each N treatment, i. e., three seedlings for each replication, were chosen and placed into holes fastened by spongy material and then transferred into a plastic tank (660 mm × 480 mm × 280 mm) containing 20 L of nutrient solution in an environmentally controlled room (16℃ in day time with a light intensity of 400 μmol m^{-2} s^{-1} photosynthetically active radiation and 13℃ at night, and a relative humidity of 70%). The solution was changed every 3 days. Plants were harvested after 10 days and stored in 30% ethanol prior to imaging.

Table 1   Nutrient solution ingredients for wheat seedling growth

Ingredient	Concentration (mmol L^{-1})	Ingredient	Concentration (mmol L^{-1})
K_2SO_4	0.75	$MnSO_4$ · H_2O	0.001
KH_2PO_4	0.25	$ZnSO_4$ · $7H_2O$	0.001
$MgSO_4$	0.60	$CuSO_4$ · $5H_2O$	0.0001
FeEDTA	0.04	Na_2MoO_4 · $2H_2O$	0.000005
H_3BO_3	0.001	Ca $(NO_3)_2$ · $4H_2O$	0/0.05/2.0
KCl	0.10	$CaCl_2$ · $2H_2O$	2.0/1.95/0

2.3 Trait measurements

The seedling biomass traits, *viz*., shoot dry weight (SDW) and root dry weight (RDW), were determined after oven drying at 70℃ for 72 h using 1/1 000 balances. The total dry weight (TDW) was calculated as the sum of SDW and RDW, and the ratio of root to shoot dry weight (RRS) was defined as the ratio of RDW to SDW. RSA related traits, *viz*., main root length (RL), root diameter (RD), root surface area (ROSA), root tip number (RTN), and root volume (RV), were measured using a recording scanner (Perfection V700/V750 2.80A; Epson, Chi-

na), and images were analyzed by semi-automated Software RootNav V1.7.5 (Pound *et al*. 2013).

2.4 SNP genotyping and QTL detection

The DH lines and parents were genotyped using the Wheat 660K iSelect SNP array from Affymetrix at Capital Bio Corporation (Beijing, China; http: // www. capitalbio. com). The genetic map was constructed by Wang *et al*. (2017). The linkage map of Yangmai 16/Zhongmai 895 population contained 10 242 SNP markers in 25 linkage groups, covering all 21 wheat chromosomes. Inclusive composite interval mapping (ICIM) was used to QTL, based on the best linear unbiased estimation (BLUE) values from three replicates by the Software IciMapping V4. 1 (Meng *et al.* 2015). The genotypes of Yangmai 16 and Zhoumai 985 were defined as 0 and 2, respectively. Hence, alleles from Zhongmai 895 increased trait values when the additive effects were positive. Recombination frequencies were converted into map distances using the Kosambi mapping function (Kosambi 1943). In order to ensure the existence of accurate QTLs, the LOD value of 2. 5 was used as a threshold based on the 1 000 permutation test. The QTL was computed with a walking step of 1. 0 cM, a PIN of 0. 001 and a Type I error of 0. 05 (Churchill and Doerge 1994).

2. 5 Conversion of SNP markers to KASP assay

The physical positions of markers were based on wheat genome sequences from the International Wheat Genome Sequencing Consortium (IWGSC RefSeq 1. 0, http: //www. wheatgenome. org/). Allele specific primers for KASP assays were designed using the PolyMarker (http: //polymarker. tgac. ac. uk/) following Rasheed *et al.* (2016). Thedetailed information for KASP markers corresponding to two SNPs *AX-109558906* on chromosome 6B and *AX-95025477* on chromosome 7B, respectively, is provided in Appendix A. The primer mixture included 46 μL ddH$_2$O, 30 μL common primer (100 μmol L^{-1}) and 12 μL of each tailed primer (100 μmol L^{-1}). Assays were tested in 384-well formats and set up as ~3 μL reactions (10-20 ng μL^{-1} DNA, 3 μL of 1× KASP master mixture and 0. 056 μL of primer mixture). PCR cycling was performed by the following procedure: hot start for 15 min at 95℃, 10 touchdown cycles (95℃ for 20 s; touchdown at 65℃ initially and decreasing by −1℃ per cycle for 25 s), then 30 additional cycles of denaturation and annealing/extension (95℃ for 10 s; 57℃ for 60 s). PCR was performed in a Bio-Rad CFX Real-Time PCR System, and fluorescence was detected using the Bio-Rad CFX Manage 3. 1 Software.

2. 6 Statistical analysis

Data analysis was performed using the R package (R Core Team 2013). Correlations were determined by Pearson's coefficient, and significance tests among DH lines and parents were made using a mixed linear model at $P < 0. 05$. Broad-sense heritability for each trait was estimated by:

$$h^2 = \sigma_g^2 / (\sigma_g^2 + \sigma_{gt}^2 / r + \sigma_\varepsilon^2 / rt)$$

where σ^2 is the genetic variance, σ_{gt}^2 is the genotype the genotype (line) × N treatment interaction, σ_ε^2 is the error variance, and t and r represent the numbers of N treatments and replications, respectively.

3 Results

3. 1 Phenotypic performance of DH lines for seedling biomass and RSA related traits under three N levels

The phenotypic performance of DH lines was normally distributed, and transgressive segregations were observed for seedling biomass and RSA related traits under all three N levels (Appendices B and C). Significant variations ($P < 0. 001$) were observed among the DH lines with high broad-sense heritabilities (0. 83 to 0. 98). Variations for all traits, treatments and the genotype× treatment interaction were also significant (Table 2). In particular, DH lines exhibited higher values for RRS, RL and RTN under the control and low N conditions, and higher values for SDW, TDW, RD, and RV at the high N level. Yangmai 16 had higher values of SDW, RDW, TDW, RL, and RTN under the control and low N conditions, whereas Zhongmai 895 showed higher values at the high N level for all traits except RTN (Fig. 1).

High and positive correlations between most seedling biomass and RSA traits were observed across all three N levels ($r = 0. 20$ to 0. 98; Fig. 2). RD showed significantly negative correlations with RTN ($r = -0. 45$ to $-0. 62$), and with RL at the low N ($r = -0. 29$) and high N levels ($r = -0. 40$). Negative correlations were also observed between SDW and RRS across the three N treatments.

3. 2 QTL for seedling biomass traits

Twenty-five QTLs weremapped on 12 chromosomes

for seedling biomass traits (Table 3). Ten major QTLs, accounting for >10% of the phenotypic variances, were located in the three N treatments. Twelve QTLs were detected in the high N treatment, seven in the control and six in the low N treatment. One stable QTL on 4DS was detected across the three N treatments. This corresponded to *Rht-D1* region with favorable alleles from the male parent Zhongmai 895, explaining 13.1 to 19.0% of the phenotypic variances. A QTL for RRS on 4BS was detected under both low and

high N treatments, corresponding to the *Rht-B1* region. The presence of *Rht-B1b* in Yangmai 16 and *Rht-D1b* in Zhongmai 895 and their segregation in DH population was confirmed through functional markers in a previous study (Hassan *et al*. 2019; Appendix D). At the high N level, an important pleiotropic QTL on 6BL simultaneously controlling SDW, RDW and TDW explained 10.4, 9.9 and 10.3% of the phenotypic variances, respectively.

Table 2　Significance test and heritability (h^2) values for the measured traits of doubled haploid (DH) lines in three nitrogen (N) treatments

Trait[1]	N level[2]	Min.	Max.	Mean SD	Genotype F-value	Replication F-value	Genotype× Treatment F-value	h^2
SDW (mg)	C	0.01	0.08	0.05±0.01	3.31***	0.06	1.60***	0.96
	L	0.02	0.06	0.06±0.01	3.25***	0.04*		0.95
	H	0.01	0.12	0.09±0.02	1.94***	0.12		0.92
RDW (mg)	C	0.01	0.04	0.02±0.001	2.26***	0.08	1.62***	0.95
	L	0.01	0.03	0.02±0.001	3.22***	0.05		0.96
	H	0.01	0.04	0.03±0.001	2.20***	0.10		0.94
TDW (mg)	C	0.02	0.11	0.07±0.01	2.96***	0.08	1.61***	0.95
	L	0.02	0.09	0.06±0.01	3.19***	0.04*		0.97
	H	0.01	0.16	0.09±0.02	1.99***	0.13		0.92
RRS	C	0.14	0.82	0.48±0.11	2.74***	0.01*	1.70***	0.98
	L	0.26	0.73	0.49±0.07	4.20***	0.07		0.96
	H	0.10	0.54	0.32±0.06	2.27***	0.20		0.90
RL (mm)	C	65.46	672.24	314.19±104.61	1.24*	0.19	1.23*	0.85
	L	31.83	646.11	267.51±112.17	1.51**	0.13		0.90
	H	6.55	662.63	247.31±124.77	0.95	0.04*		0.94
RD (mm³)	C	0.14	0.28	0.21±0.02	2.30***	0.12	1.88***	0.93
	L	0.13	0.24	0.19±0.02	2.37***	0.08		0.95
	H	0.09	0.36	0.24±0.04	2.18***	0.03*		0.96
RV (mm³)	C	0.01	0.25	0.10±0.04	1.15	0.21	1.13*	0.83
	L	0.01	0.17	0.07±0.03	1.35**	0.06		0.94
	H	0.01	0.24	0.11±0.05	1.04	0.01*		0.97
RTN	C	93.55	2 631.05	1053.4±465.11	1.55**	0.28	1.43*	0.83
	L	149.04	3 579.63	1675.62±582.32	2.08***	0.01*		0.97
	H	129.32	3 154.98	1177.65±588.83	1.51**	0.23		0.85
ROSA (mm²)	C	1.73	38.80	20.52±6.84	1.20*	0.08	1.06*	0.92

(continued)

Trait[1]	N level[2]	Min.	Max.	Mean SD	Genotype	Replication	Genotype×Treatment	h^2
					F-value	F-value	F-value	
ROSA （mm²）	L	1.86	36.88	15.46±6.37	1.39*	0.03*		0.96
	H	1.07	44.98	18.02±8.3	0.92	0.03*		0.95

[1] SDW, shoot dry weight; RDW, root dry weight; TDW, total dry weight; RRS, ratio of root to shoot dry weight; RL, root length; RD, root diameter; RV, root volume; RTN, root tip number; ROSA, root surface area.

[2] C, control; L, low N treatment; H, high N treatment.

***, $P<0.001$; **, $P<0.01$; *, $P<0.05$.

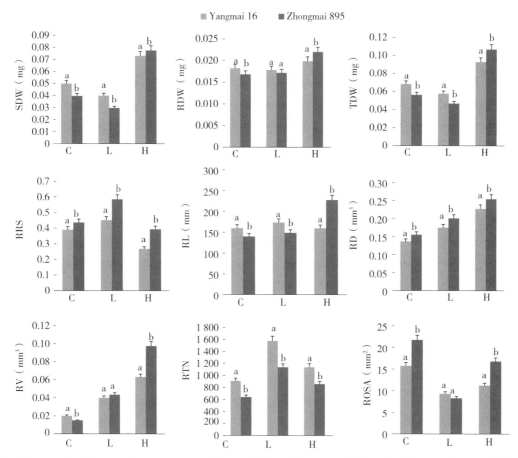

Fig. 1　Phenotypic differences between parents Yangmai 16 and Zhongmai 895 under different nitrogen （N） treatments. C, control; L, low N treatment; H, high N treatment. SDW, shoot dry weight; RDW, root dry weight; TDW, total dry weight; RRS, ratio of root to shoot dry weight; RL, root length; RD, root diameter; RV, root volume; RTN, root tip number; ROSA, root surface area. Error bars represent standard deviations for each proportion （$n=3$）; different letters indicate significant differences between the genotypes determined by Duncan's multiple range tests. *, significant at $P<0.05$.

3.3　QTL for RSA-related traits

Twenty-six QTLs were identified for all five RSA traits （Table 3）. Among them, five QTLs for RL were identified on chromosomes 1AL （2）, 2BS, 3BS, and 7DS, nine for RD on chromosomes 1AL, 1DS, 2AL, 3AS, 4AS, 4BS, 4BL, 5AL and 5BL, four for RV on chromosomes 4BL （2）, 5AS and 7DS, five for RTN on chromosomes 1DS, 2BS, 4DS, 5AL and 7DS, and three for ROSA on chromosomes 2AS, 4BL and 7BL. The phenotypic variances explained （PVE） of these QTLs ranged from 4.7 to 19.1%. Five major QTLs QRD.caas-1AL, QRD.caas-4BL, QRV.caas-4BL, QRTN.caas-5AL, and QROSA.caas-4BL with larger effects, explained 19. 1, 18. 3, 13. 3, 14. 5, and 12. 5% of phenotypic variances, respectively. Under the low N treatment, one pleiotropic QTL for RL and RTN on 2BS, and one for RL and RV on chromosome 7DS were identified in the marker intervals of *AX-108920782-AX-110463005* (2BS) and *AX-1089522259-AX-111881572* (7DS) under low N treatment, respectively. Two pleiotropic QTLs conditioning RD and RTN in the high N treatment were identified on 1DS and 5AL in the marker intervals of *AX-109849862- AX-108727857* (1DS) and *AX-109958693-AX-94700681* (5AL), respectively. These QTLs explained 5. 9 to 14. 5% of the phenotypic variances.

3. 4　QTL clusters

Twelve clusters (C1-C12) including 28 QTLs for different traits were identified on 10 chromosomes (i. e. , 1AL, 1DS (2), 2BS, 4BL (2), 4DS, 5AL, 6AL, 6BL, 7BL, and 7DS) across the three N treatments (Figs. 3 and 4). Among these, more QTL clusters under the high N treatment (5) were identified than those under the control (1) and the low N treatment (3), and the other three clusters were detected in two or three N treatments.

In the control, C1 for RL and RRS was detected on chromosome 1AL in the marker interval *AX-89541634-AX- 109280493*. In the low N treatment, three QTL clusters were identified including C3 for SDW and TDW on chromosome 1DS, C4 for RDW, RL and RTN on 2BS, and C12 for RL and RV on 7DS. In the high N condition, C8, C9, C10, and C11 were mapped on chromosomes 5AL, 6AL, 6BL, and 7BL, respectively, comprising 14 QTLs and explaining 5. 1 to 14. 5% of the phenotypic variances. In addition, C5 for RDW and RV and C6 for RDW, RV and ROSA were detected on chromosome 4BL. In addition, C7 for SDW and RRS was mapped on 4DS (AX-109861583-AX-109478820), with favorable alleles contributed by Zhongmai 895.

3. 5　KASP marker development for important loci

Two KASP assays were developed for new QTL clusters C10 and C11. The first assay for *AX-109558906-6B* locus in C10 was related to SDW, RDW and TDW, and the second one for *AX-95025477-7B* locus in C11 controlled SDW, RDW and ROSA (Appendix A). Fourteen varieties and advanced breeding lines and two parents were chosen for dissecting the characteristics of SDW, RDW, TDW, and ROSA to verify the markers using Gel-free KASP assays. At *AX-109558906- 6B* loci, the genotypes showing the blue color for "AA" allele were the same as Zhongmai 895, while the red for "GG" allele were the same as Yangmai 16. Higher SDW, RDW and TDW were observed in six varieties with the "AA" allele than those eight with the "GG" allele. At *AX-95025477-7B* locus, the genotypes showing the blue color for "GG" allele were the same as Yangmai 16, whereas those with the red color for "CC" allele were the same as Zhongmai 895. The ROSA among six varieties with the "CC" allele were greater than those eight varieties with the "GG" allele (Fig. 5; Appendix E). The genotyping results of the entire population with two KASP markers were the same as those of the corresponding SNPs from the Wheat 660K SNP assay.

4　Discussion

4. 1　Phenotypic trait variation and correlations

Considering the importance of root traits for nutrient uptake, root characteristics at seedling stage could help to predict the mature root system and important yield related traits as previously established by correlation analysis (Atkinson *et al*. 2015). Complexity in observing these " hidden half " has remained a bottleneck in understanding their genetic control and effective optimization for efficient nutrient uptake (Meister *et al*. 2014). Advanced phenotyping approaches have opened a new avenue for detecting precise and novel phenotypic information which allows for effective improvement of relevant traits.

In this study, rapid screening of a large DH population for seedling biomass and RSA related traits was accomplished to dissect their genetic basis under three

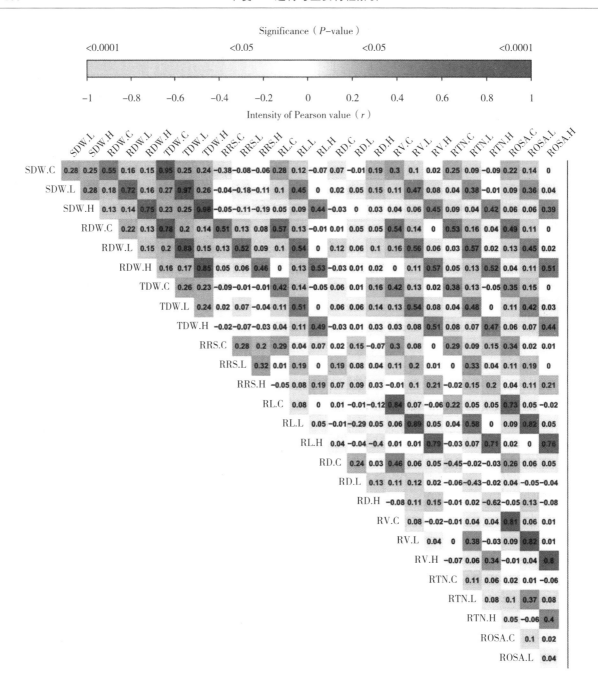

Fig. 2　Correlations between seedling biomass and root system architecture（RSA）-related traits under three nitrogen（N）treatments. Color intensity indicates the levels of positive and negative significance at $P<0.05$. Red and yellow colors indicate significantly positive and negative correlations，respectively，whereas white color indicates no significant correlation. C，control；L，low N treatment；H，high N treatment. SDW，shoot dry weight；RDW，root dry weight；TDW，total dry weight；RRS，ratio of root to shoot dry weight；RL，root length；RD，root diameter；RV，root volume；RTN，root tip number；ROSA，root surface area.

N conditions through a hydroponic culture-based pipeline. Significant phenotypic variations（$P<0.001$）and high repeatability indicated that DH lines presented high genetic diversity for root traits at seedling stage. It had great value for dissecting the genetic basis of plant growth and N uptake（An *et al.* 2006；Melino *et*

al. 2015）. The significant genotype × treatment interaction detected under different N levels demonstrated that the deep information of root traits was important，and could be significantly affected by nitrogen supply conditions. Similar results were observed in root traits under nutrient supply manipulations by Guo *et al.*

(2012) and Horn *et al*. (2016).

DH lines performed better for all traits investigated, such as higher RTN, RRS and RDW in low N condition and RV, TDW and SDW in high N condition (Appendix C). High levels of these traits in correlation with root early vigor and efficient accumulation of nutrients, and transgressive segregation indicated that DH lines could be a potential source for the identification of novel alleles (Gao *et al*. 2017). Flexibility and alteration in root traits could help adult plants by modifying the cellular mechanisms according to the influence of varied growing environments (López-Bucio *et al*. 2002; Horn *et al*. 2016). In the present study, we observed longer RL and RRS, and more RTN under the control and low N condi-

tions. Shorter RL, less RTN and higher SDW and TDW were found in the high N condition. These types of changes could improve plant stability under diverse nutrient conditions. Strong correlations among measured traits were indicative of closely linked or pleiotropic loci, and RL, RRS and RTN were crucial in N uptake in the three N levels. Zhongmai 895 showed higher values of traits at the high N level compared to Yangmai 16, because Zhongmai 895 was developed through selection by screening for high yield potential, NUE and drought resistance (He *et al*. 2014). Therefore, these traits could be selection targets in breeding, and significant variations could help to identify important QTLs for adaptive traits (Li *et al*. 2011; Yuan *et al*. 2017).

Table 3   Quantitative trait loci (QTLs) for seedling biomass and root system architecture (RSA) traits in the Yangmai 16/Zhongmai 895 doubled haploid (DH) population[1]

Trait[2]	Treatment[3]	QTL	Marker interval	Physical position (Mb)	LOD	PVE (%)	Add
SDW	L	QSDW. *caas-1DS*	AX-94979950-AX-110428402	68.44-77.34	3.52	8.72	−0.002
	H	QSDW. *caas-1BL*	AX-111590092-AX-110990632	665.05-665.06	3.36	5.37	−0.003
		QSDW. *caas-4DS*	AX-109816583-AX-109478820	16.64-30.66	3.34	5.33	−0.003
		QSDW. *caas-6AL*	AX-109373226-AX-109855287	597.86-597.92	4.38	6.90	−0.004
		QSDW. *caas-6BL*	AX-109558906-AX-110028322	675.44-675.52	5.73	10.39	−0.004
		QSDW. *caas-7BL*	AX-95025477-AX-94890497	700.83-740.04	3.09	5.07	0.003
RDW	C	QRDW. *caas-2AS*	AX-110988586-AX-110607196	21.46-27.11	3.62	7.01	0.001
	L	QRDW. *caas-2BS*	AX-108920782-AX-110463005	36.51-37.43	4.17	8.84	0.001
		QRDW. *caas-4BL*	AX-94496964-AX-109446017	587.47-601.10	3.44	7.27	−0.001
	H	QRDW. *caas-4BL*	AX-86179151-AX-109387538	526.93-534.24	5.92	11.07	−0.002
		QRDW. *caas-6BL*	AX-109558906-AX-110028322	675.44-675.52	5.41	9.94	−0.001
		QRDW. *caas-7BL*	AX-95025477-AX-94890497	700.83-740.04	4.21	7.78	0.001
TDW	L	QTDW. *caas-1DS*	AX-94979950-AX-110428402	68.44-77.34	3.95	9.21	−0.002
		QTDW. *caas-7DL*	AX-108909298-AX-109307518	453.62-456.31	2.95	6.77	−0.002
	H	QTDW. *caas-6AL*	AX-109373226-AX-109855287	597.86-597.92	4.67	7.69	−0.005
		QTDW. *caas-6BL*	AX-109558906-AX-110028322	675.44-675.52	5.52	10.29	−0.006
RRS	C	QRRS. *caas-1AL*	AX-89541634-AX-109280493	582.86-593.27	7.35	10.76	−0.027
		QRRS. *caas-1AS*	AX-110602477-AX-111172881	13.90-21.40	4.98	7.08	0.022
		QRRS. *caas-3AL*	AX-110994464-AX-109445920	607.47-612.99	3.32	4.82	0.018
		QRRS. *caas-4BS*	AX-109291938-AX-111068079	21.55-25.34	7.54	11.25	−0.028

（continued）

Trait[2]	Treatment[3]	QTL	Marker interval	Physical position (Mb)	LOD	PVE（%）	Add
		QRRS. caas-4DS	AX-109816583-AX-109478820	16. 64-30. 66	8. 62	13. 08	0. 031
	L	QRRS. caas-4BS	AX-110498886-AX-110411152	100. 69-108. 90	6. 46	10. 75	−0. 021
		QRRS. caas-4DS	AX-109816583-AX-109478820	16. 64-30. 66	10. 97	18. 99	0. 028
	H	QRRS. caas-4BS	AX-94912823-AX-111826497	166. 30-172. 72	6. 71	10. 91	−0. 017
		QRRS. caas-4DS	AX-109816583-AX-109478820	16. 64-30. 66	9. 92	16. 59	0. 021
RL	C	QRL. caas-1AL	AX-89541634-A X-109280493	582. 86-593. 27	4. 01	9. 86	−20. 98
	L	QRL. caas-1AL	AX-110573930-AX-109411284	530. 69-532. 79	3. 67	8. 28	−20. 888
		QRL. caas-2BS	AX-108920782-AX-110463005	36. 51-37. 43	4. 35	9. 46	22. 259
		QRL. caas-7DS	AX-108952259-AX-111881572	393. 94-399. 62	4. 45	9. 67	−23. 83
	H	QRL. caas-3BS	AX-86178172-AX-110962448	69. 36-71. 20	4. 28	8. 73	22. 28
RD	C	QRD. caas-2AL	AX-109922869-AX-110457187	703. 87-704. 75	3. 53	8. 04	0. 004
		QRD. caas-4AS	AX-111788010-AX-110061005	17. 03-17. 05	3. 73	7. 62	−0. 004
		QRD. caas-5BL	AX-94815980-AX-111449617	520. 89-546. 70	2. 97	5. 94	−0. 004
	L	QRD. caas-1AL	AX-94796020-AX-111736411	491. 14-493. 22	9. 66	19. 12	0. 006
		QRD. caas-4BS	AX-94816872-AX-111765045	138. 68-139. 62	4. 65	8. 54	−0. 004
	H	QRD. caas-1DS	AX-109849862-AX-108727857	4. 05-11. 53	6. 57	8. 50	−0. 009
		QRD. caas-3AS	AX-110515614-AX-108843261	45. 09-49. 97	3. 77	4. 66	−0. 007
		QRD. caas-4BL	AX-110367312-AX-110050167	619. 77-621. 71	12. 77	18. 31	−0. 013
		QRD. caas-5AL	AX-109958693-AX-94700681	578. 56-581. 36	4. 62	5. 92	0. 007
RV	L	QRV. caas-4BL	AX-86179151-AX-109387538	526. 93-534. 24	5. 03	11. 87	−0. 007
		QRV. caas-7DS	AX-108952259-AX-111881572	393. 94-399. 62	3. 41	7. 94	−0. 006
	H	QRV. caas-4BL	AX-94496964-AX-109446017	587. 47-601. 10	7. 02	13. 28	−0. 012
		QRV. caas-5AS	AX-110418367-AX-108855322	9. 65-9. 66	3. 94	7. 11	0. 008
RTN	L	QRTN. caas-2BS	AX-108920782-AX-110463005	36. 51-37. 43	3. 75	8. 91	119. 079
		QRTN. caas-7DS	AX-109036760-AX-110916001	134. 11-135. 93	3. 02	7. 04	−108. 997
	H	QRTN. caas-1DS	AX-109849862-AX-108727857	4. 05-11. 53	5. 36	8. 77	121. 538
		QRTN. caas-4DS	AX-110570496-AX-111128307	0. 45-1. 34	3. 46	5. 52	96. 49
		QRTN. caas-5AL	AX-109958693-AX-94700681	578. 56-581. 36	8. 51	14. 47	−155. 655
ROSA	L	QROSA. caas-4BL	AX-94496964-AX-109446017	587. 47-601. 10	4. 82	12. 51	−1. 466
	H	QROSA. caas-2AS	AX-110430307-AX-108748677	1. 31-1. 44	3. 81	6. 66	1. 318
		QROSA. caas-7BL	AX-95025477-AX-94890497	700. 83-740. 04	4. 68	7. 78	1. 448

[1] Physical positions of SNP markers based on wheat genome sequences from the International Wheat Genome Sequencing Consortium (IWGSC，http：//www. wheatgenome. org/）；LOD，LOD value of each QTL；PVE，phenotypic variance explained by QTL；Add，a positive sign means increased effect contributed by Zhongmai 895；a negative sign indicates increased effect contributed by Yangmai 16.

[2] SDW，shoot dry weight；RDW，root dry weight；TDW，total dry weight；RRS，ratio of root to shoot dry weight；RL，root length；RD，root diameter；RV，root volume；RTN，root tip number；ROSA，root surface area.

[3] C，control；L，low N treatment；H，high N treatment.

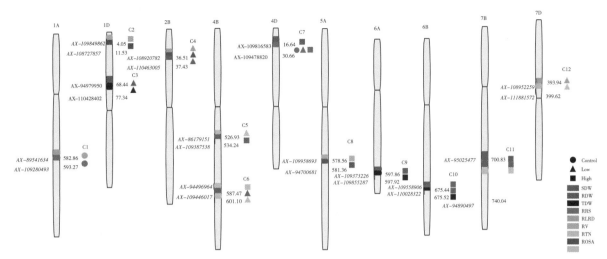

Fig. 3   Twelve QTL clusters for seedling biomass and root system architecture (RSA) traits. Different colors indicate traits and shapes indicate the three N treatments. C, control; C1-12, QTL clusters 1-12; L, low N treatment; H, high N treatment. SDW, shoot dry weight; RDW, root dry weight; TDW, total dry weight; RRS, ratio of root to shoot dry weight; RL, root length; RD, root diameter; RV, root volume; RTN, root tip number; ROSA, root surface area.

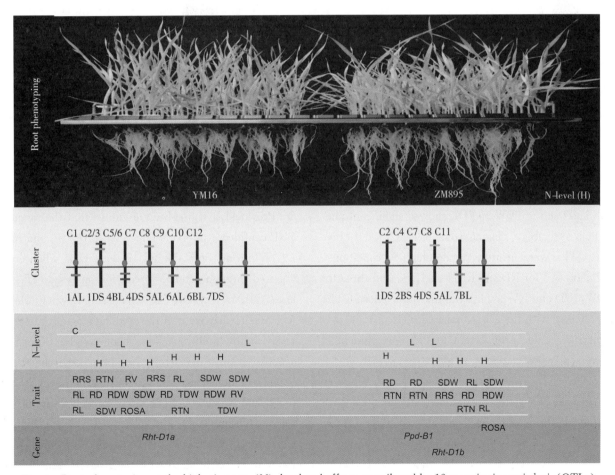

Fig. 4   Root phenotyping at the high nitrogen (N) level and effects contributed by 12 quantitative trait loci (QTLs) clusters from Yangmai 16 and Zhongmai 895. YM16, Yangmai 16; ZM895, Zhongmai 895. C, control; C1-12, QTL clusters 1-12; L, low N treatment; H, high N treatment. SDW, shoot dry weight; RDW, root dry weight; TDW, total dry weight; RRS, ratio of root to shoot dry weight; RL, root length; RD, root diameter; RV, root volume; RTN, root tip number; ROSA, root surface area.

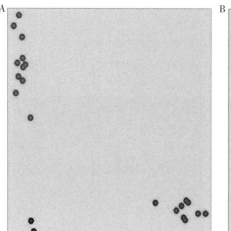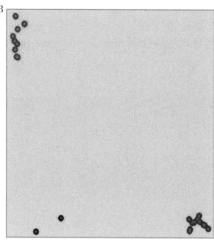

Fig. 5　Scatter plots for two kompetitive allele-specifc PCR（KASP）marker assays in 16 wheat varieties. A，KASP assay for *AX- 109558906-6B*，the blue color for "AA" allele corresponded to the Zhongmai 895 genotype，whereas the red color for "GG" allele corresponds to the Yangmai 16 genotype. B，KASP assay for *AX-95025477-7B*，the blue color for "GG" allele corresponds to the Yangmai 16 genotype，and those with the red color for "CC" allele correspond to the Zhongmai 895 genotype. Scatter plot for KASP assays showing clustering of genotypes on the *X*-（FAM）and *Y*-（HEX）axes. Blue genotypes have the FAM-type allele；red genotypes have the HEX-type allele；black dots indicate the NTC（non-template control）.

4. 2　QTL identified for nitrogen supply

Efficient N accumulation and utilization are crucial for plant growth in fulfilling N requirements（Ryan *et al*. 2015；Horn *et al*. 2016）. In present study，51 QTLs were identified under three N treatments. In low N treatment，two new QTLs on 4BL for RDW and ROSA，three new QTLs for RL，RV and RTN on 7DS and one for TDW on 7DL showed significant phenotypic variances，ranging from 6. 1 to 12. 5%. Five new QTLs were identified under high N conditions，including two for RD and RTN on 1DS，and three for RDW，RD and RV on chromosome 4BL. These QTLs were linked with the genes that might affect root early vigor and would be important for the development of N-efficient wheat varieties（Table 3）.

Fourteen QTLs were co-localized with previously reported loci for seedling and yield-related traits（An *et al*. 2006；Li *et al*. 2011；Yuan *et al*. 2017；Guan *et al*. 2018）. Among them，the most stable QTL，*QRRS. caas-4DS* for *Rht-D1b*，was confirmed through its functional marker by Hassan *et al*. （2019），and Zhongmai 895 contributed the favorable allele. This could provide a useful selection criterion in N-deficient and sufficient conditions，by using a gene-specific marker for *Rht-D1b*（Rasheed *et al*. 2016），which could help in evaluating root phenotypes and might be used for marker- assisted selection. Moreover，a major QTL with significant phenotypic impact was detected on chromosome 4BS（for RRS and RD）corresponding to *Rht-B1* on 4BS at three different confidence intervals，with a favorable allele from Yangmai 16. Previously，it had been discussed that the range of SNP confidence intervals linked with *Rht-B1* and *Rht-D1* on chromosome 4D and 4B were wide when cross checked through genomic prediction analysis（Hassan *et al*. 2019）. These QTLs have been shown to play a key role in early vigor at the seedling stage by affecting root length，and root and shoot weight（Horn *et al*. 2016；Gao *et al*. 2017）. Notably，Laperche *et al*. （2007）demonstrated the role of dwarfing gene *Rht* in increasing N content（g N m^{-2}）as well as grain protein content in plants. The QTL on 2BS showed significant effects on RDW，RL and RTN，which were likely to be co-localized with the *Ppd-B1* gene and an important locus，*qTaLRO-B1*，which is involved in shoot height fluctuations，seedling biomass accumulation and mineralization of N at later growth stages（Cao *et al*. 2014）. In our results，the presence of dwarfing alleles of plant height on 4BS, 4DS and 5AL and the photoperiodism gene on 2BS had significant influences in biomass accumulation, root development and N accumulation (Laperche *et al.* 2007; Quraishi *et al.* 2011).

At the high N level, 10 QTLs were located on chromosomes 5AL, 6AL, 6BL, and 7BL. Among them, 5AL had significant influences on RD and RTN. It also had roles in increasing the duration of organ development for nitrogen accumulation, and grain yield under rain-fed conditions (Quraishi *et al.* 2011; Gahlaut *et al.* 2017; Guan *et al.* 2018). Three QTLs on 6BL for SDW, RDW and TDW were likely in the same region as a previously reported gene for NUE (Guo *et al.* 2012), and thousand-grain weight and grain yield (Yuan *et al.* 2017). In addition, an important QTL on chromosome 7B was reported to have a strong relationship with kernel number per spike (Guan *et al.* 2018). Three QTLs on 7BL for SDW, RDW and ROSA were detected, with favorable alleles from Zhongmai 895. These important QTL for seedling biomass and RSA traits could be useful for root morphogenesis and early vigor, and for high degrees of development of mature plant organs and yield stability under varied N conditions.

4. 3　QTL clusters

The presence of favorable genetic relationships between the traits and stable loci in QTL clusters are more reliableunder varied environments (Gong *et al.* 2016). In the present study, 12 QTL clusters were identified for closely associated traits (Fig. 3). C1 on 1AL showed 9. 9 to 10. 8% of PVE for RL and RRS under control treatment, and C2 and 3 on 1DS exhibited 8. 5 to 8. 8% of PVE for SDW, RD and RTN under low and high N levels. These are likely to be new QTL for the alteration of root and biomass-related seedling traits. C4 on chromosome 2BS was likely to be linked with *Ppd-B1* which has been cloned as an important gene for harvest index and NUE. C7 was the most stable cluster for RRS and SDW detected in all three N treatments (*AX-109816583-AX-109478820*)

on chromosome 4DS, with a significant phenotypic effect ranging from 5. 3 to 19%, and all of the favorable alleles from C7 were provided by Zhongmai 895; also, it is likely to be the reduced height allele *Rht-D1b*, affecting root and shoot biomass (Ryan *et al.* 2015; Horn *et al.* 2016). *Rht* loci have effects on increasing N content, thousand-grain weight and grain yield. Therefore, the presence of *Rht1* and *Rht2* genes in DH lines seems to have a great influence in N uptake. Besides the clusters containing dwarfing and photoperiod related genes, the important QTLs cluster of C10 on chromosome 6BL (*AX-109558906-AX-110028322*) significantly increased SDW, RDW and TDW with 9. 9 to 10. 4% PVE, and C11 on chromosome 7BL (*AX-95025477-AX-95121748*) enhanced SDW, RDW and ROSA with 5. 1 to 7. 8% of PVE at the high N level. KASP markers for C10 and C11 on chromosomes 6BL and 7BL were developed and validated in the study. Previously, the QTL in cluster of C10 was reported for N accumulation and thousand-grain weight in RIL populations under field conditions (Gahlaut *et al.* 2017; Yuan *et al.* 2017). While the QTL identified in C11 were likely to be new for fluctuations of seedling traits in response to N, these QTLs have been reported for kernel number per spike in a previous study (Zhang H *et al.* 2016). Thus, the above QTL clusters, particularly C10 and C11, could provide important information of genetic background for early vigor selection of N uptake efficiency genotypes.

5　Conclusion

Fifty-one QTLs for seedling biomass and RSA-related traits were detected across three N treatments using a hydroponic experiment for high-throughput genotyping in wheat. Twelve new QTLs on chromosomes 1AL, 1DS (2), 4BL (5), 7DS (3) and 7DL influenced the root biomass traits for N uptake. The QTL on chromosome 4DS for RRS with high phenotypic effects was most stable across the N treatments. Among 12 QTL clusters, C10 on 6BL for SDW, RDW and TDW, and C11 on 7BL for SDW, RDW and ROSA were new loci under high N levels，and these were also reported to have links with yield-related traits. These QTLs and clusters associated with seedling biomass and RSA-related traits，together with KASP markers developed for C10 and C11，could help to genetically improve the N uptake efficiency in wheat.

❖ Acknowledgements

We thank Prof. R. A. McIntosh from Plant Breeding Institute，University of Sydney，for review of this manuscript. This work was funded by the National Key R&D Program of China（2016YFD0101804-6），the National Natural Science Foundation of China（31671691）and the International Science & Technology Cooperation Program of China（2016YFE0108600）.

❖ References

An D G，Su J Y，Liu Q Y，Zhu Y G，Tong Y P，Li J M，Jing R L，Li B，Li Z S. 2006. Mapping QTLs for nitrogen uptake in relation to the early growth of wheat（*Triticum aestivum* L.）. *Plant and Soil*，284，73-84.

Atkinson J A，Wingen L U，Griffiths M，Pound M P，Gaju O，Foulkes M J，Gouis J L，Griffiths S，Bennett M J，King J，Wells D N. 2015. Phenotyping pipeline reveals major seedling root growth QTL in hexaploid wheat. *Journal of Experimental Botany*，66，2283-2292.

Aziz M M，Palta J A，Siddique K H M，Sadras V O. 2017. Five decades of selection for yield reduced root length density and increased nitrogen uptake per unit root length in Australian wheat varieties. *Plant and Soil*，413，181-192.

Bai C，Liang Y，Hawkesford M J. 2013. Identification of QTL associated with seedling root traits and their correlation with plant height in wheat. *Journal of Experimental Botany*，64，1745-1753.

Barraclough P B，Kuhlmann H，Weir A H. 1989. The Effects of prolonged drought and nitrogen fertilizer on root and shoot growth and water uptake by winter wheat. *Journal of Agronomy and Crop Science*，163，352-360.

Bettembourg M，Dardou A，Audebert A，Thomas E，Frouin J，Guiderdoni E，Ahmadi N，Perin C，Dievart A，Courtois B. 20017. Genome-wide association mapping for root cone angle in rice. *Rice*，10，45.

Cao P，Ren Y，Zhang K，Teng W，Zhao X，Dong Z，Liu X，Qin H，Li Z，Wang D，Tong Y. 2014. Further genetic analysis of a major quantitative trait locus controlling root length and relatedtraits in common wheat. *Molecular Breeding*，33，975-985.

Churchill G A，Doerge R W. 1994. Empirical threshold values for quantitative trait mapping. *Genetics*，138，963-971.

Cormier F，Foulkes J，Hirel B，Gouache D，Moënne-Loccoz Y，Gouis J. 2016. Breeding for increased nitrogen-use efficiency：A review for wheat（*T. aestivum* L.）. *Plant Breeding*，135，255-278.

Cui F，Fan X L，Zhao C H，Zhang W，Chen M，Ji J，Li J M. 2014. A novel genetic map of wheat：utility for mapping QTL for yield under different nitrogen treatments. *BMC Genetics*，130，1235-1252.

Cui Z，Zhang H，Chen X，Zhang C，Ma W，Huang C，Zhang W F，Mi G，Miao Y，Li X，Gao Q，Yang J，Wang Z，Ye Y，Guo S，Lu J，Huang J，Lv S，Sun Y，Liu Y，*et al*. 2018. Pursuing sustainable productivity with millions of smallholder farmers，*Nature*，555，363-366.

Diaz R J，Rosenberg R. 2008. Spreading dead zones and consequences for marine ecosystems. *Science*，321，926-929.

FAOSTAT. 2020. Food and Agriculture Organization of the United Nations. ［2020-11-08］. http：//www. fao. org/3/i6895e/i6895e. pdf

Forde B G. 2014. Nitrogen signalling pathways shaping root system architecture：An update. *Current Opinion in Plant Biology*，21，30-36.

Gahlaut V，Jaiswal V，Tyagi B S，Singh G，Sareen S，Balyan H S，Gupta P K. 2017. QTL mapping for nine drought-responsive agronomic traits in bread wheat under irrigated and rainfed environments. *PLoS ONE*，12，e0182857.

Gao F，Ma D，Yin G，Rasheed A，Dong Y，Xiao Y，Xia X，Wu X，He Z. 2017. Genetic progress in grain yield and physiological traits in Chinese wheat cultivars of southern Yellow and Huai Valley since 1950. *Crop Science*，57，760-773.

Gong X，Wheeler R，Bovill W D，McDonald G K. 2016. QTL mapping of grain yield and phosphorus efficiency in barley in a Mediterranean-like environment. *Theoretical and Applied Genetics*，129，1657-1672.

Guan P，Lu L，Jia L，Kabir M R，Zhang J，Lan T，Zhao Y，Xin M，Hu Z，Yao Y，Ni Z，Sun Q，Peng

H. 2018. Global QTL analysis identifies genomic regions on chromosomes 4A and 4B harboring stable loci for yield-related traits across different environments in wheat (*Triticum aestivum* L.). *Frontiers in Plant Science*, 9, 529.

Guo J, Liu X, Zhang Y, Shen J, Han W, Zhang W, Christie P, Goulding K, Vitousek P, Zhang F. 2010. Significant acidification in major Chinese croplands. *Science*, 327, 1008-1010.

Guo Y, Kong F M, Xu Y F, Zhao Y, Liang X, Wang Y Y, An D G, Li S S. 2012. QTL mapping for seedling traits in wheat grown under varying concentrations of N, P and K nutrients. *Theoretical and Applied Genetics*, 124, 851-865.

Hassan M A, Yang M, Fu L, Rasheed A, Zheng B, Xia X, Xiao Y, He Z. 2019. Accuracy assessment of plant height using an unmanned aerial vehicle for quantitative genomic a-nalysis in bread wheat. *Plant Methods*, 15, 37.

He Z, Yan J, Zhang Y. 2014. A new wheat variety Zhong-mai895 with high yield and wide adaptability, dwarf and lodge resistance. *China Seed Industry*, 6, 76-77. (in Chinese)

Hermans C, Hammond J P, White P J, Verbruggen N. 2006. How do plants respond to nutrient shortage by biomass allocation? *Trends in Plant Science*, 11, 610-617.

Hoagland D R, Arono D I. 1950. The water-culture method for growing plants without soil. *California Agricultural Experiment Station Circular*, 347, 1-32.

Horn R, Wingen L U, Snape J W, Dolan L. 2016. Mapping of quantitative trait loci for root hair length in wheat identifies loci that co-locate with loci for yield components. *Journal of Experimental Botany*, 67, 4535-4543.

Hund A, Ruta N, Liedgens M. 2009. Rooting depth and water use efficiency of tropical maize inbred lines, differing in drought tolerance. *Plant and Soil*, 318, 311-325.

Kosambi D D. 1943. The estimation of map distances from recombination values. *Annals of Human Genetics*, 12, 172-175.

Laperche A, Brancourt-Hulmel M, Heumez E, Gardet O, Hanocq E, Devienne-Barret F, Le Gouis J. 2007. Using genotype × nitrogen interaction variables to evaluate the QTL involved in wheat tolerance to nitrogen con-straints. *Theoretical and Applied Genetics*, 115, 399-415.

Li F, Xiao Y, Jing S, Xia X, Chen X, Wang H, He Z. 2015. Genetic analysis of nitrogen and phosphorus effi-ciency related traits at seedling stage of Jing 411 and its derivatives. *Journal of Triticeae Crops*, 35 737-746. (in Chinese)

Li H, Bradbury P, Ersoz E, Buckler E S, Wang J. 2011. Joint QTL linkage mapping for multiple-cross mating design sharing one common parent. *PLoS ONE*, 6, e17573.

Li P C, Chen F J, Cai H G, Liu J C, Pan Q C, Liu Z G, Gu R L, Mi G H, Zhang F S, Yuan L X. 2015. A genetic relationship between nitrogen use efficiency and seedling root traits in maize as revealed by QTL analysis. *Journal of Ex-perimental Botany*, 66, 3175-3188.

López-Bucio J, Hernández-Abreu E, Sánchez-Calderón L, Nieto-Jacobo M F, Simpson J, Herrera-Estrella L. 2002. Phosphate availability alters architecture and causes changes in hormone sensitivity in the Arabidopsis root system. *Plant Physiology*, 129, 244-256.

Lynch J P. 2007. Roots of the second green revolu-tion. *Australian Journal of Botany*, 55, 493-512.

Meister R, Rajani M S, Ruzicka D, Schachtman D P. 2014. Challenges of modifying root traits in crops for agriculture. *Trends in Plant Science*, 19, 779-788.

Melino V J, Fiene G, Enju A, Cai J, Buchner P, Heuer S. 2015. Genetic diversity for root plasticity and nitrogen uptake in wheat seedlings. *Functional Plant Biology*, 42, 942-956.

MengL, Li H, Zhang L, Wang J. 2015. QTL IciMapping: Integrated software for genetic linkage map construction and quantitative trait locus mapping in biparental popula-tions. *The Crop Journal*, 3, 269-283.

Moore F C, Lobell D B. 2015. The fingerprint of climate trends on European crop yields. *Proceedings of the Na-tional Academy of Sciences of the United States of A-merica*, 112, 2670-2675.

Narayanan S, Prasad P V. 2014. Characterization of a spring wheat association mapping panel for root traits. *Agronomy Journal*, 106, 1593-1604.

Pound M P, French A P, Atkinson J A, Wells D M, Ben-nett M J, Pridmore T. 2013. RootNav: Navigating images of complex root architectures. *Plant Physiology*, 162, 1802-1814.

QuraishiU M, Abrouk M, Murat F, Pont C, Foucrier S, Desmaizieres G, Confolent C, Rivière N, Charmet G, Paux E, Murigneux A, Guerreiro L, Lafarge S, Gouis J L, Feuillet C, Salse J. 2011. Cross-genome map based dissection of a nitrogen use efficiency ortho-meta QTL in bread wheat unravels concerted cereal genome evolution. *The Plant Journal*, 65, 745-756.

R Core Team. 2013. *R: A Language and Environment for Statistical Computing*. R Foundation for Statistical Computing, Vienna, Austria.

Rasheed A, Wen W, Gao F, Zhai S, Jin H, liu J, Guo Q, Zhang Y, Dreisigacker S, Xia X, He Z. 2016. Development and validation of KASP assays for genes underpinning key economic traits in bread wheat. *Theoretical and Applied Genetics*, 129, 1843-1860.

Rengel Z, Marschner P. 2005. Nutrient availability and management in the rhizosphere: Exploiting genotypic differences. *New Phytologist*, 168, 305-312.

Reynolds M, Dreccer F, Trethowan R. 2007. Drought-adaptive traits derived from wheat wild relatives and landraces. *Journal of Experimental Botany*, 58, 177-186.

Robinson D. 2001. Root proliferation, nitrate inflow and their carbon costs during nitrogen capture by competing plants in patchy soil. *Plant and Soil*, 232, 41-50.

Ryan P R, Liao M, Delhaize E, Rebetzke G J, Weligama C, Spielmeyer W, James R A. 2015. Early vigour improvesphosphate uptake in wheat. *Journal of Experimental Botany*, 66, 7089-7100.

Subira J, Ammar K, Álvaro F, del Moral L F G, Dreisigacker S, Royo C. 2016. Changes in durum wheat root and aerial biomass caused by the introduction of the *Rht-B1b* dwarfing allele and their effects on yield formation. *Plant and Soil*, 403, 291-304.

Wang J P. 2017. Genetic analysis and functional marker development of key traits in wheat. Ph D thesis, Shandong Agricultural University, China. (in Chinese)

Wasson A P, Richards R A, Chatrath R, Misra S C, Prasad S V, Rebetzke G J, Watt M. 2012. Traits and selection strategies to improve root systems and water uptake in water-limited wheat crops. *Journal of Experimental Botany*, 63, 3485-3498.

Yuan Y, Gao M, Zhang M, Zheng H, Zhou X, Guo Y, Zhao Y, Kong F, Li S. 2017. QTL mapping for phosphorus efficiency and morphological traits at seedling and maturity stages in wheat. *Frontiers in Plant Science*, 8, 614.

Zhang H, Chen J, Li R, Deng Z, Zhang K, Liu B, Tian J. 2016. Conditional QTL mapping of three yield components in common wheat (*Triticum aestivum* L.). *The Crop Journal*, 4, 220-228.

Zhang H, Wang H. 2015. QTL mapping for traits related to P-deficient tolerance using three related RIL populations in wheat. *Euphytica*, 203, 505-520.

Zhang W, Cao G, Li X, Zhang H, Wang C, Liu Q, Chen X, Cui Z, Shen J, Jiang R, Mi G, Miao Y, Zhang F, Dou Z. 2016. Closing yield gaps in China by empowering smallholder farmers. *Nature*, 537, 671-674.

QTL mapping of root traits in wheat under different phosphorus levels using hydroponic culture

Mengjiao Yang[1], Cairong Wang[1,2], Muhammad Adeel Hassan[1],
Faji Li[1], Xianchun Xia[1], Shubing Shi[3], Yonggui Xiao[1*] and Zhonghu He[1,4]

[1] Institute of Crop Sciences, National Wheat Improvement Centre, Chinese Academy of Agricultural Sciences (CAAS), Beijing 100081, China.

[2] Agricultural Research Institute of Yili, Yili 835000, Xinjiang, China.

[3] College of Agriculture, Xinjiang Agricultural University, Urumqi 830052, Xinjiang, China.

[4] International Maize and Wheat Improvement Centre (CIMMYT) China Office, c/o CAAS, Beijing 100081, China.

* Correpondence: xiaoyonggui@caas.cn

Background: Phosphorus (P) is an important in ensuring plant morphogenesis and grain quality, therefore an efficient root system is crucial for P-uptake. Identification of useful loci for root morphological and P uptake related traits at seedling stage is important for wheat breeding. The aims of this study were to evaluate phenotypic diversity of Yangmai 16/Zhongmai 895 derived doubled haploid (DH) population for root system architecture (RSA) and biomass related traits (BRT) in different P treatments at seedling stage using hydroponic culture, and to identify QTL using 660 K SNP array based high-density genetic map.

Results: All traits showed significant variations among the DH lines with high heritabilities (0.76 to 0.91) and high correlations ($r = 0.59$ to 0.98) among all traits. Inclusive composite interval mapping (ICIM) identified 34 QTL with 4.64-20.41% of the phenotypic variances individually, and the log of odds (LOD) values ranging from 2.59 to 10.43. Seven QTL clusters (C1 to C7) were mapped on chromosomes 3DL, 4BS, 4DS, 6BL, 7AS, 7AL and 7BL, cluster C5 on chromosome 7AS (*AX-109955164-AX-109445593*) with pleiotropic effect played key role in modulating root length (RL), root tips number (RTN) and root surface area (ROSA) under low P condition, with the favorable allele from Zhongmai 895.

Conclusions: This study carried out an imaging pipeline-based rapid phenotyping of RSA and BRT traits in hydroponic culture. It is an efficient approach for screening of large populations under different nutrient conditions. Four QTL on chromosomes 6BL (2) and 7AL (2) identified in low P treatment showed positive additive effects contributed by Zhongmai 895, indicating that Zhongmai 895 could be used as parent for P-deficient breeding. The most stable QTL *QRRS. caas-4DS* for ratio of root to shoot dry weight (RRS) harbored the stable genetic region with high phenotypic effect, and QTL clusters on 7A might be used for speedy selection of genotypes for P-uptake. SNPs closely linked to QTLs and clusters could be used to improve nutrient-use efficiency.

Keywords: Genome-wide linkage mapping, Quantitative trait loci, Root traits, SNP array

Background

Phosphorus (P) is an important macro-element forensuring plant development, productivity, and grain quality[1]. P deficiency causes abnormal physiological and bio-chemical metabolism during critical plant growth stages and resulted yield losses[2]. P as phosphate is immobile in most of the soil types that make its application on the soil surface less beneficial for plants. While efficient up-take of P from deep soil depends on plant's underground organs[3]. In crops, an efficient root system is crucial for P-uptake. For example, increase in root to shoot ratio in most of the elite cultivars helps to up-take P from deep soil or by growing longer root hairs to exploit the spatial characteristics of soil for maximum nutrient storage in shoots[3,4]. Therefore, optimization of root and biomass related attributes such as root length, root width, root tips number, root diameter, root biomass and shoot biomass at seedling stage could provide a promising avenue to explore early variations correlated with high P uptake. Genetic diversity for root-related traits under different nutrient conditions has been also considered very important for grain yield enhancement[5,6]. Therefore, improvement of nutrient up-take through useful variations in seedling root and biomass traits under varied growth conditions could provide a sustainable solution for developing elite cultivars[2,7,8].

In wheat, QTLs have been detected for root traits under different P treatments across the 21 chromosomes[9,10]. But despite the many genetic interactions which have been determined for the seedling biomass and root system architecture traits, still few loci were reported with major effects[11,12]. Accurate phenotyping of root traits under normal field conditions is difficult, whereas traditional methods such as soil columns and soil cores are time-consuming and laborious for screening of large populations[13,14]. Artificial systems like sand, germination paper and hydroponic based cultures have been used as proxies for characterization of root traits[4,15]. Hydroponic culture with digital imaging has given new opportunities to detect number of root traits with different aspects of root development compared with sand culture and germination paper techniques[11]. Moreover, hydroponic technique can be easily applied for rapid and precise screening of large populations to bridge the phenome to genome knowledge gaps. Several common QTL related to root biomass and root system architecture traits have been reported in both hydroponic and field trials conditions[11,12,16]. But, there is no report regarding cloning of QTL for P-uptake related root traits or P uptake efficiency yet.

Nowadays, construction of high-density genetic maps has accelerated the accuracy of quantitative genomic analysis. Therefore, it could increase the chance for identification of true loci for complex traits[17]. The 660 K SNP array in wheat has greatly improved the density of genetic maps for QTL analysis compared with the earlier 90 K array[18,19]. In this study, we have used hydroponic culture-based image pipeline for root phenotyping and 660 K SNPs array for QTL and joint QTL analysis. This will identify useful loci and can help to understand their pleiotropic effect for multiple root morphological and P-uptake related traits during selection. The aims of this study were (1) to evaluate phenotypic diversity of Yangmai 16/Zhongmai 895 derived doubled haploid (DH) population for seedling root and biomass related traits in different P treatments, (2) to identify QTLs for root traits and biomass at seedling stage under low and high P conditions using 660 K SNP array based high-density genetic map and (3). to detect QTL with pleiotropic effect for multiple traits.

Methods

Plant materials

The panel of 198 DH lines from the cross of Yangmai 16/Zhongmai 895 were evaluated for root-related traits in hydroponic culture. The female parental Yangmai 16 is a spring wheat cultivar with drought resistance attributes and cover the largest planting area in the middle and lower Yangtze River Region. Zhongmai 895 is a facultative cultivar and widely cultivated in southern parts of the Yellow and Huai Valleys. This has been characterized as high yielding with drought and heat resistance ability, strong roots and early vigor.

Hydroponic culture and experimental design

Hoagland's nutrient solution was used[20], and three P levels were kept at zero (control), low and high P con-tents (KH_2PO_4 0, 0.005, 0.25 mmol/L, respectively). Whereas, levels of KCl in solution were at 0.35, 0.345 and 0.10 mmol/L, respectively in the three treatments to maintain a common nutrient concentration across treatments[21] (Table 1). The experiments were conducted in randomized complete blocks (RCBD), with three replications from March 15 to April 28, 2017, and 30 healthy seeds of each DH line were used for each treatment.

The seeds of each DH line were sterilized for 15-20 min in a 10% H_2O_2 solution. After rinsing 5-6 times in sterilized water, seeds were placed on moisturized germination paper in glass Petri dishes, crease-side down and left for 36 h in darkness to initiate germination. After the early appearance of germination, seeds were transferred to sand (2 mm diameter) box in a dark environment for seedling growth at 24℃ for 72 h. The sand box was kept under a constant environment room (12 h photo-period: 16℃ day and 13 ℃ night, light intensity at 400 μmol m^{-2} s^{-1} PAR and relative humidity at 70%). Three uniformly sprouted seeds with ～5 mm inroots length for each replication were transferred to holes in trays (the seedling was holed with a sponge), and were placed on plastic tanks (660×480×280 mm) containing 20 L of nutrient solution. The solution was renewed after every 3 days. After 10 days (two-leaf stage) plants were harvested and placed in 30% ethanol prior to imagery for phenotyping.

Table 1 Nutrient solution ingredients for wheat seedling growth

Ingredient	Concentration (mmol/L)	Ingredient	Concentration (mmol/L)
K_2SO_4	0.75	$MnSO_4 \cdot H_2O$	0.001
$Ca (NO_3)_2 \cdot 4H_2O$	0.25	$ZnSO_4 \cdot 7H_2O$	0.001
$MgSO_4$	0.60	$CuSO_4 \cdot 5H_2O$	0.0001
FeEDTA	0.04	$Na_2MoO_4 \cdot 2H_2O$	0.000005
H_3BO_3	0.001	KH_2PO_4	0/0.005/0.25
		KCl	0.35/0.345/0.10

Trait measurements

Five root system architecture (RSA) traits, viz. root length (RL), root volume (RV), root diameter (RD), root tip number (RTN), root surface area (ROSA) were captured through images using a scanner (Perfection V700/V750 2.80A; Epson, China). Images were analyzed by using a software RootNav V1.7.5, which was operationally semi-automated[22]. Root biomass-related traits (BRT), including shoot dry weight (SDW) and root dry weight (RDW) were measured after oven-drying for 72 h at 70 ℃ (mg/plant). Total dry weight (TDW)

was estimated as the sum of SDW and RDW, and ratio of root to shoot dry weight (RRS) was measured as ratio between RDW and SDW.

SNP genotyping and QTL analysis

The DH lines and parent cultivars were genotyped through Wheat 660 K SNP array synthesized by Affymetrix and commercially available from Capital Bio Corporation (Beijing, China; http://www.capitalbio.com). Dataset is available in an online repository named "Data-set Yang et al. at Dryad data bank (please see data sharing link in availability of data and materials section). Genetic map was contacted by Wang et al. [23] from our lab. Briefly，markers with no polymorphisms between parents，severely distorted segregations，and missing rate greater than 20% were removed in the subsequent linkage analysis. Finally，10,242 markers each representing a bin site were selected to construct the linkage map of Yangmai 16/Zhongmai 895 population. Map was comprised 25 linkage groups， covering all 21 wheat chromosomes. Among them，chromosomes 1B，2B，4A and 7A consisted of two linkage groups，and the remaining chromosomes were with only one linkage group. Inclusive composite-interval mapping was used for QTL analysis in IciMappingV4. 1 software[24]，and averaged data from the three replicates were used for QTL detection. The SNP genotypes of Yangmai 16 were defined as A，and those of Zhongmai 895 as B. Alleles from Yangmai 16 reduced trait values when the additive effects were negative. Kosambi mapping approach was used to convert recombinant frequencies into distance map[25]. Locations of QTLs for the root traits were detected by inclusive composite interval mapping-additive（ICIM-ADD）by using same software as for the QTL analysis. The threshold for significant QTL of each trait was demarcated by 1000-permutations at $P = 0.05$[26]，and minimum LOD score at 2.5 with walking speed at 1.0 cM. Joint QTL analysis for closely linked pleiotropic QTL was performed using the general linear model scripted in lm package of R software. Phenotypic variance explained by each QTL was calculated as demonstrated in Li et al. [27].

Identification of putative candidate genes

The genes located in the physical intervals of RSA and BRT-associated genomic regions were screened based on the annotations in the wheat reference genome（CS RefSeq v1. 0；IWGSC 2018），and those related to growth，development and nutrient mobilization were considered as candidate genes. Gene annotation was retrieved using EmsemblPlant and EMBL-EBI（http：//www. ebi. ac. uk/interpro）databases. Gene annotation for putative proteins was performed using BLAST2GO（https：//www. blast2go. com/）.

Statistical analysis

Pearson's correlations analysis among the traits were estimated using averaged data from each replicate. Significance of variances among DH lines，treatments and interactions between genotypes and treatments（G×T）was calculated using following mixed linear model and considered significant at $P<0.05$.

$$Y = XB + Z\mu + ge + \varepsilon \qquad (1)$$

where Y is demonstrated as the response from fixed（β）and random（μ）effects with random error（ε），ge is the genotype×environment effect while X and Z illustrate fixed and random effects，respectively. Broad sense heritabilities for all traits were estimated using genotypes as a random effect following[28].

$$h^2 = \sigma_g^2 / (\sigma_g^2 + \sigma_{gt}^2/r + \sigma_\varepsilon^2/rt) \qquad (2)$$

where σ_g^2，σ_{gt}^2 and σ_ε^2 represent genotype，genotype（DH line）× P treatment interaction and error variances，respectively，while t is indicated P treatments and r is replicates. The R package was used for all statistical analyses[29].

Results

Phenotypic variation and correlations among traits under P treatments

Data for all nine traits were normally distributed across the three P treatments（Figure S1）. Under low P condition， Zhongmai 895 showed higher RL，RV，ROSA，RDW，TDW and RRS，but lower RD，RTN，and SDW than Yangmai 16. Except for RD and RRS in the high P treatment，Yangmai 16 had higher RL，RV，RTN，ROSA，SDW，RDW and TDW than Zhongmai 895（Fig. 1）.

Phenotypic variances among the DH lines were significant（$P < 0.0001$）. Transgressive segregations across P treatments were observed for most of traits（Figs. 1，2；Table 2）. The average values of the DH lines for RL，RV，RTN，and RRS were higher than parents in the low P treatment， indicating positiveeffects for root vigor from both parents（Figs. 1 and 2）. Broad sense heritabilities of nine traits were high ranged from 0.76 to 0.91（Table

2). RSA and BRT traits were correlated significantly ($r = 0.59$ to 0.98 at all three P levels) with each other. However, RD was negatively correlated with RL, RTN and ROSA ($r = -0.23$ to -0.69) (Fig. 3).

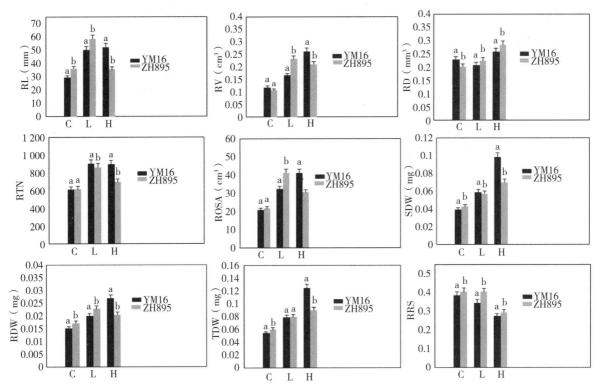

Fig. 1   The phenotypic difference between parents Yangmai 16 and Zhongmai 895 under P treatments. Error bars represent standards deviation for each proportion; letters indicate significant differences between the genotypes determined by Duncan tests. Abbreviations: ZM895, Zhongmai 895; YM16, Yangmai 16; C, control; L, low P treatment; H, high P treatment; RL, root length; RV, root volume; RD, root diameter; RTN, root tip number; ROSA, root surface area; SDW, shoot dry weight; RDW, root dry weight; TDW, total dry weight; RRS, ratio of root to shoot dry weight

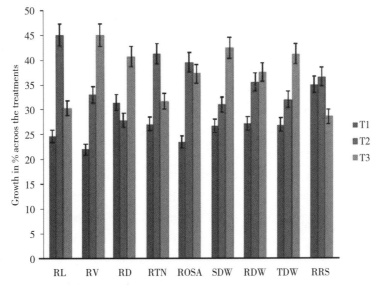

Fig. 2   Effects of phosphorus treatment on traits in DH lines. Error bars represent standard deviation for each trait under P treatments. Abbreviations: C, control; L, low P treatment; H, high P treatment; RL, root length; RV, root volume; RD, root diameter; RTN, root tip number; ROSA, root surface area; SDW, shoot dry weight; RDW, root dry weight; TDW, total dry weight; RRS, ratio of root to shoot dry weight

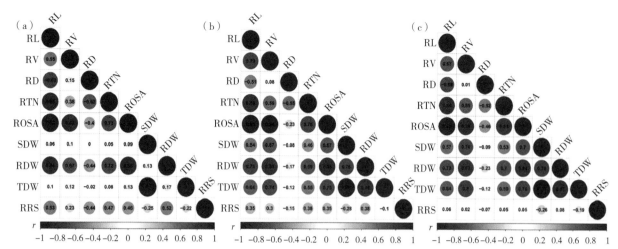

Fig. 3　Correlation analysis among the traits under （a）control，（b）low and （c）high P treatment. Intensities of blue colour show degrees of positive and red colour shows degrees of negative correlations；circle sizes indicate low to high significance. Abbreviations：RL，root length；RV，root volume；RD，root diameter；RTN，root tip number；ROSA，root surface area；SDW，shoot dry weight；RDW，root dry weight；TDW，total dry weight；RRS，ratio of root to shoot dry weight

QTL for root system architecture traits

Nineteen QTL from the three P treatments were identified for RSA traits i. e. RL，RV，RTN and ROSA on chro-mosomes 1BL，2BL，2DL，3DL，4BS，6AL，6BL （2），6DS，7AS （5），7AL （3） and 7BL （2） （Table 3）. Among them 6 QTL were detected for the zero control，and 7 in the low and 6 in the high P treatments with the phenotypic variances explained of 45. 1，48. 0，53. 7%，respectively. Seventeen QTL on chromosomes 2BL，2DL，3DL （2），4BS （3），6AL，6BL （3），6DS （2），7AS （2），7AL （2） conferred positive additive effects contributed by Zhongmai 895 （Table 3）. In the low P treatment，three QTL were identified for each of RL and RTN，explaining 20. 7 and 21. 7% of phenotypic variances，respectively. RL and RTN were co-located on chromosomes 6BL，7AL and 7BL. Zhongmai 895 possessed the positive alleles on 6BL and 7AL for increased phenotypic values （Table 3）.

QTL for root biomass-related traits

Fifteen QTL for BRT traits were identified on chromosomes 3AS （3），3DL，4BS （5），4DS （4），6BL and 6DS （Table3）. Two，4 and 9 QTL were identified in the zero，low and high P treatments，and explained 34. 0，42. 8 and 67. 6% of the phenotypic

variances，respectively. A stable QTL （QRDW. caas-4BS） was identified under both low and high P conditions，explaining 8. 1 to 17. 7% of the phenotypic variances for RDW. A pleiotropic QTL on chromosome 4DS in interval AX-109816583 -AX-109478820 （16. 64-30. 66 Mb） detected in the low P treatment for RDW co-located with 3 QTL in all three P treatments for RRS explained 7. 1 to 20. 4% of the phenotypic variances （Table 3）.

QTL clusters

Fifteen QTL were grouped in seven clusters （C1 to C7） for both RSA and BRT traits in all three P treatments. These clusters were identified in the same or close marker intervals （Fig. 4）. Genetic regions of these clusters were located on 3DL，4BS，4DS，6BL，7AS，7AL and 7BS chromosomes，respectively. Of these，C1 was on 3DL for SDW and RTN under high P treatment. C2 for RV and RRS on 4BS was detected in a 0. 31 Mb interval between SNPs AX-109491270 （21. 79 Mb） and AX-108815849 （21. 42 Mb）. C3 on 4DS comprised QTL for RDW and RRS. QTL in clusters on 4B （16. 64-30. 66 Mb） and 4D （32. 42-37. 86 Mb） were at or very near to reduced plant height gene loci Rht-B1 （30. 8 Mb） and Rht-D1 （18. 9 Mb），respectively. C4，C6 and C7 involved the same traits （RL and RTN） on chromosomes 6BL, 7AL and 7BL. C5 on 7AS （ AX-109955164-AX-109445593） affected RL, RTN and ROSA, with the favorable allele contributed by Zhongmai 895 （Fig. 4）. Joint QTL analysis reveal that, QTLs presented in four clusters. i. e. C3 （RDW and RRS）, C4 （RL and RTN）, C5 （RL, RTN, ROSA） and C7 （RL and RTN） were identified as pleiotropic QTL under different P levels （Fig. 4 and Table 4）.

Discussion

Significant variations and correlations observed for RSA and BRT

Hydroponic-based rapid phenotyping approach wasused to measure for BRT and RSA related traits in DH population. Previous studies identified strong correlations between such laboratory-based experiments and field data[30,31]. Therefore, use of digital dataset on BRT and RSA can be used to explore phenotypic variations among the DH population under varied nutrient levels. Most of the phenotypes were higher under the high P treatment compared with zero and low P treatments （Fig. 2）. High heritabilities and significant genetic variances for root system architecture and biomass traits indicated that these traits could be used as primary selection criteria for enhancement of P uptake and to identify underlying genetics[12,17] （ Table 2 ）. The present results corroborated earlier findings for genetic variances of root-related traits under different nutrient treatments[6,32,33].

P deprivation restricts the growth of main roots, while increases the lateral roots elongation with high numbers of root hairs[34]. This phenomenon leads higher ratio of root to shoot that significantly changes the root architecture for greater nutrient up-take[10]. In this study, RSA traits i. e. RL, RV, RTN and ROSA, and BRT traits i. e. RDW and RRS showed higher growth and highly positive correlation between each other at the low P level （Figs. 2 and 3）. This trend was contributed by Zhongmai 895, which also had longer roots, high root tip number and high RRS in the low P treatment. High accumulation of SDW under high P conditions and greater RRS under low P were observed in Yangmai 16 （Fig. 1）. Zhongmai 895 performed higher as compared to Yangmai 16 across the P treatments, as previously reported elite for high root growth at seedling stage and field-based nitrogen use efficiency[35,36]. SDW and RRS were negatively correlated across treatments （ Fig. 3 ）, although previous work found that low P led to reduced root bio-mass and had a significant impact on root system1 architecture traits[4,17,37]. A significantly negative correlations ranging from −0. 23 to −0. 69 of RD with RL, RTN and ROSA in all three P treatments indicated a negative association of RD with high P uptake among DH lines （Fig. 3）. A similar trend also observed for Zhongmai 895 under high P treatment （ Fig. 1 ）. These kind of variations among the population for complex root behavior could be important for genetic dissection of useful loci.

Table 2   Phenotypic variances for three P treatments in DH lines

Trait	P Level	Min	Max	Mean±SD	h²	G F. value	T F. value	G×T F. value
RL （cm）	C	18. 13	56. 86	33. 07 ± 10. 72	0. 90	2. 89**	706. 75**	1. 39**
	L	22. 21	98. 83	60. 51 ± 21. 07	0. 83	1. 53**		
	H	2. 01	63. 37	40. 00 ± 12	0. 77	1. 09		
RV （cm³）	C	0. 05	0. 23	0. 11 ± 0. 03	0. 86	2. 06**	1252. 17**	1. 46**
	L	0. 04	0. 34	0. 17 ± 0. 05	0. 81	1. 45**		
	H	0. 08	0. 42	0. 22 ± 0. 008	0. 87	2. 28**		
RD （cm³）	C	0. 16	0. 35	0. 21 ± 0. 03	0. 87	2. 27**	2274. 54**	1. 43**
	L	0. 14	0. 28	0. 19 ± 0. 02	0. 86	2. 01**		

（continued）

Trait	P Level	Min	Max	Mean±SD	h²	G F. value	T F. value	G×T F. value
	H	0.08	0.42	0.23 ± 0.06	0.88	2.54**		
RTN	C	209	1822	716.25 ± 246.7	0.91	3.19**	403.58**	1.34**
	L	183	2477	1090.1 ± 352.41	0.84	1.76**		
	H	184	1593	836.63 ± 341.52	0.82	1.56**		
ROSA （cm³）	C	7.25	39.36	21.44 ± 5.24	0.88	2.45**	700.51**	1.36**
	L	7.48	66.38	36.07 ± 10.12	0.82	1.48**		
	H	9.78	66.83	33.99 ± 9.32	0.83	1.68**		
SDW （mg）	C	0.02	2.06	0.04 ± 0.009	0.76	1.04	46.72**	1.01
	L	0.02	0.09	0.05 ± 0.01	0.86	2.09**		
	H	0.005	0.12	0.071 ± 0.009	0.88	2.54**		
RDW （mg）	C	0.01	0.03	0.02 ± 0.001	0.89	2.69**	451.45**	1.41**
	L	0.01	0.04	0.02 ± 0.001	0.84	1.72**		
	H	0.01	0.12	0.07 ± 0.02	0.89	2.67**		
TDW （mg）	C	0.01	2.08	0.06 ± 0.008	0.76	1.05	65.22**	1.03
	L	0.01	0.12	0.07 ± 0.02	0.85	1.88**		
	H	0.01	0.04	0.02 ± 0.001	0.88	2.42**		
RRS	C	0.01	0.71	0.38 ± 0.07	0.91	3.34**	180.38**	1.03
	L	0.18	1.04	0.39 ± 0.07	0.83	1.68**		
	H	0.01	0.16	0.09 ± 0.02	0.88	2.50**		

* Significant at $P<0.05$， ** Significant at $P<0.001$

Abbreviations：C，control；L，low；P，treatment；H，high；RL，root length；RV，root volume；RD，root diameter；RTN，root tip number；$ROSA$，root surface area；SDW，shoot dry weight；RDW，root dry weight；TDW，total dry weight；RRS，ratio of root to shoot dry weight

QTL identified under phosphorus treatments

Significant variances among genotypes allowed to explore genomic regions associated with the observed traits[4]. Several QTL were identified previously using hydroponic culture，which showed a vital role in nutrient uptake and a positive correlation with yield-related morphological traits. Here we identified 34 QTL on 1B，2B，2D，3A，3D，4B，4D，6A，6B，6D，7A and 7B chromosomes across the P treatments （Table 3）. Some QTL were already reported for root system architecture and root biomass-related traits[12,38]. New loci closely linked with *AX-109273188* （16.63 Mb） on 3AS，*AX-10934618* （18.67 Mb） on 6DS，*AX-109955164* （116.62 Mb） on 7AS，*AX-109109966788* （725.54 Mb） on 7AL and *AX-109289805* （705.84 Mb） on 7BL showed

significant influence for root growth under low P-condition.

The genetic diversity for root vigor under low P conditions could potentially improve P acquisition efficiency and described some QTL for BRT under low P conditions[11]. We also detected QTL on 2BL，4BS，4DS，6BL，7AL and 7BL showing high phenotypic variances for RSA traits i.e. RTN，RRS and RL under low P. （Table2）. Previously，Su et al. [9,37] evaluated two DH populations in pot and field experiments and reported a common QTL on chromosome 4B associated with shoot biomass and tiller number，was potentially important for P efficiency. Here，QTL in same genomic region of the 4B chromosome were detected for RDW and RRS （Table 3）. This QTL was nearly co-located with *Rht* genes for plant height. The *Rht-1* locus on 4B, which was a major factor against lodging as part of the green revolution. Interactions between root-related growth traits and *Rht* genes were reported as important in early vigor and nutrient uptake[11,39-41]. These reports had demonstrated contrasting impact of *Rht* genes on seedling traits under different nutrient conditions as compare to our findings. In our results, both QTL mapped on 4B (16. 64-30. 66 Mb) and 4D (32. 42-37. 86 Mb) in low P corresponding to *Rht-B1b* (30. 9 Mb) and *Rht-D1b* (18. 9 Mb), respectively, had negative impact on RRS among DH lines in contrast with a previous report[40].

QTL on 2DL and 6BL for RTN, RL and SDW under low P condition were likely to be those which were demonstrated for root related traits under P sufficiency in pot trials[37] and hydroponic culture[4]. Therefore, identification of those QTL influencing root traits for high nutrient uptake under varied conditions could be important for genotypic selection. RL, RTN and ROSA are important traits for nutrient uptake and early seedling vigor. Five QTL on chromosome 7A explained higher phenotypic variances of 4. 6 to 11. 9% for RL, RTN and ROSA. Four of these QTL detected under high P conditions and negatively influenced the root elongation, RTN and ROSA. Whereas, the QTL on 7AL controlling RTN had a positive role by increasing RTN in low P condition (Table 3). This QTL could play a vital for future crop breeding to improve the P-uptake. Closely linked QTL *QRL. caas-7BL* and *QRTN. caas-7BL* near *AX-109289805* (705. 84 Mb) could also be important for early vigor under P deficient conditions, whereas their strong relationship with kernel number per spike was earlier demonstrated by Zhang et al. [42]. SNPs linked with QTL for these traits on 7A and 7B stably detected in different nutrient conditions might be of great value for wheat breeding.

Putative candidate genes

Putative candidate genes for new QTL identified on 3AS for SDW, RDW and TDW was linked with E3-Ubiquitin ligase gene (TraesCS3A01G030100 14Kb)

(Table5). It had been reported that E3 ubiquitin ligases are involved in lateral root development in monocots and dicots through regulating plant phytohormone biosynthesis, transport and signaling pathways or cell cycle progression. Whereas, Ubiquitin-mediated proteolysis also has a pivotal role in root development, flowering time control and hypocotyl elongation[43]. Based on putative analysis, flanking sequence of SNP linked with QTL on 7AS identified for RL, RTN and ROSA was associated with gene for Gb protein. Previous reports had shown significant role of Gb protein in stress tolerance and plant growth[44]. Furthermore, a gene for transmembrane domain containing protein family was associated with QTL on 7BL for RL and RTN. These genes have been reported for nutrient import through cell membrane, stress tolerance and vegetative growth in *Arabidopsis* [45]. These QTL could play an important role for high uptake of P through alteration in root traits under varied P conditions.

QTL clusters

Previously high-density physical mapping results demonstrated about 85% of gene expression and also has many cluster distributions in wheat genome that covered 5 to 10% of chromosome regions[46]. In wheat, many QTL clusters were reported in several studies[47-51]. In the present study, seven clusters were identified in three different P treatments (Fig. 4). Clusters C3, C4, C5 and C7 had pleiotropic QTLs for root system architecture and biomass-related traits under different P treatments, whereas QTLs in other clusters contained closely linked multiple genes without pleiotropic effects. Clusters C2 and C3 on chromosomes 4BS and 4DS were linked with reduced plant height genes. Cluster C4 on 6BL had QTLs controlling RL and RTN and previously were re-ported for thousand grain weight[17]. Cluster C5 on chromosome 7AS (*AX-109955164 -AX-109445593*) affecting RL, RTN and ROSA was identified for the first time in the present study and SNP linked with this cluster can be used for future wheat improvement. Whereas cluster C7 on chromosome 7BL containing QTLs for RL and RTN was reported for RL and SDW under P-deficient condition. Similar QTL were also identified on 7B chromosome in previous studies[4,17], but it is not confirmed that QTL in our results are new or co-localization with previously identified loci. (Fig. 4). These results also indicated that QTL identified at seedling stage could be reliable for the selection of yield-related traits measured at maturity[12]. The SNPs tightly linked to QTLs or QTL clusters identified in the present study can be converted to KASP assays and effectively used for MAS to improve nutrient-use efficiency in wheat breeding.

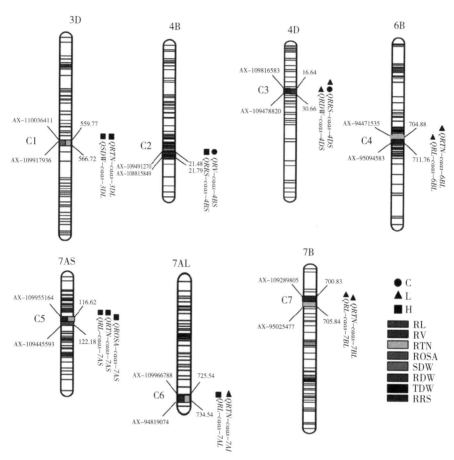

Fig. 4　Seven QTL clusters containing pleiotropic loci with closely linked markers and physical positions. Abbreviations: C, control; L, low; P, treatment; H, high; P, treatment; RL, root length; RV, root volume; RTN, root tip number; ROSA, root surface area; SDW, shoot dry weight; RDW, root dry weight; TDW, total dry weight; RRS, ratio of root to shoot dry weight

Conclusions

Thirty-four QTL with significant phenotypic variations for root system architecture and biomassrelated traits were identified using high-density genetic map constructed from 660k SNP array and cost-effective hydroponic-based phenotyping pipe-line. Four QTLs on chromosomes 6BL (2) and 7AL (2) identified in low P treatment with positive additive effects from Zhongmai 895, indicating that it could be used as parent for P-deficient breeding. A stable QTL *QRRS. caas-4DS* (16. 64-30. 66 Mb) was also detected across the three P levels, accounted for 8. 4 to 20. 4% of the phenotypic variances, which could be used for speedy selection of genotypes for P-uptake. Among the seven QTLs clusters, C5 identified on chromosome 7AS (*AX-109955164-AX-109445593*) affecting RL, RTN and ROSA could be vital for improving the nutrient use efficiency of roots under high P condition, whereas, C6 on 7AL under low P condition. Identified chromosome regions, particularly the QTL clusters could be used for molecular marker-assisted selection in future breeding.

Table 3   QTLs for root system architecture and root biomass-related traits identified under three P treatments

Trait	Treatment	QTL[a]	Marker interval	Physical interval[b] (Mb)	Genetic position[c] (cM)	LOD[d]	PVE[e] (%)	Add[f]
RL	C	QRL. caas-1BL	AX-10928674-AX-94446430	671. 33-688. 38	2. 5	3. 13	5. 56	−18. 89
		QRL. caas-7AS. 1	AX-110432090-AX-109345074	20. 95-22. 87	0. 97	3. 51	6. 38	19. 78
		QRL. caas-6AL	AX-86165298-AX-109431293	612. 11-613. 49	1. 17	4. 53	8. 20	23. 51
		QRL. caas-7AL	AX-109966788-AX-94819074	725. 54-734. 54	4. 84	3. 37	6. 22	19. 54
	L	QRL. caas-6BL	AX-109368729-AX-95094583	699. 99-711. 76	4	4. 90	8. 61	45. 43
		QRL. caas-7BL	AX-95025477-AX-109289805	700. 83-705. 84	2. 35	3. 74	6. 57	−38. 04
		QRL. caas-7AL	AX-109966788-AX-94819074	725. 54-734. 54	4. 84	3. 46	5. 50	35. 19
	H	QRL. caas-7AS. 2	AX-109955164-AX-109445593	116. 62-122. 18	1. 72	4. 72	11. 89	30. 81
RV	C	QRV. caas-6DS	AX-10934618-AX-94913665	18. 67-25. 66	3. 45	3. 98	8. 41	0. 01
		QRV. caas-4BS	AX-108815849-AX-109491270	21. 48-21. 79	2. 5	4. 90	10. 32	0. 01
	H	QRV. caas-7AS	AX-108763612-AX-111263013	85. 44-85. 71	1. 95	3. 49	8. 82	0. 01
RTN	L	QRTN. caas-6BL	AX-109368729-AX-95094583	699. 99-711. 76	4	4. 21	7. 08	73. 56
		QRTN. caas-2BL	AX-86176885-AX-94933613	647. 32-647. 71	0. 35	2. 86	4. 79	57. 39
		QRTN. caas-7BL	AX-95025477-AX-109289805	700. 83-705. 84	2. 35	5. 52	9. 78	−81. 68
		QRTN. caas-7AL	AX-109966788-AX-94819074	725. 54-734. 54	4. 84	2. 77	4. 64	56. 97
	H	QRTN. caas-2DL	AX-94959623-AX-94821426	79. 54-82. 22	1. 26	2. 74	7. 78	45. 97
		QRTN. caas-7AS	AX-109955164-AX-109445593	116. 62-122. 18	1. 72	3. 84	8. 17	48. 20
		QRTN. caas-3DL	AX-110036411-AX-109917936	559. 77-566. 72	2. 79	2. 99	6. 21	42. 26
ROSA	H	QROSA. caas-7AS	AX-109955164-AX-109445593	116. 62-122. 18	1. 72	4. 43	10. 80	2. 24
SDW	H	QSDW. caas-6BL	AX-109368729-AX-95094583	699. 99-711. 76	4	3. 44	6. 80	0. 00
		QSDW. caas-6DS	AX-109346183-AX-94913665	18. 67-25. 66	3. 45	2. 59	4. 95	0. 00
		QSDW. caas-3AS	AX-111507145-AX-109273188	15. 80-16. 63	1. 62	3. 01	5. 99	0. 00
		QSDW. caas-3DL	AX-110036411-AX-109917936	559. 77-566. 72	2. 79	4. 17	8. 21	0. 00
RDW	L	QRDW. caas-4BS	AX-111068079-AX-111164540	32. 42-37. 86	0. 85	7. 70	17. 70	0. 00
		QRDW. caas-4DS	AX-109816583-AX-109478820	16. 64-30. 66	3. 6	3. 12	7. 14	0. 00
	H	QRDW. caas-3AS	AX-111507145-AX-109273188	15. 80-16. 63	1. 62	3. 08	7. 06	0. 00
		QRDW. caas-4BS	AX-111068079-AX-111164540	32. 42-37. 86	0. 85	3. 59	8. 12	0. 00
TDW	H	QTDW. caas-3AS	AX-111507145-AX-109273188	15. 80-16. 63	1. 62	3. 11	7. 68	0. 00
RRS	C	QRRS. caas-4BS. 1	AX-109494015-AX-108991675	97. 25-103. 06	0. 48	6. 78	13. 59	0. 01
		QRRS. caas-4DS	AX-109816583-AX-109478820	16. 64-30. 66	3. 6	10. 43	20. 41	−0. 01
	L	QRRS. caas-4BS. 1	AX-109494015-AX-108991675	97. 25-103. 06	0. 48	2. 94	6. 25	0. 01
		QRRS. caas-4DS	AX-109816583 -AX-109478820	16. 64-30. 66	3. 6	5. 55	11. 71	−0. 01
	H	QRRS. caas-4BS. 2	AX-108815849-AX-109491270	21. 48-21. 79	2. 5	4. 33	9. 35	0. 01
		QRRS. caas-4DS	AX-109816583-AX-109478820	16. 64-30. 66	3. 6	3. 85	8. 30	−0. 01

[a] Quantitative trait loci

[b] Physical positions of SNP markers based on wheat genome sequences from the International Wheat Genome Sequencing Consortium (IWGSC, http: //www. wheatgenome. org/)

[c] Genetic position of closest marker to the identified QTL on linkage map

[d] LOD value of each QTL

[e] Phenotypic variance explained by QTL

[f] A positive sign means the increased effect contributed by Zhongmai 895, whereas the negative effect was contributed by Yangmai 16

Abbreviations: C, control; L, low; P, treatment; H, high; RL, root length; RV, root volume; RD, root diameter; RTN, root tip number; ROSA, root surface area; SDW, shoot dry weight; RDW, root dry weight; TDW, total dry weight; RRS, ratio of root to shoot dry weight (Q=QTL) + (caas=Chinese Academy of Agricultural Sciences) + Chromosome=long arm (L) or short arm (S)

Table 4　Joint QTL analysis results for pleiotropic QTL in clusters

Cluster	Trait	P Level	Chromosome	Marker interval	Physical interval[a] (Mb)	Pr $(>\mid t \mid)$[b]	Adjusted R^2
C1	SDW	H	3D	AX-110036411-AX-109917936	559.77-566.72	0.02*	0.38
	RTN	H					
C2	RV	C	4B	AX-109491270-AX-108815849	21.48-21.79	0.05*	0.12
	RRS	H					
C3	RDW	L	4D	AX-109816583-AX-109478820	16.64-30.66	0.15, 0.18	0.18, 0.12
	RRS	C, L					
C4	RL	L	6B	AX-94471535-AX-95094583	704.88-711.76	0.7	0.75
	RTN	L					
C5	RL	H	7A	AX-109955164-AX-109445593	116.62-122.18	0.68, 0.55, 0.71	0.76, 0.90, 0.74
	RTN	H					
	ROSA	H					
C6	RL	H	7A	AX-109966788-AX-94819074	725.54-134.54	0.037*	0.12
	RTN	L					
C7	RL	L	7B	AX-109289805-AX-95025477	700.83-705.84	0.8	0.75
	RTN	L					

* Indicates significance at $P < 0.05$

[a] Physical positions of SNP markers based on wheat genome sequences from the International Wheat Genome Sequencing Consortium (IWGSC, http://www.wheatgenome.org/)

[b] Significance of t-values at $P < 0.05$

Abbreviations: C, control; L, low; P, treatment; H, high P treatment; RL, root length; RV, root volume; RD, root diameter; RTN, root tip number; ROSA, root surface area; SDW, shoot dry weight; RDW, root dry weight; TDW, total dry weight; RRS, ratio of root to shoot dry weight

Table 5　QTL with corresponded candidate genes and putative proteins

Trait	eatme	QTL[a]	Marker interval	Physical interval[b] (Mb)	Closest SNP	Seq of SNP markers	Wheat Genes	Distance from SNP	Putative protein
RL	C	QRL.caas-1BL	AX-109286742—AX-94446430	671.33—688.38	AX-109286742	GAAGGCATTATATTGTAAGGTCAAGAAAATAGCAA[A/G]GATCACTTGCTTGTTATGCCATATTCCTTCTGCAT	TraesCS1B02G457000	28kb	Biotype:protein
		QRL.caas-7AS.1	AX-110432090—AX-109345074	20.95—22.87	AX-110432090	CCTTGTACGGAAGAGCATGTACGATCTGTCATTTT[A/G]TACAATGTCACTTGTGGTGGTCAGTAAATAACATT	TraesCS7A01G046100	28Kb	Disease resistance protein RPM1
		QRL.caas-6AL	AX-86165298—AX-109431293	612.11—613.49	AX-86165298	AACAAAAGTAACCAACACTTCCAACAATGTTAAAT[C/T]AAGGAGTTGAAGTGTTGGAAATCAGCTTGTCCTCA	TraesCS6A02G406900	72b	Biotype:protein
		QRL.caas-7AL	AX-109966788—AX-94819074	725.54—734.54	AX-109966788	CATTTTCTGTACAAACGGTGATTCCGGACCTGATC[C/T]TTCCTTGTGTGGCCGCACATCCATCTCAACATGCA	TraesCS7A01G785700L	2.87kb	Biotype:protein
	L	QRL.caas-6BL	AX-109368729—AX-95094583	699.99—711.76	AX-109368729	TCATATAAAATCTTTTGTAATAATGATGAAAG[C/T]AGTTTGAAGCTTATACCTGAAGTCTGTTAACTCTC	TraesCS6B01G770800L	284.6kb	Biotype:protein
		QRL.caas-7BL	AX-95025477—AX-109289805	700.83—705.84	AX-95025477	CTCAGCGAGTCGATCTTCTCCAGCGCGCACAGCCG[C/G]GGCGAGGAACACCACGAGGACGAACAGGATGAGCAG	TraesCS7B01G433400	72b	Biotype:protein
		QRL.caas-7AL	AX-109966788—AX-94819074	725.54—734.54	AX-109966788	CATTTTCTGTACAAACGGTGATTCCGGACCTGATC[C/T]TTCCTTGTGTGGCCGCACATCCATCTCAACATGCA	TraesCS7A01G785700L	2.87kb	Transmembrane domain containing protein
	H	QRL.caas-7AS.2	AX-109955164—AX-109445593	116.62—122.18	AX-109955164	GCTTGTAAACCTAAAGAACATTTGTGACAAAAAGA[A/C]AAATAAAGAAAAATACCACATAGCTTCCTTAAAAA	TraesCS7A01G166800	72b	Similar to Gb protein (Os06t0206100-01)
RV	C	QRV.caas-4BS	AX-108815849—AX-109491270	21.48—21.79	AX-108815849	AGCGGCAACAATTACGTGTCAGTGGCTCTTTCTCA[C/T]GGCTGTAGTGGTTGTTCAGTGGTTTAGAGACATTG	TraesCS4B01G044700L	14kb	Biotype:protein
	H	QRV.caas-4DS	AX-109341618—AX-94931665	18.67—25.66	AX-109341618	CCATATATCATTACTCCTGTTATTCATAGGACCCA[C/G]TTTTCTATAAGGTCACAACATCAAAGGTCATAGCA	TraesCS4D01G044700L	72b	Biotype:protein
RTN	L	QRTN.caas-6BL	AX-109368729—AX-95094583	699.99—711.76	AX-109368729	TCATATAAAATCTTTTGTAATAATGATGAAAG[C/T]AGTTTGAAGCTTATACCTGAAGTCTGTTAACTCTC	TraesCS6B01G770800L	284.6kb	Biotype:protein
		QRTN.caas-2BL	AX-86176885—AX-94933613	647.32—647.71	AX-86176885	CGGGTTATTGAATTTGCAACACATATGTAACTTAGTT[G/T]TCCATAGGTTGTTGCTTTCATGGGGCTCTCCTGGA	TraesCS2B01G453500	72b	Alpha-1,2-Mannosidase
		QRTN.caas-7BL	AX-95025477—AX-109289805	700.83—705.84	AX-95025477	CTCAGCGAGTCGATCTTCTCCAGCGCGCACAGCCG[C/G]GGCGAGGAACACCACGAGGACGAACAGGATGAGCAG	TraesCS7B01G433400	72b	Transmembrane domain containing protein (Os06t0644700-01)
		QRTN.caas-7AL	AX-109966788—AX-94819074	725.54—734.54	AX-109966788	CATTTTCTGTACAAACGGTGATTCCGGACCTGATC[C/T]TTCCTTGTGTGGCCGCACATCCATCTCAACATGCA	TraesCS7A01G785700L	2.87kb	Transmembrane domain containing protein
	H	QRTN.caas-3DL	AX-94959625—AX-109445593	79.54—82.22	AX-94959623	GGATAGACTCTTCACCGCCCACTTCCGCCTTAAGCA[C/T]ATATAGGTTTAATCTCGTATGACTTTGTGTGCGAG	TraesCS3D01G135300	72b	Biotype:protein
		QRTN.caas-7AS	AX-109955164—AX-109445593	116.62—122.18	AX-109955164	GCTTGTAAACCTAAAGAACATTTGTGACAAAAAGA[A/C]AAATAAAGAAAAATACCACATAGCTTCCTTAAAAA	TraesCS7A01G166800	72b	Similar to Gb protein (Os06t0206100-02)
		QRTN.caas-3DL	AX-110036411—AX-109917916	559.77—566.72	AX-110036411	GGTAAAAAATTTAGCGCTTGTTGCAGAAGCCTAAC[C/T]GAAAGGCCGCCTTGAAACTGCATCCTCAAAGTA	TraesCS3A01G030100	284kb	Biotype:protein
ROSA	H	QROSA.caas-7AS	AX-109955164—AX-109445593	116.62—122.18	AX-109955164	GCTTGTAAACCTAAAGAACATTTGTGACAAAAAGA[A/C]AAATAAAGAAAAATACCACATAGCTTCCTTAAAAA	TraesCS7A01G166800	72b	Similar to Gb protein (Os06t0206100-02)
SDW	H	QSDW.caas-6BL	AX-109368729—AX-95094583	699.99—711.76	AX-109368729	TCATATAAAATCTTTTTGTAATAATATGATGAAAG[C/T]AGTTTGAAGCTTATACCTGAAGTCTGTTAACTCTC	TraesCS6B01G770800L	284.6kb	Biotype:protein
		QSDW.caas-3AS	AX-11507145—AX-109273188	15.80—16.63	AX-11507145	AATAGGCGTCTTCTTCTTTTGTGGTTGTTGGTTGCGTG[A/G]JTCTCATTGATGTCTTACTTCGATGGTTGCTCCT	TraesCS3A01G133200	14kb	Biotype:protein
		QSDW.caas-3DL	AX-110036411—AX-109917916	559.77—566.72	AX-110036411	GGTAAAAAATTTAGCGCTTGTTGCAGAAGCCTAAC[C/T]GAAAGGCCGCCTTGAAACTGCATCATCTCCAAAGTA	TraesCS3A01G452000	284kb	E3 ubiquitin-protein ligase
RDW	L	QRDW.caas-4BS	AX-11068079—AX-111164540	32.42—37.86	AX-11068079	TGTGTTTCGCAGCCGCCACTGCTAGCGCTGCCTGC[A/G]GTC]TCACTTCCTTGGCGCGGATCAGCAAGATCAACTCA	TraesCS4B01G034600	71.7kb	Hypothetical protein
		QRDW.caas-4DS	AX-109816583—AX-109478820	16.64—30.66	AX-109816583	GATAGAATTCGGCAGGAATTCCATCTACTCCATGA[A/G]TCCAATTATTGACCATCTTATAAAAAGCAAACCTA	TraesCS4D01G038700	1.14kb	Biotype:protein
		QRDW.caas-3AS	AX-11507145—AX-109273188	15.80—16.63	AX-11507145	AATAGGCGTCTTCTTCTTTTGTGGTTGTTGGTTGCGTG[A/G]TCTCATTGATGTCTTACTGCTCGATGGTTGCTCCT	TraesCS3A01G034100	14kb	E3 ubiquitin-protein ligase
	H	QRDW.caas-4BS	AX-11507145—AX-109273188	32.42—37.86	AX-11068079	TGTGTTTCGCAGCCGCCACTGCTAGCGCTGCCTGC[A/G]JTCACTTCCTTGGCGCGGATCAGCAAGATCAACTCA	TraesCS4B01G034600	71.7kb	Hypothetical protein
TDW	H	QTDW.caas-3AS	AX-11507145—AX-109273188	15.80—16.63	AX-11507145	AATAGGCGTCTTCTTCTTTTGTGGTTGTTGCGTG[A/G]TCTCATTGATGTCTTACTGCTCGATGGTTGCTCCT	TraesCS3A01G030100	14kb	E3 ubiquitin-protein ligase
RRS	C	QRRS.caas-4BS.1	AX-109494015—AX-108991675	97.25—103.06	AX-109494015	TGACATTTCTAAAATTTATGATAAATATACAAAGA[A]TGCCAGATTTGAAATGCGTCGACGCTGTGGATATG	TraesCS4B01G094000	72b	Biotype:protein
		QRRS.caas-4DS	AX-109816583—AX-109478820	16.64—30.66	AX-109816583	GATAGAATTCGGCAGGAATTCCATCTACTCCATGA[A/G]TCCAATTATTGACCATCTTATAAAAAGCAAACCTA	TraesCS4D01G038700	1.14kb	Biotype:protein
	L	QRRS.caas-4BS.1	AX-109494015—AX-108991675	97.25—103.06	AX-109494015	TGACATTTCTAAAATTTATGATAAATATACAAAGA[A/T]TGCCAGATTTGAAATGCGTCGACGCTGTGGATATG	TraesCS4B01G094000	72b	Biotype:protein
		QRRS.caas-4DS	AX-109816583—AX-109478820	21.48—21.79	AX-108815849	AGCGGCAACAATTACGTGTCAGTGGCTCTTTCTCA[C/T]GGCTGTAGTGGTTGTTCAGTGGTTTAGAGACATTG	TraesCS4B01G029200	2.87kb	Biotype:protein
	H	QRRS.caas-4DS	AX-109816583—AX-109478820	16.64—30.66	AX-109816583	GATAGAATTCGGCAGGAATTCCATCTACTCCATGA[A/G]TCCAATTATTGACCATCTTATAAAAAGCAAACCTA	TraesCS4D01G038700	1.14kb	Biotype:protein

[a] Quantitative trait loci

[b] Physical positions of SNP markers based on wheat genome sequences from the International Wheat Genome Sequencing Consortium (IWGSC, http://www.wheatgenome.org/)

Abbreviations: C, control; L, low; P, treatment; H, high; P, treatment; RL, Root length; RV, Root volume; RD, Root diameter; RTN, root tip number; ROSA, root surface area; SDW, shoot dry weight; RDW, root dry weight; TDW, total dry weight; RRS, ratio of root to shoot dry weight

manuscript.

❖ Acknowledgements

We are thankful to Dr. R. A. McIntosh at the Plant Breeding Institute, University of Sydney, and Awais Rasheed at CIMMYT China office for reviewing this

❖ References

[1] Malhi SS, Vera CL, Brandt SA. Seed yield potential of five wheat species/cultivars without and with

phosphorus fertilizer application on a p-deficient soil in northeastern Saskatchewan. Agri Sci. 2015; 6: 224-31.

[2] James RA, Weligama C, Verbyla K, Ryan PR, Rebetzke GJ, Rattey A, Richardson AE, Delhaize E. Rhizosheaths on wheat grown in acid soils: phosphorus acquisition efficiency and genetic control. J Exp Bot. 2016; 67: 3709-18.

[3] Wang W, Ding GD, White PJ, Wang XH, Jin KM, Xu FS, Shi L. Mapping and cloning of quantitative trait loci for phosphorus efficiency in crops: opportunities and challenges. Plant Soil. 2019; 439: 91-112.

[4] Guo Y, Kong FM, Xu YF, Zhao Y, Liang X, Wang YY, An DG, Li SS. QTL mapping for seedling traits in wheat grown under varying concentrations of N, P and K nutrients. Theor Appl Genet. 2012; 124: 851-65.

[5] Carvalho P, Azam-Ali S, Foulkes MJ. Quantifying relationships between rooting traits and water uptake under drought in Mediterranean barley and durum wheat. J Integr Plant Biol. 2014; 56: 455-69.

[6] Liu Z, Gao K, Shan S, Gu R, Wang Z, Craft EJ, Mi G, Yuan L, Chen F. Comparative analysis of root traits and the associated QTL for maize seedlings grown in paper roll, hydroponics and vermiculite culture system. Front Plant Sci. 2017; 8: 436.

[7] Clárk RB. Plant genotype differences in the uptake, translocation, accumulation, and use of mineral elements required for plant growth. Plant Soil. 1983; 72: 175-96.

[8] Habash DZ, Bernard S, Schondelmaier J, Weyen J, Quarrie SA. The genetics of nitrogen use in hexaploid wheat: N utilisation, development and yield. Theor Appl Genet. 2007; 114: 403-19.

[9] Su JY, Xiao YM, Li M, Liu QY, Li B, Tong YP, Jia JJ, Li ZS. Mapping QTL for phosphorus-deficiency tolerance at wheat seedling stage. Plant Soil. 2006; 281: 25-36.

[10] Ayalew H, Ma X, Yan G. Screening wheat (Triticum spp.) genotypes for root length under contrasting water regimes: potential sources of variability for drought resistance breeding. J Agron Crop Sci. 2015; 201: 189-94.

[11] Ryan PR, Liao M, Delhaize E, Rebetzke GJ, Weligama C, Spielmeyer W, James RA. Early vigour improves phosphate uptake in wheat. J Exp Bot. 2015; 66: 7089-100.

[12] Yuan Y, Gao M, Zhang M, Zheng H, Zhou X,

Guo Y, Zhao Y, Kong F, Li S. QTL mapping for phosphorus efficiency and morphological traits at seedling and maturity stages in wheat. Front Plant Sci. 2017; 8: 614.

[13] Bovill WD, Huang CY, McDonald GK. Genetic approaches to enhancing phosphorus-use efficiency (PUE) in crops: challenges and directions. Crop Pasture Sci. 2013; 64: 179-98.

[14] Gong X, McDonald GK. QTL mapping of root traits in phosphorus-deficient soils reveals important genomic regions for improving NDVI and grain yield in barley. Theor Appl Genet. 2017; 130: 1885-902.

[15] Atkinson JA, Wingen LU, Griffiths M, Pound MP, Gaju O, Foulkes MJ, Le Gouis J, Griffiths S, Bennett MJ, King J, Wells DM. Phenotyping pipeline reveals major seedling root growth QTL in hexaploid wheat. J Exp Bot. 2015; 66: 2283-92.

[16] Kabir MR, Liu G, Guan P, Wang F, Khan AA, Ni Z, Yao Y, Hu Z, Xin M, Peng H, Sun Q. Mapping QTL associated with root traits using two different populations in wheat (Triticum aestivum L.). Euphytica. 2015; 206: 175-90.

[17] Zhang H, Wang HG. QTL mapping for traits related to P-deficient tolerance using three related RIL populations in wheat. Euphytica. 2015; 203: 505-20.

[18] Cui F, Zhang N, Fan X, Zhang W, Zhao C, Yang L, Pan R, Chen M, Han J, Zhao X, Ji J, Tong Y, Zhang H, Jia J, Zhao G, Li M. Utilization of a wheat 660K SNP array-derived high-density genetic map for high-resolution mapping of a major QTL for kernel number. Sci Rep. 2017; 7: 3788.

[19] Jin H, Wen W, Liu J, Zhai S, Zhang Y, Yan J, Liu Z, Xia X, He Z. Genome-wide QTL mapping for wheat processing quality parameters in a Gaocheng 8901/Zhoumai 16 recombinant inbred line population. Front Plant Sci. 2016; 7: 1032.

[20] Hoagland DR, Arnon DI. The water culture method for growing plants without soil. Calif Agric Exp Stn Bull. 1938; 347: 36-9.

[21] Li F, Xiao Y, Jing S, Xia X, Chen X, Wang H, He Z. Genetic analysis of nitrogen and phosphorus utilization efficiency related traits at seedling stage of Jing 411 and its derivatives. J Triticeae Crops. 2015; 35: 737-46.

[22] Pound MP, French AP, Atkinson J, Wells DM, Bennett MJ, Pridmore TP. RootNav: navigating images of complex root architectures. Plant

Physiol. 2013; 162; 1802-14.

[23] Wang JP. Genetic analysis and functional marker development of key traits in wheat. Tai´an Shandong Province, China; Dissertation, Shandong Agricultural University; 2017.

[24] Meng L, Li H, Zhang L, Wang J. QTL IciMapping; integrated software for genetic linkage map construction and quantitative trait locus mapping in biparental populations. Crop J. 2015; 3; 269-83.

[25] Kosambi DD. The estimation of map distances from recombination values. Annu Eugen. 1944; 12; 172-5.

[26] Churchill GA, Doerge RW. Empirical threshold values for quantitative trait mapping. Genetics. 1994; 138; 963-71.

[27] Li H, Bradbury P, Ersoz E, Buckler ES, Wang J. Joint QTL linkage mapping for multiple-cross mating design sharing one common parent. PLoS One. 2011; 6; e17573.

[28] Sehgal D, Skot L, Singh R, Srivastava RK, Das SP, Taunk J, Sharma PC, Pal R, Raj B, Hash CT, Yadav RS. Exploring potential of pearl millet germplasm association panel for association mapping of drought tolerance traits. PLoS One. 2015; 10; e0122165.

[29] R Core Team. R; A language and environment for statistical computing. Vienna; R Foundation for Statistical Computing; 2013.

[30] Ren Y, Qian Y, Xu Y, Zou C, Liu D, Zhao X, Zhang A, Tong Y. Characterization of QTL for root traits of wheat grown under different nitrogen and phosphorus supply levels. Front Plant Sci. 2017; 8; 2096.

[31] Xie Q, Fernando KM, Mayes S, Sparkes DL. Identifying seedling root architectural traits associated with yield and yield components in wheat. Ann Bot. 2017; 119; 1115-29.

[32] Ao J, Fu J, Tian J, Yan X, Liao H. Genetic variability for root morpho-architecture traits and root growth dynamics as related to phosphorus efficiency in soybean. Funct Plant Biol. 2010; 37; 304-12.

[33] Bayuelo-Jiménez JS, Gallardo-Valdéz M, Pérez-Decelis VA, Magdaleno-ArmasL, Ochoa I, Lynch JP. Genotypic variances for root traits of maize (Zea mays L.) from the Purhepecha plateau under contrasting phosphorus availability. Field Crops Res. 2011; 121; 350-62.

[34] Gu R, Chen F, Long L, Cai H, Liu Z, Yang J, Wang L, Li H, Li J, Liu W, Mi G, Zhang F,

Yuan L. Enhancing phosphorus uptake efficiency through QTL-based selection for root system architecture in maize. J Genet Genomics. 2016; 43; 663-72.

[35] Yang M, Hassan MA, Xu K, Zheng C, Rasheed A, Zhang Y, Jin X, Xia X, Xiao Y, He Z. Assessment of water and nitrogen use efficiencies through UAV-based multispectral phenotyping in winter Wheat. Front Plant Sci. 2020; 11; 0927.

[36] Yang M, Wang C, Hassan MA, Wu Y, Xia X, Shi S, Xiao Y, He Z. QTL mapping of seedling biomass and root traits under different nitrogen conditions in bread wheat (Triticum aestivum L.). J Integrat Agric. 2020; 19; 2-14.

[37] Su JY, Zheng Q, Li HW, Li B, Jing RL, Tong YP, Li ZS. Detection of QTL for phosphorus use efficiency in relation to agronomic performance of wheat grown under phosphorus sufficient and limited conditions. Plant Sci. 2009; 176; 824-36.

[38] Sun JJ, Guo Y, Zhang GZ, Gao MG, Zhang GH, Kong FM, Zhao Y, Li SS. QTL mapping for seedling traits under different nitrogen forms in wheat. Euphytica. 2013; 191; 317-31.

[39] Wojciechowski T, Gooding MJ, Ramsay L, Gregory PJ. The effects of dwarfing genes on seedling root growth of wheat. J Exp Bot. 2009; 60; 2565-73.

[40] Bai C, Liang Y, Hawkesford MJ. Identification of QTL associated with seedling root traits and their correlation with plant height in wheat. J Exp Bot. 2013; 64; 1745-53.

[41] Narayanan S, Vara Prasad PV. Characterization of a spring wheat association mapping panel for root traits. Agron J. 2014; 106; 1593-604.

[42] Zhang H, Chen J, Li R, Deng Z, Zhang K, Liu B, Tian J. Conditional QTL mapping of three yield components in common wheat (Triticum aestivum L.). Crop J. 2016; 4; 220-8.

[43] Shu K, Yang W. E3 Ubiquitin Ligases; ubiquitous actors in plant development and abiotic stress responses. Plant Cell Physiol. 2017; 58; 1461-76.

[44] Giri J. Glycinebetaine and abiotic stress tolerance in plants. Plant Signal Behav. 2011; 6; 1746-51.

[45] Deng Y, Srivastava R, Howell SH. Protein kinase and ribonuclease domains of IRE1 confer stress tolerance, vegetative growth, and reproductive development in "Arabidopsis". Proceed Natl Acad Sci. 2013; 110; 19633-8.

[46] Cui F, Ding A, Li J, Zhao C, Li X, Feng D, Wang X, Wang L, Gao J, Wang H. Wheat kernel dimensions: how do they contribute to kernel weight at an individual QTL level? J Genet. 2011; 90: 409-25.

[47] Quarrie SA, Pekic Quarrie S, Radosevic R, Rancic D, Kaminska A, Barnes JD, Leverington M, Ceoloni C, Dodig D. Dissecting a wheat QTL for yield present in a range of environments: from the QTL to candidate genes. J Exp Bot. 2006; 57: 2627-37.

[48] Crossa J, Burgueño J, Dreisigacker S, Vargas M, Herrera-Foessel SA, Lillemo M, Singh RP, Trethowan R, Warburton M, Franco J, Reynolds M, Crouch JH, Ortiz R. Association analysis of historical bread wheat germplasm using additive genetic covariance of relatives and population structure. Genetics. 2007; 177: 1889-913.

[49] Kong FM, Guo Y, Liang X, Wu CH, Wang YY, Zhao Y, Li SS. Potassium (K) effects and QTL mapping for K efficiency traits at seedling and adult stages in wheat. Plant Soil. 2013; 373: 877-92.

[50] Zhao Y, Li XY, Zhang SH, Wang J, Yang XF, Tian JC, Hai Y, Yang XJ. Mapping QTL for potassium-deficiency tolerance at the seedling stage in wheat (*Triticum aestivum* L.). Euphytica. 2014; 198: 185-98.

[51] Gong X, Wheeler R, Bovill WD, McDonald GK. QTL mapping of grain yield and phosphorus efficiency in barley in a Mediterranean-like environment. Theor Appl Genet. 2016; 129: 1657-72.

Genome-wide association mapping of root system architecture traits in common wheat（*Triticum aestivum* L.）

Peng Liu. Yirong Jin. Jindong Liu. Caiyun Liu. Hongping Yao.
Fuyi Luo. Zhihui Guo. Xianchun Xia. Zhonghu He

P. Liu • Y. Jin • C. Liu • Z. Guo

Dezhou Academy of Agricultural Sciences，926 Dexing Road，Dezhou 253015，Shandong，China

J. Liu

Agricultural Genomics Institute at Shenzhen，Chinese Academy of Agricultural Sciences（CAAS），Pengfei Road，Shenzhen 518124，Guangdong，China

e-mail：liujindong _ 1990@163. com

H. Yao • F. Luo

Dezhou Bureau of Agriculture，2 Qinglong Road，Dezhou 253016，Shandong，China

X. Xia • Z. He

Institute of Crop Sciences，National Wheat Improvement Center，Chinese Academy of Agricultural Sciences（CAAS），12 Zhongguancun South Street，Beijing 100081，China

e-mail：hezhonghu02@caas. com

X. Xia • Z. He

International Maize and Wheat Improvement Center（CIMMYT）China Office，c/o CAAS，12 Zhongguancun South Street，Beijing 100081，China

Abstract：Uncovering the genetic basis and optimization of root system architecture（RSA）traits are crucial for modern wheat breeding. Genome-wide association mapping has become a powerful approach to dissect the genetic architecture of complex quantitative traits. In the present study，RSA traits，viz. total root length（TRL），total root area，average root diameter and number of root tips in a diverse panel of 165 elite wheat cultivars from the Yellow and Huai River Valley Facultative Wheat Region of China were evaluated as seedlings in hydroponic culture and in the field to identify loci significantly associated with those traits. The diverse panel was genotyped using the wheat 90 K and 660 K SNP arrays，and a genome-wide association study using a mixed linear model identified 28 and 4 loci significantly associated with RSA traits in hydroponic culture and in the field，explaining 8. 8-15. 6% and 8. 9-12. 6% of phenotypic variances，respectively. Seven loci for RSA traits colocated with known genes or quantitative trait loci（QTL），whereas the other 22 were potentially new. Linear regression between favorable alleles and RSA traits suggested that QTL pyramiding should be effective in optimizing root systems. Two candidate genes for RSA traits were identified，including genes encoding calcium dependent protein kinase and E3-Ubiquitin protein ligase. This study provides novel insights into the genetic architecture of RSA traits.

Published in Euphytica，2019，215：121.

Keywords: Bread wheat • GWAS • Marker-assisted selection • RSA traits • SNP arrays

Introduction

Common wheat (*Triticum aestivum* L.) is one of the most important staple food crops in the world (http: //www. fao. org/). Wheat productivity is largely limited by abiotic stresses, thus development of cultivars with high yield and good stability under abiotic stress has become an important objective in wheat breeding programs. Root traits are key to wheat cultivar performance and are not only important for the acquisition of nutrients and water from soil but also for tolerance to environmental stress (Osmont et al. 2007; Smith and De Smet 2012; Bishopp and Lynch 2015).

Root system architecture (RSA) traits define the shape of the root system and play important roles in plant growth and development, particularly under water and nutrient-deficient conditions (de Dorlodot et al. 2007; Hawkesford 2014; Bishopp and Lynch 2015; Mickelbart et al. 2015; Paez-Garcia et al. 2015). Root system morphology includes basic traits such as root length, root surface area, root volume, root diameter, and number of root tips (Kabir et al. 2015; Maccaferri et al. 2016). Root length, surface area and volume affect the spatial arrangement of the root structure, and are significantly associated with the uptake of nutrients and water (Maccaferri et al. 2016). Root diameter is associated with drought tolerance; thicker roots have large xylem vessels and were more efficient in deeply penetrating soil layers to extract water and nutrients (Maccaferri et al. 2016). In addition, fine roots constituting the major component of root systems and number of root tips increase the absorption efficiency of water and nutrients by increasing root surface area and volume (Osmont et al. 2007; Bishopp and Lynch 2015).

Most previous breeding programs have focusedon above-ground components, particularly increased harvest index, disease resistance and decreased plant height, whereas optimization of RSA traits was ignored (Osmont et al. 2007; Bishopp and Lynch 2015; Maccaferri et al. 2016), mainly due to the difficulty of high-throughput and cost-effective evaluation of RSA traits under field conditions (Mace et al. 2012; Maccaferri et al. 2016). However, several recent studies have dissected the genetic basis of RSA traits for rice (*Oryza sativa* L.)(Biscarini et al. 2016; Deshmukh et al. 2018), maize (*Zea mays* L.) (Pace et al. 2015; Sanchez et al. 2018), sorghum (*Sorghum bicolor* L.) (Mace et al. 2012) and wheat (Sanguineti et al. 2007; Rahnama et al. 2011; Ren et al. 2012; Bai et al. 2013; Liu et al. 2013; Cane et al. 2014; Narayanan et al. 2014; Kabir et al. 2015; Maccaferri et al. 2016). RSA traits are controlled by minor genes and are influenced by environmental conditions (Rahnama et al. 2011; Maccaferri et al. 2016). Although several studies reported the development and physiology of RSA traits, the genetic mechanisms controlling RSA traits remain limited. Clearly, it is necessary to uncover the genetic basis of RSA traits by linkage or association mapping studies in order to facilitate the optimization of RSA traits.

With the development of high-throughputgeno-typing, particularly the wheat 90 K (Wang et al. 2014) and 660 K (Cui et al. 2017) SNP assays, genome-wide association study (GWAS) has become a powerful approach for genetic dissection of complex quantitative traits in common wheat (Collard and Mackill 2008; Zhu et al. 2008), such as enduse quality (Marcotuli et al. 2015), disease resistance (Zegeye et al. 2014; Liu et al. 2017) and yield components (Rasheed et al. 2014; Sun et al. 2017). GWAS is an efficient and reliable tool for deciphering the molecular basis of complex quantitative traits.

In the present study, we performed a GWASfor RSA traits using the 90 K and 660 K SNP assays on a panel of 165 cultivars originating from the Yellow and Huai

Valley Region. Our objective was to dissect their genetic architecture and identify loci and molecular markers associated with the traits, and to identify candidate genes for RSA improvement.

Materials and methods

Plant materials

One hundred and sixty-five common wheat cultivars collected from the Yellow andHuai River Valley Facultative Wheat Region of China, were used for GWAS to dissect the genetic basis of RSA traits (Table S1).

A hydroponic experiment was arranged in randomized complete blocks with three replicates in a greenhouse at the Dezhou Academy of Agricultural Sciences. Before planting, 20 seeds of each cultivar were surface sterilized by soaking in 10% H_2O_2 for 10 min, followed by thrice washing with fresh water. Surface-sterilized seeds were put in Petri dishes with moist filter paper to allow germination. When coleoptiles were about 2 cm in length, seedlings were transferred to plastic trays (53×27 cm) containing Hoagland's nutrient solution (Hoagland and Arnon 1950). Plastic trays were kept in a growth chamber with a cycle of 16 h light and 8 h darkness (22-25 ℃). The nutrient solution was replaced by fresh solution every third day. After three weeks, roots were rinsed with sterile deionized water prior to recording RSA traits.

All 165 cultivars were grown in the field atthe Dezhou Academy of Agricultural Sciences Experimental Station (36. 24°N, 115. 45°E) in 2016-2017. Field trials were performed in randomized complete blocks with three replicates. Each plot contained one 2 m row with 25cm between rows and 10cm between plants. After 3 weeks, roots were removed from the soil, rinsed with fresh water before measuring RSA traits.

Phenotype evaluation and statistical analysis

Four RSA traits were evaluated, including total root length (TRL), total root area (TRA), average root diameter (ARD), and number of root tips (NRP). The WinRHIZO root analysis system (LA6400XL) (Regent Instruments Inc., CITY, Canada) consists of both image acquisition hardware and root analysis soft-ware. Roots were isolated from the plant and placed on a dish in an orderly manner to avoid excessive overlapping. The image was obtained through the compatible scanner Expression 11000XL and the resolution was set to 400 dpi. During analysis, the root images were imported into WinRHIZO software (www. regentinstruments. com) and analysed using a fixed threshold parameter of 40. Total root length calculated for all the roots and can be measured with the following formula: Length= (number of pixels in the root skeleton) × (pixel size). WinRhizo estimates average diameter from the total projected root area and length. Average diameter is calculated with the following formula: $Diam_{avg} =$ Projected area/ Total length. The formula is based on the assumption that roots are round. From this hypothesis, root height and surface area can be calculated from the measured diameter with trigonometric. Each trait was scored on five plants in each accession, and means of three replicates were used for subsequent statistical analysis and GWAS.

Basic statistical analyses were performed using Office Excel 2016 and SAS v9. 3 (http: //www. sas. com). Frequency distributions were calculated by SPSS Statistics 17. 0 software.

Genome-wide association analysis and prediction of candidate genes

Genetic diversity, population structure and linkage disequilibrium analysis of data for the panel followed Liu et al. (2017). Previous reports indicated that the 165 accessions could be divided into three subpopulations, which were largely consistent with pedigrees and geographic origins (Liu et al. 2017). The presence of population structure in an association panel could lead to spurious marker-trait associations (MTAs) (Pritchard et al. 2000; Yu and Buckler

2006). To eliminate spurious MTAs, associations among phenotypic data and genotypic factors were analyzed using the mixed linear model (MLM) (Q + K) in Tassel v5.0 (Bradbury et al. 2007). The Q (structure) and K (kinship) matrix were calculated by Structure v2.3.4 (Evanno et al. 2005) and Tassel v5.0, respectively. In this study, the Bonferroni-Holm correction for multiple testing (alpha = 0.05) was too conservative and no significant MTAs were detected. Thus, markers with an adjusted $-\log_{10}$ (P value) $\geqslant$3.0 were regarded as significantly associated (Houston et al. 2014; Gurung et al. 2014; Bellucci et al. 2015; Liu et al. 2017). Manhattan plots and Quantile-quantile (Q-Q) plots were drawn using the CMplot package (https://github.com/YinLiLin/R-CMplot) in R.

Alleles with positive effects leading to higher values of RSA traits were described as "favorable alleles", whereas those resulted with values were "unfavorable alleles". Candidate genes for RSA traits were identified according to Liu et al. (2017).

Results

Phenotypic evaluation

All five RSA traits in hydroponic culture and in the field exhibited continuous and significantly wide variation across the 165 wheat accessions (Figs. S1-S4). The mean values of TRL, TRA, ARD and NRT were higher in hydroponic culture (233.6, 25.6, 0.23 and 255.0) than in the field (116.2, 16.1, 0.18 and 83.5). The coefficients of variation (CV) for all RSA traits ranged from 26% to 34% for hydroponic culture and from 20% to 26% for the field (Table 1). To analyze the relationships among all RSA traits, correlation coefficients (r^2) between traits in the hydroponic culture (Table S2) and in the field (Table S3) were calculated. Significantly positive correlations were observed among TRL, TRA, ARD and NRT both in hydroponic culture and the field. Strong correlations between RSA traits in hydroponic culture and in the field indicated reliability of the phenotypic data.

Marker-trait associations

After removing SNPs with minor allele frequencies (MAF) <5% and missing data points>20%, 259,922 SNPs from the 90 K and 660 K SNP arrays were used for GWAS of RSA traits (Fig. S5). Marker density for the whole genome was 18.5 SNPs/Mb, indicating potential reliability for detecting MTAs based on LD decay (Liu et al. (2017).

Thirty-eight loci (159 MTAs) detected forRSA traits in hydroponic culture distributed across all chromosomes except 3D, 4D and 7A explained 9.3-25.8% of the phenotypic variances (Tables 2, S4; Figs. 1, S6). Among these loci, 11 for TRL were mapped to 1B, 2B, 2D, 3B, 4A, 5B and 7B, explaining 10.5-15.6% of the phenotypic variances. Twenty-four loci significantly associated with TRA on chromosomes 1A, 1B, 1D, 2B, 2D, 3A, 3B, 4A, 4B, 5B, 5D, 6B, 7B and 7D, accounted for 8.9-15.6% of the phenotypic variances. Several loci for ARD were located on chromosomes 1A, 1B, 1D, 2A, 2B, 2D, 3A, 4A, 4B, 5B, 5D, 6B, 7B and 7D, accounting for 8.8-15.6% of the phenotypic variances. Seven loci for NRT on chromosomes 1B, 2B, 2D and 5B explained 10.3-15.6% of the phenotypic variances.

Of these loci, two on chromosome 2B (4.5 Mb, 239.8 Mb) showed pleiotropic effects on TRL, TRA, ARD and NRT; 4 on chromosomes 1B (676.2 Mb), 2B (568.8 Mb), and 2D (145.7 Mb), and 5B (607.5 Mb) were pleiotropic for NRT, TRL, TRA and ARD; 3 on chromosomes 4B (1.1 Mb), 5B (654.9 Mb), and 5D (542.7 Mb) had pleiotropic effects on TRA and ARD; 3 on chromosomes 1B (193.1 Mb), 2D (285.4 Mb) and 4A (46.1 Mb) were significantly associated with TRA, ARD and TRL; TRA and ARD shared 8 common loci on chromosomes 1A (498.5 Mb), 1B (466.3 Mb), 1D (35.3 Mb), 2D (16.7 Mb), 3A (659.8 Mb), 6B (604.1 Mb), 7B (186.9 Mb), and 7D (181.3 Mb); and one on chromosome 3B (255.7 Mb) was a pleiotropic locus for TRL and TRA.

Four loci (68 MTAs) were detected for RSA traits in the field, distributing on chromosomes 1B, 2A, 2B and 2D and 5A, and explained 8.9-15.8% of the phenotypic variances (Tables3 and S5; Figs.2 and S7). Three loci for TRA on chromosomes 1B, 2B and 2D explained 8.9-12.6% of the phenotypic variances. Three on chromosomes 1B, 2A and 2D significantly associated with ARD accounted for 8.9-12.4% of the phenotypic variances. Two loci on chromosomes 1B (550.42 Mb) and 2D (15.965 Mb) showed a pleiotropic effect on ARD and TRA.

Several loci identified in hydroponic culture colocated with those detected in the field; for example, loci on chromosomes 1B (550.42 Mb) and 2B (788.14 Mb) significantly associated with TRA, and one locus on chromosome 2D (16.7 Mb) was pleiotropic for TRA and ARD.

Table 1 Basic statistics of root system architecture (RSA) traits for 165 wheat accessions grown in hydroponic culture and the field

Environment	Trait	Min	Max	Average	Standard deviation	Coefficient of variation (%)
Hydroponic culture	TRL (cm)	135.7	465.3	233.6	73.9	0.31
	TRA (cm^2)	8.2	47.4	25.6	7.4	0.29
	ARD (mm)	0.072	0.397	0.226	0.059	0.26
	NRT	54.0	451.4	225.0	77.1	0.34
Field	TRL (cm)	48.4	216.3	116.2	29.3	0.25
	TRA (cm^2)	7.3	29.8	16.1	4.1	0.25
	ARD (mm)	0.081	0.326	0.179	0.046	0.26
	NRT	46.5	133.5	83.5	16.6	0.20

TRL, total root length; *TRA*, total root area; *ARD*, average root diameter; *NRT*, number of root tips

Table 2 Markers significantly associated with root system architecture (RSA) traits in 165 wheat accessions grown in hydroponic culture

Trait	Marker[a]	Chromosome	Interval[b] (Mb)	SNP[c]	*P*-value	R^{2d} (%)
TRL	*AX_109924351*	1B	677.9-682.4	C	2.1E-05	14.8
	AX_111653240	2B	2.6-4.8	A	1.9E-05	14.0
	AX_111251784	2B	199.5	A	7.7E-05	12.2
	AX_94405934	2B	239.8	A	9.1E-05	11.9
	AX_108756976	2B	568.8	C	6.9E-06	15.6
	AX_110594265	2D	144.3-175.5	T	8.4E-06	15.5
	AX_111505152	2D	285.4	C	6.0E-05	10.5
	AX_109948487	3B	255.7	C	7.5E-05	12.1
	AX_111706221	4A	46.1	A	8.0E-05	12.0
	AX_109410506	5B	607.5	G	3.8E-05	13.1
	AX_108912470	7B	332.3	A	9.5E-05	11.9
TRA	*AX_110979981*	1A	498.5	A	8.6E-05	12.8
	AX_89723417	1B	466.3	C	5.4E-05	12.9
	AX_109875446	1B	550.7-555.1	A	1.1E-05	15.0
	AX_109924351	1B	675.8-682.4	C	2.1E-05	14.8

(continued)

Trait	Marker[a]	Chromosome	Interval[b] (Mb)	SNP[c]	*P*-value	R^{2d} (%)
	AX _ 110062330	1D	35. 3	C	7. 0E-05	12. 5
	AX _ 111653240	2B	2. 6-4. 8	A	1. 9E-05	14. 0
	AX _ 108881774	2B	193. 1-199. 5	A	6. 7E-05	12. 7
	AX _ 94405934	2B	239. 8	A	9. 1E-05	11. 9
	AX _ 108756976	2B	568. 8	C	6. 9E-06	15. 6
	AX _ 110977006	2B	793. 2	C	8. 2E-04	9. 6
	AX _ 111220010	2D	16. 7	A	6. 7E-04	9. 3
	AX _ 110594265	2D	145. 7-175. 5	T	8. 4E-06	15. 5
	AX _ 111505152	2D	285. 4	C	6. 0E-05	10. 5
	AX _ 111617187	3A	659. 8-660. 9	C	5. 6E-05	12. 7
	AX _ 109948487	3B	255. 7	C	7. 5E-05	12. 1
	AX _ 111706221	4A	46. 1	A	8. 0E-05	12. 0
	AX _ 108830894	4B	1. 1	A	5. 4E-05	12. 9
	AX _ 109410506	5B	604. 1-607. 9	G	3. 8E-05	13. 1
	AX _ 110914792	5B	654. 9	C	9. 2E-05	12. 2
	AX _ 110196801	5D	545. 7	A	8. 8E-05	12. 0
	AX _ 109584945	6B	604. 1	A	5. 6E-05	13. 4
	AX _ 95684378	7B	186. 9	C	4. 0E-05	13. 3
	AX _ 108912470	7B	332. 3	A	9. 5E-05	11. 9
	AX _ 111157278	7D	181. 3	A	5. 4E-05	13. 0
ARD	*AX _ 110979981*	1A	498. 5-503. 4	A	8. 6E-05	12. 8
	AX _ 89723417	1B	466. 3	C	5. 4E-05	12. 9
	AX _ 109924351	1B	675. 8-682. 4	C	2. 1E-05	14. 8
	AX _ 110062330	1D	35. 3-36. 5	C	7. 0E-05	12. 5
	AX _ 111107192	2A	43. 3-43. 5	C	8. 8E-05	12. 4
	AX _ 111653240	2B	2. 6-4. 8	A	1. 9E-05	14. 0
	AX _ 108881774	2B	193. 1-199. 5	A	6. 7E-05	12. 7
	AX _ 109897011	2B	238. 2-239. 8	C	8. 4E-05	12. 3
	AX _ 108756976	2B	568. 8	C	6. 9E-06	15. 6
	AX _ 111220010	2D	16. 7	A	6. 7E-04	9. 3
	AX _ 110594265	2D	145. 7-147. 2	T	8. 4E-06	15. 5
	AX _ 111505152	2D	285. 4	C	6. 0E-05	10. 5
	AX _ 111617187	3A	659. 8-660. 9	C	5. 6E-05	12. 7
	AX _ 111706221	4A	46. 1	A	8. 0E-05	12. 0
	AX _ 108830894	4B	1. 1-5. 0	A	5. 4E-05	12. 9
	AX _ 111047107	5B	438. 9	C	9. 2E-05	13. 5
	AX _ 109410506	5B	593. 4-607. 9	G	3. 8E-05	13. 1
	AX _ 110914792	5B	654. 9	C	9. 2E-05	12. 2
	AX _ 110196801	5D	545. 7	A	8. 8E-05	12. 0
	AX _ 109584945	6B	604. 1-604. 7	A	5. 6E-05	13. 4
	AX _ 95684378	7B	186. 9	C	4. 0E-05	13. 3
	AX _ 108912470	7B	332. 3	A	9. 5E-05	11. 9

（continued）

Trait	Marker[a]	Chromosome	Interval[b]（Mb）	SNP[c]	P-value	R^{2d}（%）
	AX_111157278	7D	181.3	A	5.4E-05	13.0
NRT	AX_108944064	1B	675.8-677.9	A	3.2E-05	13.6
	AX_111653240	2B	2.6-6.4	A	1.9E-05	14.0
	AX_94405934	2B	239.8	A	9.1E-05	11.9
	AX_108756976	2B	568.8	C	6.9E-06	15.6
	AX_110594265	2D	144.3-150.7	T	8.4E-06	15.5
	AX_108834280	2D	175.9	T	8.3E-05	12.4
	AX_109410506	5B	607.5	G	3.8E-05	13.1

TRL，total root length；TRA，total root area；ARD，average root diameter；NRT，number of root tips

[a]Representative marker at the resistance locus

[b]Physical position of SNP marker based on wheat genome sequences from the International Wheat Genome Sequencing Consortium（IWGSC，http：//www.wheatgenome.org/）

[c]Favorable allele（SNP）is underlined

[d]Percentage of phenotypic variance explained by the MTA from the results of Tassel v5.0

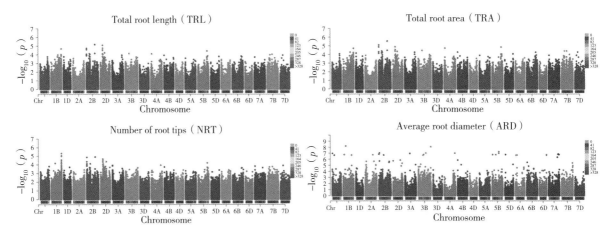

Fig.1　Manhattan plots for root system architecture（RSA）traits in 165 wheat accessions grown in hydroponic culture and analyzed by the mixed linear model（MLM）in Tassel v5.0. $-\log_{10}$（P）values from a genome-wide scan are plotted against positions on each of the 21 chromosomes

Table 3　Markers significantly associated with root system architecture（RSA）traits in 165 wheat accessions grown in the field

Trait[a]	Marker[b]	Chromosome	Position[c]（Mb）	SNP[d]	P-value	R^{e}（%）
TRA	AX_111531930	1B	549.9-550.9	C	9.5E-04	11.5
	AX_111174916	2B	788.1	C	8.8E-04	11.6
	AX_108872097	2D	15.6-18.6	C	5.0E-04	12.6
ARD	AX_110375130	1B	549.9-550.9	A	6.6E-04	12.3
	AX_95014398	2A	32.2	T	9.3E-04	11.5
	AX_109447682	2D	16.2	C	5.3E-04	12.4

[a]TRA，total root area；ARD，Average root diameter

[b]Representative marker at the resistance locus

[c]Physical position of SNP marker based on wheat genome sequences from the International Wheat Genome Sequencing Consortium（IWGSC，http：//www.wheatgenome.org/）

[d]Favorable allele（SNP）is underlined

[e]Percentage of phenotypic variance explained by the MTA from the results of Tassel v5.0

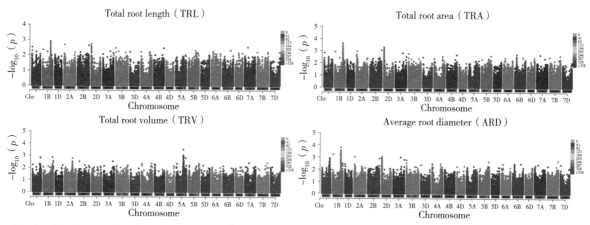

Fig. 2　Manhattan plots for root system architecture (RSA) traits in 165 wheat accessions grown in the field and analyzed by the mixed linear model (MLM) in Tassel v5.0. $-\log_{10}$ (*P*) values from a genome-wide scan are plotted against positions on each of the 21 chromosomes

Relationship between RSA traits and number of favorable alleles

To further understand the additive effects of alleles on RSA traits, we calculated the number of favorable and unfavorable alleles in each accession. In hydroponic culture, the numbers of favorable alleles in single accessions ranged from 0 to 9, 4 to 19, 3 to 16 and 0 to 7 for TRL, TRA, ARD and NRT, respectively (Fig. S8). Based on the MTA from the field trial, the number of favorable alleles in single accessions ranged from 0 to 3, and 1 to 4 for TRA and ARD, respectively (Fig. S9). Linear regression analysis identified a significant correlation between RSA traits and number of favorable alleles, with $r^2 = 0.83$, 0.88, 0.42 and 0.96 for TRL, TRA, ARD and NRT, respectively, in hydroponic culture (Fig. S8), and $r^2 = 0.87$ and 0.86 for TRA and ARD, respectively, in the field. Thus, accessions with higher numbers of favorable alleles and lower number of unfavorable alleles have better root systems.

To verify the reliable of the GWAS results, MTA from hydroponic were used to verify corresponding traits from field and vice versa. Linear regression analysis identified a certain extent between RSA traits from hydroponic and number of favorable alleles from field, with $r^2 = 0.59$ and 0.59 for TRA and ARD, respectively (Fig. S10). A certain extent between

RSA traits from field and number of favorable alleles from hydroponic were also founded, with $r^2 = 0.42$, 0.55, 0.51 and 0.48 for TRL, TRA, ARD and NRT (Fig. S11), respectively.

Candidate genes for RSA traits

Two candidate genes for RSA traits were identified (Table 4). Bioinformatics analysis indicated that *AX-110092021* on chromosome 5BL (604.1 Mb) corresponded to a calcium dependent protein kinase (CDPK), whereas *AX-95681859* on chromosome 1BL (676.2 Mb) corresponds to a E3-Ubiquitin protein ligase.

Discussion

Although optimization of crop root systems hasbeen proposed for a long time, genetic dissection and improvement of roots is rarely attempted (Bishopp and Lynch 2015; Maccaferri et al. 2016). A deep and comprehensive understanding of the genetic basis of RSA traits would help to improve root systems in high yielding wheat lines under conditions of water and/or nutrients deficiency. High-throughput genotyping and association mapping have paved the way for better understanding the genetic basis of complex quantitative traits like RSA (Flint-Garcia et al. 2003; Wang et al. 2014; Xu and Crouch 2008; Cui et al. 2017).

In this study, 38 and 5 loci, significantly associated

with RSA traits were respectively identified in hydroponic culture and the field（Fig. S3）. Comparisons of loci detected in this study and those reported previously based on the physical map of IWGSC V1.0 and integrated map constructed by Maccaferri et al. （2015）revealed several shared loci.

Total root length（TRL）

Loci involved in TRL were reported in common wheat （Ren et al. 2012; Cane et al. 2014; Kabir et al. 2015; Petrarulo et al. 2015; Maccaferri et al. 2016; Kabir et al. 2015）. Kabir et al. （2015）identified a QTL for TRL on chromosome 2B（*QTrl-2B*, *Xgwm501*）in a Nongda 3331 × Jindong 6 recombinant inbred line （RIL）population; this QTL overlapped with a locus identified on chromosome 2B（*AX_111251784*）in the present study. Another locus on chromosome 2B （*AX_108944064*）co-located with *qTRL-2B*（*Xgwm210*）identified in a Xiaoyan 54 × Jing 411 population by Ren et al. （2012）, and *QPRL-2B* （*Xbarc183*）detected by Cane et al. （2014）in a GWAS panel. The remaining eight loci located on chromosomes 1B（*AX_109824789*）, 2B（*AX_108756976*; *IWA1799*）, 2D（*AX_110594265*; *AX_111505152*）, 3B（*AX_109948487*）, 4A （*AX_111706221*）, 5B（*AX_109410506*）and 7B （*AX_108912470*）are likely to be new.

Total root area（TRA）

Maccaferri et al. （2016）identified four loci associated with TRA on chromosomes 2B（*QTrs. ubo-2B.1*, *wPt-7970*; *QTrs. ubo-2B*, *Xwmc441*）and 4B （*QTrs. ubo-4B*, *Xgwm6*）. These overlapped with present loci on chromosomes 2B（*AX_108881774*, *AX_108756976*）and 4B（*AX_108830894*）. Cane et al. （2014）detected QTL for TRA, explaining 4.5 to 9.4% of the phenotypic variances on chromosomes 1B（*QSRA1-1B*, *wPt-0655*）and 3A（*QSRA2-3A*, *Xbarc1177*）, corresponding to two loci identified in our study（1B, *AX_109875446* and 3A, *AX_111617187*）. The locus on chromosome 3A also overlapped with *QRSA-3A.3*（*IWB30296*）reported by Kabir et al. （2015）. The remaining 20 loci located

on chromosomes 1A, 1B（2）, 1D, 2B（4）, 2D （3）, 3B, 4A, 5B（2）, 5D, 6B, 7B（2）and 7D are likely to be new.

Average root diameter（ARD）

Maccaferri et al. （2016）found one locus for ARD on chromosomes 4B（*QTrd. ubo-4B*, *IWB74227*）in a Meridiano 9 Claudio RIL population and one locus on 6B（*QTrd. ubo-6B*, *IWB73246*）in a Colosseo 9 Lloyd population. *QTrd. ubo-4B* and *QTrd. ubo-6B* co-located with loci on chromosomes 4B（*AX_108830894*）and 6B（*AX_109584945*）in this study. Besides, Ren et al. （2012）and Cane et al. （2014）detected eight loci for ARD on chromosomes 2B, 3A, 3B and 5A. However, no overlapping was identified between those loci and the loci identified in our study. Accordingly, the other 22 loci may be new.

Number of root tips（NRT）

Iannucci et al. （2017）reported a locus associated with the NRT on chromosome 1B（*CAP11_c6406_104*, *655.2 Mb*）in a Simeto × Molise Colli RIL population, corresponding to our locus on chromosome 1B（*AX_108944064*, *676.2 Mb*）. Maccaferri et al. （2016）also found 5 loci for NRP on 2B, 3A, 4B and 5A in the Colosseo 9 Lloyd population, and 5 loci on 2A, 2B, 3A, 4B and 6B in the Meridiano 9 Claudio population, but there was no overlapping with loci identified in the present study. The six loci on chromosomes 2B（*AX_111653240*, *AX_94405934*, *AX_108756976*）, 2D（*AX_110594265*, *AX_108834280*）and 5D（*AX_109410506*）are likely to be new.

Although several MTAs identified in both hydroponic culture and field, most of the MTAS are different. The differences for MTAs between hydroponic and filed maybe due to RSA traits are highly affected by environment. However, the linear regression analysis between favorable alleles in hydroponic culture and RSA traits in field and favorable alleles in field and RSA traits in hydroponic culture indicated the GWAS results provide a relatively reliable basis for the RSA traits in wheat and could be used in wheat breeding. Besides, although the $-\log_{10}$ (P-value) $\geqslant$ 3.0 were set for the identification of MTA in thisstudy as previous studies (Houston et al. 2014; Gurung et al. 2014; Bellucci et al. 2015; Liu et al. 2017), this constitutes a high risk of identifying false positives. Thus, GWAS based on haplotype analysis were conducted in the future.

Genomic regions for RSA traits and agronomic traits

Previous studies reported that optimized RSA traits were important for agronomic performance ofcrops, such as soybean (Song et al. 2016), rice (Ju et al. 2015), common wheat (Maccaferri et al. 2016), and maize (Mu et al. 2015). The coincidence of genomic regions for RSA and other agronomic traits indicated the effects of RSA traits in agronomic performance. We also performed GWAS for agronomic traits in the present wheat accessions (Li et al. unpublished) and identified several genomic regions associated with both RSA traits and agronomic

traits. The loci for RSA traits on chromosomes 1B (*AX _ 108944064*, 676.2 Mb), 2A (*AX _ 111107192*, 27.3 Mb;) and 4A (*AX _ 111706221*, 46.1 Mb) were co-located with regions affecting plant height in 1B (*AX _ 94564150*, 539.6 Mb; 668.1 Mb), 2A (*AX _ 110988136*, 30.3 Mb; *AX _ 94494373*, 715.3 Mb), and 4A (*AX _ 109840724*, 41.1 Mb). Two loci associated with RSA traits on chromosomes 1B (*AX _ 108944064*, 675.8 Mb) and 2A (*AX _ 111107192*, 27.3 Mb), overlapped with loci affecting grain yield (GY) in 1B (*AX _ 109820171*, 673.6 Mb) and 2A (*AX _ 94546135*, 30.3 Mb). Genomic regions related to RSA traits on chromosomes 5B (*AX _ 109410506*, 593.4 Mb) and 6B (*AX _ 109584945*, 604.1 Mb) were co-located with loci affecting grain-related traits, including kernel length and spike number per square meter on chromosomes 5B (*AX _ 95632082*, 594.2 Mb) and 6B (*IWB51629*, 605.8 Mb). These results indicated that the loci for RSA traits could also be selected to enhance yield potential and stability by marker assisted selection (MAS).

Table 4 Candidate genes for SNPs significantly associated with root system architecture (RSA) traits

Marker	Chromosome	Position (Mb)	Candidate gene	Sequence coverage (%)	Quality parameter
AX-95681859	1BL	676.2	E3-Ubiquitin protein ligase	97	7 E^{-08}
AX-110092021	5BL	448.3	Calcium dependent protein kinase	97	7 E^{-07}

The evaluation of RSA traits in hydroponic culture and the field

The measurement of root traits under field conditions is extremely difficult. Various laboratory methods have been applied to studies of RSA traits; these include hydroponic culture (Ren et al. 2012), gelchamber-based observation systems (Liu et al. 2013), and a 'cigar roll' system (Zhu et al. 2005). Of these, hydroponic culture has been widely used to study RSA traits in crop species, such as rice (Biscarini et al. 2016), maize (Zhu et al. 2005; Pace et al. 2015) and wheat (Ren et al. 2012; Liu et al. 2013; Maccaferri et al. 2016). Previous studies were conducted to correlate laboratory results from seedlings with field performance (Tuberosa et al. 2002; An et al. 2006; Su et al. 2009). Tuberosa et

al. (2002) reported that some loci associated with RSA traits detected in hydroponic culture overlapped with those affecting grain yield in maize. An et al. (2006) and Su et al. (2009) indicated that loci for RSA traits detected in the hydroponic culture coincided with those affecting water and nutrient uptake in wheat under field conditions. Bai et al. (2013) found that QTL for RSA traits identified in seedlings grown in hydroponic culture corresponded with genes for plant height detected in the field. For these reasons we used a hydroponic system to measure RSA traits in order to improve selection efficiency in wheat breeding.

Candidate genes for RSA traits

Two candidate genes for RSA traits were identified (Table4). Bioinformatics analysis indicated that SNP *AX-110092021* on chromosome 5BL (604.1 Mb) corresponded to a calcium dependent protein kinase (CDPK) in *Triticum urartu*. CDPKs in plants are involved in various signaling pathways and play a critical role in root growth and development (Ludwig et al. 2004; Gargantini et al. 2006; Asano et al. 2012). Ivashuta et al. (2005) reported that silencing of CDPK in *Medicago truncatula* significantly reduced root hair growth and root cell length. *AX-95681859* on chromosome 1BL (676.2 Mb) corresponds to a E3-Ubiquitin protein ligase. E3-Ubiquitin protein ligases are a large protein family that is important in plant growth and development (Cyr et al. 2002; Craig et al. 2009). Park et al. (2010) indicated that E3-Ubiquitin protein ligase are essential to the growth of primary roots and shoot development.

Application of MTAs to optimize RSA traits in breeding

It is impossible to select root traits in conventional wheat breeding because of difficulties in measuring RSA traits in the field. Although conventional breeding has led to improved root systems on wheat, selection is time-consuming and not very efficient (Holland 2007). Molecular markers associated with RSA detected in this study should facilitate MAS. Significant additive effects were identified between RSA traits and a number of favorable alleles, indicating that pyramiding favorable alleles will improve RSA traits. Loci with pleiotropic and consistent effects in both hydroponic culture and field trials should be amenable to MAS. In addition, accessions with superior RSA traits, high numbers of favorable alleles and suitable agronomic traits, such as Bainong 64, Jimai 22, Jinan 17, Yannong 15, Yumai 34, Yumai 47 and Zhengyin 1 are recommended as parental lines for improvement of RSA traits in wheat breeding.

Conclusion

In the present study, 38 and 5 loci significantly associated with RSA traits were identified in hydroponic culture and the field, respectively, confirming the reliability and efficiency of GWAS for identification of loci affecting quantitative traits. Markers significantly associated with RSA traits and cultivars already having superior RSA traits can be used to make further advances in varietal improvement. Our study provides a comprehensive survey of the loci for RSA traits and provides a basis to initiate gene cloning and improvement of wheat root systems.

❖ Acknowledgements

We are grateful to Prof. R. A. McIntosh, Plant Breeding Institute, University of Sydney, for critical review of this manuscript. This work was implemented bythe Key Research and Development Plan of Shandong Province (2017GNC10107), the Key Laboratory of Biology, Genetics & Breeding for Triticeae Crops, Ministry of Agriculture and Rural Affairs, the Natural Science Foundation of Shandong Province (ZR2017BC015), and the National Natural Science Foundation of China (31671691).

❖ References

An DG, Su JY, Liu QY, Li B, Jing RL, Li JY, Li ZS (2006) Mapping QTLs for nitrogen uptake in relation to the early growth of wheat (*Triticum aestivum* L.). Plant Soil 284: 73-84.

Asano T, Hayashi N, Kikuchi S, Ohsugi R (2012) CDPK-mediated abiotic stress signaling. Plant Signal Behav 7: 817-821.

Bai C, Liang Y, Hawkesford MJ (2013) Identification of QTLs associated with seedling root traits and their correlation with plant height in wheat. J Exp Bot 64: 1745-1753.

Bellucci A, Torp AM, Bruun S, Magid J, Andersen SB, Ras-mussen SK (2015) Association mapping in scandinavian winter wheat for yield, plant height, and traits important for second-generation bioethanol production. Front Plant Sci 6: 1046.

Biscarini F, Cozzi P, Casella L, Riccardi P, Vattari A, Orasen G, Perrini R, Tacconi G, Tondelli A, Biselli C, Cattivelli L, Spindel J, McCouch S, Abbruscato P, Vale' G, Piffanelli P, Greco R (2016) Genome-wide association study for traits related to plant and grain morphology, and root architec-ture in temperate rice accessions. PLoS ONE 11: e0155425.

Bishopp A, Lynch JP（2015）The hidden half of crop yields. Nat Plants 1: 15117.

Bradbury PJ, Zhang Z, Kroon DE, Casstevens TM, Ramdoss Y, Buckler ES（2007）TASSEL. Software for association mapping of complex traits in diverse samples. Bioinformatics 23: 2633-2635.

Cane MA, Maccaferri M, Nazemi G, Salvi S, Francia R, Colalongo C, Tuberosa R（2014）Association mapping for root architectural traits in durum wheat seedlings as related to agronomic performance. Mol Breeding 34: 1629-1645.

Collard BC, Mackill DJ（2008）Marker-assisted selection: an approach for precision plant breeding in the twenty-first century. Philos Trans R Soc B 363: 557-572.

Craig A, Ewan R, Mesmar J, Gudipati V, Sadanandom A（2009）E3 ubiquitin ligases and plant innate immunity. J Exp Bot 60: 1123-1132.

Cui F, Zhang N, Fan XL, Zhang W, Zhao CH, Yang LJ, Pan RQ, Chen M, Han J, Zhao XQ, Ji J, Tong YP, Zhang HX, Jia JZ, Zhao GY, Li JM（2017）Utilization of a wheat 660 K SNP array-derived high-density genetic map for high-resolution mapping of a major QTL for kernel number. Sci Rep-UK 7: 3788.

Cyr DM, Höhfeld J, Patterson C（2002）Protein quality control: U-box-containing E3 ubiquitin ligases join the fold. Trends Biochem Sci 27: 368-375.

de Dorlodot S, Forster B, Pagès L, Price A, Tuberosa R, Draye X（2007）Root system architecture: opportunities and constraints for genetic improvement of crops. Trends Plant Sci 12: 474-481.

Deshmukh V, Mankar SP, Muthukumar C, Divahar P, Bharathi A, Thomas HB, Ashish R, Reena S, Poornima R, Senthivel S, Chandra BR（2018）Genome-wide consistent molecular markers associated with phenology, plant production and root traits in diverse rice（*Oryza sativa* L.）accessions under drought in rainfed target populations of the environment. Curr Sci India 114: 329-340.

Evanno G, Regnaut S, Goudet J（2005）Detecting the number of clusters of individuals using the software STRUCTURE: a simulation study. Mol Ecol 14: 2611-2620.

Flint-Garcia SA, Thornsberry JM, Buckler ES（2003）Structure of linkage disequilibrium in plants. Annu Rev Plant Biol 54: 357-374.

Gargantini PR, Gonzalez-Rizzo S, Chinchilla S, Raices M, Giammaria V, Ulloa RM, Frugier F, Crespi MD（2006）A CDPK isoform participates in the regulation of nodule number in *Medicago truncatula*. Plant J 48: 843-856.

Gurung S, Mamidi S, Bonman JM, Xiong M, Brown-Guedira G, Adhikari TB（2014）Genome-wide association study reveals novel quantitative trait loci associated with resistance to multiple leaf spot diseases of spring wheat. PLoS ONE 9: e108179.

Hawkesford MJ（2014）Reducing the reliance on nitrogen fer-tilizer for wheat production. J Cereal Sci 59: 276-283.

Hoagland DR, Arnon DI（1950）The water-culture method for growing plants without soil. Calif Agric Exp Stn Circ 347: 1-32.

Holland JB（2007）Genetic architecture of complex traits in plants. Curr Opin Plant Biol 10: 156-161.

Houston K, Russell J, Schreiber M, Halpin C, Oakey H, Washington JM, Booth A, Shirley N, Burton RA, Fincher GB, Waugh R（2014）A genome wide association scan for（1, 3; 1, 4）-b-glucan content in the grain of contemporary 2-row Spring and Winter barleys. BMC Genom 15: 907.

Iannucci A, Marone D, Russo MA, Vita PD, Miullo V, Ferragonio P, Blanco A, Gadaleta A, Mastrangelo AM（2017）Mapping QTL for root and shoot morphological traits in a durum wheat × *T. dicoccum* segregating population at seedling stage. Int J Genom 2017: 6876393.

Ivashuta S, Liu J, Lohar DP（2005）RNA interference identifies a calcium-dependent protein kinase involved in *Medicago truncatula* root development. Plant Cell 17: 2911-2921.

Ju C, Buresh RJ, Wang Z, Zhang H, Liu L, Yang J, Zhang J（2015）Root and shoot traits for rice varieties with higher grain yield and higher nitrogen use efficiency at lower nitrogen rates application. Field Crops Res 175: 47-55.

Kabir MR, Liu G, Guan PF, Wang F, Khan AA, Ni ZF, Yao YY, Hu ZR, Xin MM, Peng HR（2015）Mapping QTLs associated with root traits using two different populations in wheat（*Triticum aestivum* L.）. Euphytica 206: 175-190.

Liu X, Li R, Chang X, Jing R（2013）Mapping QTLs for seedling root traits in a doubled haploid wheat population under different water regimes. Euphytica 189: 51-66.

Liu JD, He ZH, Rasheed A, Wen WE, Yan J, Zhang PZ, Wan YX, Zhang Y, Xie CJ, Xia XC（2017）Genome-wide association mapping of black point reaction in common wheat（*Triticum aestivum* L.）. BMC Plant

Biol 17：220.

Ludwig AA，Romeis T，Jones JD（2004）CDPK-mediated signaling pathways：specificity and cross-talk. J Exp Bot 55：181-188.

Maccaferri M，Zhang J，Bulli P，Abate Z，Chao S，Cantu D，Bossolini E，Chen X，Pumphrey M，Dubcovsky J（2015）A genome-wide association study of resistance to stripe rust（Puccinia striiformis f. sp. tritici）in a worldwide collection of hexaploid spring wheat（*Triticum aestivum* L.）. G3（Bethesda）5（3）：449-465.

Maccaferri M，El-Feki W，Nazemi G，Salvi S，Cane` MA，Colalongo MC，Stefanelli S，Tuberosa R（2016）Prioritizing quantitative trait loci for root system architecture in tetraploid wheat. J Exp Bot 67：1161-1178.

Mace ES，Singh V，VanOosterom EJ，Hammer GL，Hunt CH，Jordan DR（2012）QTL for nodal root angle in sorghum（*Sorghum bicolor* L.）co-locate with QTL for traits associated with drought adaptation. Theor Appl Genet 124：97-109.

Marcotuli I，Houston K，Waugh R，Fincher GB，Burton RA，Blanco A，Gadaleta A（2015）Genome-wide association mapping for arabinoxylan content in a collection of tetraploid wheats. PLoS ONE 10：0132787.

Mickelbart MV，Hasegawa PM，Bailey-Serres J（2015）Genetic mechanisms of abiotic stress tolerance that translate to crop yield stability. Nat Rev Genet 16：237-251.

Mu X，Chen F，Wu Q，Chen Q，Wang J，Yuan L，Mi G（2015）Genetic improvement of root growth increases maize yield.

via enhanced post-silking nitrogen uptake. Eur J Agron 63：55-61.

Narayanan S，Mohan A，Gill KS，Prasad PV（2014）Variabilityof root traits in spring wheat germplasm. PLoS ONE 9：e100317.

Osmont KS，Sibout R，Hardtke CS（2007）Hidden branches：developments in root system architecture. Annu Rev Plant Biol 58：93-113.

Pace J，Gardner C，Romay C，Ganapathysubramanian B，Lu¨bberstedt T（2015）Genome-wide association analysis of seedling root development in maize（*Zea mays* L.）. BMC Genom 16：47.

Paez-Garcia A，Motes CM，Scheible WR，Chen R，Blancaflor EB，Monteros MJ（2015）Root traits and phenotyping strategies for plant improvement. Plants 4：334-355.

Park GG，Park JJ，Yoon JM，Yu SN，An GH（2010）A RING finger E3 ligase gene，*Oryza sativa* delayed seed germination 1（*OsDSG*1），controls seed germination and stress responses in rice. Plant Mol Biol 74：467-478.

Petrarulo M，Marone D，Ferragonio P，Cattivelli L，Rubiales D，Vita PD，Mastrangelo AM（2015）Genetic analysis of root morphological traits in wheat. Mol Genet Genom 290：785-806.

Pritchard JK，Stephens M，Rosenberg NA，Donnelly P（2000）Association mapping in structured populations. Am J Hum Genet 67：170-181.

Rahnama A，Munns R，Poustini K，Watt M（2011）A screening method to identify genetic variation in root growth response to a salinity gradient. J Exp Bot 62：69-77.

Rasheed A，Xia XC，Ogbonnaya F，Mahmood T，Zhang Z，Mujeeb-Kazi AM，He ZH（2014）Genome-wide association for grain morphology in synthetic hexaploid wheats using digital imaging analysis. BMC Plant Biol 14：128.

Ren R，He X，Liu D，Li J，Zhao X，Li B，Li Z（2012）Majorquantitative trait loci for seminal root morphology of wheat seedlings. Mol Breed 30：139-148.

Sanchez DL，Liu S，Ibrahim R，Blanco M，Lu¨bberstedt T（2018）Genome-wide association studies of doubled haploid exotic introgression lines for root system architecture traits in maize（*Zea mays* L.）. Plant Sci 268：30-38.

Sanguineti MC，Li S，Maccaferri M，Corneti S，Rotondo F，Chiari T，Tuberosa R（2007）Genetic dissection of seminal root architecture in elite durum wheat germplasm. Ann Appl Biol 151：291-305.

Smith S，Smet ID（2012）Root system architecture：insights from Arabidopsis and cereal crops. Philos Trans R Soc B 367：1441-1452.

Song L，Prince S，Valliyodan B，Joshi T，Santos JV，Wang JJ，Lin L，Wan JR，Wang YQ，Xu D，Nguyen HT（2016）Genome-wide transcriptome analysis of soybean primary root under varying water-deficit conditions. BMC Genom 17：57.

Su YH，Mcgrath SP，Zhao FJ（2009）Rice is more efficient in arsenite uptake and translocation than wheat and barley. Plant Soil 328：27-34.

Sun CW，Zhang FY，Yan XF，Zhang XF，Dong ZD，Cui DQ，Chen F（2017）Genome-wide association study for 13 agronomic traits reveals distribution of superior alleles in bread wheat from the yellow andHuai Valley of China. Plant Biotechnol J 8：953-969.

Tuberosa R, Sanguineti MC, Landi P, Giuliani MM, Salvi S, Conti S (2002) Identification of QTLs for root character-istics in maize grown in hydroponics and analysis of the overlap with QTLs for grain yield in the field at two water regimes. Plant Mol Biol 48: 697-712.

Wang SC, Wong D, Forrest K, Allen A, Chao S, HuangBE, Maccaferri M, Salvi S, Milner SG, Cattivelli L, Mastrangelo AM, Whan A, Stephen S, Barker G, Wieseke R, Plieske J, Lillemo M, Mather D, Appels R, Dolferus R, Guedira GB, Korol A, Akhunova AR, Feuillet C, Salse J, Morgante M, Pozniak C, Luo MC, Dvorak J, Morell M, Dubcovsky J, Ganal M, Tuberosa R, Lawley C, Mikoulitch I, Cavanagh C, Edwards KJ, Hayden M, Akhunov E (2014) Characterization of polyploidy wheat genomic diversity using a high-density 90000 single nucleotide polymorphism array. Plant Biotechnol J 12: 787-796.

Xu Y, Crouch JH (2008) Marker-assisted selection inplant breeding: from publications to practice. Crop Sci 48: 391-407.

Yu J, Buckler ES (2006) Genetic association mappingand genome organization of maize. Curr Opin Biotechnol 17: 155-160.

Zegeye H, Rasheed A, Makdis F, Badebo A, Ogbonnaya FC (2014) Genome-wide association mapping for seedling and adult plant resistance to stripe rust in synthetic hexaploid wheat. PLoS ONE 9: e105593.

Zhu J, Kaeppler SM, Lynch JP (2005) Mapping of QTLs for lateral root branching and length in maize (*Zea mays* L.) under differential phosphorus supply. Theor Appl Genet 111: 688-695.

Zhu CS, Gore M, Buckler ES, Status YJM (2008) Prospects of association mapping in plants. Plant Genome 1: 5-20.

抗病性与黑胚

Molecular mapping of quantitative trait loci for *fusarium* head blight resistance in a doubled haploid population of Chinese bread wheat

Zhanwang Zhu,[1,2] Xiaoting Xu,[1] Luping Fu,[1] Fengju Wang,[1]
Yachao Dong,[1] Zhengwu Fang,[3] Wenxue Wang,[2] Yanping Chen,[3]
Chunbao Gao,[2] Zhonghu He,[1,4] Xianchun Xia,[1] and Yuanfeng Hao[1,†]

[1] Institute of Crop Sciences, Chinese Academy of Agricultural Sciences, Beijing 100081, China

[2] Hubei Key Laboratory of Food Crop Germplasm and Genetic Improvement, Wheat Disease Biology Research Station for Central China, Food Crops Institute, Hubei Academy of Agricultural Sciences, Wuhan, Hubei 430064, China

[3] College of Agriculture, Yangtze University, Jingzhou, Hubei 434000, China

[4] International Maize and Wheat Improvement Center (CIMMYT)-China Office, Beijing 100081, China

† Corresponding author: haoyuanfeng@caas.cn

Abstract: Fusarium head blight (FHB) is a destructive disease of wheat worldwide, particularly in China. To map genetic loci underlying FHB resistance, a doubled haploid (DH) population consisting of 174 lines was developed from a cross between widely grown Chinese cultivars Yangmai 16 and Zhongmai 895. The DH population and parents were evaluated in field nurseries at Wuhan in 2016 to 2017 and 2017 to 2018 crop seasons with both spray inoculation and natural infection, and at Jingzhou in 2017 to 2018 crop season with grain-spawn inoculation. The DH lines were genotyped with a wheat 660K SNP array. The FHB index, plant height, anther extrusion, and days to anthesis were recorded and used for quantitative trait loci (QTL) analysis. Seven QTL for FHB resistance were mapped to chromosome arms 3BL, 4AS, 4BS, 4DS, 5AL, 6AL, and 6BS in at least two environments. *QFhb. caas-4BS* and *QFhb. caas-4DS* co-located with semi-dwarfing alleles *Rht-B1b* and *Rht-D1b*, respectively, and were associated with anther extrusion. The other five QTL were genetically independent of the agronomic traits, indicating their potential value when breeding for FHB resistance. Based on correlations between FHB indices and agronomic traits in this population, we concluded that increasing plant height to some extent would enhance FHB resistance, that anther extrusion had a more important role in environments with less severe FHB, and that days to anthesis were independent of the FHB response when viewed across years. PCR-based markers were developed for the 3BL and 5AL QTL, which were detected in more than three environments. The InDel marker *InDel _ AX-89588684* for *QFhb. caas-5AL* was also validated on a wheat panel, confirming its effectiveness for marker-assisted breeding for improvements in FHB resistance.

Keywords: *Fusarium graminearum*, QTL mapping, *Triticum aestivum*

Published in Plant Disease, 2021, 105 (05): 1339-1345. (doi. org/10. 1094/PDIS-06-20-1186-RE)

Fusarium head blight (FHB), or scab, mainly caused by the fungus *Fusarium graminearum* Schwabe, is an economically important disease of wheat (*Triticum aestivum* L.) and other cereals particularly in warm and humid regions (Bai et al. 2018). Epidemics of FHB reduce grain yield, processing quality, and seed quality (Bai and Shaner 1994; Dexter et al. 1996). Infected grains contain toxins, such as deoxynivalenol (DON) and 3-O-acetyl-DON (3-AcDON), which can make the grain unsuitable for food and feed (Foroud et al. 2019; McMullen et al. 1997; Xu et al. 2019). In China, FHB is a long-term, major wheat disease in the Middle and Lower Yangtze River Valleys, and it has recently become prevalent in the Yellow and Huai River Valleys, which are the most important wheat-producing areas in China, because of the predominant wheat-maize cropping system, stubble retention with reduced tillage, and effects of climate change (Zhu et al. 2018, 2019). The area severely affected by FHB in China each year now exceeds 5 million hectares (Chen et al. 2017).

Wheat resistance to FHB is a quantitative trait controlled by minor quantitative trait loci (QTL), and it is largely affected by the environment (Bai and Shaner 1994). More than 400 QTL for FHB resistance were allocated to 44 chromosomal regions (Buerstmayr et al. 2009, 2020; Liu et al. 2009; Ma et al. 2020). However, only *Fhb1* and *Fhb7* have been cloned; *Fhb1* was shown to be a histidine-rich calcium-binding (*TaHRC* or *His*) gene (Li et al. 2019; Su et al. 2019), and *Fhb7* is a glutathione S-transferase (GST) gene (Wang et al. 2020). The FHB response is often confounded by morphological and phenological traits of the host plant (Bai et al. 2018). Taller plants and higher anther extrusion (AE) rates usually lead to better resistance. Approximately 40% of the QTL for plant height (PH) and 60% of the QTL for AE overlap with QTL for the FHB response (Buerstmayr et al. 2020). Flowering time is a key stage for FHB infection, and its effect is highly variable (Buerstmayr et al. 2020).

Opportunities for agronomists and farmers to take advantage of these variables are very limited; taller genotypes are more prone to lodging, which is already a serious problem in China, and flowering dates are restricted to narrow bands to meet environmental or double cropping system limitations.

Wheat germplasm resistant to FHB has been identified inEast Asia, South America, North America, and Europe (Liu and Anderson 2003; Yu et al. 2008a). Among Chinese sources, many landraces and historical cultivars and some modern cultivars are characterized as FHB-resistant (Yu et al. 2006; Zhu et al. 2019). Extensive genetic studies have been conducted on landraces, such as Wangshuibai (Yu et al. 2008b), Haiyanzhong (Li et al. 2011), Huangcandou (Cai and Bai 2014), Baishanyuehuang (Zhang et al. 2012), and Huangfangzhu (Li et al. 2012), and historical wheat cultivars or lines, such as Sumai 3 (Waldron et al. 1999), Ning 7840 (Bai et al. 1999), Wuhan 1 (Somers et al. 2003), CJ9306 (Jiang et al. 2007a, b), Huapei 57-2 (Bourdoncle and Ohm 2003), W14 (Chen et al. 2006), and AQ 24788-83 (Ren et al. 2019). However, genetic studies of Chinese modern cultivars remain limited.

Because of the many additive genes involved in the FHB response in wheat and the wide range of phenotypic responses observed during trials, it is highly likely that breeders could select transgressive segregants with improved resistance in crosses of unrelated elite parents with intermediate FHB responses. For example, Yangmai 16 and Zhongmai 895 are widely grown wheat cultivars that show moderately resistant and moderately susceptible responses to FHB, respectively. To test the prediction of transgressive segregation, a doubled haploid (DH) population developed from a cross of Yangmai 16 × Zhongmai 895 was assessed to determine the FHB response during field trials involving different methods of disease induction. The agronomic traits, including PH, AE, and days to anthesis (DTA), previously shown to be related to the FHB response were also investigated.

Materials and Methods

Plant materials. Yangmai 16 (*Rht-B1b/Rht-D1a*, *vrn-A1/vrn-B1/Vrn-D1*, *Ppd-A1a/Ppd-B1a/Ppd-D1a*), a spring wheat cultivar released in 2004, is a major cultivar in the Middle and Lower Yangtze River Valleys. Zhongmai 895 (*Rht-B1a/Rht-D1b*, *vrn-A1/vrn-B1/vrn-D1*, *Ppd-A1a/Ppd-B1a/Ppd-D1a*), a facultative wheat cultivar released in 2012, is widely grown in the Yellow and Huai River Valleys. The *Rht*, *Vrn*, and *Ppd* genotypes have been reported by Rasheed et al. (2016). In the present study, 174 DH lines from Yangmai 16 × Zhongmai 895 were used. Cultivars Sumai 3, Zhengmai 9023, and Zhoumai 22 were included as resistant, moderately resistant, and susceptible checks, respectively.

A wheat association panel for scab (WAPS) research comprising 192 geographically diverse cultivars and elite lines was used to validate the QTL identified in the Yangmai 16 × Zhongmai 895 population. Detailed information about this panel has been reported previously (Zhu et al. 2020).

Field trials and agronomic traits measurement. The DH population and parents were planted at Nanhu Experimental Station, Hubei Academy of Agricultural Sciences in Wuhan (30.48° N; annual rainfall, 1, 270 mm) during the 2016 to 2017 and 2017 to 2018 crop seasons, and at Yangtze University in Jingzhou (Hubei province)(30.36°N; annual rainfall, 1, 200 mm) during the 2018 to 2019 crop season. Both sites have favorable conditions for FHB development. All trials were arranged in randomized complete blocks with two replications. Each plot comprised 2-m×1-m rows with 25-cm spacing.

Agronomic traits were only recorded at Wuhan. PH was recorded approximately 2 weeks after anthesis, and an average value was obtained from two measures. The AE was visually rated using a linear scale from 0 (no extrusion) to 9 (full extrusion) (Skinnes et al. 2010) at approximately 3 days after anthesis on a whole-plot basis. DTA was recorded when 50% of spikes per plot were flowering.

Pathogen inoculation and FHB scoring. An adequate level of disease is required for a genetic study of disease response. Therefore, three different inoculation methods were used to test the DH population and parents. Both spray inoculation and natural infection methods were used at Wuhan, and the corresponding environments were named 2017 Wuhan-spray, 2017 Wuhan-natural, 2018 Wuhan-spray, and 2018 Wuhan-natural. Two isolates of *F. graminearum*, Huanggang 1 and Fg 5035, were mixed for the spray inoculation, and the inocula were prepared according to the methods of Zhu et al. (2020). Briefly, mycelia preserved on potato dextrose agar (PDA) medium at − 20℃ were inoculated in a 5% mung bean broth medium with shaking at 180 rpm under 28℃ for 5 to 7 days to produce conidia. Ten spikes at anthesis within approximately 25 cm per plot were labeled and inoculated with approximately 30 ml *F. graminearum*-water suspension at a concentration of 50, 000 spores/ ml. To favor infection and disease development, both trials were equipped with an overhead misting system of 1.5-m-high microsprinklers spaced at 1.5 m × 1.5 m and operated automatically by a programable timer from 9: 00 AM to 7: 00 PM, with 2 minutes of misting per hour. The third method used at Jingzhou in 2018 (2018 Jingzhou) was grain-spawn inoculation by scattering scabby wheat grains producing ascospores on the soil surface at Zadoks growth stage 33 (Zadoks et al. 1974) approximately 1 month before anthesis. The FHB indices under five environments were collected and used for subsequent analyses.

For spray-inoculated trials, the labeled spikes were assessed21 days after inoculation to determine FHB incidence and severity. The FHB index was calculated using the following formula:

index (%) = (incidence×severity) /100

(Stack and McMullen 1994), where incidence is the

percentage of FHB-infected spikes and severity is the average percentage of diseased spikelets per infected spike. During natural infection and grain-spawn inoculated trials, the FHB index for each plot was scored visually 24 days after anthesis (Zhu et al. 2020). The FHB indices of wheat accessions in the WAPS population were evaluated in field nurseries at Wuhan during the 2013 to 2014, 2014 to 2015, and 2015 to 2016 crop seasons with spray inoculation using the same protocol as that mentioned here.

Statistical analysis. An analysis of variance (ANOVA), correlation coefficients, and t tests were performed using SAS 9.4 software (SAS Institute, NC). The ANOVA was performed with the PROC MIXED, with environment considered as a fixed effect and genotype, genotype × environment interaction, replication, and residual error considered random. Broad-sense heritability (H^2) for the FHB index was calculated using the following formula:

$$H^2 = \sigma_G^2 / (\sigma_G^2 + \frac{\sigma_{GE}^2}{e} + \frac{\sigma_\varepsilon^2}{re})$$

where σ_G^2 is the genotypic effect, σ_{GE}^2 is the genotype × environment effect, σ_ε^2 is the residual error, and e and r are the numbers of environments and replicates, respectively. The best linear unbiased estimator (BLUE) values were estimated using the ANOVA function in IciMapping v4.1 (Meng et al. 2015).

Genotyping, genetic map construction, and QTL mapping. The DH population and parents were genotyped with the wheat 660K SNP array and 13 polymorphic kompetitive allele-specific PCR (KASP) markers for known genes, including *Rht-B1 _ SNP* for *Rht-B1* and *Rht-D1 _ SNP* for *Rht-D1* (Rasheed et al. 2016). The genetic linkage maps used for QTL analyses were constructed by Xu et al. (2020b) with updates of the QTL target regions in the present study.

Inclusive composite interval mapping (ICIM) was used to detect QTL with IciMapping v4.1. Lines with missing phenotypic data were deleted. The walk speed was set as 1.0 centimorgan (cM). The largest P-

value for variables in the stepwise regression of phenotypes on marker variables was set as 0.001. The logarithm of the odds (LOD) thresholds were calculated by a permutation test with 1 000 replications with a type I error of 0.05. For FHB resistance, loci with LOD values exceeding the thresholds or exceeding 2.5 under two or more environments were reported. The confidence interval of the QTL was calculated by a one-LOD drop from the estimated QTL position. Genetic maps covering QTL regions were drawn using MapChart v2.32 (Voorrips 2002). QTL situated within the same or overlapping confidence intervals were considered the same, and those identified in two or more environments were defined as stable.

Both physical and genetic positions were used to compare the QTL identified in the current study with genes or QTL previously reported. First, the physical positions of linked markers were compared. Then, the physical positions of previously reported QTL were used to choose representative single nucleotide polymorphisms (SNPs) from the present linkage maps. Next, the genetic positions of the representative SNPs were compared with flanking SNPs of currently mapped QTL. The markers' physical positions were obtained from the Chinese Spring reference genome sequences RefSeq v2.0 (https://wheat-urgi.versailles.inra.fr/Seq-Repository/Assemblies).

Prediction of candidate genes. The annotations of high-confidence genes referring to IWGSC RefSeq Annotation v1.1 (International Wheat Genome Sequencing Consortium, IWGSC 2018) in the QTL confidence intervals were examined to determine their candidacy for FHB resistance based on omics or targeted transgenic studies (Ma et al. 2020; Walter et al. 2009). Their potential functions were compared with those of homologous genes in Arabidopsis, maize, and rice through UniProt (https://www.uniprot.org/).

PCR-based marker development and validation. The SNPs closely linked to mapped QTL were converted to

KASP or InDel markers to facilitate their application in marker-assisted selection (MAS). KASP markers were designed following protocols described previously (Xu et al. 2020b). Primers for designing the InDel marker were based on chromosome-specific polymorphisms among homologous sequences in the target region obtained from the EnsemblPlants database (http: // plants. ensembl. org/index. html). The PCR-based markers were validated in the DH population by comparing the amplification results with the chip data. Developed markers were also tested in the WAPS population to compare the FHB indices of two groups with different alleles using the t test.

Results

Resistance to FHB and its relationship with agronomic traits. Pearson's correlation coefficients of FHB indices ranged from 0. 65 to 0. 76 between replications and from 0. 42 to 0. 87 across environments of the mapping population (Supplementary Table S1). The ANOVA showed significant variations for genotypes, environments, and the genotype $\times$ environment interaction (Supplementary Table S2). Broad-sense heritability (H^2) of the FHB index in this population was estimated to be 0. 86.

The Sumai 3, Zhengmai 9023, and Zhoumai 22 checks had average FHB indices of 3. 5%, 26. 5%, and 53. 9%, respectively. The BLUE values of FHB indices for Yangmai 16 and Zhongmai 895 were 13. 3% and 38. 7%, respectively, indicating their moderately resistant and moderately susceptible responses. Apparent transgressive segregation of FHB resistance in the DH population indicated that both parents contributed to FHB resistance (Supplementary Fig. S1). FHB at Wuhan in 2018 was less severe than that in 2017, which might be attributable to the relatively lower temperature during that season. However, at Jingzhou in 2018, the disease was relatively severe because of the humid environment and more suitable conditions for FHB development.

Significant and negative correlations ($r = -0.73$) were observed between plant height and the FHB index in the

mapping population across 2 years at Wuhan (Table 1). AE was negatively correlated with the FHB index, but in different degrees, from 2017 ($r = -0.26$) to 2018 ($r = -0.58$), suggesting that AE is probably more important during years when FHB is less severe, such as 2018. The DTA showed no correlation with the FHB index across years ($r = 0.01$).

Table 1 Pearson's correlation coefficients among traits days to anthesis (DTA), anther extrusion (AE), plant height (PH), and Fusarium head blight (FHB) index of the Yangmai 16/Zhongmai 895 population at Wuhan

Year	Trait	FHBindex_natural	DTA	AE
2017	DTA	−0. 22**		
	AE	−0. 26**	−0. 23**	
	PH	−0. 63**	−0. 02ns	0. 37**
2018	DTA	0. 35**		
	AE	−0. 58**	0. 18*	
	PH	−0. 59**	0. 01ns	0. 60**
Across years	DTA	0. 01ns		
	AE	−0. 61**	−0. 06ns	
	PH	−0. 73**	0. 003ns	0. 59**

ns: not significant, *: significant at $P < 0.05$, **: significant at $P < 0.01$.

Resistance QTL for FHB, agronomic traits, and their correlations. The QTL significance thresholds for FHB indices under five environments and the BLUE values ranged from 3. 62 to 3. 84. The threshold ranges for PH, AE, and DTA were 3. 78 to 3. 80, 3. 55 to 3. 90, and 3. 72 to 3. 75, respectively. Seven QTL for FHB resistance, four for PH, two for AE, and two for DTA were detected by ICIM analysis (Table 2 and Supplementary Table S3). FHB resistance at four loci on chromosome arms 3BL, 4DS, 6AL, and 6BS were contributed by Yangmai 16, whereas resistance QTL on 4AS, 4BS, and 5AL came from Zhongmai 895 (Table 2 and Fig. 1). *QFhb. caas-4BS* and *QFhb. caas-4DS* situated at *Rht-B1* and *Rht-D1* loci had the largest effects on resistance in this population and were significant in all five environments and BLUE values; they accounted for 8. 6 to 21. 7% and 13. 0 to 20. 8% of the total phenotypic variation explained (PVE), respectively. *QFhb. caas-5AL* (572. 5 to

574. 8 Mb) with resistance contributed by Zhongmai 895 was identified in four environments and BLUE, explaining 2. 4 to 9. 0% of the phenotypic variation. *QFhb. caas-3BL* (832. 8 to 834. 9 Mb) and *QFhb. caas-6BS* (26. 1 to 29. 4 Mb) were each detected in three environments, accounting for 6. 8 and 4. 7% of PVE in BLUE values, respectively. The remaining two QTL on 4AS (12. 6 to 12. 9 Mb) and 6AL (616. 7 to 617. 0 Mb) were each detected in two environments (Table 2).

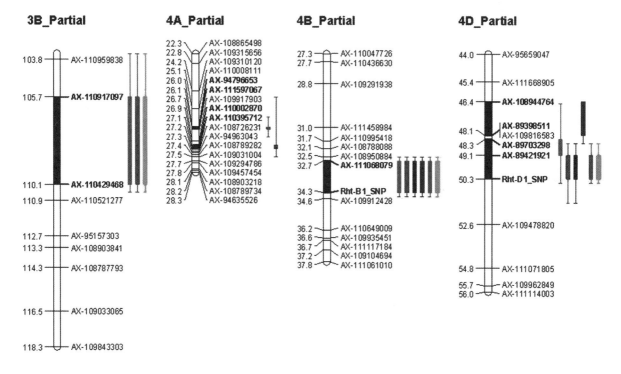

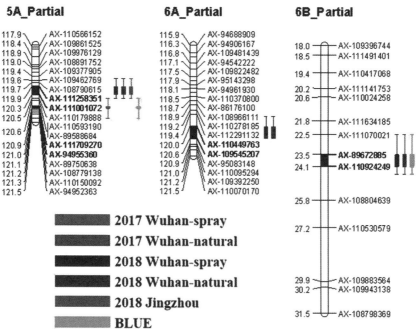

Fig. 1　Quantitative trait loci (QTL) for Fusarium head blight resistance identified in the Yangmai 16/Zhongmai 895 population. Marker names are shown on the right and genetic positions in cM are on the left. QTL flanking markers are in bold. The solid boxes are defined by QTL flanking markers. Confidence intervals calculated by 1-LOD drop from the estimated QTL positions are shown by whiskers.

There were four QTL for PH, including *Rht-B1* and *Rht-D1* (Supplementary Table S3), that explained 29. 7 and 40. 3% of the variation and co-locating in repulsion with *QFhb. caas-4BS* and *QFhb. caas-4DS*, respectively. Two QTL corresponding to AE were associated with *Rht-B1* and *Rht-D1*, respectively. No DTA QTL was associated with the PH or AE. The QTL at the *Vrn-D1* locus on chromosome 5D explained the highest variation (45. 9%) in BLUE values across years (Supplementary Table S3).

Regression of QTL number and FHB resistance. Closely linked markers were used as proxies for the five stable QTL (*QFhb. caas-3BL*, *QFhb. caas-4AS*, *QFhb. caas-5AL*, *QFhb. caas-6AL*, and *QFhb. caas-6BS*), excluding *QFhb. caas-4BS* (*Rht-B1*) and *QFhb. caas-4DS* (*Rht-D1*), to assess their additive effects on the FHB response. The number of resistance alleles in DH lines ranged from 0 to 4, and the corresponding average FHB index decreased from 47. 1 to 20. 3%. The average FHB indices of each group of the DH lines were regressed with the numbers of resistance alleles. Linear regression ($r^2 = 0. 98$) showed increasing FHB resistance with the number of favorable alleles (Supplementary Fig. S2).

PCR-based marker development and validation of *QFhb. caas-5AL*. The SNP *AX-110917097* associated with

QFhb. caas-3BL was converted to the KASP marker *KASP _ AX-110917097* (protocol in Supplementary Table S4). Genotypes of *KASP _ AX-110917097* in 170 DH lines were identical to the chip data, indicating the reliability of the marker. We identified 107 and 64 wheat accessions from the WAPS population containing resistance and susceptibility alleles of *QFhb. caas-3BL*, respectively (Supplementary Fig. S3 and Supplementary Table S5). However, the mean FHB indices of the two groups were not significantly different ($P=0. 27$).

SNP *AX-89588684* associated with *QFhb. caas-5AL* was converted to the InDel marker *InDel _ AX-89588684* (protocol in Supplementary Table S4). The primer pair P8684 amplified the flanking sequences of *AX-89588684*, yielding 756 bp and 909 bp fragments from Zhongmai895 and Yangmai16, respectively (Fig. 2). After sequencing, one 152-bp InDel, one 1-bp InDel, and five SNPs were found between Zhongmai 895 and Yangmai16 (data not shown). Genotypic data for 173 of the 174 DH lines detected by *InDel _ AX-89588684* were identical to the chip data, except one missing genotype, indicating the reliability of the InDel marker. For the WAPS population, 72 accessions carried the *QFhb. caas-5AL* resistance allele (Supplementary Table S5), with an average FHB index of 52. 3% compared with 63. 9% for the mean of lines with the alternative allele ($P < 0. 01$) (Fig. 2).

Table 2 Quantitative trait loci (QTL) for Fusarium head blight resistance mapped in the Yangmai 16/Zhongmai 895 population using the inclusive composite interval mapping in IciMapping v4. 1

QTL	Environment	Physical position (Mb)[a]	Marker interval	LOD	PVE (%)[b]	Add[c]
QTL with resistance allele contributed by Yangmai 16						
QFhb. caas-3BL	2017 Wuhan-spray	832. 8-834. 9	*AX-110917097-AX-110429468*	3. 8	9. 8	−7. 7
	2017 Wuhan-natural	832. 8-834. 9	*AX-110917097-AX-110429468*	2. 8	7. 2	−6. 6
	BLUE	832. 8-834. 9	*AX-110917097-AX-110429468*	2. 6	6. 8	−3. 5
QFhb. caas-4DS	2017 Wuhan-spray	17. 3-19. 7	*AX-89703298-AX-89421921*	7. 9	19. 1	−10. 9
	2017 Wuhan-natural	19. 2-19. 7	*Rht-D1 _ SNP-AX-89421921*	5. 9	14. 3	−9. 5
	2018 Wuhan-spray	19. 2-19. 7	*Rht-D1 _ SNP-AX-89421921*	5. 3	13. 3	−5. 0
	2018 Wuhan-natural	16. 3-17. 5	*AX-108944764-AX-89398511*	5. 2	13. 0	−5. 0
	2018Jingzhou	19. 2-19. 7	*Rht-D1 _ SNP-AX-89421921*	5. 6	13. 6	−7. 1
	BLUE	19. 2-19. 7	*Rht-D1 _ SNP-AX-89421921*	9. 0	20. 8	−6. 5

(continued)

QTL	Environment	Physical position (Mb)[a]	Marker interval	LOD	PVE (%)[b]	Add[c]
QFhb. caas-6AL	2018 Wuhan-natural	616. 7-617. 0	*AX-109545207-AX-110449763*	3. 5	3. 0	−3. 9
	2018 Wuhan-spray	616. 7-617. 0	*AX-109545207-AX-110449763*	3. 0	2. 6	−2. 5
QFhb. caas-6BS	2018 Wuhan-spray	26. 1-29. 4	*AX-89672885-AX-110924249*	2. 7	8. 1	−3. 6
	2018 Wuhan-natural	26. 1-29. 4	*AX-89672885-AX-110924249*	2. 6	7. 7	−3. 6
	BLUE	26. 1-29. 4	*AX-89672885-AX-110924249*	1. 8	4. 7	−3. 0
QTL with resistance allele contributed by Zhongmai 895						
QFhb. caas-4AS	2017 Wuhan-natural	12. 6-12. 9	*AX-94796653-AX-111597067*	3. 1	8. 0	7. 0
	2018 Wuhan-natural	12. 6-12. 8	*AX-110395712-AX-110002870*	2. 0	3. 2	4. 3
QFhb. caas-4BS	2017 Wuhan-spray	28. 1-33. 6	*AX-111068079-Rht-B1 _ SNP*	4. 1	10. 4	8. 3
	2017 Wuhan-natural	28. 1-33. 6	*AX-111068079-Rht-B1 _ SNP*	6. 5	15. 2	10. 2
	2018 Wuhan-spray	28. 1-33. 6	*AX-111068079-Rht-B1 _ SNP*	8. 1	19. 3	6. 1
	2018 Wuhan-natural	28. 1-33. 6	*AX-111068079-Rht-B1 _ SNP*	8. 1	20. 7	6. 3
	2018Jingzhou	28. 1-33. 6	*AX-111068079-Rht-B1 _ SNP*	3. 2	8. 6	5. 5
	BLUE	28. 1-33. 6	*AX-111068079-Rht-B1 _ SNP*	9. 2	21. 7	6. 6
QFhb. caas-5AL	2017 Wuhan-spray	572. 5-574. 8	*AX-111709270-AX-94955360*	4. 1	3. 9	5. 8
	2018 Wuhan-spray	571. 0-571. 1	*AX-111001072-AX-111258351*	9. 5	9. 0	4. 6
	2018 Wuhan-natural	571. 0-571. 1	*AX-111001072-AX-111258351*	2. 8	2. 4	3. 5
	2018Jingzhou	571. 0-571. 1	*AX-111001072-AX-111258351*	2. 6	3. 9	4. 2
	BLUE	572. 5-574. 8	*AX-111709270-AX-94955360*	7. 8	6. 7	3. 8

[a] The physical positions （Mb） of the SNPs are referred to Chinese Spring reference genome sequences IWGSC RefSeq v2. 0.

[b]PVE，phenotypic variation explained.

[c] Add，estimated additive effects of QTL at the current scanning position. Positive and negative values indicate that the resistance alleles are inherited from Zhong-mai 895 and Yangmai 16，respectively.

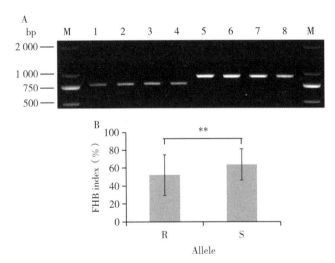

Fig. 2　Amplification profiles of marker *InDel _ AX-89588684* and its association with Fusarium head blight resistance in the wheat association panel for scab research. （a）M：DL2000 DNA marker；1：Zhongmai 895；2 to 4：doubled haploid lines with the *QFhb. caas-5AL* resistance allele；5：Yangmai 16；and 6 to 8：doubled haploid lines with the susceptibility allele. （b）Average Fusarium head blight indices of the wheat association panel for scab research accessions with the resistance （R） and susceptibility （S） alleles of *QFhb. caas-5AL* detected by *InDel _ AX-89588684*.

Candidate gene prediction. Three genes encoding cytochrome P450 and three encoding kinase-like protein were identified in the region of *QFhb. caas-3BL* （Supplementary Table S6）. One cytochrome P450 gene was identified in the region of *QFhb. caas-4AS*. Genes encoding kinase-like protein，protein with a leucine-rich repeat domain， glutathione S-transferase, and WRKY transcription factor might be candidates for *QFhb. caas-5AL*. One gene encoding protein with a leucine-rich repeat domain and eight genes encoding kinase-like protein were identified at the *QFhb. caas-6AL* locus. Five genes for kinase-like protein, three for protein with a leucine-rich repeat domain, and two cytochrome P450 genes might be related to *QFhb. caas-6BS*.

Discussion

Semi-dwarfing alleles *Rht-B1b* and *Rht-D1b* were reported to affect FHB resistance (Srinivasachary et al. 2009). Three possible mechanisms, disease escape, pleiotropy, or close linkage, were proposed previously (Mao et al. 2010). Although increased PH could reduce FHB, tall plants are prone to lodging, which is already a significant problem in China. An alternative approach might be other sources of reduced height that have no influence on the FHB response. The importance of AE on the FHB response has been emphasized by Buerstmayr et al. (2020). Again, the contribution of the semi-dwarfing alleles *Rht-B1b* and *Rht-D1b* to lower AE was evident in the present study and has been reported by He et al. (2016b) and Xu et al. (2020a). Alternate sources of height reduction might permit higher AE.

Yangmai 16 was developed in an FHB-prevalent region by phenotypic selection. Its parentage includes St1472/506 (Italy), Mentana (Italy), Funo (Italy), and Triumph (United States), all of which are susceptible or moderately susceptible to FHB (Zhu et al. 2019) but likely carry minor QTL for FHB resistance. In addition to *QFhb. caas-4DS* at the *Rht-D1* locus, we identified three other FHB resistance QTL, *QFhb. caas-3BL*, *QFhb. caas-6AL*, and *QFhb. caas-6BS*, with resistance alleles from Yangmai 16; each had minor or moderate effects on the FHB response. These QTL could partially explain the moderately resistant response of Yangmai 16.

We identified two minor resistance QTL, *QFhb. caas-4AS* and *QFhb. caas-5AL*, from the moderately

susceptible parent Zhongmai 895 beside *QFhb. caas-4BS* (*Rht-B1a*). Zhongmai 895 was selected from the Zhoumai 16 × Liken 4 cross, in which Liken 4 was the donor of resistance detected by the marker *InDel _ AX-89588684*. Liken 4, which is moderately susceptible to FHB, is from Shaanxi province, where FHB epidemics are more frequent than they are in other provinces in the Yellow and Huai River Valleys (Zhuang 2003). We concluded that the minor QTL contribute to the FHB resistance in both Yangmai 16 and Zhongmai 895. This clearly demonstrates that pyramiding minor QTL from regionally adapted elite germplasm can be an effective way to breed for improved FHB resistance. Pyramiding these minor resistance loci with major QTL, such as *Fhb1* and *Fhb7*, is valuable for pursuing high resistance to FHB.

Bourdoncle and Ohm (2003) mapped a QTL for type II resistance on chromosome 3BL from Chinese wheat Huapei 57-2. The physical position of its closest marker *Xgwm247* was at 847. 7 Mb based on the Chinese Spring reference sequence IWGSC RefSeq v2. 0 (Supplementary Table S7), which is ≈14 Mb apart from the markers flanking *QFhb. caas-3BL*, namely *AX-110917097* (832. 8 Mb) and *AX-110429468* (834. 9 Mb) (Table 2). SNPs with physical positions near *Xgwm247* were ≈10 cM apart from *QFhb. caas-3BL* based on the linkage map of the present population. Huapei 57-2 was developed at Huazhong Agricultural University at Wuhan. Yangmai 16 was developed at Yangzhou, Jiangsu province. The two places are located in the same agro-ecological wheat zone. It is reasonable to speculate that the 3BL QTL in Huapei 57-2 and Yangmai 16 might be the same.

Paillard et al. (2004) identified a minor QTL on chromosome 6AL in the Swiss winter wheat cultivar Forno (Supplementary Table S7). The QTL named *QFhs. fal-6AL*, flanked by *Xgwm169* (598. 2 Mb) and *Xpsr966b*, is approximately 19 Mb away from *QFhb. caas-6AL* (616. 7 to 617. 0 Mb) (Table 2). SNPs from the wheat 660K SNP array with physical positions near *Xgwm169* were approximately 16 cM apart from *QFhb. caas-6AL* based on the linkage map of the present population. This genetic distance indicates that *QFhb. caas-6AL* may be different from *QFhs. fal-6AL*.

Although three QTL were reported on chromosome 4AL (Holzapfel et al. 2008; Paillard et al. 2004), no QTL for FHB resistance has been reported on 4AS (Supplementary Table S7). *QFhb. caas-4AS* appears to be a new locus. *Rht-A1* on chromosome 4AL, the homolog of reduced plant height genes *Rht-B1b* and *Rht-D1b*, is at the position of 581. 4 Mb on the long arm (Cao et al. 2020), indicating that *QFhb. caas-4AS* has no relation with *Rht-A1*.

A QTL on chromosome 5AL was linked with *Xgwm410* in European cultivars Pirat and Apache (Holzapfel et al. 2008). In Swiss wheat Arina, *QFhs. fal-5AL. 1* was mapped between *Xgwm291* and *Xglk348c* (Paillard et al. 2004). The physical positions of *Xgwm410* and *Xgwm291* were at 680. 1 Mb and 700. 5 Mb, respectively (Supplementary Table S7), whereas *QFhb. caas-5AL* was mapped at an interval of 572. 5 to 574. 8 Mb. SNPs in similar physical positions with the two SSR markers were far from *QFhb. caas-5AL* on the linkage map in the present population. The different positions indicate that *QFhb. caas-5AL* is new. *Vrn-A1* at 589. 3 Mb on chromosome 5A was reported to be possibly related to FHB response (He et al. 2016a), but both parents of the present DH population were monomorphic with *vrn-A1*. Furthermore, no QTL for DTA was detected at this locus, indicating that *QFhb. caas-5AL* is independent of the variation at the *Vrn-A1* locus.

Many QTL (Supplementary Table S7) were detected previously on chromosome 6BS in Chinese germplasm or derivatives (Buerstmayr et al. 2009; Häberle et al. 2009; Lin et al. 2004; Liu et al. 2009; Semagn et al. 2007; Shen et al. 2003; Yang et al. 2005). These loci might share the same genomic region according to their linked and common SSR markers. The locus that was finely mapped later was designated as *Fhb2* and located

in the centromeric region of chromosome 6B (224. 1 to 233. 3 Mb) (Cuthbert et al. 2007) (Supplementary Table S7). *QFhb. caas-6BS* detected in this study was located in the distal region of 6BS with a physical position of 26. 1 to 29. 4 Mb. Flanking markers of *QFhb. caas-6BS* were 25 cM apart from SNPs with similar physical positions of *Fhb2*-linked SSR markers. The differences indicate that *QFhb. caas-6BS* is not *Fhb2*. Therefore, *QFhb. caas-6BS* is likely a new QTL for FHB resistance.

Yangmai16 and Zhongmai895 are both leading cultivars and good parents for breeding. Except for *QFhb. caas-4BS* (*Rht-B1*) and *QFhb. caas-4DS* (*Rht-D1*), the other five stable QTL were genetically independent of plant height and DTA. Because these QTL are unlikely to have linkage drag in Chinese breeding programs as their hosts are adapted modern cultivars, they have high value in breeding for FHB resistance. In particular, the PCR-based marker *InDel _ AX-89588684* will facilitate MAS of *QFhb. caas-5AL* from Zhongmai 895.

❖ Acknowledgements

We are grateful to Prof. R. A. McIntosh, Plant Breeding Institute, University of Sydney, for English editing of the manuscript.

❖ Literature Cited

Bai, G., Kolb, F. L., Shaner, G., and Domier, L. L. 1999. Amplified fragment length polymorphism markers linked to a major quantitative trait locus controlling scab resistance in wheat. Phytopathology 89: 343-348.

Bai, G., and Shaner, G. 1994. Scab of wheat: prospects for control. Plant Dis. 78: 760-766.

Bai, G., Su, Z., and Cai, J. 2018. Wheat resistance to Fusarium head blight. Can. J. Plant Pathol. 40: 336-346.

Bourdoncle, W., and Ohm, H. W. 2003. Quantitative trait loci for resistance to Fusarium head blight in recombinant inbred wheat lines from the cross Huapei 57-2/Patterson. Euphytica 131: 131-136.

Buerstmayr, H., Ban, T., and Anderson, J. A. 2009. QTL mapping and marker-assisted selection for *Fusarium* head blight resistance in wheat: a review. Plant Breed. 128: 1-26.

Buerstmayr, M., Steiner, B., and Buerstmayr, H. 2020. Breeding for Fusarium head blight resistance in wheat-Progress and challenge. Plant Breed. 139: 429-454.

Cai, J., and Bai, G. 2014. Quantitative trait loci for Fusarium head blight resistance in Huangcandou × 'Jagger' wheat population. Crop Sci. 54: 2520-2528.

Cao, S., Xu, D., Hanif, M., Xia, X., and He, Z. 2020. Genetic architecture underpinning yield component traits in wheat. Theor. Appl. Genet. 133: 1811-1823.

Chen, J., Griffey, C. A., Maroof, M. A. S., Stromberg, E. L., and Zeng, Z. 2006. Validation of two major quantitative trait loci for Fusarium head blight resistance in Chinese wheat line W14. Plant Breed. 125: 99-101.

Chen, Y., Wang, J., Yang, R., and Ma, Z. 2017. Current situation and management strategies of Fusarium head blight in China. Plant Prot. 43: 11-17 (in Chinese with English abstract).

Cuthbert, P. A., Somers, D. J., and Brulé-Babel, A. 2007. Mapping of *Fhb2*, on chromosome 6BS: a gene controlling Fusarium head blight field resistance in bread wheat (*Triticum aestivum* L.). Theor. Appl. Genet. 114: 429-437.

Dexter, J. E., Clear, R. M., and Preston, K. R. 1996. Fusarium head blight: effect on the milling and baking of some Canadian wheats. Cereal Chem. 73: 695-701.

Foroud, N. A., Baines, D., Gagkaeva, T. Y., Thakor, N., Badea, A., Steiner, B., Buŕstmayr, M., and Bü rstmayr, H. 2019. Trichothecenes in cereal grains-an update. Toxins (Basel) 11: 634.

Häberle, J., Schweizer, G., Schondelmaier, J., Zimmermann, G., and Hartl, L. 2009. Mapping of QTL for resistance against Fusarium head blight in the winter wheat population Pelikan//Bussard/

Ning8026. Plant Breed. 128: 27-35.

He, X., Lillemo, M., Shi, J., Wu, J., Bjørnstad, Å., Belova, T., Dreisigacker, S., Duveiller, E., and Singh, P. 2016a. QTL characterization of Fusarium head blight resistance in CIMMYT bread wheat line Soru♯1. PLoS One 11: e0158052.

He, X., Singh, S., Singh, P. K., Duveiller, E., Dreisigacker, S., and Lillemo, M. 2016b. Dwarfing genes *Rht-B1b* and *Rht-D1b* are associated with both Type I FHB susceptibility and low anther extrusion in two bread wheat populations. PLoS One 11: e0162499.

Holzapfel, J., Voss, H. H., Miedaner, T., Korzun, V., Häberle, J., Schweizer, G., Mohler, V., Zimmermann, G., and Hartl, L. 2008. Inheritance of resistance to Fusarium head blight in three European winter wheat populations. Theor. Appl. Genet. 117: 1119-1128.

International Wheat Genome Sequencing Consortium (IWGSC) 2018. Shifting the limits in wheat research and breeding using a fully annotated reference genome. Science 361: eaar7191.

Jiang, G., Dong, Y., Shi, J., and Ward, R. W. 2007a. QTL analysis of resistance to Fusarium head blight in the novel wheat germplasm CJ9306. II. Resistance to deoxynivalenol accumulation and grain yield loss. Theor. Appl. Genet. 115: 1043-1052.

Jiang, G., Shi, J., and Ward, R. W. 2007b. QTL analysis of resistance to Fusarium head blight in the novel wheat germplasm CJ9306. I. Resistance to fungal spread. Theor. Appl. Genet. 116: 3-13.

Li, G., Zhou, J., Jia, H., Gao, Z., Fan, M., Luo, Y., Zhao, P., Xue, S., Li, N., Yuan, Y., Ma, S., Kong, Z., Jia, L., An, X., Jiang, G., Liu, W., Cao, W., Zhang, R., Fan, J., Xu, X., Liu, Y., Kong, Q., Zheng, D., Wang, Y., Qin, B., Cao, S., Ding, Y., Shi, J., Yan, H., Wang, X., Ran, C., and Ma, Z. 2019. Mutation of a histidine-rich calcium-binding-protein gene in wheat confers resistanceto Fusarium head blight. Nat. Genet. 51: 1106-1112.

Li, T., Bai, G., Wu, S., and Gu, S. 2011. Quantitative trait loci for resistance to Fusarium head blight in a Chinese wheat landrace Haiyanzhong. Theor. Appl. Genet. 122: 1497-1502.

Li, T., Bai, G., Wu, S., and Gu, S. 2012. Quantitative trait loci for resistance to Fusarium head blight in the Chinese wheat landrace Huangfangzhu. Euphytica 185: 93-102.

Lin, F., Kong, Z. X., Zhu, H. L., Xue, J. Z., Wu, D. G., Tian, J. B., Wei, C. Q., Zhang, Z. Q., and Ma, Z. Q. 2004. Mapping QTL associated with resistance to Fusarium head blight in the Nanda2419 × Wangshuibai population. I. Type II resistance. Theor. Appl. Genet. 109: 1504-1511.

Liu, S., and Anderson, J. A. 2003. Marker assisted evaluation of Fusarium head blight resistant wheat germplasm. Crop Sci. 43: 760-766.

Liu, S., Hall, M. D., Griffey, C. A., and Mckendry, A. L. 2009. Meta-analysis of QTL associated with Fusarium head blight resistance in wheat. Crop Sci. 49: 1955-1968.

Ma, Z., Xie, Q., Li, G., Jia, H., Zhou, J., Kong, Z., Li, N., and Yuan, Y. 2020. Germplasms, genetics and genomics for better control of disastrous wheat Fusarium head blight. Theor. Appl. Genet. 133: 1541-1568.

Mao, S. L., Wei, Y. M., Cao, W., Lan, X. J., Yu, M., Chen, Z. M., Chen, G. Y., and Zheng, Y. L. 2010. Confirmation of the relationship between plant height and Fusarium head blight resistance in wheat (Triticum aestivum L.) by QTL meta-analysis. Euphytica 174: 343-356.

McMullen, M., Jones, R., andGallenberg, D. 1997. Scab of wheat and barley: a re-emerging disease of devastating impact. Plant Dis. 81: 1340-1348.

Meng, L., Li, H., Zhang, L., and Wang, J. 2015. QTLIciMapping: Integrated software for genetic linkage map construction and quantitative trait locus mapping in biparental populations. Crop J. 3: 269-283.

Paillard, S., Schnurbusch, T., Tiwari, R., Messmer, M., Winzeler, M., Keller, B., and Schachermayr, G. 2004. QTL analysis of resistance to Fusarium head blight in Swiss winter wheat (Triticum aestivum L.). Theor. Appl. Genet. 109: 323-332.

Rasheed, A., Wen, W., Gao, F., Zhai, S., Jin, H., Liu, J., Guo, Q., Zhang, Y., Dreisigacker, S., Xia, X., and He, Z. 2016. Development and validation of KASP assays for genes underpinning key economic traits in bread wheat. Theor. Appl. Genet. 129: 1843-1860.

Ren, J., Wang, Z., Du, Z., Che, M., Zhang, Y., Quan, W., Wang, Y., Jiang, X., and Zhang, Z. 2019. Detection and validation of a novel major QTL for resistance to Fusarium head blight from Triticum aestivum in the terminal region of chromosome 7DL. Theor. Appl. Genet. 132: 241-255.

Semagn, K., Skinnes, H., Bjørnstad, Å., Marøy, A. G., and Tarkegne, Y. 2007. Quantitative trait loci controlling Fusarium head blight resistance and low deoxynivalenol content in hexaploid wheat population from 'Arina' and NK93604. Crop Sci. 47: 294-303.

Shen, X., Zhou, M., Lu, W., and Ohm, H. 2003. Detection of Fusarium head blight resistance QTL in a wheat population using bulked segregant analysis. Theor. Appl. Genet. 106: 1041-1047.

Skinnes, H., Semagn, K., Tarkegne, Y., Marøy, A. G., and Bjørnstad, Å. 2010. The inheritance of anther extrusion in hexaploid wheat and its relationship to Fusarium head blight resistance and deoxynivalenol content. Plant Breed. 129: 149-155.

Somers, D. J., Fedak, G., and Savard, M. 2003. Molecular mapping of novel genes controlling Fusarium head blight resistance and deoxynivalenol accumulation in spring wheat. Genome 46: 555-564.

Srinivasachary, S., Gosman, N., Steed, A., Hollins, T. W., Bayles, R., Jennings, P., and Nicholson, P. 2009. Semi-dwarfing Rht-B1 and Rht-D1 loci of wheat differ significantly in their influence on resistance to Fusarium head blight. Theor. Appl. Genet. 118: 695-702.

Stack, R. W., and McMullen, M. P. 1994. A visual scale to estimate severity of Fusarium head blight in wheat. Page 1095 North Dakota State University Extension Service, Fargo, ND.

Su, Z., Bernardo, A., Tian, B., Chen, H.,

Wang, S., Ma, H., Cai, S., Liu, D., Zhang, D., Li, T., Trick, H., St. Amand, P., Yu, J., Zhang, Z., and Bai, G. 2019. A deletion mutation in *TaHRC* confers *Fhb1* resistance to Fusarium head blight in wheat. Nat. Genet. 51: 1099-1105.

Voorrips, R. E. 2002. MapChart: Software for the graphical presentation of linkage maps and QTLs. J. Hered. 93: 77-78.

Waldron, B. L., Moreno-Sevilla, B., Anderson, J. A., Stack, R. W., and Frohberg, R. C. 1999. RFLP mapping of QTL for Fusarium head blight resistance in wheat. Crop Sci. 39: 805-811.

Walter, S., Nicholson, P., and Doohan, F. M. 2009. Action and reaction of host and pathogen during *Fusarium* head blight disease. New Phytol. 185: 54-66.

Wang, H., Sun, S., Ge, W., Zhao, L., Hou, B., Wang, K., Lyu, Z., Chen, L., Xu, S., Guo, J., Li, M., Su, P., Li, X., Wang, G., Bo, C., Fang, X., Zhuang, W., Cheng, X., Wu, J., Dong, L., Chen, W., Li, W., Xiao, G., Zhao, J., Hao, Y., Xu, Y., Gao, Y., Liu, W., Liu, Y., Yin, H., Li, J., Li, X., Zhao, Y., Wang, X., Ni, F., Ma, X., Li, A., Xu, S.S., Bai, G., Nevo, E., Gao, C., Ohm, H., and Kong, L. 2020. Horizontal gene transfer of Fhb7 from fungus underlies Fusarium head blight resistance in wheat. Science 368: eaba5435.

Xu, K., He, X., Dreisigacker, S., He, Z., and Singh, P. K. 2020a. Anther extrusion and its association with Fusarium head blight in CIMMYT wheat germplasm. Agronomy (Basel) 10: 47.

Xu, W., Han, X., and Li, F. 2019. Co-occurrence of multi-mycotoxins in wheat grains harvested in Anhui province, China. Food Control 96: 180-185.

Xu, X., Zhu, Z., Jia, A., Wang, F., Wang, J., Zhang, Y., Fu, C., Fu, L., Bai, G., Xia, X., Hao, Y., and He, Z. 2020b. Mapping of QTL for partial resistance to powdery mildew in two Chinese common wheat cultivars. Euphytica 216: 3.

Yang, Z., Gilbert, J., Fedak, G., and Somers, D. J. 2005. Genetic characterization of QTL associated with resistance to Fusarium head blight in a doubled-haploid spring wheat population. Genome 48: 187-196.

Yu, J., Bai, G., Cai, S., and Ban, T. 2006. Marker-assisted characterization of Asian wheat lines for resistance to Fusarium head blight. Theor. Appl. Genet. 113: 308-320.

Yu, J., Bai, G., Cai, S., Dong, Y., and Ban, T. 2008a. New Fusarium head blight-resistant sources from Asian wheat germplasm. Crop Sci. 48: 1090-1097.

Yu, J., Bai, G., Zhou, W., Dong, Y., and Kolb, F. L. 2008b. Quantitative trait loci for Fusarium head blight resistance in a recombinant inbred population of Wangshuibai/Wheaton. Phytopathology 98: 87-94.

Zadoks, J.C., Chang, T.T., and Konzak, C. F. 1974. A decimal code for the growth stages of cereals. Weed Res. 14: 415-421.

Zhang, X., Pan, H., and Bai, G. 2012. Quantitative trait loci responsible for Fusarium head blight resistance in Chinese landraceBaishanyuehuang. Theor. Appl. Genet. 125: 495-502.

Zhu, Z., Chen, L., Zhang, W., Yang, L., Li, J., Liu, Y., Tong, H., Fu, L., Liu, J., Rasheed, A., Xia, X., He, Z., Hao, Y., and Gao, C. 2020. Genome-wide association analysis of Fusarium head blight resistance in Chinese elite wheat lines. Front. Plant Sci. 11: 206.

Zhu, Z., Hao, Y., Mergoum, M., Bai, G., Humphreys, G., Cloutier, S., Xia, X., and He, Z. 2019. Breeding wheat for resistance to Fusarium head blight in the Global North: China, USA, and Canada. Crop J. 7: 730-738.

Zhu, Z., Xu, D., Cheng, S., Gao, C., Xia, X., Hao, Y., and He, Z. 2018. Characterization of Fusarium head blight resistance gene *Fhb1* and its putative ancestor in Chinese wheat germplasm. Acta Agron. Sin. 44: 473-482.

Zhuang, Q. S. 2003. Chinese wheat improvement and pedigree analysis. Page 108 in: China Agriculture Press, Beijing. (in Chinese)

Harnessing wheat *Fhb1* for *Fusarium* resistance

Yuanfeng Hao,[1] Awais Rasheed,[1,2,3] Zhanwang Zhu,[1]
Brande B. H. Wulff,[4] and Zhonghu He[1,2,*]

[1] Institute of Crop Sciences, Chinese Academy of Agricultural Sciences (CAAS), 12 Zhongguancun South Street, Beijing 100081, China

[2] International Maize and Wheat Improvement Centre (CIMMYT), c/o CAAS, 12 Zhongguancun South Street, Beijing 100081, China

[3] Department of Plant Sciences, Quaid-i-Azam University, Islamabad 45320, Pakistan

[4] John Innes Centre, Norwich Research Park, Norwich, NR4 7UH, UK

[*] Correspondence: z. he@cgiar. org https: //doi. org/10. 1016/j. tplants. 2019. 10. 006

Fusarium head blight (FHB), caused by the fungus *Fusarium graminearum*, is an economically devastating disease of wheat worldwide. *Fhb1*, a widely used genetic source of FHB resistance, originated in East Asia. The recent cloning of *Fhb1* opens a new avenue to improve FHB resistance in wheat and potentially other crops.

Role of *Fhb1* Carrier in Breeding Wheat for FHB Resistance

FHB is a fungal disease of cereals, mainly affecting wheat and barley. Since the first epidemics reported on wheat in 1884 in the UK, FHB has become a major disease in most wheat-producing areas. It causes significant reductions in grain yield and quality, costing farmers billions of dol-lars each year worldwide[1]. Despite intensive searches, no completely resistant germplasm has been identified. Sumai 3, a Chinese wheat cultivar and known *Fhb1* carrier, is recognized as the best source of FHB resistance and has been widely used as a parent in many breeding programs[1].

The first Sumai 3-derived cultivar in North America was McVey, released in Minnesota in 1999 (Figure 1). Since then, more than 20 hard red spring (HRS) wheat cultivars with Sumai 3 in their pedigrees have been extensively grown in the northern USA and Canada, where severe FHB epidemics occur on a regular basis[2]. Indeed, Sumai 3-derived cultivars occupied a respective 48%, 36%, and 83% of the HRS wheat areas in North Dakota (2017), South Dakota (2016), and Manitoba (2018), totaling 3-4 million ha each year (https: //www. nass. usda. gov; https: //www. masc. mb. ca). Cultivar AAC Brandon alone occupied 2. 4 million ha in the Canadian prairies in 2018 (https: //www. masc. mb. ca). The wide acceptance of these cultivars suggests that FHB resistance has become an important criterion for choice of cultivars by producers. In contrast to the successful

Published in Trends in Plant Science, 2020, 25 (1): 1-3. (doi org/10. 1016/J. tplants. 2019. 10. 006)

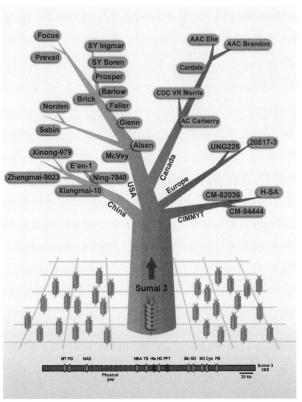

Trends in Plant Science

Figure 1   Significance of Sumai 3 to World Wheat for Fusarium Head Blight (FHB) Resistance. Top: Sumai 3 derivatives distributed in the USA, Canada, China, Europe, and by CIMMYT. Derivatives include the leading cultivars Alsen (led in 2002-2006), Glenn (2007-2011), Barlow (2012-2015), SY Soren (2016), and SY Ingmar (2017-2018) in North Dakota, Faller (2009-2012) and Prosper (2013-2015) in Minnesota, Prevail (2016-2017) in South Dakota, USA, and AAC Brandon (2016-2018) in Manitoba, Canada. Middle: highly resistant accessions, including Sumai 3 and many landraces identified from East Asia; green spikelets indicate genotypes resistant to FHB and orange indicates susceptibility. Bottom: physical map of the *Fhb1* region of Sumai 3 containing 13 open reading frames (arrows) and their respective positions. Red arrows indicate the two candidate proteins contributing to *Fhb1* resistance. Sourced from Rawat *et al.* [3], with minor revisions. Abbreviations: Cys, Cystatin; FB, F-box domain containing; HC, HCBT-like defense response protein; His, histidine-rich calcium-binding protein; MT, tRNA methyl transferase; NAD, oxidoreductase NAD binding; NBA, Nb-ARC domain containing; PFT, pore-forming toxin-like; PG, polygalacturonase; SG, SGNH plant lipase; Sin, Sina superfamily; TS, terpene synthase.

deployment of Sumai 3 derivatives in North America, the use of Sumai 3 in other wheat breeding programs has been slower (Figure 1). This may be due to different choices of breeders. For instance, many Japanese cultivars can trace their FHB resistance back to another *Fhb1* carrier, Shinchunaga, while breeders in China have accidently used Norin 129, a derivative of Shinchunaga, as major *Fhb1* donor[2]. Although many Chinese landraces contained *Fhb1*, their direct use as resistant parents was not successful due to poor agronomic traits[1]. In CIMMYT (International Maize and Wheat Improvement Centre), breeders have preferentially selected stem rust resistance gene *Sr2*, thus excluding *Fhb1* since the two genes are in very close repulsion linkage.

Molecular Identification of *Fhb1*

Among several quantitative trait loci for FHB resistance identified in Sumai 3, only *Fhb1* consistently shows a major effect on FHB resistance, making it attractive for wheat breeding[1]. The molecular isolation of the gene was hampered by the complexity of FHB resistance and hexaploid nature of the wheat genome. In 2016, Rawat *et al.* [3] reported that a pore-forming toxin-like (*PFT*) gene was the candidate for *Fhb1* among 13 putative genes (Figure 1) using positional cloning, mutation analysis, gene silencing, and transgenic over-expression strategies. However, *PFT* was subsequently found to be present in many FHB-susceptible cultivars, casting doubt on the role of the gene[4,5].

In two recent studies, Su *et al.* [6] and Li *et al.* [7] identified a histidine-rich calcium-binding (*TaHRC* or *His*) gene adjacent to *PFT* from Ning 7840 and Wangshuibai, respectively, as a new *Fhb1* candidate. Despite agreement on *His* candidacy, the two groups reached contrasting conclusions regarding the gene's function. Su *et al.* [6] proposed that the *Fhb1* product is a susceptibility factor and FHB resistance results from loss-of-function. However, Li *et al.* [7] concluded that *Fhb1* is a gain-of-function gene and that the newly generated protein acts as a regulator of host immunity. A thorough comparison between the two studies has been performed[8] and we will not readdress this here. The dominant-negative model proposed[8] could partially explain the differences between the two studies, but the nature of *Fhb1* resistance remains elusive.

Even though discrepancies exist, all three *Fhb1* cloning studies agree that the *PFT* gene is expressed at low level after anthesis and that expression is further lowered after FHB inoculation[3,6,7]. *PFT*, which encodes a chimeric lectin, has been proposed to participate in recognition of fungus-specific carbohydrates during early phases of infection and to confer type I-like (initial pathogen penetration) resistance by causing toxicity to the fungus[3,9]. Its expression is very high in pre-emergent and pre-anthesis spikes[3] and this may contribute to a certain level of FHB resistance, in particular when the gene was overexpressed[3]. In this context, it is important to note that Rawat *et al.*[3], Su *et al.*[6], and Li *et al.*[7] used different developmental stages, namely pre-anthesis, early anthesis, of *Fhb1*, probably plays a leading role for resistance to disease spread. Other genes, including those in the *Fhb1* interval (Figure 1), may be involved in the molecular network of FHB resistance.

In conclusion, the underlying genetic basis of *Fhb1* resistance might involve more than one gene, orchestrated by a spatiotemporal expression pattern targeting different stages of infection. More research and collaborative efforts are required to uncover the hidden mystery of *Fhb1*.

Practical Breeding Strategies

Although the precise genetic basis of *Fhb1* remains uncertain, scientists have developed diagnostic molecular markers that are useful for selection in regions where FHB is less prevalent and phenotypic selection is challenging[4,10]. The markers allow breeders to predict the presence of *Fhb1* in leading cultivars (Figure 1) and to select the best *Fhb1* carrying genotypes. Marker-assisted selection (MAS) underpinned by 'seed chipping' and barcoding technologies

combined with speed breeding that allows up to six generations of wheat per year, could more than halve the time taken to introduce *Fhb1* into an elite wheat line[11].

Breeders at CIMMYT have successfully transferred *Fhb1* from H-SA (Hartog/Sumai3, an *Fhb1-Sr2* recombinant) into Quaiu, Munal, Super 152, and many other elite lines by shuttle breeding and MAS. Breeders at Chinese Academy of Agricultural Sciences (CAAS) have also invested considerable effort in introducing *Fhb1* into Jimai 22, Zhoumai 16, and other leading cultivars through MAS, shuttle breeding, and doubled and anthesis, respectively, for FHB inoculation. Inoculation at different times might be partially responsible for the differences among the three studies. Once *Fusarium* infection is established, *His*, as a major determinant haploid strategies. Once the gene has been fixed in elite backgrounds, breeders can focus mainly on selection of agronomic and quality-related traits other than FHB resistance.

Enabling Technologies for Accelerating Genetic Improvement

Using conventional approaches, breeders can now easily transfer *Fhb1* into elite wheat, but it may be difficult to remove linkage drag in certain genetic backgrounds resulting from suppressed recombination in the target region[9]. However, transgenesis can overcome this limitation[12]. Since *Fhb1* could be a gene complex, transformation of a multigene cassette with *His*, *PFT*, and other genes, might confer more stable FHB resistance than the transfer of a single gene. Gene editing is perhaps another option. Indeed, a CRISPR-Cas9-mediated 1-bp insertion in the *His* gene significantly increased FHB resistance in the susceptible cultivar Bobwhite, indicating the effectiveness of this approach in certain genetic backgrounds[6].

Fhb1 bioengineering may be applicable to other crops such as maize, which is affected by ear or stalk rot caused by the same fungus. The intensive maize-wheat rotation in China, the USA, central Europe, and elsewhere is an important reason for increased prevalence of FHB in wheat in recent years. Therefore, control of maize ear and stalk rot through transgenesis or gene editing could reduce the amount of *Fusarium* inoculum, with huge bene-fits for protection of both crops.

❖ Acknowledgements

We are grateful to Dr Robert McIntosh (University of Sydney) for useful suggestions. We acknowledge financial support from the National Key Research and Development Program of China (2016YFD0101802, 2016YFE0108600), Agricultural Science and Technology Innovation Program of CAAS, and the BBSRC cross-institutional strategic program Designing Future Wheat (BB/P016855/1).

❖ References

[1] Bai, G. *et al.* (2018) Wheat resistance to Fusarium head blight. *Can. J. Plant Pathol.* 40, 336-346.

[2] Zhu, Z. *et al.* (2019) Breeding wheat for resistance to Fusarium head blight in the Global North: China, USA and Canada. *Crop J.* Published online July 19, 2019. https: //doi. org/10. 1016/j. cj. 2019. 06. 003

[3] Rawat, N. *et al.* (2016) Wheat *Fhb1* encodes a chimeric lectin with agglutinin domains and a pore-forming toxin-like domain conferring resistance to Fusarium head blight. *Nat. Genet.* 48, 1576-1580.

[4] Zhu, Z. *et al.* (2018) Characterization of *Fusarium* head blight resistance gene *Fhb1* and its putative ancestor in Chinese wheat germplasm. *Acta Agronomica Sinica* 44, 473-482.

[5] He, Y. *et al.* (2018) Molecular characterization and expression of *PFT*, an FHB resistance gene at the *Fhb1* QTL in wheat. *Phytopathology* 108, 730-736.

[6] Su, Z. *et al.* (2019) A deletion mutation in *TaHRC* confers *Fhb1* resistance to Fusarium head blight in wheat. *Nat. Genet.* 51, 1099-1105.

[7] Li, G. *et al.* (2019) Mutation of a histidine-rich calcium-binding-protein gene in wheat confers resistance to Fusarium head blight. *Nat. Genet.* 51, 1106-1112.

[8] Lagudah, E. S. and Krattinger, S. G. (2019) A new player contributing to durable *Fusarium* resistance. *Nat. Genet.* 51, 1070-1071.

[9] Schweiger, W. *et al.* (2016) Suppressed recombination and unique candidate genes in the divergent haplotype encoding *Fhb1*, a major Fusarium head blight resistance locus in wheat. *Theor. Appl. Genet.* 129, 1607-1623.

[10] Su, Z. *et al.* (2018) Development and validation of diagnostic markers for *Fhb1* region, a major QTL for *Fusarium* head blight resistance in wheat. *Theor. Appl. Genet.* 131, 2371-2380.

[11] Hickey, L. T. *et al.* (2019) Breeding crops to feed 10 billion. *Nat. Biotechnol.* 37, 744-754.

[12] Wulff, B. B. H. and Dhugga, K. S. (2018) Wheat—the cereal abandoned by GM. *Science* 361, 451-452.

Mapping of QTL for partial resistance to powdery mildew in two Chinese common wheat cultivars

Xiaoting Xu. Zhanwang Zhu. Aolin Jia. Fengju Wang. Jinping Wang. Yelun Zhang. Chao Fu. Luping Fu. Guihua Bai. Xianchun Xia. Yuanfeng Hao. Zhonghu He

X. Xu   Z. Zhu   A. Jia   F. Wang   J. Wang
C. Fu   L. Fu   X. Xia   Y. Hao   Z. He
Institute of Crop Sciences, Chinese Academy of Agricultural Sciences (CAAS), 12 Zhongguancun South Street, Beijing 100081, China
e-mail: haoyuanfeng@caas. cn
Z. He
e-mail: zhhecaas@163. com
Z. He
CIMMYT-China Office, c/o CAAS, 12 Zhongguancun South Street, Beijing 100081, China
Y. Zhang
Institute of Cereal and Oil Crops, Hebei Academy of Agriculture and Forestry Sciences, Shijiazhuang 050035, China
G. Bai
USDA-ARS, Hard Winter Wheat Genetics Research Unit, Manhattan, KS 66506, USA

Abstract: The increasing severity and prevalence of powdery mildew aided by extensive use of semi-dwarf cultivars and high levels of nitrogenous fertilizers are causing significant yield losses in wheat. Resistant cultivars are the most cost-effective and environmentally friendly approach to manage the disease. The objective of this study was to identify quantitative trait loci (QTL) for powdery mildew resistance in a doubled haploid (DH) population from a cross between leading Chinese cultivars, Yangmai 16 and Zhongmai 895. A high-density genetic map comprising of 14, 480 non-redundant markers (equal to 148, 179 SNPs) in 21 wheat chromosomes was constructed by genotyping the population with the Wheat 660 K SNP array. The DH population was phenotyped for powdery mildew resistance at the adult plant stage in multiple field trials, including four environments in the 2016-2017 cropping season and two environments in $2017 - 2018$. Composite interval mapping detected six stable QTL explaining 3. 8-23. 6% of the phenotypic variance across environments. *QPmyz. caas-5DS*, *QPmyz. caas-6BL* and *QPmyz. caas-7BS*, are probably new QTL for powdery mildew resistance. One SNP marker closely linked to *QPmyz. caas-6BL*, the most stable QTL, was converted into a Kompetitive Allele-Specific PCR marker (*K _ AX-94973433*) and validated on 103 commercial wheat cultivars. Significantly lower maximum disease severities of cultivars with the resistance-associated allele than those with the susceptibility-associated allele at *QPmyz. caas-6BL* in some environments indicated partial effectiveness of the marker. The novel QTL and their closely linked markers identified in the present study should facilitate development of cultivars with improved powdery mildew resistance.

Keywords: *Blumeria graminis* f. sp. *tritici* • High-density linkage map • QTL mapping • *Triticum aestivum*

Introduction

Common wheat (*Triticum aestivum* L.) is one of the most important staple food crops. Its yield and quality are threatened by many biotic and abiotic factors (Figueroa et al. 2018). Powdery mildew, caused by *Blumeria graminis* f. sp. *tritici* (*Bgt*), is one of the most damaging fungal diseases for wheat production. Yield losses from powdery mildew range from 5 to 40%, and can exceed 62% following severe epidemics (Singh et al. 2016). Powdery mildew has occurred on an average of over seven million ha in China during the past 5 years, and was particularly damaging in high yield areas such as the Yellow and Huai River Valleys (https://www.natesc.org.cn/). Although fungicides can reduce disease losses, resistant cultivars are a more cost-effective and environmentally friendly means of control (Huang and Röder 2004).

More than 90 powdery mildew resistance alleles have been designated at 60 loci (*Pm1-Pm64*) (McIntosh et al. 2013, 2018; Zhang et al. 2019). Most resistance genes are race-specific but a smaller number conferring adult-plant resistance (APR) such as *Pm38*, *Pm39*, *Pm46*, and *Pm62* (Lillemo et al. 2008; Moore et al. 2015; Zhang et al. 2018) are race non-specific with a response that is usually quantitative and more durable (Lillemo et al. 2008). Although individual APR genes confer small to intermediate effects, combination of 4-5 such genes can be highly effective (Singh et al. 2000). Therefore, pyramiding quantitative trait genes (QTL) by means of marker-assisted selection (MAS) is considered an important strategy in breeding cultivars with durable resistance.

Chip-based genotyping platforms such as the 9K, 15 K, 35 K, 50 K (*Triticum* TraitBreed array), 90 K, 660 K and 820 K Wheat SNP arrays have been rapidly developed in common wheat in recent years (Cavanagh et al. 2013; Wang et al. 2014; Winfield et al. 2016; Allen et al. 2017; Cui et al. 2017; Rasheed and Xia 2019). Cui et al. (2017) mapped 4959 non-redundant markers from the Wheat 660 K SNP array using a recombination inbred line (RIL) population and identified a major QTL for kernel number. Rimbert et al. (2018) developed a high-throughput genotyping array TaBW280 K from whole-genome resequencing data and constructed a high-density genetic map comprising 83, 721 markers (2616 nonredundant SNPs). New genotyping platforms such as KASP (Kompetitive Allele-Specific PCR) assay were established to facilitate application of SNP markers in breeding and to achieve both flexibility and high-throughput (Semagn et al. 2014). Seventy KASP assays have been validated and the number exceeded 150 with continued effort (Rasheed et al. 2016, unpublished data) based on the expectation that KASP assays will be extensively applied in MAS by wheat breeding programs.

Yangmai 16 has been a leading cultivar in the Middle and Lower Yangtze River Valleys, and cumulatively covered more than 5 million ha; Zhongmai 895 has been planted mainly in the Yellow and Huai River Valleys with total planting area about three million ha since it was released in 2012 (Ministry of Agriculture and Rural Development, China). Both cultivars are documented as susceptible to the prevalent *Bgt* isolates E09 and E20 as seedlings but exhibit moderate resistance at the adult plant stage. Here, a doubled haploid (DH) population from a cross between

Yangmai 16 and Zhongmai 895 was used to determine the underlying QTL for powdery mildew resistance.

Materials and methods

Plant material and field trials

The mapping population developed from the Yangmai 16/Zhongmai 895 cross comprised 198 DH lines. Yangmai 16 (*Rht-B1b*; *Rht-D1a*) was developed by the Lixiahe Agricultural Research Institute in Jiangsu province. Zhongmai 895 (*Rht-B1a*; *Rht-D1b*) with pedigree Zhoumai 16/Liken 4 was jointly released by the Institute of Crop Sciences and Institute of Cotton Research, Chinese Academy of Agricultural Sciences. The 103 commercial cultivars for QTL validation were the same as those used in a previous study (Jia et al. 2018). These cultivars were tested for powdery mildew resistance in the field at Beijing, Xingyang and Zhengzhou in 2017.

The DH lines and parents were planted at Gaoyi (Hebei province), Shijiazhuang (Hebei), Zhengzhou (Henan) and Xinxiang (Henan) in the 2016-2017 cropping season, and at Gaoyi and Xinxiang in 2017-2018, hereafter referred to as 2017 and 2018, respectively. Field trials were arranged in randomized complete blocks with two or three replications. Plots consisted of 1.5 m single rows spaced 20-30 cm apart. Approximately 50 seeds were sown in each row. The parents were planted after every 18 rows as checks. To ensure ample powdery mildew inoculum in the field, highly susceptible wheat cultivars Jingshuang 16, Shi 4185 or their mixture were planted around the test plots.

Jingshuang 16 planted in plastic pots in greenhouse was first inoculated with mixture of two highly virulent isolates, E09 and E20, at 3-4 leaves stage. After about a week, the fully infected Jingshuang 16 seedlings were then transplanted to the field and planted closely to highly susceptible checks. All plants were inoculated at growth stage 51-69 in the field (Zadoks et al. 1974). Maximum disease severities (MDS) were evaluated as the average percentage of leaf area covered by powdery mildew when the susceptible controls reached maximum severity about 6 weeks after inoculation.

Plant height was measured from the ground to spike (awns excluded) at grain-filling. Three representative primary tillers of different plants were selected to measure in the 2017 Xinxiang and 2018 Xinxiang, and the mean value was used for subsequent analysis.

Statistical analysis

MDS correlation coefficients, analysis of variance (ANOVA) and the best linear unbiased prediction (BLUP) values were obtained using SAS 9.4 software (SAS Institute, North Carolina, USA). Broad-sense heritability (h^2) for MDS was calculated as:

$$h^2 = \sigma_g^2 / (\sigma_g^2 + \frac{\sigma_{ge}^2}{r} + \frac{\sigma_\varepsilon^2}{re}),$$

where σ_g^2 is the genotypic variance, σ_{ge}^2 is the genotype by environment variance, σ_ε^2 is the residual error, e is the number of environments and r is the number of replicates (Nyquist and Baker 1991).

Construction of genetic linkage map

Genomic DNA of the DH lines and parents was extracted from fresh leaf tissues using the CTAB method (Doyle and Doyle 1987), and they were genotyped with the Wheat 660 K SNP array containing 630,517 SNP markers at Capital-Bio, Beijing, China (http://www.capitalbiotech.com/). SNPs were called using Axiom Analysis Suite v.1.1.1 (Affymetrix, Waltham, MA) as described (Wang 2017). KASP markers for known functional genes were also included in this study (Rasheed et al. 2016).

All SNPs with low minor allele frequencies (<0.3) and high missing rates (>10%) were excluded from analysis using Tassel v5.0 (Bradbury et al. 2007). Marker redundancy was eliminated by the BIN function in IciMapping v4.1 (Meng et al. 2015). According to the integrated genetic map (Guangyao Zhao, personal communication) and the Chinese Spring reference genome sequence RefSeq v1.0 (International Wheat

Genome Sequencing Consortium (IWGSC) 2018), 461 markers were anchored into 21 wheat chromosomes. Genetic maps were con-structed jointly using MSTmap (Wu et al. 2008) and Joinmap v4. 0 (Stam 1993).

QTL analysis

QTL were identified by composite interval mapping (CIM) using Windows QTL Cartographer v2. 5 (Wang et al. 2012). A logarithm of odds (LOD) threshold of 2. 5 was set for QTL mapping. QTL situated within the same interval or overlapping confidence regions were considered to be identical, and those identified in two or more environments and BLUP were defined as stable. The plot of phenotypes against genotypes at selected loci was made using the 'effectplot' function in R/qtl (Broman et al. 2003). Physical positions of the QTL were determined by closely linked markers based on RefSeq v1. 0 (International Wheat Genome Sequencing Consortium (IWGSC) 2018). MapChart v2. 3 was used to draw the genetic maps of the regions where stable QTL resided (Voorrips 2002).

KASP marker design and application

The SNP closely linked to the 6BL QTL was converted to a KASP assay. The KASP assay were designed following Ramirez-Gonzalez et al. (2015b) and comprised forward primer A: 5′-GAAGGTGACCA AGTTCATGCTactgctggagaggattgcA -3′, forward primer B: 5′-GAAGGTCGGAGTCAACGGATTact gctg-gagaggattgc G-3′ and reverse primer C: 5′-ggattgcaggtcccattatctc-3′. The KASP marker was validated for its genetic positions in the DH population and tested for trait association in 103 wheat cultivars. PCR was performed in a 384-well plate format and reaction mixtures comprising 60-100 ng dry genomic DNA, 2. 0 μL of 2 × KASP master mix (V4.0, LGC Genomics), 0. 045 μL of primer mix (12 μM of primer A, B and 30 μM of primer C) and 1. 924 μL of ddH2O (Ramirez-Gonzalez et al. 2015a). PCR cycling was followed Rasheed et al. (2016). End-point fluorescence was analyzed using PHERAstar[plus] SNP (BMG

LABTECH) and Klustercaller Version 3. 4. 1. 39 (LGC, Hoddesdon, UK).

Results

The DH population and its genetic map

The Yangmai 16/Zhongmai 895 DH populationoriginally included 198 lines, but after removal of 24 lines due to low DNA quality, 174 lines remained for genetic map construction and QTL analysis. A total of 151, 286 SNPs (24.0%) were polymorphic and assigned to 14, 782 bins, in which14, 467 non-redundant SNPs (equal to 148, 166 SNPs before redundancy) were mapped on 21 wheat chromosomes. Thirteen gene-specific KASP markers were also mapped (data not shown). The final map with 14, 480 markers covered 3681. 7 cM with an average marker interval of 0. 25 cM.

Powdery mildew responses of the parents and DH population

MDS were significantly correlated ($P<0.01$) among the six environments with Pearson's correlation coef-ficients ranging from 0. 34 to 0. 75 (Table S1). Both Yangmai 16 and Zhongmai 895 were moderately resistant to powdery mildew with average MDS of 27. 2 and 28. 2%, respectively. The frequency distributions of for each of the six environments showed continuous variation, confirming quantitative inheritance (Figure S1). The broad-sense heritability of powdery mildew resistance across the environments was 0. 86. ANOVA showed significant variation among DH lines, environments, replicates, and line × environment interactions (Table S2).

QTL for powdery mildew resistance

Six stable QTL for powdery mildew resistancewere detected on wheat chromosomes 2DL, 4BS, 4DS, 5DS, 6BL, and 7BS, explaining 3. 8-23. 6% of the phenotypic variance (PVE) (Table 1, Fig. 1). QPmyz. -caas-2DLwas mapped to the distal end of 2DL accounting for 6. 6-13. 5% of the PVE across four environments and the BLUP. QPmyz. caas-4BS and QPmyz. caas-4DS overlapped with the semi-dwarfing genes Rht-B1 and *Rht-D1* based on gene-specific markers for those loci，and contributed 7. 4-23. 6％ and 3. 8-22. 7％ of the PVE，respectively. *QPmyz. caas-5DS* explained 3. 9-7. 1％ of the PVE for resistance. *QPmyz. caas-6BL* was flanked by *AX-94973433* and *AX-94516327*，and was detected in four environments and the BLUP，explaining 6. 5-10. 3％ of the PVE. *QPmyz. caas-7BS* was mapped to the distal region of chromosome 7BS and explained 5. 7-11. 0％ of the PVE. The dot plots clearly reflect the differences in effect of the Yangmai 16 and Zhongmai 895 alleles at each of the six loci（Fig. 2）. Zhongmai 895 contributed the resistance alleles on 2DL，4BS，6BL and 7BS，and Yangmai 16 contributed the resistance alleles on 4DS and 5DS QTL.

Validation of *QPmyz. caas-6BL*

The most stable QTL，*QPmyz. caas-6BL*，was closely linked with a SNP marker *AX-94973433*. This marker was transferred into a KASP assay designated as *K ＿ AX-94973433*. It was assigned to the same genetic position as the original SNP marker confirming the success of the conversion. In 2017 Beijing the powdery mildew response of lines with the favorable allele was 30. 9％ lower than that of lines with the alterative allele（*P*＜0. 01，Table S3）. In the other two environments，2017 Xingyang and 2017 Zhengzhou，however，corresponding difference in mean MDS was not significant.

Relationship between powdery mildew resistance and plant height

Map locations of the 4BS and 4DS QTL corresponded to that of the *Rht-B1* and *Rht-D1* loci，respectively. Lines with resistance alleles of *QPmyz. caas-4BS* and *QPmyz. caas-4DS* were，respectively，an average of 18. 7 and 23. 4 cm taller than those with the contrasting alleles（*P*＜0. 001）. In regard to each of the other four QTL，there was no significant difference in plant height between the two genotypes indicating their independence of plant height（Fig. 3）. These four QTL could be used more effectively in wheat breeding for powdery mildew resistance.

To make a further understanding of the relationship between QTL for powdery mildew on 4BS and 4DS，and semi-dwarfing genes *Rht-B1* and *Rht-D1*. Six makers closely linked to the QTL were used to find recombination lines. Among 174 DH lines，only two lines，XY027 and XY048，were found the same in other four QTL and only have one recombination in *Rht-B1* and *Rht-D1* gene-specific KASP markers（Figure S2）. However，the BLUP value of XY027 was significantly higher than that of XY048，which suggested the 4BS and 4DS QTL for powdery mildew resistance was not linked but the same with the semi-dwarfing genes.

Table 1　QTL results by composite interval mapping in the Yangmai 16/Zhongmai 895 DH population，showing positions and effects of QTL for powdery mildew resistance in six environments and the BLUP

QTL	Environment	Position (cM)	Flanking marker	Physical interval (Mb)	LOD	PVE (％)[a]	Add[b]
QPmyz. caas-2DL	BLUP	95. 23	*AX-111501111-AX-111886292*	639. 36-640. 39	3. 43	8. 5	3. 03
	2017 Gaoyi	96. 59	*AX-108961748-AX-111440263*	635. 78-635. 94	5. 35	11. 7	6. 01
	2017 Shijiazhuang	95. 21	*AX-111501111-AX-111886292*	639. 36-640. 39	5. 63	13. 5	5. 38
	2018 Gaoyi	96. 52	*AX-110581180-AX-108961748*	635. 78-639. 47	5. 14	10. 9	5. 81
	2018 Xinxiang	95. 23	*AX-111501111-AX-111886292*	639. 36-640. 39	2. 51	6. 6	3. 18
QPmyz. caas-4BS	BLUP	28. 25	*AX-94407750-Rht-B1*	30. 86-38. 28	10. 39	15. 2	4. 09
	2017 Gaoyi	28. 25	*AX-94407750-Rht-B1*	30. 86-38. 28	6. 95	10. 8	6. 06
	2017 Zhengzhou	28. 25	*AX-94407750-Rht-B1*	30. 86-38. 28	16. 96	23. 6	10. 85
	2017 Xinxiang	28. 25	*AX-94407750-Rht-B1*	30. 86-38. 28	3. 88	7. 4	4. 92

(continued)

QTL	Environment	Position (cM)	Flanking marker	Physical interval (Mb)	LOD	PVE (%)[a]	Add[b]
	2018 Gaoyi	28.25	AX-94407750-Rht-B1	30.86-38.28	7.03	12.1	6.11
QPmyz. caas-4DS	BLUP	54.76	AX-89421921-Rht-D1	18.78-19.29	11.68	16.2	−4.17
	2017 Gaoyi	54.76	AX-89421921-Rht-D1	18.78-19.29	9.11	14.1	−6.55
	2017 Shijiazhuang	54.76	AX-89421921-Rht-D1	18.78-19.29	8.61	14.0	−5.64
	2017 Zhengzhou	54.76	AX-89421921-Rht-D1	18.78-19.29	15.35	22.7	−9.87
	2017 Xinxiang	54.76	AX-89421921-Rht-D1	18.78-19.29	5.88	9.8	−5.78
	2018 Gaoyi	54.76	AX-89421921-Rht-D1	18.78-19.29	2.57	3.8	−3.49
QPmyz. caas-5DS	BLUP	5.56	AX-111464164-AX-94924969	1.89-5.56	2.71	4.7	−2.30
	2017 Zhengzhou	5.56	AX-111464164-AX-94924969	1.89-5.56	4.54	7.1	−5.56
	2018 Gaoyi	5.56	AX-111464164-AX-94924969	1.89-5.56	2.18[c]	3.9	−3.84
QPmyz. caas-6BL	BLUP	56.45	AX-94973433-AX-94516327	475.27-477.80	3.23	7.2	3.20
	2017 Gaoyi	56.45	AX-94973433-AX-94516327	475.27-477.80	2.54	6.5	4.42
	2017 Shijiazhuang	56.45	AX-94973433-AX-94516327	475.27-477.80	3.22	8.1	5.03
	2017 Zhengzhou	56.45	AX-94973433-AX-94516327	475.27-477.80	4.05	9.8	6.75
	2017 Xinxiang	56.45	AX-94973433-AX-94516327	475.27-477.80	4.21	10.3	8.01
QPmyz. caas-7BS	BLUP	9.88	AX-110953411-AX-108848791	5.93-8.97	4.80	9.9	3.38
	2017 Gaoyi	9.88	AX-110953411-AX-108848791	5.93-8.97	5.16	11.0	5.91
	2017 Xinxiang	9.88	AX-110953411-AX-108848791	5.93-8.97	2.81	5.7	4.39
	2018 Xinxiang	9.88	AX-110953411-AX-108848791	5.93-8.97	3.59	7.5	3.78

[a] Percentages of phenotypic variance explained by individual QTL

[b] Additive effect of resistance allele, positive and negative values indicate that the alleles were inherited from Zhongmai 895 and Yangmai 16, respectively

[c] The LOD value is lower than 2.5 but higher than 2.0

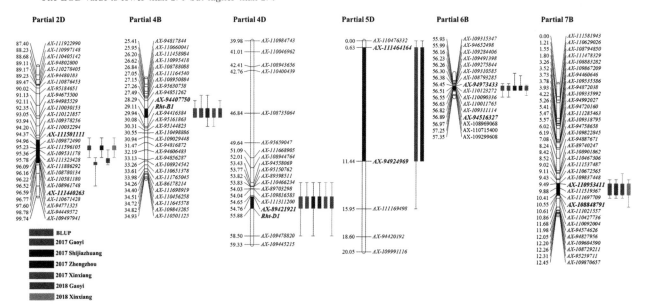

Fig. 1　Six stable QTL for powdery mildew resistance identified in six environments and the BLUP of a DH population from theYangmai 16/Zhongmai 895 cross; QTL flanking markers are in bold

Pyramiding various number of QTL

Except for the QTL located at *Rht-B1b* and *Rht-D1b*, four stable QTL detected in present study. To further understand the combined effects of QTL on reaction to powdery mildew, we examined the number of QTL in 161 DH lines without any missing genotype in the QTL closest markers. The lines contained 0-4 QTL with the powdery mildew severity decreasing linearly from 39.6 to 22.0% in the BLUP and the same trend was observed in all the six environments (Figure S3).

Discussion

Yangmai 16 and Zhongmai 895 were leading cultivars in different wheat-growing regions in China. They showed a high level of polymorphism in DNA markers and both had moderate resistance to powdery mildew; therefore, a very high-density genetic map was constructed with the Wheat 660 K SNP array. Using this map, we identified six stable QTL for wheat powdery mildew resistance.

QPmyz. caas-2DL

QPmyz. caas-2DL was flanked by SNPs *AX-111501111* and *AX-111440263* at a 635.78-640.39 Mb interval based on the Chinese Spring reference genome, RefSeq v1.0 (IWGSC 2018). One formally named gene (*Pm43*) and one temporarily designated gene (*PmSE5785*) were previously reported on this chromosome arm. *Pm43* was putatively introgressed from *Thinopyrum intermedium* (accession Z1141) into common wheat line CH5025, and flanked by *Xbarc11* and *Xwmc41* located at 467 and 577 Mb, respectively (Fig. 4. He et al. 2009). *PmSE5785* from 'SE5785' was identified in a synthetic hexaploid derivative and mapped very close to SSR marker *Xgwm539* at physical position 513.10 Mb (Wang et al. 2016). Thus, *QPmyz. caas-2DL* is likely different from *Pm43* and *PmSE5785*. In addition, seven QTL on chromosome 2DL were identified *QPm. sfr-2D* in Oberkulmer (Keller et al. 1999), *Qpm. ipk-2D* in W7984 (Börner et al. 2002), *QPm. inra-2D* in RE9001 (Bougot et al. 2006), *QPm.-caas-2DL* in Lumai 21 (Lan et al. 2010), *QPm. umb-2DL* in Folke (Lillemo et al. 2012), *2DLc* in Naxos (Lu et al. 2012), and *2DL* in Naxos (Windju et al. 2017). Of them two QTL were mapped far from *QPmyz. caas-2DL*. *Qpm. caas-2DL* was flanked by *Xwmc18* and *Xcfd233* at position 131 and 561 Mb, and 2DL by *Xmag3616* and *Xwmc41* at 287 and 577 Mb, respectively. *QPm. sfr-2D*, *Qpm. ipk-2D* and *QPm. inra-2D* were corresponds to *QPm. umb-2DL* as well as *2DL*. All the five QTL are partial resistance to powdery mildew and were located possibly at an overlap genetic region around 632-644 Mb. Therefore, the five QTL are probably the same with *QPmyz. caas-2DL*.

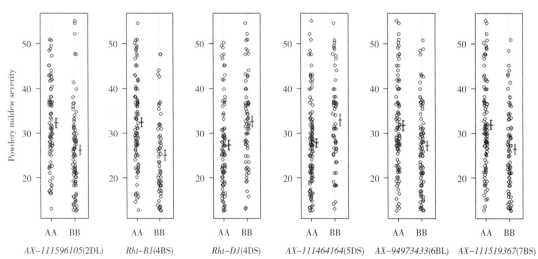

Fig. 2 Dot plots showing allelic effects of six markers closely linked to QTL for powdery mildew resistance in the Yangmai 16/ Zhongmai 895 DH population; *AA*, the allele from Yangmai 16; *BB*, the allele from Zhongmai 895; the estimated mean of BLUP associated with AA and BB alleles were shown on blue and red marks. (Color figure online)

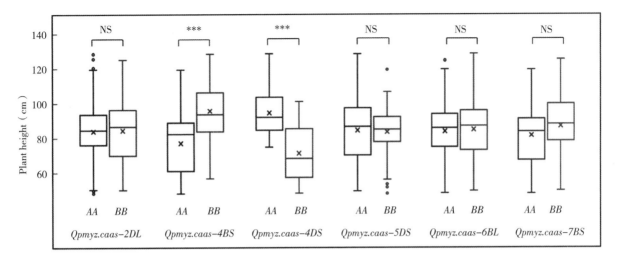

Fig. 3   Boxplots showing plant heights of different alleles in the six markers closely linked to QTL for powdery mildew resistance; AA, the allele from Yangmai 16, BB, the allele from Zhongmai 895; NS, not statistically significant; ***$P<0.001$

QPmyz. caas-4BS and *QPmyz. caas-4DS*

The 4BS and 4DS QTL were assigned to the *Rht-B1* and *Rht-D1* loci, respectively. Although very few studies report a relationship between powdery mildew resistance and plant height at these loci, many Fusarium head blight (FHB) studies indicated that semi-dwarfing alleles enhanced FHB susceptibility (Mesterházy 1995; Voss et al. 2008; Lu et al. 2013). Three possible mechanisms, including height per se, linkage and pleiotropy, have been proposed. Reduced height affected on disease susceptibility and the correlation remained significant in the sub-populations with homozygous *Rht-B1* and *Rht-D1* alleles (He et al. 2016). However, some researchers regarded linkage and pleiotropy as the most likely causes because not all plant height QTL are coincident with those for disease susceptibility (Draeger et al. 2007). Some observed close proximity of the peaks of FHB resistance QTL and the *Rht-D1* locus (Srinivasachary et al. 2008), while others reports suggested that *Rht-B1a* and *Rht-D1a* increased susceptibility to biotrophic pathogens in wheat and barley (Saville et al. 2012). Thus, no conclusive result has been obtained so far. In our study, two recombinant lines, XY027 and XY048, supported pleiotropy, while transgenic materials or near inbred lines still need to be created to confirm that in further researches.

QPmyz. caas-5DS

QPmyz. caas-5DS was physically mapped on the distal end of chromosome arm 5DS (1.89-5.56 Mb). This QTL was detected in the 2017 Zhengzhou and the BLUP, and was also detected in the 2018 Gaoyi when the LOD threshold was set at 2.0. Nine genes or alleles including *Pm2a* (Qiu et al. 2006), *Pm2b* (Ma et al. 2015), *Pm2c* (Xu et al. 2015), *Pm48* (Gao et al. 2012), *MlBrock* (Li et al. 2009), *PmD57-5D* (Ma et al. 2011), *PmLX66* (Huang et al. 2012), *PmX3986-2* (Ma et al. 2014), *PmW14* (Sun et al. 2015), and one QTL *QPm. inra-5D* (Bougot et al. 2006) were mapped on this chromosome arm. Of them all *Pm48* flanked by *Xgwm205* and *Xcfd18* was the nearest to 5DS distal end at physical position between 35 and 50 Mb. Thus, these genes were at least 29 Mb from *QPmyz. caas-5DS*, suggesting that they are different from *QPmyz.-caas-5DS* in consideration of high recombination rates at the end of Chromosomes.

QPmyz. caas-6BL

Three formally (*Pm20*, *Pm27*, *Pm54*) and two temporarily (*PmG3M*, *PmY150*) named genes, and four QTL were mapped on 6BL (Fig. 4). *Pm20* occurs in a T6BS. 6RL translocation from *Secale cereal* (Friebe et al. 1994), *Pm54* in common wheat line AGS 2000 (Hao et al. 2015) and *PmG3M* in *Triticum dicoccoides* (Xie et al. 2012), and of the four QTL, *CP3* and *CP4* were from Pedroso (Marone et al. 2013); *Qpm-caas. 6BL. 1* and *Qpm-caas. 6BL. 2* were from Chinese landrace Huixianhong (Asad et al. 2012). All the seven genes were located around 569-698 Mb, apparently different from *QPmyz. caas-6BL* (475. 27-477. 80 Mb) for about 100 Mb space. The other two genes *Pm27* (Järve et al. 2000) was cosegregation with *Xpsp3131* located at 554 Mb and *PmY150* (Zhou et al. 2005) was linked with *Xwmc397* at 437 Mb. It is possible that *PmY150* overlaps with *QPmyz. caas-6BL*. However, *Pm27* from *Triticum timopheevii* and *PmY150* from the translocation of *Aegilops longissima* were both dominant and race specific genes. Therefore, *QPmyz. -caas-6BL* is most likely a new QTL that is different from any previously reported genes for powdery mildew resistance on 6BL.

A SNP marker closely linked to *QPmyz. caas-6BL* was converted into a KASP assay. Analysis of the KASP marker in the panel of 103 cultivars showed that most of the cultivars (79. 6%) carried the resistance allele (Table S4). The allelic effect was observed only in the 2017 Beijing and not in the other two environments. This might be due to: (1) Under-estimation of the actual marker to QTL distance. (2) Although isolates E09 and E20 were inoculated, some other isolates were likely present in the fields as we have observed the severity was more serious in the 2017 Beijing. (3) Major seedling resistance gene (s) in this panel may be effective against E09 and E20 in the 2017 Xingyang and 2017 Zhengzhou, and have confounded the results. This will particularly be the case when the unfavorable allele frequency is extremely low and the validated QTL confers a small effect. Therefore, the KASP marker for *QPmyz. caas-6BL* was only useful for MAS in certain genetic backgrounds and further fine mapping work is necessary to develop more diagnostic markers.

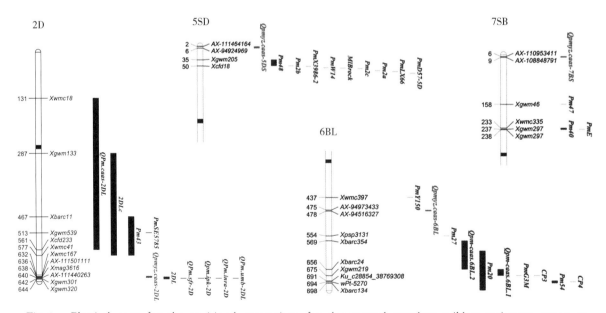

Fig. 4　Physical map for the positional comparison for documented powdery mildew resistance genes on chromosome 2DL, 5DS, 6BL and 7BS. The approximate position of centromeres was showed in black bars referring to IWGSC (2018). Genes or QTL linked markers are shown on the right and physical position on the left. QTL detected in the present study were in red, others in blue. (Color figure online)

QPmyz. caas-7BS

QPmyz. caas-7BS was physically assigned to 5. 93-8. 97 Mb on the distal end of 7BS. Three genes (*Pm40*, *Pm47* and *PmE*) were reported on 7BS (Fig. 4). *Pm40* putatively from *Thinopyrum intermedium* was introduced into bread wheat line CRY19, and is flanked by *BF291338* and *Xwmc335* at the physical position of 227. 30-233. 16 Mb (Zhong et al. 2016). *Pm47* in the Chinese landrace Hongyanglazi was closely linked with the marker *Xgwm46* at 158. 08 Mb (Xiao et al. 2013). *PmE* was also derived from *Thinopyrum intermedium* and is located close to *Pm40* (Ma et al. 2007). The three genes were at least 140 Mb from *QPmyz. caas-7BS* and no QTL for powdery mildew resistance was reported in this chromosome; therefore; *QPmyz. caas-7BS* is novel.

Conclusions

Using a high-density genetic map, we identifiedsix stable QTL that explained 3. 8-23. 6% of the phenotypic variance in powdery mildew resistance. Three of them, *QPmyz. caas-5DS*, *QPmyz. caas-6BL*, and *QPmyz. caas-7BS*, are likely new QTL and can be used for improvement of wheat powdery mildew resistance in breeding programs.

❖ Acknowledgements

The authors are grateful to Prof. R. A. McIntosh, Plant Breeding Institute, University of Sydney, for critical review of this manuscript. This study was supported bythe National Key R&D Program of China (2016YFE0108600 and 2016YFD0101802) and CAAS Science and Technology Innovation Program.

❖ References

Allen AM, Winfield MO, Burridge AJ et al (2017) Characterization of a Wheat Breeders' Array suitable forhigh-throughput SNP genotyping of global accessions of hexaploid bread wheat (*Triticum aestivum*). Plant Biotechnol J 15: 390-401. https: //doi. org/10. 1111/pbi. 12635

Asad MA, Bai B, Lan CX et al (2012) Molecular mapping of quantitative trait loci for adult-plant resistance to powdery mildew in Italian wheat cultivar Libellula. Crop Pasture Sci 63: 539-546. https: //doi. org/10. 1071/cp12174

Börner A, Schumann E, Fürste A et al (2002) Mapping of quantitative trait loci determining agronomic important characters in hexaploid wheat (*Triticum aestivum* L.). Theor Appl Genet 105: 921-936. https: //doi. org/10. 1007/s00122-002-0994-1

Bougot Y, Lemoine J, Pavoine MT et al (2006) A major QTL effect controlling resistance to powdery mildew in winterwheat at the adult plant stage. Plant Breed 125: 550-556. https: //doi. org/10. 1111/j. 1439-0523. 2006. 01308. x

Bradbury P, Zhang Z, Kroon D et al (2007) TASSEL: software for association mapping of complex traits in diverse samples. Bioinformatics 23: 2633-2635. https: //doi. org/10. 1093/bioinformatics/btm308

Broman KW, Hao W, Saunak S et al (2003) R/qtl: QTL mapping in experimental crosses. Bioinformatics 19: 889-890. https: //doi. org/10. 1093/bioinformatics/btg112

Cavanagh CR, Shiaoman C, Shichen W et al (2013) Genome-wide comparative diversity uncovers multiple targets of selection for improvement in hexaploid wheat landraces and cultivars. Proc Natl Acad Sci USA 110: 8057-8062. https: //doi. org/10. 1073/pnas. 1217133110

CuiF, Zhang N, Fan XL et al (2017) Utilization of a Wheat 660 K SNP array-derived high-density genetic map for high-resolution mapping of a major QTL for kernel number. Sci Rep 7: 3788. https: //doi. org/10. 1038/s41598-017-04028-6

Doyle J, Doyle JL (1987) A rapid DNA isolation procedure from small quantities of fresh leaf tissues. Phytochem Bull 19: 11-15.

Draeger R, Gosman N, Steed A et al (2007) Identification ofQTLs for resistance to Fusarium head blight, DON accumulation and associated traits in the winter wheat variety Arina. Theor Appl Genet 115: 617-625. https: //doi. org/10. 1007/s00122-007-0592-3

Figueroa M, Hammond-Kosack KE, Solomon PS (2018) A review of wheat diseases: a field perspective. Mol Plant Pathol 19: 1523-1536. https: //doi. org/10. 1111/mpp. 12618

Friebe B, Heun M, Tuleen N et al (1994) Cytogenetically monitored transfer of powdery mildew resistance from rye into wheat. Crop Sci 34：621-625. https：//doi. org/10. 2135/cropsci1994. 0011183X003400030003x

Gao H，Zhu F，Jiang Y et al（2012）Genetic analysis and molecular mapping of a new powdery mildew resistantgene *Pm*46 in common wheat. Theor Appl Genet 125：967-973. https：//doi. org/10. 1007/s00122-012-1886-7

Hao Y，Parks R，Cowger C et al（2015）Molecular characterization of a new powdery mildew resistance gene *Pm*54 in soft red winter wheat. Theor Appl Genet 128：465-476. https：//doi. org/10. 1007/s00122-014-2445-1

He R，Chang Z，Yang Z et al（2009）Inheritance and mapping of powdery mildew resistance gene *Pm43* introgressed from *Thinopyrum intermedium* into wheat. Theor Appl Genet 118：1173-1180. https：//doi. org/10. 1007/s00122-009-0971-z

He X，Singh PK，Dreisigacker S et al（2016）Dwarfinggenes *Rht-B1b* and *Rht-D1b* are associated with both type I FHB susceptibility and low anther extrusion in two bread wheat populations. PLoS ONE 11：e0162499. https：//doi. org/10. 1371/journal. pone. 0162499

Huang XQ，Röder MS（2004）Molecular mapping of powdery mildew resistance genes in wheat：a review. Euphytica 137：203-223. https：//doi. org/10. 1023/b：euph. 0000041576. 74566. d7

Huang J，Zhao Z，Song F et al（2012）Molecular detection of a gene effective against powdery mildew in the wheat cul-tivar Liangxing 66. Mol Breed 30：1737-1745. https：//doi. org/10. 1007/s11032-012-9757-0

International Wheat Genome Sequencing Consortium（IWGSC）（2018）Shifting the limits in wheat research and breeding using a fully annotated reference genome. Science 361：eaar7191. https：//doi. org/10. 1126/science. aar7191

Järve K，Peusha H，Tsymbalova J et al（2000）Chromosomal location of a *Triticum timopheevii*-derived powdery mildew resistance gene transferred to common wheat. Genome 43：377-381.

Jia A，Ren Y，Gao F et al（2018）Mapping and validation of a new QTL for adult-plant resistance to powdery mildew in Chinese elite bread wheat line Zhou8425B. Theor Appl Genet 131：1063-1071. https：//doi. org/10. 1007/s00122-018-3058-x

Keller M，Keller B，Schachermayr G et al（1999）Quantitative trait loci for resistance against powdery mildew in a segregating wheat×spelt population. Theor Appl Genet 98：903-912. https：//doi. org/10. 1007/s001220051149

Lan C，Ni X，Yan J et al（2010）Quantitative trait loci mapping of adult-plant resistance to powdery mildew in

Chinese wheat cultivar Lumai 21. Mol Breed 25：615-622. https：//doi. org/10. 1007/s11032-009-9358-8

Li G，Fang T，Zhu J et al（2009）Molecular identification of a powdery mildew resistance gene from common wheat cultivar Brock. Acta Agron Sin 35：1613-1619. https：//doi. org/10. 3724/SP. J. 1006. 2009. 01613

Lillemo M，Asalf B，Singh RP et al（2008）The adult plant rust resistance loci *Lr34/Yr18* and *Lr46/Yr29* are important determinants of partial resistance to powdery mildew in bread wheat line Saar. Theor Appl Genet 116：1155-1166. https：//doi. org/10. 1007/s00122-008-0743-1

Lillemo M，Bjørnstad Å，Skinnes H（2012）Molecular mapping of partial resistance to powdery mildew in winter wheat cultivar Folke. Euphytica 185：47-59. https：//doi. org/10. 1007/s10681-011-0620-x

Lu Q，Bjornstad A，Ren Y et al（2012）Partial resistanceto powdery mildew in German spring wheat 'Naxos' is based on multiple genes with stable effects in diverse environments. Theor Appl Genet 125：297-309. https：//doi. org/10. 1007/s00122-012-1834-6

Lu Q，Lillemo M，Skinnes H et al（2013）Anther extrusion and plant height are associated with Type I resistance to Fusarium head blight in bread wheat line 'Shanghai-3/Catbird'. Theor Appl Genet 126：317-334. https：//doi. org/10. 1007/s00122-012-1981-9

Ma Q，Luo PG，Ren ZL et al（2007）Genetic analysis and chromosomal location of two new genes for resistance to powdery mildew in wheat（*Triticum aestivum* L. ）. Acta Agron Sin 33：1-8.

Ma H，Kong Z，Fu B et al（2011）Identification and mapping of a new powdery mildew resistance gene on chromosome 6D of common wheat. Theor Appl Genet 123：1099-1106. https：//doi. org/10. 1007/s00122-011-1651-3

Ma P，Xu H，Luo Q et al（2014）Inheritance and genetic mapping of a gene for seedling resistance to powdery mildew in wheat line X3986-2. Euphytica 200：149-157. https：//doi. org/10. 1007/s10681-014-1178-1

Ma P，Xu H，Xu Y et al（2015）Molecular mapping of a new powdery mildew resistance gene *Pm2b* in Chinese breeding line KM2939. Theor Appl Genet 128：613-622. https：//doi. org/10. 1007/s00122-015-2457-5

Marone D，Russo MA，Laidò G et al（2013）Genetic basis of qualitative and quantitative resistance to powderymildew in wheat：from consensus regions to candidate genes. BMC Genom 14：562. https：//doi. org/10. 1186/1471-2164-14-562

McIntosh RA, Yamazaki Y, Dubcovsky J et al (2013) Gene symbol: catalogue of gene symbols for wheat. In: Proceedings of 12th International wheat genetics symposium, vol 57, 8-13 Sept, Yokohama, Japan, pp 303-321.

McIntosh RA, Dubcovsky J, Rogers WJ et al (2018) Catalogueof gene symbols for wheat: 2018 supplement. Annu Wheat Newsl 63: 73-93.

Meng L, Li H, Zhang L et al (2015) QTL IciMapping: integrated software for genetic linkage map construction and quantitative trait locus mapping in biparental populations. Crop J 3: 269-283. https://doi.org/10.1016/j.cj.2015.01.001

Mesterházy A (1995) Types and components of resistance to Fusarium head blight of wheat. Plant Breed 114: 377-386. https://doi.org/10.1111/j.1439-0523.1995.tb00816.x

Moore JW, Herrera-Foessel S, Lan C et al (2015) A recently evolved hexose transporter variant confers resistance tomultiple pathogens in wheat. Nat Genet 47: 1494-1498. https://doi.org/10.1038/ng.3439

Nyquist WE, Baker RJ (1991) Estimation of heritability and prediction of selection response in plant populations. CritRev Plant Sci 10: 235-322. https://doi.org/10.1080/07352689109382313

Qiu YC, Sun XL, Zhou RH et al (2006) Identification of microsatellite markers linked to powdery mildew resistance gene *Pm2* in wheat. Cereal Res Commun 34: 1267-1273. https://doi.org/10.1556/CRC.34.2006.4.268

Ramirez-Gonzalez RH, Segovia V, Bird N et al (2015a) RNA-Seq bulked segregant analysis enables the identification of high-resolution genetic markers for breeding in hexaploid wheat. Plant Biotechnol J 13: 613-624. https://doi.org/10.1111/pbi.12281

Ramirez-Gonzalez RH, Uauy C, Caccamo M (2015b) Poly-Marker: a fast polyploid primer design pipeline. Bioinformatics 31: 2038-2039. https://doi.org/10.1093/bioinformatics/btv069

Rasheed A, Xia X (2019) From markers to genome-basedbreeding in wheat. Theor Appl Genet. https://doi.org/10.1007/s00122-019-03286-4

Rasheed A, Wen W, Gao F et al (2016) Development and validation of KASP assays for genes underpinning key economic traits in bread wheat. Theor Appl Genet 129: 1843-1860. https://doi.org/10.1007/s00122-016-2743-x

Rimbert H, Darrier B, Navarro J et al (2018) High throughput SNP discovery and genotyping in hexaploid wheat. PLoS ONE 13: e0186329. https://doi.org/10.1371/journal.pone.0186329

Saville RJ, Gosman N, Burt CJ et al (2012) The 'Green Revolution' dwarfing genes play a role in disease resistance in *Triticum aestivum* and *Hordeum vulgare*. J Exp Bot 63: 1271-1283. https://doi.org/10.1093/jxb/err350

Semagn K, Babu R, Hearne S et al (2014) Single nucleotide polymorphism genotyping using Kompetitive Allele Specific PCR (KASP): overview of the technology and its application in crop improvement. Mol Breed 33: 1-14. https://doi.org/10.1007/s11032-013-9917-x

Singh RP, Huerta-Espino J, Rajaram S (2000) Achieving near-immunity to leaf and stripe rusts in wheat by combining slow rusting resistance genes. Acta Phytopathol Entomol Hung 35: 133-139.

Singh RP, Singh PK, Rutkoski J et al (2016) Disease impact on wheat yield potential and prospects of genetic control. Annu Rev Phytopathol 54: 303-322. https://doi.org/10.1146/annurev-phyto-080615-095835

Srinivasachary GN, Steed A et al (2008) Susceptibilityto Fusarium head blight is associated with the *Rht-D1b* semidwarfing allele in wheat. Theor Appl Genet 116: 1145-1153. https://doi.org/10.1007/s00122-008-0742-2

Stam P (1993) Construction of integrated genetic linkage maps by means of a new computer package: JOINMAP. Plant J 3: 739-744. https://doi.org/10.1111/j.1365-313X.1993.00739.x

Sun Y, Zou J, Sun H et al (2015) *PmLX66* and *PmW14*: new alleles of *Pm2* for resistance to powdery mildew in the Chinese winter wheat cultivars Liangxing 66 and Wennong 14. Plant Dis 99: 1118-1124. https://doi.org/10.1094/PDIS-10-14-1079-RE

Voorrips RE (2002) MapChart: software for the graphical presentation of linkage maps and QTLs. J Hered 93: 77-78. https://doi.org/10.1093/jhered/93.1.77

Voss HH, Holzapfel J, Hartl L et al (2008) Effect of the *Rht-D1* dwarfing locus on Fusarium head blight rating in three segregating populations of winter wheat. Plant Breed 127: 333-339. https://doi.org/10.1111/j.1439-0523.2008.01518.x

Wang JP (2017) Genetic dissection of key traits and functional marker development in common wheat. Ph.D, Shandong Agricultural University

Wang S, Basten CJ, Zeng ZB (2012) Windows QTL Cartographer 2.5. Department of Statistics, North Carolina State University, Raleigh, NC. http://

statgen. ncsu. edu/qtlcart/WQTLCart. htm

Wang S，Wong D，Forrest K et al（2014）Characterization of polyploid wheat genomic diversity using a high-density 90，000 single nucleotide polymorphism array. Plant Biotechnol J 12：787-796. https：//doi. org/10. 1111/pbi. 12183

Wang Y，Wang C，Quan W et al（2016）Identification and mapping of *PmSE* 5785，a new recessive powdery mildew resistance locus，in synthetic hexaploid wheat. Euphytica 207：619-626. https：//doi. org/10. 1007/s10681-015-1560-7

Windju SS，Malla K，Belova T et al（2017）Mapping and validation of powdery mildew resistance loci from spring wheat cv. Naxos with SNP markers. Mol Breed 37：61. https：//doi. org/10. 1007/s11032-017-0655-3

Winfield MO，Allen AM，Burridge AJ et al（2016）High-density SNP genotyping array for hexaploid wheat and its secondary and tertiary gene pool. Plant Biotechnol J 14：1195-1206. https：//doi. org/10. 1111/pbi. 12485

Wu Y，Bhat PR，Close TJ et al（2008）Efficient and accurateconstruction of genetic linkage maps from the minimum spanning tree of a graph. PLoS Genet 4：e1000212. https：//doi. org/10. 1371/journal. pgen. 1000212

Xiao M，Song F，Jiao J et al（2013）Identification of the gene Pm47 on chromosome 7BS conferring resistance to powdery mildew in the Chinese wheat landrace Hongyanglazi. Theor Appl Genet 126：1397-1403. https：//doi. org/10. 1007/s00122-013-2060-6

Xie W，Ben-David R，Zeng B et al（2012）Identification and characterization of a novel powdery mildew resistancegene *PmG3M* derived from wild emmer wheat，*Triticum dicoccoides*. Theor Appl Genet 124：911-922. https：//doi. org/10. 1007/s00122-011-1756-8

Xu H，Yi Y，Ma P et al（2015）Molecular tagging of a new broad-spectrum powdery mildew resistance allele *Pm2c* in Chinese wheat landrace Niaomai. Theor Appl Genet128：2077-2084. https：//doi. org/10. 1007/s00122-015-2568-z

Zadoks JC，Chang TT，Konzak CF（1974）A decimal code forthe growth stages of cereals. Weed Res 14：415-421. https：//doi. org/10. 1111/j. 1365-3180. 1974. tb01084. x

Zhang R，Fan Y，Kong L et al（2018）*Pm62*，an adult-plant powdery mildew resistance gene introgressed from.

Dasypyrum villosum chromosome arm 2VL into wheat. Theor Appl Genet 131：2613-2620. https：//doi. org/10. 1007/s00122-018-3176-5

Zhang D，Zhu K，Dong L et al（2019）Wheat powdery mildew resistance gene *Pm64* derived from wild emmer （*Triticum turgidum var. dicoccoides*）is tightly linked in repulsion with stripe rust resistance gene *Yr5*. Crop J. https：//doi. org/10. 1016/j. cj. 2019. 03. 003

Zhong S，Ma L，Fatima SA et al（2016）Collinearity analysis and high-density genetic mapping of the wheat powdery mil-dew resistance gene *Pm40* in PI 672538. PLoS ONE 11：e0164815. https：//doi. org/10. 1371/journal. pone. 0164815

Zhou R，Zhu Z，Kong X et al（2005）Development of wheat near-isogenic lines for powdery mildew resistance. Theor Appl Genet 110：640-648. https：//doi. org/10. 1007/s00122-004-1889-0

Fine mapping of *QPm. caas-3BS*，a stable QTL for adult-plant resistance to powdery mildew in wheat (*Triticum aestivum* L.)

Yan Dong[1] • Dengan Xu[2] • Xiaowan Xu[3] • Yan Ren[4] • Fengmei Gao[5] •
Jie Song[1] • Aolin Jia[1] • Yuanfeng Hao[1] • Zhonghu He[1,6] • Xianchun Xia[1]

[1] Institute of Crop Sciences，National Wheat Improvement Center，Chinese Academy of Agricultural Sciences (CAAS)，12 Zhongguancun South Street，Beijing 100081，China

[2] Shandong Provincial Key Laboratory of Dryland Farming Technology，College of Agronomy，Qingdao Agricultural University，Qingdao 266109，Shandong，China

[3] College of Agronomy，Northwest A & F University，Yangling 712100，Shaanxi，China

[4] College of Agronomy，Henan Agricultural University，63 Agricultural Road，Zhengzhou 450002，Henan，China

[5] Institute of Crop Germplasm Resources，Heilongjiang Academy of Agricultural Sciences，Harbin 150086，Heilongjiang，China

[6] International Maize and Wheat Improvement Center (CIMMYT) China Office，c/o CAAS，12 Zhongguancun South Street，Beijing 100081，China

Key message: A stable QTL *QPm. caas-3BS* for adult-plant resistance to powdery mildew was mapped in an interval of 431 kb，and candidate genes were predicted based on gene sequences and expression profiles.

Abstract: Powdery mildew is a devastating foliar disease occurring in most wheat-growing areas. Characterization and fine mapping of genes for powdery mildew resistance can benefit marker-assisted breeding. We previously identified a stable quan-titative trait locus (QTL) *QPm. caas-3BS* for adult-plant resistance to powdery mildew in a recombinant inbred line population of Zhou8425B/Chinese Spring by phenotyping across four environments. Using 11 heterozygous recombinants and high-density molecular markers，*QPm. caas-3BS* was delimited in a physical interval of approximately 3. 91 Mb. Based on re-sequenced data and expression profiles，three genes *TraesCS3B02G014800*，*TraesCS3B02 G016800* and *TraesCS3B02G019900* were associated with the powdery mildew resistance locus. Three gene-specific kompetitive allele-specific PCR (KASP) markers were developed from these genes and validated in the Zhou8425B derivatives and Zhou8425B/Chinese Spring population in which the resistance gene was mapped to a 0. 3 cM interval flanked by *KASP14800* and *snp _ 50465*，corresponding to a 431 kb region at the distal end of chromosome 3BS. Within the interval，*TraesCS3B02G014800* was the most likely candidate gene for *QPm. caas-3BS*，but *TraesCS3B02G016300* and *TraesCS3B02G016400* were less likely candidates based on gene annotations and sequence variation between the parents. These results

Published in Theoretical and Applied Genetics，2022，135 (3)：1083-1099. (doi. org/10. 1007/s00122-021-04019-2)

not only offer high-throughput KASP markers for improvement of powdery mildew resistance but also pave the way to map-based cloning of the resistance gene.

Abbreviations：

APR	Adult-plant resistance
ASR	All-stage resistance
Bgt	*Blumeria graminis* f. sp. *Tritici*
BLUE	Best linear unbiased estimate
FPKM	Fragments per kilobase per million reads
KASP	Kompetitive allele-specific PCR
MDS	Maximum disease severity
NIL	Near-isogenic line
PCR	Polymerase chain reaction
qPCR	Quantitative real-time PCR
QTL	Quantitative trait locus
RIL	Recombinant inbred line
SNP	Single nucleotide polymorphism

Introduction

Bread wheat as one of the most important staple crops provides nearly one-fifth of dietary calories for human beings (Reynolds et al. 2012). Biotic and abiotic stresses remain a large constraint to wheat production (Gao et al. 2018). Powdery mildew, caused by *Blumeria graminis* f. sp. *tritici* (*Bgt*), is one of the most devastating fungal diseases in wheat across many regions of the world (Juroszek and von Tiedemann 2013). Breeding resistant wheat cultivars is a more effective, economic and environment-friendly strategy to control powdery mildew than is the use of fungicides.

Plants evolved with several mechanisms to defend against pathogens. Disease resistance in wheat has been classified as all-stage resistance (ASR) and adult-plant resistance (APR) (Chen 2005; Lin and Chen 2007). ASR usually confers a hypersensitive response following recognition of pathogen-derived effector molecules by host receptors (Bent and Mackey 2007), whereas APR seems to be mechanistically distinct from ASR whereby individual resistance genes are race non-specific and act quantitatively to provide partial resistance at post-seedling growth stages.

To date, more than 100 powdery mildew (*Pm*) resistance alleles at 65 loci (*Pm1-Pm68*) have been catalogued in bread wheat and its close relatives (McIntosh et al. 2017; Zhang et al. 2019; He et al. 2020). Twelve ASR alleles, viz. *Pm1a* (Hewitt et al. 2020), *Pm2* (Sánchez-Martín et al. 2016), *Pm3* (Yahiaoui et al. 2004), *Pm4* (Sánchez-Martín et al. 2021), *Pm5e* (Xie et al. 2020), *Pm8* (Hurni et al. 2013), *Pm17* (Singh et al. 2018), *Pm21* (He et al. 2018; Xing et al. 2018), *Pm24* (Lu et al. 2020), *Pm41* (Li et al. 2020), *Pm60* (Zou et al. 2018) and *WTK4* (Gaurav et al. 2021), have been cloned and characterized. Among them, *Pm4* encodes a MCTP kinase (multiple C2-domains and transmembrane region kinase protein), *Pm24* (*WTK3*) and *WTK4* encode a wheat tandem kinase protein and the other nine *Pm* genes encode coiled-coil nucleotide-binding site leucine-rich repeat proteins (Sánchez-Martín and Keller 2021). Due to ease of selection and high effectiveness, ASR genes have been used more widely than APR genes in wheat breeding. However, excessive use of single resistance genes leads to rapid evolution of new virulent *Bgt* races and consequent huge economic losses (Zeng et al. 2014; Singh et al. 2016). In contrast, race-non-specific APR genes tend to provide broad-spectrum resistance to one or multiple diseases as exemplified by genes multiply named as *Pm39/Lr46/Yr29/Sr58* (Singh et al. 1998), *Pmx/Lr27/Yr30/Sr2* (Singh et al. 2000b; Mago et al. 2011b), *Pm38/Lr34/Yr18/Sr57* (hereafter referred as *Lr34*) (Krattinger et al. 2009) and *Pm46/Lr67/Yr46/Sr55* (hereafter referred as *Lr67*) (Moore et al. 2015). Previous studies revealed that pyramiding 4-5 APR genes can achieve high resistance or near immunity to leaf rust and stripe rust (Singh et al. 2000a, 2005; Lu et al. 2009). Thus, development of wheat cultivars that combine several quantitative resistance loci should be important breeding objectives.

Although nearly 200 quantitative trait loci (QTL) for powdery mildew resistance have been reported (McIntosh et al. 2014, 2016, 2017; Liu et al. 2017, 2021; Jia et al. 2018; Kang et al. 2020; Xu et al. 2020), most of them were detected with linked rather than co-segregating markers. Therefore, fine mapping of resistance QTL with minor to moderate effect and investigation of the underlying mechanism will greatly assist the deployment of varieties with durable resistance based on multiple genes.

Due to the large complex hexaploid wheat genome with many repetitive sequences, gene cloning is a huge challenge, particularly for genes with minor effect. The cloned APR genes for powdery mildew resistance at present are *Pm38* and *Pm46*. Advances in next-generation sequencing technology and increasingly released genomic and transcriptomic sequence data greatly facilitate fine mapping and cloning of wheat genes (Qi et al. 2004; Brenchley et al. 2012; IWGSC 2018; Ramírez-González et al. 2018).

QPm. caas-3BS was previously mapped between SNP markers *IWB21064* and *IWB64002* on chromosome 3BS in a Zhou8425B/Chinese Spring population, explaining 4.4-9.1% of the phenotypic variances across four environments (Jia et al. 2018). Two markers flanking *QPm. caas-3BS* were at 4.59 Mb and 11.08 Mb,

respectively, overlap-ping the genomic region of *Pmx/Lr27/Yr30/Sr2* located at 5.80-6.56 Mb (Mago et al. 2011b). Therefore, fine mapping of *QPm. caas-3BS* and characterization of its relationship with *Pmx* will be important in understanding this potentially durable resistance gene. The aims of the present study were to: (1) fine map *QPm. caas-3BS* and predict candidate genes based on annotated gene sequences and expression profiles and (2) develop kompetitive allele-specific PCR (KASP) markers for marker-assisted breeding.

Materials and methods

Whole-genome re-sequencing of Zhou8425B

The DNA library of Zhou8425B was constructed using NEB Next © Ultra™ DNA Library Prep Kit for Illumina (NEB, USA) following the manufacturer's recommendations. The DNA library was sequenced on an Illumina Highseq platform at Beijing Novogene Bioinformatics Company Ltd. (Beijing, http://www.novogene.com), generating 150 bp paired-end reads. The original image data generated by the sequencer were converted into sequence data via base calling (Illumina pipeline CASAVA v1.8.2) and then subjected to a quality control procedure to remove unusable reads, including those containing the Illumina library construction adapters or more than 10% of unknown bases (N bases), and those with one end having more than 50% of low-quality bases (sequencing quality value ≤ 5). After quality control, the reads were aligned to the Chinese Spring reference genome (ftp://ftp.ensemblgenomes.org/pub/release-41/plants/fasta/triticum_aestivum/dna/Triticum aestivum.IWGSC.dna.toplevel.fa.gz) using BWA 0.7.8-r455 with default parameters (Li and Durbin 2009). Duplications were removed using SAMtools (Li et al. 2009) and Picard (http://picard.sourceforge.net). The raw SNP/InDel sets were called by SAM-tools with the parameters set as "-q 1 -C 50 -m 2 -F 0.002 -d 1000." The data sets were then filtered following the criteria of the mapping quality > 20, and the depth of the variate position > 4. ANNOVAR (Wang et al. 2010) was used for functional annotation of variants. Know genes datasets available at the University of California Santa Cruz (UCSC) (Hsu et al. 2006) were used for gene annotations.

Marker development based on re-sequenced data

The physical interval delimiting *QPm. caas-3BS* was from 4. 59 to 11. 08 Mb on chromosome 3BS (Jia et al. 2018); 40 markers were developed in the current study based on a com-parison of re-sequenced data for Zhou8425B and genome sequences of Chinese Spring (IWGSC 2018; https://urgi. versailles. inra. fr/ blast_iwgsc/). Corresponding sequences of SNPs for targeted regions were submitted to PolyMarker (https://www. polymarker. info/) for development of KASP markers (Ramirez-Gonzalez et al. 2015). Chromosome 3B-specific primers were preferentially chosen based on results obtained from the PolyMarker website. Alternatively, primers were selected if they were chromosome 3A and chromosome 3B-specific or chromosome 3D and chromo-some 3B-specific. Primers tailed with FAM and HEX overhangs were synthesized by Sangon Biotech (https://www. sangon. com/ integral). All the synthesized KASP primers were diluted to 100 μM, and the primer mixture consisted of 46 μl of ddH$_2$O, 30 μl of common reverse primer and 12 μl of each of the FAM and HEX primers. KASP assays were performed in 384-well format in an approximately 3 μl mixture comprising 1. 5 μl of 2 $\times$ KASP mix (LGC, UK), 1. 5 μl of ddH$_2$O and 0. 0336 μl of primer mixture. PCR was run for an initial denaturation of 15 min at 94℃, followed by 10 touchdown cycles (94℃ for 20 s, touchdown at 65℃ initially and decreasing by 0. 8℃ per cycle for 60 s), and 35 additional cycles of annealing and extension (94℃ for 20 s, 57℃ for 60 s). Genotyping of the markers followed Wu et al. (2020). Three to six more cycles of annealing and extension were used in the PCR program if the genotyping results were not good enough. Development of primers for Sanger sequencing to confirm the presence of SNPs was based on re-sequenced data for Zhou8425B. PCR products amplified in Chinese Spring and Zhou8425B were used for Sanger sequencing. Primers that were polymorphic

between Chinese Spring and Zhou8425B were then used to genotype recombinant lines screened from two BC$_1$F$_2$ (g93/2 $\times$ g16 and g115/2 $\times$ g5) populations. Sequences of KASP markers and primers for Sanger sequencing are shown in Tables S1 and S2.

Secondary population construction

Six F$_{2:8}$ RILs (g5, g16, g37, g93, g115 and g160) from the Zhou8425B/Chinese Spring cross previously studied by Jia et al. (2018) were chosen to construct secondary mapping populations (Table S3) in four crosses (hereafter named Crosses 1-4, Table S4). The reasons for constructing secondary populations were that the parents for each cross had different alleles only within *QPm. caas-3BS*, while other loci were identical based on genotyping with wheat 90 K SNP array. Then, lines with the susceptible allele within *QPm. caas-3BS* were used as recurrent parents. The disease severities of the six lines and F$_1$ of Crosses 1-4 were evaluated, and their genotypes were also diagnosed with marker *3B5476* (at 5. 67 Mb).

Identification of recombinants in the *QPm. caas-3BS* interval

A total of 2027 BC$_1$F$_2$ plants from two crosses (Crosses 2 and 4) were screened to identify recombinants in the *QPm. caas-3BS* interval. DNA was extracted from leaves and screened with flanking KASP markers *KASP173*, *KASP3BH41*, *KASP3 BS10* and *KASP9058*. One hundred and three recombinants were identified and confirmed by Sanger sequencing. The recombinants were then genotyped with four additional markers (*KASP573*, *3B5000*, *3B6510* and *KASP9007*) to further distinguish recombinant types. To obtain near-isogenic lines (NILs), six kinds of heterozygous recombinants were sown in 16 $\times$ 8 plug trays. After genotyping all plants with markers *KASP3BH41*, *KASP573*, *KASP3B21*, *KASP9058*, *KASP3BS30* and *KASP3BH9*, two contrasting homozygous progenies from each heterozygous recombinant were grown to recover seeds of NILs.

Field trials and scoring powdery mildew response

The six lines selected from the RIL population and F$_1$

of Crosses 1-4 were planted at Beijing during the 2018-2019 cropping season. The BC_1 F_2 populations of Crosses 2 and 4 were sown at Beijing and Zhengzhou during the 2019-2020 growing season. Eleven kinds of heterozygous recombinants identified from the two BC_1 F_2 populations were sown in 1.5 m rows with 0.3 m spacing at 16 seeds per row at Beijing during 2020-2021. These 11 recombinants were used to generate segregating families (designated as L1-L11) (Fig. 1b). Within each family, homozygous non-recombinant lines with Zhou8425B and Chinese Spring alleles were designated as 3B+ NILs and 3B- NILs, respectively. Among the progeny of the 11 recombinants, 11-49 and 9-54 plants from individual recombinants with 3B+ and 3B- genotypes, respectively, were used to determine phenotypic differences between contrasting genotypes (Fig. 1; Table 1). Six sets of NILs derived from L3, 5, 6, 7, 10 and 11 were phenotyped at Beijing during 2020-2021, grown in 1 m rows spaced 0.3 m apart and arranged in randomized complete blocks with three replications. NILs with 3B- and 3B+ genotypes derived from L6, L5 and L7 were designated as NIL-S1 and NIL-R1, NIL-S2 and NIL-R2, NIL-S3 and NIL-R3, respectively. A panel of 103 Zhou8425B derivatives was grown at Beijing, Zhengzhou and Xingyang during the 2016-2017 cropping season and used for validating the functional markers (Jia et al. 2018).

The highly susceptible cultivar Jingshuang16 used as an inoculum spreader was grownperpendicular and adjacent to the test rows and planted at every tenth row as a control. All the materials were covered with plastic sheets to prevent freezing when winter temperatures were below zero, with the plastic sheets being removed in the following spring. Seedlings of Zhou8425B, Chinese Spring and 240 RILs were susceptible to isolates E09 and E20 according to Jia et al. (2018). Thus, a mixture of E09 and E20 was used to infect the materials at the jointing stage. The mean MDS (maximum disease severity) scores were evaluated according to the Cobb scale (Peterson et al. 1948) with minor modification when disease severities on the susceptible control exceeded 85%.

Statistical analysis

All statistical analyses were performed using SAS version 9.2. Phenotypic differences between homozygous lines with contrasting alleles were compared by Student's *t* tests. The difference in relative expression between parents (Chinese Spring vs. Zhou8425B, g5 *vs.* g115 and g16 *vs.* g93) and between NILs (NIL-S1 *vs.* NIL-R1, NIL-S2 *vs.* NIL-R2 and NIL-S3 *vs.* NIL-R3) was also compared by Student's *t* tests.

RNA sequencing and relative gene expression analysis by quantitative real-time PCR (qPCR)

QPm. caas-3BS is an adult-plant resistance gene that is effective in adult plants during the grain-filling stage. One piece of flag leaf from individual plants was sampled at three weeks post-anthesis for extraction of RNA and used for RNA-Seq and qPCR analysis. Twelve homozygous plants with 3B+ and 12 with 3B-were chosen from each set of NILs from L5, L6 and L7, respectively, and formed six bulks that were sent to Novogene (http://www.novogene.com/) for RNA sequencing. qPCR was performed on a BioRad CFX system with the ABclonal 2 × Universal SYBR Green Fast qPCR mix (ABclonal, Wuhan). Total RNA from three pairs of parents (Zhou8425B and Chinese Spring, g5 and g115, g16 and g93) was extracted from leaves by a TransGen plant RNA kit. One μg of RNA from each sample was reverse-transcribed to first-strand cDNA synthesized by a Prime-Script RT regent kit (Takara, Dalian). Genes in the *QPm. caas-3BS* interval with missense mutations between Chinese Spring and Zhou8425B and with differential expression between pairs of NILs from L5, L6 and L7, respectively, based on RNA-Seq, were validated by qPCR. Gene-specific qPCR primers were designed with Geneious software (https://www.geneious.com/) according to the Chinese Spring genome sequence RefSeq v1.1 (Table S5). Relative expression levels were normalized to actin gene expression by the $2^{-\Delta\Delta CT}$ method (Livak and Schmittgen 2001).

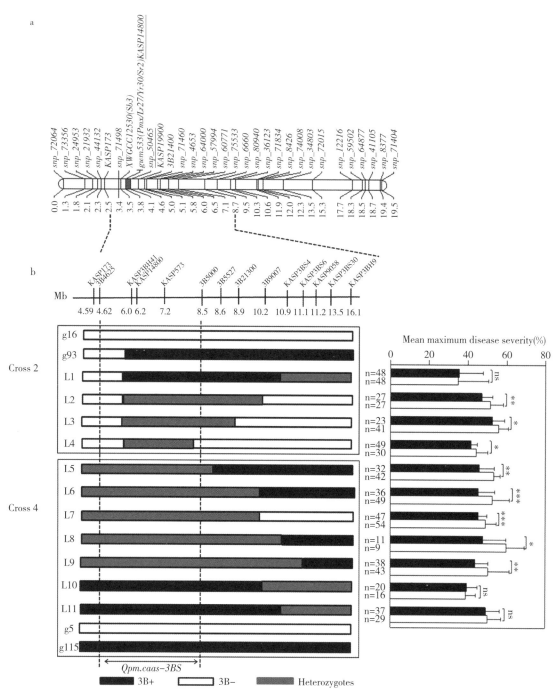

Fig. 1　Molecular mapping of *QPm. caas-3BS*.（a）Genetic map of chromosome 3BS shows QTL for maximum disease severity（MDS）in the Zhou8425B/Chinese Spring RIL population；the QTL mapping region is labeled green with flanking markers *KASP14800* and *snp _ 50465* underlined. （b）Physical map of 11. 5 Mb interval on chromosome 3BS. Nine markers（topside）were used to screen recombinants，and graphical genotypes are shown for 11 recombinant lines（L1-L11）；black，white and gray rectangles represent the genotypes of Zhou8425B（3B+genotype），Chinese Spring（3B−genotype）and heterozygotes，respectively. Statistical analysis of MDS between 3B+ and 3B−genotypes from self-pollinated progenies of individual heterozygous recombinants is shown at the right. ＊，＊＊，＊＊＊ and ns，significant at $P<0.05$，$P<0.01$，$P<0.001$ and non-significant，respectively（color figure online）

Table 1 Mean maximum disease severities (MDS) of 3B+ and 3B − progenies from individual recombinant lines L1-L11

Line	Genotype [a]	No. [b]	Mean MDS (%)
L1	3B+	48	35. 5±12. 3[c]
	3B−	48	34. 9±15. 7
			0. 2
L2	3B+	27	47. 1±5. 6
	3B−	27	51. 4±6. 8
			−2. 5[d] *
L3	3B+	23	52. 6±6. 4
	3B−	41	55. 8±5. 0
			−2. 2*
L4	3B+	49	41. 5±3. 5
	3B−	30	44. 3±6. 0
			−2. 2*
L5	3B+	32	45. 9±10. 0
	3B−	42	53. 2±11. 4
			−2. 8*
L6	3B+	36	45. 2±8. 7
	3B−	49	52. 6±8. 8
			−3. 8***
L7	3B+	47	45. 3±4. 6
	3B−	54	49. 1±5. 6
			−3. 8**
L8	3B+	11	47. 7±12. 5
	3B−	9	59. 8±9. 9
			−2. 3*
L9	3B+	38	43. 8±6. 8
	3B−	43	49. 4±11. 3
			−2. 6*
L10	3B+	20	39. 5±5. 5
	3B−	16	39. 1±5. 3
			0. 2
L11	3B+	37	49. 2±7. 3
	3B−	29	50. 1±7. 2
			−0. 5

Asterisks indicate significant differences determined by Student's *t* test

*, ** and ***Significant at $P < 0. 05$, $P < 0. 01$ and $P < 0. 001$, respectively

[a]3B+ and 3B − indicate lines with Zhou8425B and Chinese Spring alleles, respectively

[b]Number of lines for corresponding genotypes

[c]Data are means ± SD (standard deviation)

[d]Phenotypic difference between the mean MDS of genotypes 3B+ and 3B−

Genetic map construction and QTL analysis

A genetic map of chromosome 3BS was constructed by Join Map 4. 0 (Stam 1993) based on the wheat 90 K SNP array, three gene-specific markers (*KASP14800*, *KASP19900* and *3B21400*) and two SSR markers tightly linked to *Pmx/Lr27/Yr30/Sr2* (*gwm533*) and *Sb3* (*XWGCC12530*) (Mago et al. 2011b; Lu et al. 2016). QTL mapping was performed with ICIM (inclusive composite interval mapping) in IciMapping 4. 2 software (https: //www. isbreeding. net) with a LOD (logarithm of odds) threshold of 2. 5 based on 2000 permutation tests at $P < 0. 01$. Considering the influence of environments on disease severity, the MET (multi-environment trials) function in IciMapping was applied to QTL mapping. The phenotypic data from the RIL population used in QTL analysis were reported in Jia et al. (2018).

Results

Analysis of re-sequenced data from Zhou8425B

A total of 379. 6 G clean sequenced data with 150 bp paired-end reads were generated from the Zhou8425B genome. These data were mapped to the Chinese Spring reference genome, with mapped reads, total reads and mapping rate of 2, 497, 793, 240, 2, 530, 340, 764 and 98. 71%, respectively. A total of 25, 832, 296 SNPs were detected between Chinese Spring and Zhou8425B, among which non-synonymous, stop-gain and stop-loss SNPs were 75, 194, 1076 and 96, respectively. The SNPs in the mapped interval of *QPm. caas-3BS* from 4 to 17 Mb were used to develop molecular markers. Genes with non-synonymous SNPs in the interval of *QPm. caas-3BS* were accepted as reference points that were likely associated with powdery mildew resistance.

Evaluation of genotypes and phenotypes of Crosses 1 to 4

The genotypes and phenotypes of Crosses 1-4 are provided in Table S4. The genotypes of parents g16 and g160 and F_1 in Cross 1 were 2, 0 and 1, representing the Chinese Spring, Zhou8425B and F_1 genotypes, respectively, and their mean MDS were 50%, 55% and 50%, respectively. The mean MDS of g160 (with the Zhou8425B allele) was higher than that of g16 (with the Chinese Spring allele). For Cross 3, the genotypes of parents g160 and g37 and the F_1 were 0, segregating 0/2 and segregating 0/1, respectively. The recurrent parent (g37) was heterozygous when tested with marker 3B5476 (at 5.67 Mb), thus accounting for the segregation in the F_1 generation. As a consequence, Crosses 1 and 3 were not used in subsequent analyses. The genotypes of parents g93 and g16 and F_1 in Cross 2 were 0, 2 and 1, with mean MDS of 3%, 55% and 30%, respectively. For Cross 4, the genotypes of parents g5 and g115 and F_1 were 2, 0 and 1, respectively, with mean MDS of 30%, 7% and 30%, respectively. As the genotypes and phenotypes of Crosses 2 and 4 met our expectation, they were used in backcrossing and construction of advanced segregating populations.

Generation of secondary mapping populations

F_1 plants in Crosses 2 and 4 were backcrossed with lines g16 and g5, respectively. The genotypes of two $BC_1 F_1$ groups were characterized, and heterozygous plants were self-pollinated for screening heterozygous recombinants in the 2019-2020 cropping season. KASP markers KASP173, KASP573, KASP9026 and KASP9058 were used to identify recombinants in the two $BC_1 F_2$ populations. One hundred and three recombinants were identified from 2027 $BC_1 F_2$ plants, including 87 heterozygous and 16 homozygous recombinants. These recombinants were classified into 17 recombinant types using new molecular markers in the interval flanked by KASP573 and KASP9026 (Fig. S1).

Mapping of *QPm. caas-3BS* using secondary populations and NILs

For mapping a candidate gene, 28 new polymorphic markers between Chinese Spring and Zhou8425B were developed based on the 90 K SNP array and re-sequenced data (Tables S1 and S2). To narrow down the interval of *QPm. caas-3BS*, 11 candidate

recombinants were grown in the 2020-2021 cropping season. L1-4 and L5-11 came from Crosses 2 and 4, respectively. All the progeny plants were genotyped with six markers (KASP3BH41, KASP573, KASP3B21, KASP9058, KASP3BS30 and KASP3BH9) and evaluated for powdery mildew severity in the field. There were significant differences in mean severity between 3B+ and 3B− NILs within families L2-L9 ($P < 0.05$), whereas there were non-significant differences between contrasting genotypes within families L1, L10 and L11 (Fig. 1b; Table 1). These results indicated that the candidate gene was in a physical interval of approximately 3.91 Mb flanked by markers 3B4625 and 3B5000 (Fig. 1b; Table 1).

To further validate the results from the $BC_1 F_2$ populations, six pairs of NILs from each of L3, L5, L6, L7, L10 and L11 were used to verify the powdery mildew reactions. The averaged MDS of 3B+ NILs was 3.2 to 7.4% lower than those of 3B− NILs derived from L3, L5, L6 and L7, whereas differences in MDS between 3B+ and 3B− NIL pairs from L10 and L11 were not significant (Fig. 2; Table 2). These results were consistent with those of the $BC_1 F_2$ populations.

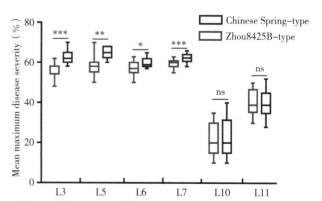

Fig. 2 Phenotypic differences in the 3B+ and 3B− near-isogenic lines (NILs) derived from recombinants L3, L5, L6, L7, L10 and L11, respectively. Chinese Spring-type and Zhou8425B-type mean lines with Chinese Spring and Zhou8425B alleles, respectively. *, **, *** and ns, significant at $P < 0.05$, $P < 0.01$, $P < 0.001$ and non-significant, respectively

Table 2   Mean maximum disease severities (MDS) of near-isogenic lines (NILs) 3B+ and 3B− from individual recombinant lines L3, L5, L6, L7, L10 and L11

Line	Genotype [a]	No. [b]	Mean MDS (%) (%) (g)
L3	3B+	32	56.3±3.7[c]
	3B−	42	62.4±2.9
			−7.9***[d]
L5	3B+	12	58.3±4.7
	3B−	9	64.5±2.8
			−3.7**
L6	3B+	23	57.5±3.4
	3B−	14	59.8±2.7
			−2.2*
L7	3B+	18	59.5±2.1
	3B−	14	62.4±2.3
			−3.7**
L10	3B+	13	22.2±8.8
	3B−	12	23.1±10.1
			−0.2
L11	3B+	12	40.3±6.2
	3B−	15	40.1±6.7
			0.1

*, ** and ***Significant at $P<0.05$, $P<0.01$ and $P<0.001$, respectively

[a] 3B+ and 3B− indicate lines with Zhou8425B and Chinese Spring alleles, respectively

[b] Number of lines for corresponding genotypes

[c] Data are means ± SD (standard deviation)

[d] Phenotypic differences between the means of genotypes 3B+ and 3B−; asterisks indicate significant differences determined by Student's *t* test

Prediction of candidate genes within *QPm. caas-3BS*

The *QPm. caas-3BS* interval contained 98 high-confidence genes based on annotated Chinese Spring reference genome RefSeq v1.1 (IWGSC 2018; https://urgi.versailles.inra.fr/blast_iwgsc/). The re-sequenced data from Zhou8425B indicated that there were 48 genes with missense mutations between Zhou8425B and Chinese Spring (Table S6). Analysis of expression patterns using RNA-Seq results (Table S7) indicated that 21 were not expressed in both 3B+ and 3B− NILs; 13 showed a lower expression level in 3B+ NILs than in 3B− NILs; FPKM(fragments per kilobase per million reads) of six genes in both genotypes were below five; and two genes showed no differential expression between contrasting genotypes.

Six genes that had missense mutations and differential expression patterns between the 3B+ and 3B− NILs were associated with powdery mildew resistance (Tables S6 and S7). Among these six genes, the FPKM of *TraesCS3B02G014800* were highest followed by *TraesCS3B02G019900* (Fig. 3a). qPCR assays were performed on three pairs of parents (Chinese Spring and Zhou8425B, g5 and g115, g16 and g93) and three pairs of NILs (NIL-S1 and NIL-R1, NIL-S2 and NIL-R2, NIL-S3 and NIL-R3) to validate the RNA-Seq results for the six genes (Fig. 3b; Tables 3 and S8). Considering that the 3B+ genotypes could have higher expression levels than 3B− genotypes, there were inconsistencies between genotypes and relative expression levels in *TraesCS3B02G013100*, *TraesCS3B02G017400* and *TraesCS3B02G018400*. In contrast, the genotypes and relative expression levels were consistent for *TraesCS3B02G014800*, *TraesCS3B02G016800* and *TraesCS3B02G019900*, whereby 3B+ genotypes showed higher relative expression than 3B− genotypes in most 3B+/3B− pairs, and no 3B− genotype had higher expression than its 3B+ counterpart for any NIL pair. Therefore, *TraesCS3B02G014800*, *TraesCS3B02G016800* and *TraesCS3B02G019900* were considered to be associated with powdery mildew resistance.

Development of KASP markers and fine mapping of causal genes

Gene-specific KASP markers *KASP14800*, *KASP16800* and *KASP19900* were developed based on the sequences of *TraesCS3B02G014800*, *TraesCS3B02G016800* and *TraesCS3B02G019900* and validated in the Zhou8425B/Chinese Spring population. Simultaneously, 13 markers reported in Mago et al. (2011a, b) and Lu et al. (2016) were screened in Zhou8425B and Chinese Spring, among which *gwm533* (located 1.6 cM from *Pmx/Lr27/Yr30/Sr2*; Spielmeyer et al. 2003) and *XWGCC12530* (0.07 cM from *Sb3*; Lu

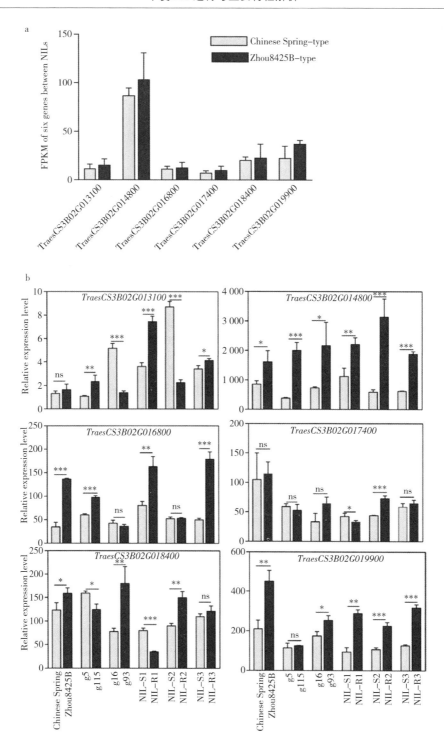

Fig. 3　Comparison of FPKM（fragments per kilobase per million reads）and relative expression levels between parents（Chinese Spring and Zhou8425B，g5 and g115，g16 and g93）and between NILs（NIL-S1 and NIL-R1，NIL-S2 and NIL-R2，NIL-S3 and NIL-R3）.（a）DEGs，differential expression genes. NIL，near-isogenic lines. FPKM values of *TraesCS3B02G013100*，*TraesCS3B02G014800*，*TraesCS3B02G016800*，*TraesCS3B02G017400*，*TraesCS3B02G018400* and *TraesCS3B02G019900* in three sets of NILs based on transcriptomic data. For each gene，the FPKM value of Chinese Spring-type was the average FPKM of NIL-S1，NIL-S2 and NIL-S3，and the FPKM value of Zhou8425B-type was the average of NIL-R1，NIL-R2 and NIL-R3. Values are represented as means ± SD（standard deviation）.（b）Relative expression levels of six genes. From left to right are Chinese Spring and Zhou8425B，g5 and g115，g16 and g93，NIL-S1 and NIL-R1，NIL-S2 and NIL-R2，NIL-S3 and NIL-R3，respectively. Values are represented as means ± SD（standard deviation）from three replicates；*，**，*** and ns，significant at *P*<0.05，*P*<0.01，*P*<0.001 and non-significant，respectively

et al. 2016) were polymorphic between Chinese Spring and Zhou8425B and validated in the RIL population. The powdery mildew resistance gene was then mapped to a 0.3 cM interval flanked by *KASP14800* (at 6.24 Mb) from *TraesCS3B02G 014800* and *snp _ 50465* (at 6.67 Mb) from a SNP in the wheat 90 K SNP array based on a new linkage map for the Zhou8425B/Chinese Spring population (Figs. 1a and 4). The physical interval spanning *KASP14800* and *snp _ 50465* was 431 kb, including 16 high-confidence genes, among which nine had missense mutations between Zhou8425B and Chinese Spring. Among the 16 genes, *TraesCS3B02G014800* showed much higher relative expression levels in 3B+ genotypes than in 3B− genotypes and was considered the most likely candidate gene. However, *TraesCS3B02G016300* and *TraesCS3B02G016400* were also possible candidates for *QPm. caas-3BS* based on gene annotations and sequence variations between the parents (Table S6).

Three KASP markers and two SSR markers were also validated in Zhou8425B derivatives. For *TraesCS3 B02G 014800* and *TraesCS3B02G019900*, the Zhou 8425B genotypes showed significantly lower mean MDS than Chinese Spring genotypes in 103 Zhou8425B derivatives, reducing mean MDS by 4.7% and 4.6%, respectively (Fig. 5; Tables 4 and S9). For *TraesCS3B02G016800*, *gwm533* and *XWGCC12530*, no significant difference in mean MDS was observed between contrasting genotypes, but the average MDS of Zhou8425B genotypes was lower than Chinese Spring genotypes.

Effects of *QPm. caas-3BS* on response to stripe rust and powdery mildew

The resistance allele proposed here to be *Pmx* reduced the powdery mildew MDS by 3.2-7.4% and 4.6-4.7% in NILs and Zhou8425B derivatives, respectively. Three gene-specific *KASP markers KASP14800*, *KASP16800* and *KASP19900* were also genotyped in the RIL population derived from Jingshuang16/ Bainong64. The QTL for stripe rust resistance is closely linked with *KASP14800*, explaining 8.5% of the phenotypic variance for stripe rust resistance (Fig. S2). The resistance allele was derived from Jingshuang16 that also has the resistance allele at *QPm. caas-3BS*. In addition, the mean MDS of powdery mildew of Jingshuang16 genotype was 3.1% lower than that of Bainong64 genotype in the Jingshuang16/Bainong64 population.

Discussion

Strategy for fine mapping of minor-effect QTL

Marker density, exchange frequency and accurate scoring of recombinant phenotypes are essential for fine mapping of causal genes in minor-effect QTL (Yang et al. 2012). Highdensity linkage maps and advanced genome sequencing technology pave the way for this to occur (Holland 2007; Daetwyler et al. 2010). NIL and CSSL (chromosome segment substitution lines) are frequently used to narrow down target QTL regions (Wissuwa et al. 2002; Ashikari et al. 2005; Schmalenbach et al. 2011; Yang et al. 2012). However, construction of ideal NIL and CSSL is laborious and time-consuming and is more suitable for large-effect QTL.

No residual heterozygous recombinants were detected in the *QPm. caas-3BS* region delineated by Jia et al. (2018). We therefore developed secondary mapping populations by crossing and backcrossing lines selected from the RIL population. Apart from *QPm. caas-3BS*, three other QTL for powdery mildew resistance were detected in the Zhou8425B/Chinese Spring RIL population. We selected lines to make crosses based on genotypes identified with 90 K SNP array to construct secondary mapping populations and ensured that the contrasting parents for each cross had different genotypes only in the *QPm. caas-3BS* region with the genetic backgrounds being identical. The principle of making crosses in this way ensured that segregation occurred only within the *QPm. caas-3BS* region. In addition, we evaluated the phenotypes and genotypes of the parents and F$_1$ of four crosses. Crosses 2 and 4 with expected phenotypes and genotypes were selected

Table 3 Relative expression levels of six genes in three pairs of parents (Chinese Spring and Zhou8425B, g5 and g115, g16 and g93) and three sets of NILs (NIL-S1 and NIL-R1, NIL-S2 and NIL-R2, NIL-S3 and NIL-R3)

Gene ID	P1		P2		P3		NIL-1		NIL-2		NIL-3	
	Chinese Spring	Zhou 8425B	g5	g115	g16	g93	NIL-S1	NIL-R1	NIL-S2	NIL-R2	NIL-S3	NIL-R3
TraesCS3B02G013100	1.3±0.6a[a][b]	1.3±0.6a	1±0a	2.3±0.6b	5±0a	1.3±0.6b	3.7±0.6a	7.3±0.6b	8.7±0.6a	2.0±0b	3.3±0.6a	4±0b
TraesCS3B02G014800	854±119a	1604±391b	382±21a	2000±270b	735±33a	2169±787b	1118±283a	2207±231b	593±81a	3139±620b	622±12a	1874±9b
TraesCS3B02G016800	35±10a	137±2b	60±2a	98±4b	43±7a	36±4a	80±9a	163±21b	52±4a	53±2a	50±4a	179±1b
TraesCS3B02G017400	105±45a	115±20a	59±4a	53±10a	33±14a	65±11a	42±6a	33±2b	44±1a	73±5b	59±6a	65±6a
TraesCS3B02G018400	123±16a	159±11b	159±4a	125±11b	78±7a	180±36b	80±5a	35±1b	90±6a	150±13b	110±6a	122±1a
TraesCS3B02G019900	210±44a	451±54b	114±24a	125±1a	174±22a	253±24b	94±22a	288±19b	106±9a	225±18b	126±5a	317±17b

[a] Data are means ± SD (standard deviation).

[b] Differences of expression levels between two genotypes followed by different letters a and b are significant at $P < 0.05$.

and used to generate $BC_1 F_2$ populations. After genotyping and phenotyping progenies from 11 heterozygous recombinants, the causal resistance gene was delimited to a physical interval of approximately 3.91 Mb. Six sets of NILs were generated and evaluated in replicated trials to validate the results from the $BC_1 F_2$ populations.

We re-constructed the genetic linkage map for 3BS in the Zhou8425B/Chinese Spring population using markers developed from three genes annotated associated with powdery mildew resistance and those tightly linked to *Pmx/Lr27/Yr30/Sr2*. Finally, the resistance gene was mapped to a 0.3 cM interval flanked by *KASP14800* and *snp_50465*, corresponding to a 431 kb region (Figs. 1a and 4). This strategy confirmed the feasibility of fine mapping minor-effect genes in common wheat.

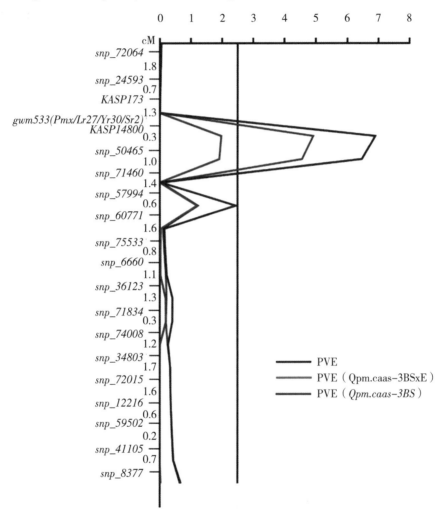

Fig. 4　LOD (logarithm of odds) contours obtained by inclusive composite interval mapping of *QPm. caas-3BS* in the Zhou8425B/Chinese Spring RIL population. Short arms are at the top. Genetic distances are shown in centiMorgans to the left of the vertical axis. LOD threshold of 2.5 is indicated by a dashed vertical line in the graphs. QTL mapping was conducted by Ici Mapping v 4.2 based on the MET (multi-environment traits) function (https://www. isbreeding. net/). Blue, green and red lines are PVE, PVE (*QPm. caas-3BS* × E) and PVE (*QPm. caas-3BS*) and indicate scores for total effect, the effect of interaction between *QPm. caas-3BS* and environment and *QPm. caas-3BS* effect, respectively (color figure online)

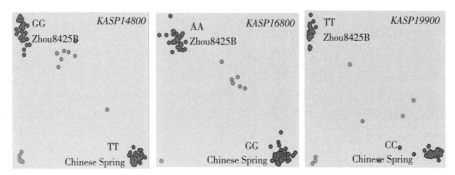

Fig. 5　Genotyping of 103 Zhou8425B derivative cultivars using three KASP markers *KASP14800*，*KASP16800* and *KASP19900*. Red，blue，green and pink dots represent Zhou8425B genotype，Chinese Spring genotype，heterozygotes and poorly genotyped DNA samples，respectively；the genotypes of the poorly DNA samples were determined by Sanger sequencing（color figure online）

Table 4　Mean maximum disease severities（MDS）of genotypes 3B＋ and 3B－ in Zhou8425B derivatives

Gene ID	Marker name	Genotype[a]	No. [b]	BLUE[c]
	XWGCC12530	3B＋	28	18.9±11.3
		3B－	71	21.0±10.6
				－0.9
	gwm533	3B＋	42	18.2±10.9
		3B－	30	20.0±8.9
				－0.7
TraesCS3B02G014800	*KASP14800*	3B＋	41	18.2±10.5
		3B－	48	22.9±11.2
				－2.0[*d]
TraesCS3B02G016800	*KASP16800*	3B＋	39	18.0±11.1
		3B－	59	22.2±10.4
				－1.9
TraesCS3B02G019900	*KASP19900*	3B＋	41	17.8±11.0
		3B－	58	22.4±10.6
				－2.1[*]

* significant at $P <0.05$.

[a] Genotypes identified using three gene-specific KASP markers，3B＋ and 3B－ indicate lines with Zhou8425B and Chinese Spring alleles，respectively.

[b] Number of cultivars for corresponding genotypes.

[c] BLUE：Best linear unbiased estimate from three environments，data are shown as means ± SD（standard deviation）.

[d] Phenotypic differences between two genotypes 3B＋ and 3B－. Asterisks indicate significant differences determined by Student's *t*-test.

Functional prediction of candidate genes in *QPm. caas-3BS*

The resistance gene was mapped to a gene-rich region in distal chromosome 3BS. There were 98 high-confidence genes in the mapping interval according to the Chinese Spring Ref-Seq v1.1 gene annotation. Following re-sequenced data and analysis of RNA-Seq profiles，genes *TraesCS3B02G014800*，*TraesCS3B02 G016800* and *TraesCS3B02G019900* were associated with powdery mildew resistance based on gene sequences and relative expression levels between three pairs of parents and three sets of NILs derived from L5，L6 and L7. *QPm. caas-3BS* was further narrowed down to a 431 kb interval that contained 16 genes based on the newly constructed linkage map using three KASP markers, two SSR markers and the wheat 90 K SNP array. *TraesCS3B02G014800* was considered the most likely candidate gene as it not only had sequence variation between Chinese Spring and Zhou8425B but also showed higher expression levels in all 3B+ genotypes than in 3B− geno-types. *TraesCS3B02G016300* and *TraesCS3B02G016400* also had sequence variations between Chinese Spring and Zhou8425B but with a lower expression level differences in three pairs of NILs. However, based on gene annotations, *TraesCS3B02G016300* and *TraesCS3B02G016400* encoding *Lr10*-like and O-acyltransferase proteins, respectively, are involved in plant immunity and were therefore potential causal genes for *QPm. caas-3BS*.

TraesCS3B02G014800 encodes an uncharacterized protein with numerous transmembrane helices. Full-length DNA and cDNA sequences of the gene were isolated from Zhou8425B and Chinese Spring (Fig. S3). Ten and 11 trans-membrane helices for Chinese Spring and Zhou8425B, respectively, were predicted based on the amino acid sequences (Figs. S4 and S5; Omasits et al. 2014). Thirty-one amino acid alterations were detected between Chinese Spring and Zhou8425B, among which there were 17 amino acids differences in the predicted transmembrane regions. Resistance gene *Lr67* encodes a predicted hexose transporter protein (LR67res) with 12 predicted transmembrane helices (Moore et al. 2015). Two amino acid changes in the transmembrane region distinguished the LR67res and LR67sus proteins (Moore et al. 2015). A tandem repeat of a 42 bp fragment in *TraesCS3B02G014800* was identified in the second exon of Chinese Spring relative to Zhou8425B. Additionally, there were many SNPs in the transmembrane helices between Chinese Spring and Zhou8425B. These sequence differences could be associated with powdery mildew resistance.

The gene annotation of *TraesCS3B02G016300* was a *Lr10*-like rust resistance kinase. *Lr10* located on chromosome 1AS in *Triticum aestivum* encodes a CC-NBS-LRR type of protein (Feuillet et al. 2003). *Lr10*

had 43% similarity with *RPM1* in *Arabidopsis thaliana* but was different from other leaf rust resistance genes such as *Lr1*, *Lr13* and *Lr21* (Huang et al. 2003; Feuillet et al. 2003; Cloutier et al. 2007; Yan et al. 2021). The amino acid sequence of *TraesCS3B02G016300* had a very low similarity with *Lr10* in common wheat but had higher similarity with *Lr10*-like homologues in *Aegilops tauschii* (XP_020171236.1), *Triticum dicoccoides* (XP_037403773.1), and *Brachypodium distachyon* (XP_003569125.1), ranging from 62 to 93%. Thus, *TraesCS3B02G016300* identified in this study was different from *Lr10* in common wheat.

TraesCS3B02G016400 encodes a O-acyltransferase WSD1 protein belonging to a plant acyltransferase family. The two known members of the acyltransferase family *Glossy2* gene in maize (Tacke et al. 1995) and *CER2* gene in *Arabidopsis thaliana* (Negruk et al. 1996; Xia et al. 1996) contribute to wax accumulation that has a role in water retention and protection from pathogen invasion (D'Auria 2006). *TraesCS3B02G016400* has a predicted wax ester synthase-like acyl-CoA acyltransferase domain that might be involved in response to pathogen infection. In the present study, the entire genome of Zhou8425B was sequenced on an Illumina Highseq platform, generating 379.6 G of clean data with 150 bp paired-end reads. The short reads of Zhou8425B were mapped to the Chinese Spring reference genome with a mapping rate of 98.71%, indicating that the re-sequenced data for Zhou8425B were high quality. We did not conduct a collinearity analysis between the genome sequences of Zhou8425B and Chinese Spring, because the genome sequence of Zhou8425B was not assembled at the chromosome level. However, we analyzed the reference genome sequence of Aikang58 (a variety derived from Zhou8425B; unpublished data provided by Professor Jizeng Jia at the institute of Crop Sciences, CAAS). The physical interval between markers *3B4625* and *3B5000* is approximately 7.5 Mb in Aikang58 and 4.0 Mb in Chinese Spring, indicating that an approximate 3.5 Mb fragment was absent in the Chinese Spring genome. We also conducted a collinearity analysis of the interval from 4. 62 to 8. 54 Mb among *T. aestivum* CS (IWGSC RefSeqv1. 1), *T. turgidum* Svevo (v1, Maccaferri et al. 2019), *T. aestivum* Zang1817 (Guo et al. 2020) and *T. aestivum* 10+Genome (Walkowiak et al. 2020). The results indicated that two fragments (approximately 0. 35 Mb and 3. 04 Mb, respectively) were absent in the Chinese Spring genome (Fig. S6, Chen et al. 2020). Fortunately, the 431 kb fragment comprising 16 genes was pre-sent in *T. aestivum* CS, Aikang58, Zang1817, Fielder (Sato et al. 2021) and 10 + Genome CDC Stanley and Julius; *T. turgidum* Svevo and *T. dicoccoides* Zavitan (v1, Avni et al. 2017). No genes in the 431 kb interval associated with plant immunity additional to those in CS were observed in the other *Triticum* genomes (Fig. S7; Table S10). The 16 genes in the 431 kb region showed high collinearity in the CS and other genomes, indicating that the candidates for *QPm. caas-3BS* predicted in this study based on CS RefSeq v1. 1 were reliable. The rapidly increasing number of assemblies of *Triticum* genome sequences greatly accelerates gene mapping and cloning in common wheat. Considering that disease resistance or other genes might be absent in the Chinese Spring reference genome sequences due to structural variation, these are valuable resources to conduct collinearity analyses and comparative genome analyses.

Genetic effect of *QPm. caas-3BS* and implication in wheat breeding

Several disease resistance genes have been mapped to chromosome 3BS. This includes the multiple resistance locus *Pmx/Lr27/Yr30/Sr2* (Mago et al. 2011b). The physical positions of five markers co-segregating with *Sr2* and *Lr27* ranged from 5. 80 to 6. 56 Mb. Lu et al. (2016) delimited spot blotch resistance gene *Sb3* to a 0. 15 cM genetic interval spanning a physical genomic region of 602 kb on chromosome 3BS. Marker *XWGGC3957* that co-segregated with *Sb3* was located at the 6. 23 Mb, thus overlapping the above interval. Zhang et al. (2017, 2021) mapped leaf rust loci

QLr. hebau-3BS in the Zhou8425B/Chinese Spring RIL population and *QLr-3BS* in a panel of 268 wheat lines, respectively, and deduced that these QTL involved the same gene. The representative SNP marker of *QLr-3BS*, *RAC875 _ c4389*, was located at 11. 08 Mb. In the present study, *QPm. caas-3BS* was firstly narrowed down to a 3. 91 Mb genomic region. We then conducted a linkage analysis of powdery mildew resistance based on SSR markers *gwm533* (putatively 1. 6 cM from *Pmx/Lr27/Yr30/Sr2*) and *XWGCC12530* (0. 07 cM from *Sb3*) and three gene-specific markers in the Zhou8425B/Chinese Spring RIL population. The peak of LOD contours of the powdery mildew resistance response was located between *KASP14800* (at 6. 24 Mb) and *snp _ 54065* (at 6. 67 Mb) (Fig. 4), indicating that the physical interval of *QPm. caas-3BS* overlapped the *Pmx/Lr27/Yr30/Sr2* (Mago et al. 2011b). Results from both the RIL population and secondary mapping populations agreed with Mago et al. (2011b), indicating that the powdery mildew resistance QTL *QPm. caas-3BS* was likely *Pmx*. The association with *Pmx/Lr27/Yr30/Sr2* was based on the location of *gwm533* (1. 6 cM from *Pmx/Lr27/Yr30/Sr2*) and consideration of pseudo-black chaff (Mago et al. 2011b). Pseudo-black chaff was observed in the Zhou8425B/Chinese Spring population but was not studied in detail. *XWGCC12530* (0. 07 cM from *Sb3*) and *gwm533* were tightly linked with *QPm. caas-3BS*, as shown by close linkage with the marker *KASP14800* designed from the candidate gene *TraesCS3B02G014800* (Fig. 1a). *Pmx/Lr27/Yr30/Sr2* confers resistance to powdery mildew, leaf rust, stripe rust and stem rust. Similarly, the candidate of *QPm. caas-3BS*, *TraesCS3B02G014800*, is within the QTL regions for powdery mildew resistance in the Zhou8425B/Chinese Spring population and for stripe rust resistance in the Jingshuang16/Bainong24 population, suggesting that *QPm. caas-3BS* also confers resistance to multiple diseases.

The use of multi-pathogen resistance genes conferring potentially durable and broad-spectrum disease resistance is a valuable strategy in wheat breeding. The cloned multiple disease resistance genes *Lr34* and *Lr67* have provided partial and durable resistance to powdery mildew, stripe rust, leaf rust and stem rust for decades (Krattinger et al. 2009; Moore et al. 2015). Transformation experiments have shown that *Lr34* not only confers resistance to multiple fungal pathogens in wheat but also in barely, rice, maize and sorghum (Risk et al. 2012, 2013; Krattinger et al. 2016; Sucher et al. 2017; Schnippenkoetter et al. 2017; Boni et al. 2018). *Lr67* encodes a hexose transporter associated with sugar uptake. The *LR67res* allele is common in landraces from the Pun-jab region of India but has not been widely used inbreeding (Moore et al. 2015). In this study, *QPm. caas-3BS* not only showed resistance to powdery mildew but also to stripe rust. Cloning and characterizing the gene candidates will not only be helpful in understanding the molecular mechanism of multiple disease resistance but might also enrich the available wheat genetic resources in breeding for durable resistance.

❖ Acknowledgements

The authors are grateful to Prof. R. A. McIntosh, Plant Breeding Institute, University of Sydney, for critical review of this manuscript. This study was supported by the National Natural Science Foundation of China (31961143007, U1904109), Henan Science and Technology Foundation (202102110016) and CAAS Science and Technology Innovation Program.

❖ References

Ashikari M, Sakakibara H, Lin SY, Yamamoto T, Takashi T, Nishimura A, Angeles ER, Qian Q, Kitano H, Matsuoka M (2005) Cytokinin oxidase regulates rice grain production. Science 309: 741-745. https://doi.org/10.1126/science.1113373

Avni R, Nave M, Barad O, Baruch K, Twardziok SO, Gundlach H, Hale I, Mascher M, Spannagl M, Wiebe K, Jordan KW, Golan G, Deek J, Ben-Zvi B, Ben-Zvi G, Himmelbach A, MacLachlan RP, Sharpe AG, Fritz A, Ben-David R, Budak H, Fahima T, Korol A, Faris JD, Hernandez A, Mikel MA, Levy AA, Steffenson B,

Maccaferri M, Tuberosa R, Cattivelli L, Faccioli P, Ceriotti A, Kashkush K, Pourkheirandish M, Komatsuda T, Eilam T, Sela H, Sharon A, Ohad N, Chamovitz DA, Mayer KFX, Stein N, Ronen G, Peleg Z, Pozniak CJ, Akhunov ED, Distelfeld A (2017) Wild emmer genome architecture and diversity elucidate wheat evolu-tion and domestication. Science 357: 93-97. https://doi.org/10.1126/science.aan0032

Bent AF, Mackey D (2007) Elicitors, effectors, and R genes: the new paradigm and a lifetime supply of questions. Annu Rev Phyto-pathol 45: 399-436. https://doi.org/10.1146/annurev.phyto.45.062806.094427

Boni R, Chauhan H, Hensel G, Roulin A, Sucher J, Kumlehn J, Brunner S, Krattinger SG, Keller B (2018) Pathogen-inducible *Ta-Lr34res* expression in heterologous barley confers disease resistance without negative pleiotropic effects. Plant Biotechnol J 16: 245-253. https://doi.org/10.1111/pbi.12765

Brenchley R, Spannagl M, Pfeifer M, Barker GL, D'Amore R, Allen AM, McKenzie N, Kramer M, Kerhornou A, Bolser D, Kay S, Waite D, Trick M, Bancroft I, Gu Y, Huo NX, Luo MC, Sehgal S, Gill B, Kianian S, Anderson O, Kersey P, Dvorak J, McCombie WR, Hall A, Mayer KFX, Edwards KJ, Bevan MW, Hall N (2012) Analysis of the bread wheat genome using whole-genome shot gun sequencing. Nature 491: 705-710. https://doi.org/10.1038/nature11650

Chen XM (2005) Epidemiology and control of stripe rust (*Puccinia striiformis* f. sp. *tritici*) on wheat. Can J Plant Pathol 27: 314-337. https://doi.org/10.1080/07060660509507230

Chen YM, Song WJ, Xie XM, Wang ZH, Guan PF, Peng HR, Jiao YN, Ni ZF, Sun QX, Guo WL (2020) A collinearity-incorporating homology inference strategy for connecting emerging assemblies in *Triticeae* tribe as a pilot practice in the plant pangenomic Era. Mol Plant 13: 1694-1708. https://doi.org/10.1016/j.molp.2020.09.019

Cloutier S, McCallum BD, Loutre C, Banks TW, Wicker T, Feuillet C, Keller B, Jordan MC (2007) Leaf rust resistance gene *Lr1*, isolated from bread wheat (*Triticum aestivum* L.) is a member of the large *psr567* gene family. Plant Mol Biol 65: 93-106. https://doi.org/10.1007/s11103-007-9201-8

D'Auria JC (2006) Acyltransferases in plants: a good time to be BAHD. Curr Opin Plant Biol 9: 331-340. https://doi.org/10.1016/j.pbi.2006.03.016

Daetwyler HD, Pong-Wong R, Villanueva B, Woolliams JA (2010) The impact of genetic architecture on genome-wide evaluation methods. Genetics 185: 1021-1031. https://doi.org/10.1534/genetics.110.116855

Feuillet C, Travella S, Stein N, Albar L, Nublat A, Keller B (2003) Map-based isolation of the leaf rust disease resistance gene *Lr10* from the hexaploid wheat (*Triticum aestivum* L.) genome. Proc Natl Acad Sci USA 100: 15253-15258. https://doi.org/10.1073/pnas.2435133100

Gao HY, Niu JS, Li SP (2018) Impacts of wheat powdery mildew on grain yield & quality and its prevention and control methods. Am J Agric Forestry 6: 141-147. https://doi.org/10.11648/j.ajaf.20180605.14

Gaurav K, Arora S, Silva P, Sanchez-Martin J, Horsnell R, Gao L, Brar GS, Widrig V, Raupp J, Singh N, Wu S (2021) Evolution of the bread wheat D-subgenome and enriching it with diversity from *Aegilops tauschii*. bioRxiv. https://doi.org/10.1101/2021.01.31.428788v1. full

Guo WL, Xin MM, Wang ZH, Yao YY, Hu ZR, Song WJ, Yu KH, Chen YM, Wang XB, Guan PF, Appels R, Peng HR, Ni ZF, Sun QX (2020) Origin and adaptation to high altitude of Tibetan semi-wild wheat. NatCommun 11: 5085. https://doi.org/10.1038/s41467-020-18738-5

He HG, Zhu SY, Zhao RH, Jiang ZN, Ji YY, Ji J, Qiu D, Li HJ, Bie TD (2018) *Pm21*, encoding a typical CC-NBS-LRR protein, confers broad-spectrum resistance to wheat powdery mildew disease. Mol Plant 11: 879-882. https://doi.org/10.1016/j.molp.2018.03.004

He HG, Liu RK, Ma PT, Du HN, Zhang HH, Wu QH, Yang LJ, Gong SJ, Liu TL, Huo NX, Gu YQ, Zhu SY (2020) Characterization of *Pm68*, a new powdery mildew resistance gene on chromosome 2BS of Greek durum wheat TRI 1796. Theor Appl Genet 134: 53-62. https://doi.org/10.1007/s00122-020-03681-2

Hewitt T, Mueller MC, Molnar I, Mascher M, Holusova K, Simkova H, Kunz L, Zhang JP, Li JB, Bhatt D, Sharma R, Schudel S, Yu GT, Steuernagel B, Periyannan S, Wulff B, Ayliffe M, McIntosh R, Keller B, Lagudah E, Zhang P (2020) A highly differentiated region of wheat chromosome 7AL encodes a *Pm1a* immune receptor that recognizes its corresponding *AvrPm1a* effector from *Blumeria graminis*. New Phytol 229: 2812-2826. https://doi.org/10.1111/nph.17075

Holland JB (2007) Genetic architecture of complex traits in plants. Curr Opin Plant Biol 10: 156-161. https://doi.org/10.1016/j.pbi.2007.01.003

Hsu F, Kent WJ, Clawson H, Kuhn RM, Diekhans M, Haussler D (2006) The UCSC known genes. Bioinformatics 22: 1036-1046. https://doi.org/10.1093/bioinformatics/btl048

Huang L, Brooks SA, Li W, Fellers JP, Trick HN, Gill BS (2003) Map-based cloning of leaf rust resistance gene *Lr21* from the large and polyploid genome of bread wheat. Genetics 164: 655-664. https://doi.org/10.1093/genetics/164.2.655

Hurni S, Brunner S, Buchmann G, Herren G, Jordan T, Krukowski P, Wicker T, Yahiaoui N, Mago R, Keller B (2013) Rye *Pm8* and wheat *Pm3* are orthologous genes and show evolutionary conservation of resistance function against powdery mildew. Plant J 76: 957-969. https://doi.org/10.1111/tpj.12345

International Wheat Genome Sequencing Consortium (IWGSC) (2018) Shifting the limits in wheat research and breeding using a fully annotated reference genome. Science 361: eaar7191. https://doi.org/10.1126/science.aar7191

Jia AL, Ren Y, Gao FM, Yin GH, Liu JD, Guo L, Zheng JZ, He ZH, Xia XC (2018) Mapping and validation of a new QTL for adult-plant resistance to powdery mildew in Chinese elite bread wheat line Zhou8425B. Theor Appl Genet 131: 1063-1071. https://doi.org/10.1007/s00122-018-3058-x

Juroszek P, von Tiedemann A (2013) Climate change and potential future risks through wheat diseases: a review. Eur J Plant Pathol 136: 21-33. https://doi.org/10.1007/s10658-012-0144-9

Kang YC, Barry K, Cao FB, Zhou MX (2020) Genome-wideassociation mapping for adult resistance to powdery mildew in common wheat. Mol Biol Rep 47: 1241-1256. https://doi.org/10.1007/s11033-019-05225-4

Krattinger SG, Lagudah ES, Spielmeyer W, Singh RP, Huerta-Espino J, McFadden H, Bossolini E, Selter LL, Keller B (2009) A puta-tive ABC transporter confers durable resistance to multiple fungal pathogens in wheat. Science 323: 1360-1363. https://doi.org/10.1126/science.1166453

Krattinger SG, Sucher J, Selter LL, Chauhan H, Zhou B, Tang MZ, Upadhyaya NM, Mieulet D, Guiderdoni E, Weidenbach D, Schaf-frath U, Lagudah ES, Keller B (2016) The wheat durable, multipathogen resistance gene *Lr34* confers partial blast resistance in rice. Plant Biotechnol J 14: 1261-1268. https://doi.org/10.1111/pbi.12491

Li H, Durbin R (2009) Fast and accurate short read alignment with Burrows-Wheeler transform. Bioinformatics 25: 1754-1760. https://doi.org/10.1093/bioinformatics/btp324

Li H, Handsaker B, Wysoker A, Fennell T, Ruan J, Homer N, Marth G, Abecasis G, Durbin R (2009) The sequence alignment/map format and SAMtools. Bioinformatics 25: 2078-2079. https://doi.org/10.1093/bioinformatics/btp352

Li MM, Dong LL, Li BB, Wang ZZ, Xie JZ, Qiu D, Li YH, Shi WQ, Yang LJ, Wu QH, Chen YX, Lu P, Guo GH, Zhang HZ, Zhang PP, Zhu KY, Li YW, Zhang Y, Wang RG, Yuan CG, Liu W, Yu DZ, Luo MC, Fahima T, Nevo E, Li HJ, Liu ZY (2020) A CNL protein in wild emmer wheat confers powdery mildew resistance. New Phytol 228: 1027-1037. https://doi.org/10.1111/nph.16761

Lin F, Chen XM (2007) Genetics and molecular mapping of genes for race-specific all-stage resistance and non-race-specific high-temperature adult-plant resistance to stripe rust in spring wheat cultivar Alpowa. Theor Appl Genet 114: 1277-1287. https://doi.org/10.1007/s00122-007-0518-0

Liu N, Bai GH, Lin M, Xu XY, Zheng WM (2017) Genome-wide association analysis of powdery mildew resistance in US winter wheat. Sci Rep 7: 11743. https://doi.org/10.1038/s41598-017-11230-z

Liu Z, Wang Q, Wan H, Yang F, Wei H, Xu Z, Ji H, Xia X, Li J, Yang W (2021) QTL mapping for adult-plant resistance to powdery mildew in Chinese elite common wheat Chuanmai104. Cereal Res Commun 49: 99108. https://doi.org/10.1007/s42976-020-00082-5

Livak KJ, Schmittgen TD (2001) Analysis of relative gene expression data using real-time quantitative PCR and the 2 (-Delta Delta C (T)) method. Methods 25: 402-408. https://doi.org/10.1006/meth. 2001. 1262

Lu YM, Lan CX, Liang SS, Zhou XC, Liu D, Zhou G, Lu LQ, Jing JX, Wang MN, Xia XC, He ZH (2009) QTL mapping for adult-plant resistance to stripe rust in Italian common wheat cultivars Libellula and Strampelli. Theor Appl Genet 119: 1349-1359. https://doi.org/10.1007/s00122-009-1139-6

LuP, Liang Y, Li DL, Wang ZZ, Li WB, Wang GX, Wang Y, Zhou SH, Wu QH, Xie JZ, Zhang DY, Chen YX, Li MM, Zhang Y, Sun QX, Han CG, Liu ZY (2016) Fine genetic mapping of spot blotch resistance gene *Sb3* in wheat (*Triticum aesti-vum*). Theor Appl Genet 129: 577-589. https://doi.org/10.1007/s00122-015-2649-z

LuP, Guo L, Wang ZZ, Li BB, Li J, Li YH, Qiu D, Shi WQ, Yang LJ, Wang N, Guo GH, Xie JZ, Wu QH, Chen YX, Li MM, Zhang HZ, Dong LL, Zhang PP, Zhu KY, Yu DZ, Zhang Y, Deal KR, Huo NX, Liu CM, Luo MC, Dvorak J, Gu YQ, Li HJ, Liu ZY (2020) A rare gain of function mutation in a wheat tandem kinase confers resistance to powdery mildew. Nat Commun 11: 680. https://doi.org/10.1038/s41467-020-14294-0

Maccaferri M, Harris NS, Twardziok SO, Pasam RK, Gundlach H, Spannagl M, Ormanbekova D, Lux T, Prade VM, Milner SG, Himmelbach A, Mascher M, Bagnaresi P, Faccioli P, Cozzi P, Lauria M, Lazzari B, Stella A, Manconi A, Gnocchi M, Moscatelli M, Avni R, Deek J, Biyiklioglu S, Frascaroli E, Corneti S, Salvi S, Sonnante G, Desiderio F, Marè C, Crosatti C, Mica E, Özkan H, Kilian B, De Vita P, Marone D, Joukhadar R, Mazzucotelli E, Nigro D, Gadaleta A, Chao S, Faris JD, Melo ATO, Pumphrey M, Pecchioni N, Milanesi L, Wiebe K, Ens J, MacLachlan RP, Clarke JM, Sharpe AG, Koh CS, Liang KYH, Taylor GJ, Knox R, Budak H, Mastrangelo AM, Xu SS, Stein N, Hale I, Distelfeld A, Hayden MJ, Tuberosa R, Walkowiak S, Mayer KFX, Ceriotti A, Pozniak CJ, Cattivelli L (2019) Durum wheat genome highlights past domestication signatures and future improvement targets. Nat Genet 51: 885-895. https://doi.org/10.1038/s41588-019-0381-3

Mago R, Simkova H, Brown-Guedira G, Dreisigacker S, Breen J, Jin Y, Singh R, Appels R, Lagudah ES, Ellis J, Dolezel J, Spielmeyer W (2011a) An accurate DNA marker assay for stem rust resistance gene *Sr2* in wheat. Theor Appl Genet 122: 735-744. https://doi.org/10.1007/s00122-010-1482-7

Mago R, Tabe L, McIntosh RA, Pretorius Z, Kota R, Paux E, Wicker T, Breen J, Lagudah ES, Ellis JG, Spielmeyer W (2011b) A multiple resistance locus on chromosome arm 3BS in wheat confers resistance to stem rust (*Sr2*), leaf rust (*Lr27*) and powdery mildew. Theor Appl Genet 123: 615-623. https://doi.org/10.1007/s00122-011-1611-y

McIntosh RA, Dubcovsky J, Rogers WJ, Morris C, Appels R, Xia XC (2014) Catalogue of gene symbols for wheat: 2013-2014 (Supplement). https://shigen.nig.ac.jp/wheat/komugi/genes/macgene/supplement2013.pdf. Accessed 6 Dec 201923

McIntosh RA, Dubcovsky J, Rogers WJ, Morris C,

Appels R, Xia XC (2016) Catalogue of gene symbols for wheat: 2015-2016 (Supplement). https://shigen.nig.ac.jp/wheat/komugi/genes/macgene/supplement 2015.pdf. Accessed 6 Dec 201924

McIntosh RA, Dubcovsky J, Rogers WJ, Morris C, Xia XC (2017) Catalogue of gene symbols for wheat: 2017 (Supplement). https://shigen.nig.ac.jp/wheat/komugi/genes/macgene/supplement2017.pdf. Accessed 6 Dec 2019

MooreJW, Herrera-Foessel S, Lan C, Schnippenkoetter W, Aylife M, Huerta-Espino J, Lillemo M, Viccars L, Milne R, Periyannan S, Kong X, Spielmeyer W, Talbot M, Bariana H, Patrick JW, Dodds P, Singh R, Lagudah E (2015) Recent evolution of a hexose transporter variant confers resistance to multiple pathogens in wheat. Nat Genet 47: 1494-1498. https://doi.org/10.1038/ng. 3439

Negruk V, Yang P, Subramanian M, McNevin JP, Lemieux B (1996) Molecular cloning and characterization of the *CER2* gene of *Arabidopsis thaliana*. Plant J 9: 137-145. https://doi.org/10.1046/j. 1365-313x. 1996. 09020137. x

Omasits U, Ahrens CH, Müller S, Wollscheid B (2014) Protter: interac-tive protein feature visualization and integration with experimen-tal proteomic data. Bioinformatics 30:884-886. https://doi.org/10.1093/bioinformatics/btt607

Peterson RF, Campbell AB, Hannah AE (1948) A diagrammatic scale for estimating rust intensity on leaves and stems of cereals. Can J Res 26: 496-500. https://doi.org/10.1139/cjr48c-033

Qi LL, Echalier B, Chao S, Lazo GR, Butler GE, Anderson OD, Akhunov ED, Dvorák J, Linkiewicz AM, Ratnasiri A, Dubcovsky J, Bermudez-Kandianis CE, Greene RA, Kantety R, La Rota CM, Munkvold JD, Sorrells SF, Sorrells ME, Dilbirligi M, Sidhu D, Erayman M, Randhawa HS, Sandhu D, Bondareva SN, Gill KS, Mahmoud AA, Ma XF, Miftahudin GJP, Conley EJ, Nduati V, Gonzalez-Hernandez JL, Anderson JA, Peng JH, Lapitan NL, Hossain KG, Kalavacharla V, Kianian SF, Pathan MS, Zhang DS, Nguyen HT, Choi DW, Fenton RD, Close TJ, McGuire PE, Qualset CO, Gill BS (2004) A chromosome bin map of 16, 000 expressed sequence tag loci and distribution of genes among the three genomes of polyploid wheat. Genetics 168: 701-712. https://doi.org/10. 1534/genetics. 104. 034868

Ramirez-Gonzalez RH, Uauy C, Caccamo M (2015) PolyMarker: a fast polyploid primer design pipeline.

Bioinformatics 31: 2038-2039. https://doi.org/10.1093/bioinformatics/btv069

Ramírez-González RH, Borrill P, Lang D, Harrington SA, Brinton J, Venturini L, Davey M, Jacobs J, van Ex F, Pasha A, Khedikar Y, Robinson SJ, Cory AT, Florio T, Concia L, Juery C, Schoonbeek H, Steuernagel B, Xiang D, Ridout CJ, Chalhoub B, Mayer KFX, Benhamed M, Latrasse D, Bendahmane A; International Wheat Genome Sequencing Consortium, Wulff BBH, Appels R, Tiwari V, Datla R, Choulet F, Pozniak CJ, Provart NJ, Sharpe AG, Paux E, Spannagl M, Bräutigam A, Uauy C (2018) The transcriptional landscape of polyploid wheat. Science 361: eaar6089. https://doi.org/10.1126/science. aar6089

Reynolds M, Foulkes J, Furbank R, Griffiths S, King J, Murchie E, Parry M, Slafer M (2012) Achieving yield gains in wheat. Plant Cell Environ 35: 1799-1823. https://doi.org/10.1111/j. 1365-3040. 2012. 02588. x

Risk JM, Selter LL, Krattinger SG, Viccars LA, Richardson TM, Bues-ing G, Herren G, Lagudah ES, Keller B (2012) Functional vari-ability of the *Lr34* durable resistance gene in transgenic wheat. Plant Biotechnol J 10: 477-487. https://doi.org/10.1111/j. 1467-7652. 2012. 00683. x

Risk JM, Selter LL, Chauhan H, Krattinger SG, Kumlehn J, Hensel G, Viccars LA, Richardon TM, Buesing G, Troller A, Lagudah ES, Keller B (2013) The wheat *Lr34* gene provides resistance against multiple fungal pathogens in barley. Plant Biotechnol J 11: 847-854. https://doi.org/10.1111/pbi. 12077

Sánchez-Martín J, Keller B (2021) NLR immune receptors and diverse types of non-NLR proteins control race-specific resistance in Trit-iceae. Curr Opin Plant Biol 62: 102053. https://doi.org/10.1016/j. pbi. 2021. 102053

Sánchez-Martín J, Steuernagel B, Ghosh S, Herren G, Hurni S, Adam-ski N, Vrana J, Kubalakova M, Krattinger SG, Wicker T, Doležel J, Keller B (2016) Rapid gene isolation in barley and wheat by mutant chromosome sequencing. Genome Biol 17: 221. https://doi.org/10.1186/s13059-016-1082-1

Sánchez-Martín J, Widrig V, Herren G, Wicker T, Zbinden H, Gronnier J, Spörri L, Praz CR, Heuberger M, Kolodziej MC, Isaksson J, Steuernagel B, Karafiátová M, Doležel J, Zipfel C, Keller B (2021) Wheat *Pm4* resistance to powdery mildew is controlled by alternative splice variants encoding chimeric proteins. Nat Plants 7: 327-341. https://doi.org/10.1038/s41477-

021-00869-2

Sato K, Abe F, Mascher M, Haberer G, Gundlach H, Spannagl M, Shirasawa K, Isobe S (2021) Chromosome-scale genome assembly of the transformation-amenable common wheat cultivar 'Fielder.' DNA Res 28: dsab008. https: //doi. org/10. 1093/dnares/dsab008

Schmalenbach I, March TJ, Bringezu T, Waugh R, Pillen K (2011) High-resolution genotyping of wild barley introgression lines and fine-mapping of the thresh ability locus thresh-1 using the Illu-mina Golden Gate Assay. G3 Genes/genomes/genetics 1: 187-196. https: //doi. org/10. 1534/g3. 111. 000182

Schnippenkoetter W, Lo C, Liu G, Dibley K, Chan WL, White J, Milne R, Zwart A, Kwong E, Keller B, Godwin I, Krattinger SG, Lagudah ES (2017) The wheat *Lr34* multipathogen resistance gene confers resistance to anthracnose and rust in sorghum. Plant Bio-technol J 15: 1387-1396. https: //doi. org/10. 1111/pbi. 12723

Singh RP, Mujeeb-Kazi A, Huerta-Espino J (1998) *Lr46*: a gene conferring slow-rusting resistance to leaf rust in wheat. Phytopathology 88: 890-894. https: //doi. org/10. 1094/PHYTO. 1998. 88. 9. 890

Singh RP, Huerta-Espino J, Rajaram S (2000a) Achieving near-immunity to leaf and stripe rusts in wheat by combining slow rusting resistance genes. Acta Phytopathologica Et Entomologica Hungarica 35: 133-139.

Singh RP, Nelson JC, Sorrells ME (2000b) Mapping *Yr28* and other genes for resistance to stripe rust in wheat. Crop Sci 40: 1148-1155. https: //doi. org/10. 1094/PHYTO-03-15-0060-R

Singh RP, Huerta-Espino J, William HM (2005) Genetics and breeding for durable resistance to leaf and stripe rusts in wheat. Turk J Agric for 29: 121-127.

SinghRP, Singh PK, Rutkoski J, Hodson DP, He XY, Jørgensen LN, Hovmøller MS, Huerta-Espino J (2016) Disease impact on wheat yield potential and prospects of genetic control. Annu Rev Phytopathol 54: 303-322. https: //doi. org/10. 1146/annur ev-phyto-080615-095835/

Singh SP, Hurni S, Ruinelli M, Brunner S, Sanchez-Martin J, Krukowski P, Peditto D, Buchmann G, Zbinden H, Keller B (2018) Evolutionary divergence of the rye *Pm17* and *Pm8* resistance genes reveals ancient diversity. Plant Mol Biol 98: 249-260. https: //doi. org/10. 1007/s11103-018-0780-3

Spielmeyer W, Sharp PJ, Lagudah ES (2003) Identification and validation of markers linked to broad-spectrum stem rust resistance gene *Sr2* in wheat (*Triticum aestivum* L.). Crop Sci 43: 333-336. https: //doi. org/10. 2135/cropsci2003. 3330

Stam P (1993) Construction of integrated genetic linkage maps by means of a new computer package: join map. Plant J 5: 739-744. https: //doi. org/10. 1111/j. 1365-313X. 1993. 00739. x

Sucher J, Boni R, Yang P, Rogowsky P, Büchner H, Kastner C, Kumlehn J, Krattinger SG, Keller B (2017) The durable wheat disease resistance gene *Lr34* confers common rust and northern corn leaf blight resistance in maize. Plant Biotechnol J 15: 489-496. https: //doi. org/10. 1111/pbi. 12647

Tacke E, Korfhage C, Michel D, Maddaloni M, Motto M, Lanzini S, Salamini F, Doring HP (1995) Transposon tagging of the maize Glossy2 locus with the transposable element En/Spm. Plant J 8: 907-917. https: //doi. org/10. 1046/j. 1365-313x. 1995. 8060907. x

Walkowiak S, Gao L, Monat C, Haberer G, Kassa MT, Brinton J, Ramirez-Gonzalez RH, Kolodziej MC, Delorean E, Thambugala D, Klymiuk V, Byrns B, Gundlach H, Bandi V, Siri JN, Nilsen K, Aquino C, Himmelbach A, Copetti D, Ban T, Venturini L, Bevan M, Clavijo B, Koo DH, Ens J, Wiebe K, N' Diaye A, Fritz AK, Gutwin C, Fiebig A, Fosker C, Fu BX, Accinelli GG, Gard-ner KA, Fradgley N, Gutierrez-Gonzalez J, Halstead-Nussloch G, Hatakeyama M, Koh CS, Deek J, Costamagna AC, Fobert P, Heavens D, Kanamori H, Kawaura K, Kobayashi F, Krasileva K, Kuo T, McKenzie N, Murata K, Nabeka Y, Paape T, Padmarasu S, Percival-Alwyn L, Kagale S, Scholz U, Sese J, Juliana P, Singh R, Shimizu-Inatsugi R, Swarbreck D, Cockram J, Budak H, Tameshige T, Tanaka T, Tsuji H, Wright J, Wu J, Steuernagel B, Small I, Cloutier S, Keeble-Gagnère G, Muehlbauer G, Tibbets J, Nasuda S, Melonek J, Hucl PJ, Sharpe AG, Clark M, Legg E, Bharti A, Langridge P, Hall A, Uauy C, Mascher M, Krattinger SG, Handa H, Shimizu KK, Distelfeld A, Chalmers K, Keller B, Mayer KFX, Poland J, Stein N, McCartney CA, Spannagl M, Wicker T, Pozniak CJ (2020) Multiple wheat genomes reveal global variation in modern breeding. Nature 588: 277-283. https://doi. org/10. 1038/s41586-020-2961-x

Wang K, Li MY, Hakonarson H (2010) ANNOVAR:

functional annotation of genetic variants from high-throughput sequencing data. Nucleic Acids Res 38：e164. https：//doi. org/10. 1093/nar/gkq603 Wissuwa M，Wegner J，Ae N，Yano M（2002）Substitution mapping of Pup1：a major QTL increasing phosphorus uptake of rice from a phosphorus-deficient soil. Theor Appl Genet 105：890-897. https：//doi. org/10. 1007/s00122-002-1051-9

Wu YY，Li M，He ZH，Dreisigacker S，Wen WE，Jin H，Zhai SN，Li FJ，Gao FM，Liu JD，Wang RG，Zhang PZ，Wan YX，Cao SH，Xia XC（2020）Development and validation of high-throughput and low-cost STARP assays for genes underpinning economically important traits in wheat. Theor Appl Genet 133：2431-2450. https：//doi. org/10. 1007/s00122-020-03609-w

Xia Y，Nikolau BJ，Schnable PS（1996）Cloning and characterization of CER2，an Arabidopsis gene that affects cuticular wax accumulation. Plant Cell 8：1291-1304. https：//doi. org/10. 1105/tpc. 8. 8. 1291

Xie JZ，Guo GH，Wang Y，Hu TZ，Wang LL，Li JT，Qiu D，Li YH，Wu QH，Lu P，Chen YX，Dong LL，Li MM，Zhang HZ，Zhang PP，Zhu KY，Li BB，Deal KR，Huo NX，Zhang Y，Luo MC，Liu SZ，Gu YQ，Li HJ，Liu ZY（2020）A rare single nucleotide variant in Pm5e confers powdery mildew resistance in common wheat. New Phytol 228：1011-1026. https：//doi. org/10. 1111/nph. 16762

Xing LP，Hu P，Liu JQ，Witek K，Zhou S，Xu JF，Zhou WH，Gao L，Huang ZP，Zhang RQ，Wang XE，Chen PD，Wang HY，Jones JDG，Karafiatova M，Vrana J，Baros J，Dolezel J，Tian YC，Wu YF，Cao AZ（2018）Pm21 from Haynaldia villosa encodes a CC-NBS-LRR protein conferring powdery mildew resistance in wheat. Mol Plant 11：874-878. https：//doi. org/10. 1016/j. molp. 2018. 02. 013

Xu XT，Zhu ZW，Jia AL，Wang FJ，Wang JP，Zhang YL，Fu C，Fu LP，Bai GH，Xia XC，Hao YF（2020）Mapping of QTL for partial resistance to powdery mildew in two Chinese common wheatcultivars. Euphytica. https：//doi. org/10. 1007/s10681-019-2537-8

Yahiaoui N，Srichumpa P，Dudler R，Keller B（2004）Genome analysis at different ploidy levels allows cloning of the powdery mildew resistance gene Pm3b from hexaploid wheat. Plant J 37：528-538. https：//doi. org/10. 1046/j. 1365-313x. 2003. 01977. x

Yan XC，Li MM，Zhang PP，Yin GH，Zhang HZ，Gebrewahid TW，Zhang JP，Dong LL，Liu DQ，Liu ZY，Li ZF（2021）High-temperature wheat leaf rust resistance gene Lr13 exhibits pleiotropic effects on hybrid necrosis. Mol Plant 14：1029-1032. https：//doi. org/10. 1016/j. molp. 2021. 05. 009

Yang Q，Zhang DF，Xu ML（2012）A sequential quantitative trait locus fine-mapping strategy using recombinant-derived progeny. JIntegr Plant Biol 54：228-237. https：//doi. org/10. 1111/j. 1744-7909. 2012. 01108. x

Zeng FS，Yang LJ，Gong SJ，Shi WQ，Zhang XJ，Wang H，Xiang LB，Xue MF，Yu DZ（2014）Virulence and diversity of Blumeria graminis f. sp. tritici populations in China. J Integr Agric 13：2424-2437. https：//doi. org/10. 1016/S2095-3119（13）60669-3

Zhang PP，Yin GH，Zhou Y，Qi AY，Gao FM，Xia XC，He ZH，Li ZF，Liu DQ（2017）QTL mapping of adult-plant resistance to leaf rust in the wheat cross Zhou8425B/Chinese spring using high-density SNP markers. Front Plant Sci 8：793. https：//doi. org/10. 3389/fpls. 2017. 00793

Zhang DY，Zhu KY，Dong LL，Liang Y，Li GQ，Fang TL，Guo GH，Wu QH，Xie JZ，Chen YX，Lu P，Li MM，Zhang HZ，Wang ZZ，Zhang Y，Sun QX，Liu ZY（2019）Wheat powdery mildew resistance gene Pm64 derived from wild emmer（Triticum turgdium var. dicoccoides）is tightly linked in repulsion with stripe rust resistance gene Yr5. Crop J 7：761-770. https：//doi. org/10. 1016/j. cj. 2019. 03. 003

Zhang PP，Yan XC，Gebrewahid TW，Zhou Y，Yang EN，Xia XC，He ZH，Li ZF，Liu DQ（2021）Genome-wide association mapping of leaf rust and stripe rust resistance in wheat accessions using the 90K SNP array. Theor Appl Genet 134：1233-1251. https：//doi. org/10. 1007/s00122-021-03769-3

Zou SH，Wang H，Li YW，Kong ZS，Tang DZ（2018）The NB-LRR genePm60 confers powdery mildew resistance in wheat. New Phytol 218：298-309. https：//doi. org/10. 1111/nph. 14964

Genome-wide association mapping of black point reaction in common wheat（*Triticum aestivum* L.）

Jindong Liu[1,2], Zhonghu He[1,3], Awais Rasheed[1], Weie Wen[1], Jun Yan[4], Pingzhi Zhang[5], Yingxiu Wan[5], Yong Zhang[1], Chaojie Xie[2] and Xianchun Xia[1]*

[1] Institute of Crop Sciences，National Wheat Improvement Center，Chinese Academy of Agricultural Sciences（CAAS），12 Zhongguancun South Street，Beijing 100081，China. [2] Department of Plant Genetics & Breeding/State Key Laboratory for Agrobiotechnology，China Agricultural University，2 Yuanmingyuan West Road，Beijing 100193，China. [3] International Maize and Wheat Improvement Center（CIMMYT）China Office，c/o CAAS，12 Zhongguancun South Street，Beijing 100081，China. [4] Institute of Cotton Research，Chinese Academy of Agricultural Sciences（CAAS），38 Huanghe Street，Anyang，Henan 455000，China. [5] Crop Research Institute，Anhui Academy of Agricultural Sciences，40 Nongke South Street，Hefei，Anhui 230001，China. * Correspondence：xiaxianchun@caas.cn

Abstract

Background：Black point is a serious threat to wheat production and can be managed by host resistance. Marker-assisted selection（MAS）has the potential to accelerate genetic improvement of black point resistance in wheat breeding. We performed a genome-wide association study（GWAS）using the high-density wheat 90 K and 660 K single nucleotide polymorphism（SNP）assays to better understand the genetic basis of black point resistance and identify associated molecular markers.

Results：Black point reactions were evaluated in 166 elite wheat cultivars in five environments. Twenty-five unique loci were identified on chromosomes 2A，2B，3A，3B（2），3D，4B（2），5A（3），5B（3），6A，6B，6D，7A（5），7B and 7D（2），respectively，explaining phenotypic variation ranging from 7.9 to 18.0%. The highest number of loci was detected in the A genome（11），followed by the B（10）and D（4）genomes. Among these，13 were identified in two or more environments. Seven loci coincided with known genes or quantitative trait locus（QTL），whereas the other 18 were potentially novel loci. Linear regression showed a clear dependence of black point scores on the number of favorable alleles，suggesting that QTL pyramiding will be an effective approach to increase resistance. In silico analysis of sequences of resistance-associated SNPs identified 6 genes possibly involved in oxidase，signal transduction and stress resistance as candidate genes involved in black point reaction.

Conclusion：SNP markers significantly associated with black point resistance and accessions with a larger number of resistance alleles can be used to further enhance black point resistance in breeding. This study provides new insights into the genetic architecture of black point reaction.

Keywords：660 K SNP array，90 K SNP array，Enzymatic browning，Favorable and unfavorable allele，GWAS

Published in BMC Plant Biology，2017，17：220.

Background

Black point, characterized by dark discoloration at the embryo end of kernels, occurs in most wheat growing regions of the world including China, USA, Australia, Canada and Serbia[1,2]. It can downgrade end-use quality of the grain due to seed discoloration[3]. Many marketing authorities have regulations on the incidence of black point, such as ≤4% in the USA, ≤5% in Australia, and ≤ 10% in Canada[4], indicating that grain with black point symptoms is more difficult to market with consequent economic losses to producers. In addition, black point can decrease the germination percentage and cause impaired seedling development[4]. It can also lead to the presence of toxic secondary metabolites, such as Alternaria mycotoxin and Alternariol monomethyl ether[5-7] that may cause oesophageal cancer[8].

Many studies indicate that black point is enhanced by abiotic stresses, as symptoms more likely occur after exposure to high humidity and extreme temperatures during grain filling[9,10]. However, the causes of black point remain unclear and contradictory. Fungi are considered as the causal agents of black point[1]; these include *Alternaria alternata* [5,11], *Bipolaris sorokiniana* [12] and *Fusarium proliferatum* [6]. However, direct association between the presence of fungi and black point development has been discounted by some workers[13-15], who pointed out that it may be caused by enzymatic browning following stress. Oxidases, such as peroxidases (POD)[15,16], polyphenol oxidase (PPO)[17,18] and lipoxygenase (LOX)[19], that catalyze oxidation of phenolic compounds to brown or black pigments (melanins and quinines)[18,20], may be triggered by high humidity during the later stages of grain filling. Susceptible varieties have higher POD[15,21] and phenylalanine ammonia-lyase (PAL)(an enzyme involved in phenolic acid biosynthesis)[21]activities.

Although several cultural, biological and chemical control strategies have been used to control black point, breeding resistant cultivars remains the most effective, economic and environmentally sustainable approachto control this disease[4,22,23]. Previous studies on the known genetic basis of black point resistance involved classical linkage-mapping methods using bi-parental populations[22-24], in which only two allelic effects can be evaluated for any single locus. Recent advances in genomics, particularly development of the wheat 90 K[25], 660 K (JZ Jia, pers. comm.) and 820 K SNP arrays[26] have made it feasible to genotype large germplasm collections with high-density SNP markers. As a result, the GWAS based on linkage disequilibrium (LD) has been widely adopted to investigate existing allelic diversity for important and complex agronomic traits. Compared with classical linkage-mapping, GWAS permits a more representative gene pool and a higher mapping resolution, because all historical meiotic events that have occurred in the ancestors of a diverse germplasm panel can be used[27]. Moreover, GWAS bypasses the expense and time of developing mapping populations, and enables the mapping of many traits in one set of genotypes, making the method more efficient and less expensive than linkage mapping[28]. Thus, GWAS has become a powerful alternative approach for linkage mapping[29]. GWAS has been applied to investigate a range of traits, including disease resistance[30,31], end-use quality[32], and yield components[33-35].

The Yellow and Huai River Valleys Facultative Wheat Region is one of the most important agricultural regions of wheat production in China with an area of 15. 3 million hectares. Black point has become one of the important diseases in this region due to increased water management and fertilizer use. Breeding for black point resistance could be greatly improved by the identification and use of closely associated molecular markers. Although GWAS has become a powerful approach to dissect the genetic architecture for many traits, it has not been used to analyze traits related to black point. In the present study, we used a diverse panel of 166 elite wheat cultivars in GWAS to (1) dissect the genetic architecture of black point resistance, （2）identify SNPs significantly associated with black point resistance, and（3）search for candidate black point resistance genes for further study.

Results

Marker coverage and genetic diversity

A total of 18,920 SNPs from the 90 K and 283,652 from the 660 K SNP array based on the consensus genetic maps and physical map（IWGSC, http：// www. wheatgen- ome. org/）were chosen for GWAS of black point reaction in 166 wheat cultivars （Additional file 1： Table S1）. After removing the SNPs with minor allele frequency（MAF）< 5% （28,935 SNPs）and missing data > 20% （13,715 SNPs）, 259,922 SNPs were employed for subsequent analysis（ Additional file 2： Table S2 ）. These markers spanned a physical distance of 14,063. 9 Mb, with an average density of 0. 054 Mb per marker. Total of 89, 519（34. 4%）, 146,270（56. 3%）and 24, 133（9. 3%）markers were from the A, B and D genomes, respectively, with corresponding map lengths of 4934. 5, 5179. 0 and 3950. 4 Mb. The marker density for the D genome（0. 202 Mb per marker）was lower than that for the A（0. 099 Mb per marker）and B（0. 042 Mb per marker） genomes. The average genetic diversity and polymorphism information content（PIC）for the whole genome were 0. 356（0. 009-0. 500）and 0. 285 （0. 009-0. 380）, respectively. Both the genetic diversity and PIC of the A（0. 365 and 0. 291）and B（0. 363 and 0. 289）genomes were higher than the D（0. 340 and 0. 265）genome. The number of markers, map length, genetic diversity and PIC for each chromosome are shown in Additional file 2： Table S2.

Population structure and linkage disequilibrium

In the plot of K against ΔK, a break in the slope was observed at K＝3 followed by flattening of the curve, indicating that this panel consists of three subgroups, which was consistent with the results of principal components analysis（ PCA ）and neighbor-joining （NJ）tree analysis（Fig. 1）. Subgroup I, the largest

group with 62 accessions, was dominated by Shandong and foreign cultivars; Subgroup II consisted of 54 accessions, mainly comprising varieties from Henan, Anhui and Shaanxi provinces; Subgroup III had 50 accessions, most of which were from Henan province （Additional file 1： Table S1）.

In total, 12,324 markers from the 90 K and 660 K SNP arrays were used to evaluate LD decay for the whole genome as well as the A, B and D genomes separately. Around 14. 3% of all pairs of loci were in significant LD（P <0. 001）with average r^2 of 0. 174 on a genome-wide level by the 90 K and 660 K SNP assays. The B genome contained the highest percentage of significant markers（44. 2%）, followed by the A （33. 6%）and D（22. 2%）genomes. The scatter plots of r^2 against physical distance（Mb）indicated a clear LD decay with increasing physical distance（Additional file 3： Figure S1）. According to[28], the critical value for significance of r^2 was evaluated at 0. 079, 0. 083, 0. 095 and 0. 082 for the A, B, D and whole genomes, respectively. The point at which the LOESS curve intercepts the critical r^2 was determined as the average LD decay of the panel[28]. Based on this criterion, LD decay distance was about 8 Mb for the whole genome. The highest LD decay was observed in the D genome（11 Mb）, followed by the A（6 Mb）and B（4 Mb） genomes（Additional file 3： Figure S1）.

Phenotypic variations for black point reaction in the field

Continuous variation was observed across five environments （Additional file 4： Figure S2; Additional file 5： Table S3）. The resulting best linear unbiased predictors（BLUPs） for black point scores across all environments ranged from 1. 6 to 80. 6% with an average of 23. 3% （Additional file 4： Figure S2; Additional file 6： Figure S3）, presenting a wide range of reactions for black point and indicating that this diversity panel was ideal for conducting GWAS. Analysis of variance （ANOVA）for black point scores revealed significant differences（ $P \leqslant$ 0. 001）among genotypes（G）, environments, and genotype $\times$ environment（G $\times$ E）

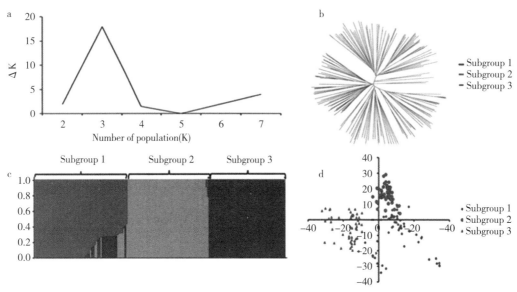

Fig. 1　Population structure analysis of 166 wheat accessions. （a） Estimated ΔK over five repeats of structure analysis； （b） three subgroups inferred by structure analysis； （c） neighbor-joining （NJ） tree； d principal components analysis （PCA） plots

interactions （Table 1）. The broad sense heritability （h^2） estimate for black point scores across all five environments was 0.62, indicating that much of the phenotypic variation was derived from genetic factors and therefore suitable for further association mapping.

Marker-trait association （MTA） analysis

The MTAs analyzed by the mixed linear model （MLM） in Tassel v5.0[36] and the FarmCPU[37] were shown in Additional file 7： Table S4 and Additional file 8： Table S5, respectively. Twenty-five loci （221 MTAs） identified by both the Tassel v5.0 and FarmCPU were considered to be more reliable （Table 2, Additional file 9： Table S6）； these were distributed on chromosomes 2A, 2B, 3A, 3B （2）, 3D, 4B （2）, 5A （3）, 5B （3）, 6A, 6B, 6D, 7A （5）, 7B and 7D （2）, respectively （Table 2, Additional file 9： Table S6）, explaining phenotypic variation ranging from 7.9 to 18.0%. Among these loci, 13 on chromosomes 2A, 2B, 3A, 3B （2）, 3D, 4B, 5A （2）, 5B, 7A, 7B and 7D were detected in two or more environments （Table 2, Additional file 9： Table S6）. The maximum number of loci were found in the A genome （11）, followed by the B genome （10）, whereas only four loci were

identified in the D genome （Table 2； Additional file 9： Table S6）.

Kinship-adjusted Manhattan plot summarizing the analysis of black point scores by Tassel v5.0 and FarmCPU are shown in Fig.2 and Fig.3, respectively. The quantile-quantile （Q-Q） plot representing expected and observed probabilities of getting associations of SNPs by Tassel v5.0 and FarmCPU are presented in Additional file 10： Figure S4 and Additional file 11： Figure S5, respectively. The LD patterns along 2A, 2B, 3A, 3B, 4B, 5A, 5B, 6A, 6B, 6D, 7A, 7B and 7D can be visualized as heatmaps in Additional file 12： Figure S6.

Table 1　Analysis of variance of black point scores in 166 wheat accessions

Source of variation	df	Mean square	F value
Replicates （nested in environments）	9	3 896	7.4***
Environments	4	499,230	480.6***
Genotypes	165	5 326	35.3***
Genotypes × Environments	660	798	4.9***
Error	1 650	153	

***significant at $P < 0.0001$

Relationship between black point reaction and the number of resistance alleles

To further understand the combined effects of alleles on reaction to black point, we examined the number of favorable alleles in each accession. The numbers of favorable alleles in single accessions ranged from 5 to 21, compared to 4 to 20 unfavorable alleles (Additional file 1:

Table S1). The relationships between black point BLUP values and numbers of favorable and unfavorable alleles estimated by linear regression showed a dependence of black point BLUP values on the number of favorable alleles with $r^2 = 0.85$ (Fig. 4a), and number of unfavorable alleles with $r^2 = 0.85$ (Fig. 4b). Thus, accessions with favorable alleles and less unfavorable alleles were more resistant to black point.

Table 2　Loci for black point resistance in 166 wheat accessions identified by both the Tassel v5. 0 and FarmCPU

Marker[a]	Chr[b]	Physical interval[c] (bp)	Environment[d]	SNP[e]	P-value[f]	R^{2g}(%)	QTL/gene[h]
IWB22408	2A	709,831,643-709,831,743	E1,E2,E3,E4,E6	T/C	1.2-9.8 E^{-04}	7.9-14.7	QBp. caas-2AL[23]; QPPO. caas-2AL [55]
PPO-A1	2A	712,188,721-712,187,200	E1,E2,E3,E4,E6	—	2.4-5.5 E^{-04}	9.9-11.6	PPO-A1[56]
AX_108,951,749	2B	714,389,068-714,388,998	E2,E3,E4,E6	T/C	2.0-7.3 E^{-04}	8.8-11.5	QBp. caas-2BL[23]
AX_111,053,669	3A	9,605,904-9,605,974	E2,E3,E4,E6	A/G	2.4-8.6 E^{-04}	8.3-10.4	
AX_108826477	3B	58,767,930-58,768,000	E1,E3	A/C	1.5-9.7 E^{-04}	7.9-11.0	
AX_108,797,097	3B	695,967,481-695,967,411	E1,E2,E3,E6	A/G	1.2-9.6 E^{-04}	8.0-11.9	QBp. caas-3BL[23]
AX_110941533	3D	4,066,092-4,066,162	E2,E3,E4,E6	A/C	4.1-9.4 E^{-04}	8.2-9.7	
AX_108983386	4B	6,961,084-6,961,154	E5	C/G	1.7-9.4 E^{-05}	8.0-10.5	
AX_111488843	4B	504,944,902-504,944,832	E1,E2,E3,E6	A/T	7.0 E^{-06}-2.6 E^{-04}	9.9-15.5	
IWB8709	5A	32,887,598-32,887,698	E3	A/G	1.9-5.3 E^{-04}	8-8.9	QBp. caas-5AS[23]; QPod. caas-5AS [57]
AX_109316564	5A	535,780,381-535,780,311	E1,E3,E6	T/G	4.8-9.9 E^{-04}	8.1-11.4	
IWA2223	5A	592,276,555-592,276,708	E1,E2,E3,E4,E6	A/G	3.4 E^{-06}-9.4 E^{-04}	8.0-18.0	
IWA5214	5B	302,177,272-302,177,428	E2	A/C	8.60 E^{-04}	8.1	
AX_110617778	5B	531,539,253-531,539,323	E2,E3	A/T	1.4-9.5 E^{-04}	8.0-11.5	
AX_110056162	5B	556,183,885-556,183,955	E5	T/C	4.2-7.3 E^{-04}	8.3-11.0	
AX_108,821,301	6A	94,2114,60-94,211,390	E4	C/G	1.4-8.0 E^{-04}	9.1-11.2	QBp. caas-6A[23]
AX_110578177	6B	676,210,414-676,210,344	E1	A/T	4.2-5.1 E^{-04}	8.6-8.9	
AX_109359792	6D	217,194,463-217,194,533	E4	A/G	6.30 E^{-04}	8.5	
AX_111086566	7A	88,862,791-88,862,721	E1,E6	T/C	2.2-8.8 E^{-04}	8.6-11.6	
AX_108743156	7A	136,398,412-136,398,482	E4,E5,E6	A/G	5.0-9.8 E^{-04}	7.9-9.2	
AX_109,311,326	7A	609,508,901-609,508,971	E2	T/C	8.50 E^{-04}	8.2	QBp. caas-7AL. 2[23]
AX_111042346	7A	670,876,731-670,876,661	E5,E6	C/G	2.6-4.9 E^{-04}	9.0-12.8	
AX_109491960	7A	70,8211,110-708,211,040	E4	A/C	1.7-9.1 E^{-04}	8.4-11.2	
AX_108870509	7B	729,224,017-729,224,087	E1,E2,E3,E6	A/G	3.0 E^{-05}-8.6 E^{-04}	8.0-13.0	
AX_109370330	7D	129,917,622-129,917,692	E1,E3	A/C	5.2-9.5 E^{-04}	8.7-9.6	
AX_109033824	7D	615,826,844-615,826,914	E5	A/C	8.30 E^{-04}	8.2	

[a]Representative markers at the resistance loci

[b]Chr: Chromosome

[c]The physical positions of SNP markers based on wheat genome sequences from the International Wheat Genome Sequencing Consortium (IWGSC, http://www. wheatgenome. org/)

[d]E1: Anyang 2013; E2: Anyang 2014; E3: Anyang 2015; E4: Suixi 2013; E5: Suixi 2014; E6: Best linear unbiased prediction (BLUP) calculated from all five environments. The data from the results of Tassel v5. 0

[e]Favorable allele (SNP) is underlined

[f]The P-values were calculated by the Tassel v5. 0

[g]Percentage of phenotypic variance explained by the MTA from the results of Tassel v5. 0

[h]The previously reported QTL or genes within the same chromosomal regions

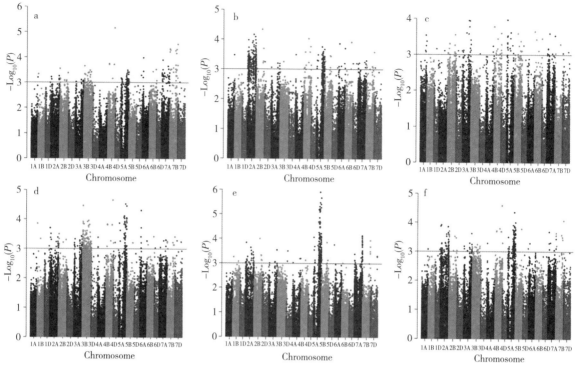

Fig. 2　Manhattan plots for black point resistance in 166 wheat accessions by the mixed linear model（MLM）in Tassel v5. 0.（a）Anyang 2013；（b）Anyang 2014；（c）Anyang 2015；（d）Suixi 2013；（e）Suixi 2014；（f）Best linear unbiased prediction（BLUP）values for black point scores calculated from all five environments. The $-\log_{10}$（P）values from a genome-wide scan are plotted against positions on each of the 21 chromosomes. Horizontal lines indicate genome-wide significance thresholds

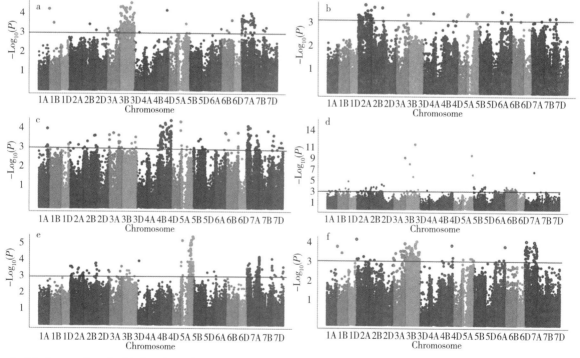

Fig. 3　Manhattan plots for black point resistance in 166 wheat accessions by the FarmCPU.（a）Anyang 2013；（b）Anyang 2014；（c）Anyang 2015；（d）Suixi 2013；（e）Suixi 2014；（f）Best linear unbiased prediction（BLUP）values for black point scores calculated from all five environments. The $-\log_{10}$（P）values from a genome-wide scan are plotted against positions on each of the 21 chromosomes. Horizontal lines indicate genome-wide significance thresholds

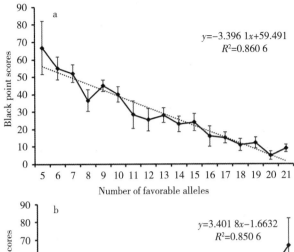

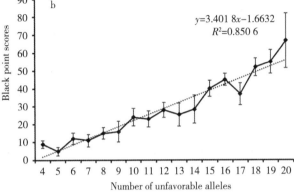

Fig. 4　Linear regression between the number of favorable alleles (a) and unfavorable alleles (b) and the BLUP values for black point scores

Discussion

The diversity panel, including released cultivars, advanced lines and landraces from different ecological regions, thus had a high genetic diversity with a wide range of reaction to black point. Our data showed that 86.1% (143) of the 166 accessions were susceptible to black point (black point score $>10\%$), indicating that black point is a considerable threat to wheat production throughout the world. However, most of the previous studies for black point were mainly conducted on pathogen identification, biological characteristics, disease cycle and control[7,13,38]. Thus, it is necessary to select cultivars highly resistant to black point and to identify markers significantly associated with resistance to facilitate breeding for resistance by MAS.

Genetic diversity, population structure and linkage disequilibrium

The mean genetic diversity and PIC of 0.356 and

0.285, respectively, indicated higher polymorphism than in previous reports[39,40]. Our diversity panel thus has high genetic diversity and approximately reflected the genetic diversity in winter wheat from the Yellow and Huai River Valleys Facultative Wheat Region. More than 56% of SNPs had PIC of 0.20-0.40, which is deemed as a suitable range for GWAS[41]. Furthermore, the A and B genomes had higher genetic diversity and PIC than the D genome, consistent with previous reports[30,40] (Additional file 2: Table S2). All results indicated that our diversity panel has high genetic diversity and was suitable for GWAS.

The diversity panel could be divided into three subgroups (Fig. 1), and the characterization of the subgroups was largely consistent with geographic origins and pedigrees. For example, Zhongmai 871, Zhongmai 875 and Zhongmai 895, which were derived from Zhoumai 16, clustered with Zhoumai 16 in group 3 (Additional file 1: Table S1). Numerous studies have shown that the lack of appropriate correction for population structure can lead to spurious MTAs[42-45]. Consequently, to eliminate spurious MTAs resulting from population structure, subpopulation data (Q matrix) were considered as fixed-effect factors, whereas the kinship matrix was considered as a random-effect factor, and a MLM implemented in Tassel v5.0 and FarmCPU were adopted for association analysis in the current study[36].

The LD decay affects the precision of GWAS and this is influenced by many factors like population structure, allele frequency, recombination rate and selection[44,46,47]. Previous studies reported that LD decay in common wheat ranged between 1.5-15 cM using SSR[28,46,48], DArT[33] or SNP[30,47] markers. In this panel, the LD decay was about 8 Mb for the whole genome (Additional file 3: Figure S1), consistent with previous reports. The LD decay of the D genome (11 Mb) was higher than the A (6 Mb) and B (4 Mb) genomes (Additional file 3: Figure S1), also consistent with previous studies[47-49], suggesting that fewer markers are needed for GWAS in the D genome than the A and B genomes. The marker densities for the A, B and D genomes were 0.099, 0.042 and 0.202 Mb/marker, and thus highly reliable for detecting MTAs with respect to LD decay in the diversity panel according to Breseghello and Sorrells[28]. The reason for the high LD of the D genome is mainly due to limited infusion of *Aegilops tauschii* in the evolutionary history of common wheat[38,49]. The average r^2 (0.174) values observed between linked loci pairs were higher than in previous studies[46,50]. Reif et al.[51] reported that LD (r^2) is expected to be higher in released cultivars than landraces. Moreover, Würschum et al.[52] indicated that QTL with small effect can be detected at higher LD (r^2), whereas only QTL with large effects can be detected at lower LD (r^2). Our results thus suggested a high mapping resolution and strong QTL detection power for black point resistance.

Comparison of the 90 K and 660 K SNP arrays

One of the key factors for GWAS is high marker density in whole genomes because sparse coverage reduces the power of marker identification[53]. Although the 90 K SNP array has emerged as a promising choice for high- density, low cost genotyping[34,54], the presence of large gaps, particularly low coverage for the D genome, reduces the power of marker identification and decreases the precision of QTL mapping. To resolve the problem, the GWAS for black point resistance was per- formed using 259, 922 markers from the 90 K and 660 K SNP arrays, providing a greater coverage of the genome. Only 8 loci were identified by the SNPs from 90 K array, whereas 23 were detected by the 660 K SNP, indicating that the 660 K SNP array with its much higher marker density had a significant advantage in GWAS.

Marker-trait associations

Some black point resistance QTL were previously identified by bi-parental linkage mapping[22,23], allowing for a comparison between loci identified in the present study and known QTL. Liu et al.[23] found seven stable black point resistance QTL on chromosomes 2AL, 2BL, 3BL, 5AS, 6A and 7AL (2) in a Linmai 2/Zhong 892 RIL popula- tion, which overlapped with loci identified in our study on chromosomes 2AL (*IWB22408*, 709.8 Mb), 2B (*AX-108951749*, 714.3 Mb), 3BL (*AX-108797097*, 695.9 Mb), 5AS (*IWB8709*, 32.8 Mb), 6A (*AX-108821301*, 94.2 Mb), and 7A (*AX-109311326*, 609.5 Mb) (Tables 2, Additional file 9: Table S6), indicating that GWAS and linkage mapping are complementary in identifying genes. Lehmensiek et al.[22] detected eight black point resistance QTL explaining 4 to 18% of the phenotypic variation on chromosomes 1D, 2A, 2B, 2D, 3D, 4A, 5A and 7A in Sunco/Tasman and Cascades/AUS1408 doubled haploid (DH) populations by SSR markers. We also identified 11 unique loci in 2A, 2B, 3D, 5A (3) and 7A (5). The loci on chromosomes 2AL (*IWB22408*, bin C-2AL1 − 0.85) and 2BL (*AX-108951749*, bin 2BL6 − 0.89-1.00) coincided with the QTL detected by Lehmensiek et al.[22] in chromosomes 2A (*Xgwm312*, bin C-2AL1 − 0.85) and 2B (*Xgwm319*, bin 2BL6 −0.89-1.00) (Table 2, Additional file 9: Table S6). However, not all of the QTL detected in linkage analysis were found in GWAS, such as *QBp. caas-3AL* and *QBp. caas-7BS*[23]. The reasons for this could be that (a) some QTL may have segregated at low frequency, or not at all in our association panel, and (b) results from the different marker platforms are difficult to align in the absence of complete genome sequences of diverse wheat cultivars.

Oxidases, such as PPO[15] and POD[17], could have enhanced the development of black point. The PPO gene (*Ppo-A1*) mapped to the long arm of chromosome 2AL in the interval *IWB59334-IWB5777* (706.2- 715.3 Mb)[55], overlapped with the loci on chromosome 2AL (*IWB22408*, 709.8 Mb) in our study. In addition, the *Ppo-A1*-specific marker *PPO18*[56] was also significantly associated with black point resistance (Table 2). Furthermore, the SNP marker *IWA5214* (302.2 Mb) on chromosome 5BL was significantly associated with both black point resistance and PPO activity (Zhai et al. unpublished data). Wei et al. [57] identified a QTL for POD activity on chromosome 5AS (15.9-36.9 Mb) using a RIL population derived from Doumai/Shi 4185, corresponding to the major loci detected on chromosome 5AS (*IWB8705*, 32.8 Mb) in this study (Additional file 9: Table S6). Shi et al. (unpublished data) identified a locus for POD activity on chromosome 2AL (*IWB59334*, 715.3 Mb) by GWAS, which overlapped with the locus on chromosome 2AL (*IWB22408*, 709.8 Mb). Thus, the GWAS results confirmed previous reports implicating phenol metabolism enzymes like PPO and POD in development of black point[15,17,18].

As the genetics of black point reaction are still poorly understood, the remaining 18 loci identified on chromosomes 3A, 3B, 3D, 4B (2), 5A (2), 5B (2), 6B, 6D, 7A (4), 7B and 7D (2) represent potentially new resistance QTL (Table 2); these may contribute to better understand of the architecture of black point reaction and provide more opportunities for resistance breeding. The above results demonstrated that GWAS was a powerful and reliable tool for identification of black point resistance genes.

Candidate genes for black point resistance

To identify candidate genes for black point resistance, the flanking sequences of SNP markers significantly associated with black point reaction were imported to Blast2Go software, and used as queries to BLAST against the National Center for Biotechnology Information (NCBI) and European Nucleotide Archive (ENA) databases; six candidate genes were identified (Table 3). Bioinformatics analysis indicated that SNP marker *AX-111518195* on chromosome 2AL corresponded to peroxisomal biogenesis factor 2, an important gene for biosynthesis of peroxidase, which can accelerate oxidation of phenolic compounds to quinones and is crucial for phenolic metabolism and melanin

synthesis[18,58]. In addition, the gene-specific marker *PPO18* for *Ppo-A1*[56] overlapping with the SNP loci on chromosome 2AL was also significantly associated with black point reaction. Fuerst et al. [18] reported that PPO catalyzes oxidation of phenolic compounds to melanins and quinines that may contribute to black point development. Thus, *Ppo-A1* is a candidate gene for this locus. Marker *AX-95684401* on chromosome 5A corresponded to a gibberellin (GA) biosynthetic process protein. GA plays an important role in modulating disease reaction throughout plant development and affects black point development by influencing seed germination[59]. Marker *IWA5463* on chromosome 2AL corresponds to an F-box repeat protein, which may affect black point development by regulating signal transduction of gibberellin[59,60]. F-box proteins have also been implicated in response to various pathogens through targeting substrates in the degradation machinery[61]. Two SNP markers (*AX-108951749* on 2B and *IWA2223* on 5AL) encode serine/threonine-protein kinases, which trigger multiple physiological and biochemical reactions in response to abiotic and biotic stresses by mediating perception and transduction of external environmental signals[62,63]. We also identified a candidate gene encoding a disease resistance RPP8-like protein (*AX-111053669* on chromosome 3A), which had been proposed to play an essential role in regulation of responses to a variety of external stimuli, including stress[64]. Bioinformatics analysis of trait-associated SNPs was proven to be an effective tool to find candidate genes for complex agronomic traits[34]. However, black point is a consequence of complicated biological processes and the mechanism of black point formation remains unclear; more detailed experimental analyses are needed to confirm the roles of candidate genes in black point resistance.

Table 3 Candidate genes for SNPs significantly associated with black point resistance

Chromosome	Marker	Candidate gene	Sequence similarity (%)	Sequence coverage (%)	Quality parameters
2AL	*IWA5463*	F-box repeat	98	98	4 E^{-36}

(continued)

Chromosome	Marker	Candidate gene	Sequence similarity (%)	Sequence coverage (%)	Quality parameters
2AL	PPO-18	Polyphenol oxidase (PPO-A1)	—	—	—
2AL	AX-111518195	Peroxisomal biogenesis factor 2	97	97	$4 E^{-12}$
2B	AX-108951749	Serine/threonine-protein kinase	97	96	$6 E^{-06}$
3A	AX-111053669	Disease resistance RPP8-like protein	97	96	$1 E^{-08}$
5AL	IWA2223	Serine/threonine-protein kinase	100	99	$8 E^{-39}$
5A	AX-95684401	Gibberellin biosynthetic process	97	100	$4 E^{-07}$

Application of MTAs for black point resistance in wheat breeding

It isdifficult to select highly resistant lines at the early stages of a breeding program in the field due to the fact that black point symptoms can be assessed only on mature seed after harvest and are highly affected by environment. A significant additive effect was identified from the linear regression between black point resistance and the number of favorable alleles, indicating that pyramiding of favorable alleles will enhance resistance. Markers significantly associated with complex traits identified by GWAS or QTL mapping can be converted into kompetitive allele-specific PCR (KASP) markers for SNP validation, MAS and QTL fine mapping[65,66]. Semi-thermal asymmetric reverse PCR (STARP) also provides a new scalable, flexible and cost-effective approach for using SNP markers in MAS[67]. QTL with consistent effects across multiple environments should be useful for MAS[68]. Thirteen of the 25 loci identified in this study were detected in two or more environments and should be suitable for MAS. Some accessions with higher black point resistance and relatively high number of resistance alleles and excellent agronomic traits, such as Kitanokaori, Norin 67, Yumai 21, Yannong19, Zhoumai19, and Zhongmai871 (Additional file 13: Table S7), should be good parental lines for breeding. Our follow-up studies will focus on validating the effects of these QTL and developing friendly, tightly linked markers that can be used in resistance breeding.

Conclusions

In the present study, a GWAS for black point resistance in a diversity panel was conducted with the 90 K and 660 K SNP arrays. Twenty-five resistance loci explained 7. 9-18. 0% of the phenotypic variations, demonstrating that GWAS can be used as a powerful and reliable tool for dissecting genes in wheat. The markers significantly associated with black point resistance and the accessions with a higher number of resistance alleles can be used as valuable markers and excellent parent material for resistance breeding. This study improves our understanding of the genetic architecture of black point resistance in common wheat.

Methods

Plant materials and field trials

The association panel used in the present study contained 166 diverse cultivars, comprising 144 accessions from the Yellow and Huai River Valley Facultative Wheat Region of China, and 22 accessions from five other countries, including Italy (9), Argentina (7), Japan (4), Australia (1) and Turkey (1) (Additional file 1: Table S1). All accessions were grown at Anyang (35°12′N, 113°37′E) in Henan province during the 2012—2013 and 2013—2014 cropping seasons, and Suixi (33°17′N, 116°23′E) in Anhui province during 2012—2013, 2013—2014 and 2014—2015. Field trials were conducted in randomized complete blocks with three replicates at all locations. Each plot contained three 2 m rows spaced 20 cm apart. Agronomic management followed local practices. All wheat accessions are deposited in the National Genebank of China, Chinese Academy of Agricultural Sciences, and available after approval.

Phenotypic evaluation and statistical analysis

After harvest and threshing three samples of 200 grains were selected from each of the three replicates of each accession, and the percentages of kernels with black point symptoms were determined and averaged. BLUPs across five environments were used as the phenotypic values for association mapping to eliminate environmental effects. BLUP estimation was calculated using the MIXED procedure (PROCMIXED) in SAS v9.3 (SAS Institute, http://www.sas.com).

ANOVA was performed using SAS v9.3 (SAS Institute, http://www.sas.com). Variance components were used to calculate broad sense heritability(h^2) of blackpoint scores as $h^2 = \sigma_g^2 / (\sigma_g^2 + \frac{\sigma_{ge}^2}{r} + \frac{\sigma_\varepsilon^2}{re})$, where σ_g^2, σ_g^2, and σ_ε^2 represent the genotype, genotype $\times$ environment interaction and residual error variances, respectively, and e and r were the numbers of environments and replicates per environment, respectively.

Genotyping and quality control

Total genomic DNA for SNP arrays was extracted from five bulked young leaves from each accession usinga modified CTAB procedure[69]. The 166 accessions were genotyped using both the Illumina wheat 90 K SNP (containing 81, 587 SNPs) and Affymetrix 660 K SNP (containing 630, 517 SNPs) arrays by Capital Bio Corporation, Beijing, China (http://www.capitalbiotech.com/). Accuracy of SNP clustering was validated visually. MAF, genetic diversity and PIC were computed by PowerMarker v3.25[70] (http://statgen.ncsu.edu/powermarker/). To avoid spurious MTAs, SNP markers with MAF< 0.05 and missing data > 20% were excluded from further analyses. The physical positions of SNP

markers from the wheat 90 K and 660 K SNP arrays were obtained from the International Wheat Genome Sequencing Consortium website (IWGSC, http:// www.wheatgenome.org/), and markers from two SNP arrays were integrated into a common physical map for GWAS.

Population structure

Population structurewas analyzed using 2000 polymorphic SNP markers from the 90 K and 660 K SNP arrays with Structure v2.3.4[41] (http://pritchardlab.stanford.edu/structure.html), which implements a model-based Bayesian cluster analysis. Five independent runs for each K value from 2 to 12 were performed based on an admixture model. Each run was carried out with 100,000 recorded Markov-Chain iterations and 10,000 burn-in periods. An adhoc quantity statistic ΔK based on the rate of change in log probability of data between successive K values[71] was used to predict the real number of subpopulations. PCA and NJ trees were also used to validate population stratification with the software Tassel v5.0[44] and PowerMarker v3.25[70] (http://www.maizegenetics.net).

Linkage disequilibrium

LD among markers was calculated using the full matrix and sliding window options in Tassel v5.0 with 12,324 evenly distributed SNP markers. The positions of these markers were based on the physical map mentioned above. Pairwise LD was measured using squared allele-frequency correlations r^2, and significance of pair-wise LD (*P*-values) was measured by Tassel v5.0 with 1000 permutations. The r^2 values were plotted against physical distance and a LOESS curve was fitted to the plot to show the association between LD decay and physical map distance. The critical value of r^2 beyond which the LD was likely to be caused by genetic linkage was determined by taking the 95th percentile in the distribution of r^2 of the selected loci[28]. The intersection of the fitted curve of r^2 values with this threshold was considered as the estimate of LD range.

Genome-wide association analysis

Associations between genotypic and phenotypic data were analyzed using the kinship matrix in a MLM by Tassel v5.0 to control background variation and eliminate the spurious MTAs. In MLM analysis, the kinship matrix (K matrix) was considered a random-effect factor, whereas the subpopulation data (Q matrix) was considered a fixed-effect factor[43]. The K matrix was calculated by the software Tassel v5.0 and the Q matrix was inferred by the program Structure v2.3.4. The P value determining whether a SNP marker was associated with the trait and the R^2 indicating the variation explained by the marker was recorded. The GWAS was also analyzed using the FarmCPU software[37] by R Language (https://www.r-project.org/). Bonferroni-Holm correction[72] for multiple testing ($\alpha=0.05$) was too conserved and no significant MTAs were detected with this criterion. Therefore, markers with an adjusted $-\log_{10}$ (P-value) $\geqslant 3.0$ were regarded as significant markers for black point reaction[73-75], as shown in Manhattan plots using the ggplot2 code in R Language. Important P value distributions (observed P values plotted against expected P values) were shown in Q-Q plots.

We checked the LD (r^2) among markers significantly associated with black point reaction on the same chromosomes to compare the resistance loci. LD block on the same chromosome were computed and visualized by Haploview v4.2[76] (www.broadinstitute.org/haploview/ haploview). To compare resistance loci identified in the present study with known genes/QTL, deletion bin information for SSR and SNP markers was obtained following[23].

The effect of favorable alleles on black point resistance

Each locus comprises two alleles based on SNP marker a single base substitution, transition or transversion. Alleles with positive effects leading to higher black point resistance are referred to as "favorable alleles", and those leading to lower resistance are "unfavorable alleles". The representative SNPs at the resistance loci were used to count the frequencies of favorable and unfavorable alleles and their allelic effects were determined (Table 2). Regression analysis between favorable, unfavorable alleles and black point scores were conducted using the line chart function in Microsoft Excel 2016.

In silico annotation of SNPs

To identify candidate genes or putative protein functions of SNP flanking-regions, the flanking sequences corresponding to the SNP markers significantly associated with black point resistance were used in BLASTn and BLASTx searches against ENA (http://www.ebi.ac.uk/ena) and NCBI (http://www.ncbi.nlm.nih.gov/) databases. Sequences were imported to Blast2Go software(https://www.blast2go.com/) in fasta formats that were blasted, mapped and annotated using standard parameters embedded in the software.

❈ Acknowledgements

We thank Prof. R.A. McIntosh, at Plant Breeding Institute, University of Sydney, for reviewing this manuscript.

❈ References

[1] Conner RL, Davidson JGN. Resistance in wheat to black point caused by *Alternaria alternata* and *Cochliobolus sativus*. Can J Plant Pathol. 1988; 68: 351-9.

[2] Sissons M, Sissons S, Egan N. The black point status of selected tetraploid species and Australian durum wheat and breeding lines. Crop Sci. 2010; 50: 1279-86.

[3] Dexter JE, Matsuo RR. Effect of smudge and black point, mildewed kernels, and ergot on durum wheat quality. Cereal Chem. 1982; 59: 63-9.

[4] Li QY, Qin Z, Jiang YM, Shen CC, Duan ZB, Niu JS. Screening wheat genotypes for resistance to black point and the effects of diseased kernels on seed germination. J Plant Dis Protect. 2014; 121: 79-88.

[5] Logrieco A, Bottalico A, Mulé G, Moretti A, Perrone G. Epidemiology of toxigenic fungi and their associated mycotoxins for some Mediterranean

crops. Eur J Plant Pathol. 2003; 109: 645-67.

[6] Desjardins AE, Busman M, Proctor RH, Stessman R. Wheat kernel black point and fumonisin contamination by *Fusarium proliferatum*. Food Addit Contam. 2007; 24: 1131-7.

[7] Busman M, Desjardins AE, Proctor RH. Analysis of fumonisin contamination and the presence of *Fusarium* in wheat with kernel black point disease in the United States. Food Addit Contam. 2012; 29: 1092-100.

[8] Palacios SA, Susca A, Haidukowski M, Stea G, Cendoya E, Ramírez ML, Chulze SN, Farnochi MC, Moretti A, Torres AM. Genetic variability and fumonisin production by *Fusarium proliferatum* isolated from durum wheat grains in Argentina. Int J Food Microbiol. 2015; 201: 35-41.

[9] Maloy OC, Spetch KL. Black point of irrigated wheat in Central Washington. Plant Dis. 1988; 72: 1031-3.

[10] Fernandez MR, Wang H, Singh AK. Impact of seed discoloration on emergence and early plant growth of durum wheat at different soil gravimetric water contents. Can J Plant Pathol. 2014; 36: 509-16.

[11] Southwell RJ, Brown JF, Wong PT. Effect of inoculum density, stage of plant growth and dew period on the incidence of black point caused by *Alternaria alternata* in durum wheat. Ann Appl Biol. 1980; 96: 29-35.

[12] Kumar J, Schäfer P, Hückelhoven R, Langen G, Baltruschat H, Stein E, Nagarajan S, Kogel KH. *Bipolaris sorokiniana*, a cereal pathogen of global concern: cytological and molecular approaches towards better control. Mol Plant Pathol. 2002; 3: 85-195.

[13] Williamson PM. Black point of wheat: in vitro production of symptoms, enzymes involved, and association with *Alternaria alternata*. Aust J Agri Res. 1997; 48: 13-9.

[14] Hudec K. Pathogenicity of fungi associated with wheat and barley seedling emergence and fungicide efficacy of seed treatment. Biologia. 2007; 62: 287-91.

[15] March TJ, Able J, Schultz C, Able AJA. Novel late embryogenesis abundant protein and peroxidase associated with black point in barley grains. Proteomics. 2007; 7: 3800-8.

[16] Mak Y, Willows RD, Roberts TH, Wrigley CW, Sharp PJ, Copeland LES. Black point is associated with reduced levels of stress, disease and defence related proteins in wheat grain. Mol Plant

Pathol. 2006; 7: 177-89.

[17] Anderson JV, Morris CF. An improved whole-seed assay for screening wheat germplasm for polyphenol oxidase activity. Crop Sci. 2001; 41: 1697-705.

[18] Fuerst EP, Okubara PA, Anderson JV, Morris CF. Polyphenol oxidase as a biochemical seed defense mechanism. Front Plant Sci. 2014; 5: 689.

[19] Porta H, Rocha-Sosa M. Plant lipoxygenases. Physiological and molecular features. Plant Physiol. 2002; 130: 15-21.

[20] Tomás-Barberán FA, Espín JC. Phenolic compounds and related enzymes as determinants of quality in fruits and vegetables. J Sci Food Agr. 2001; 81: 853-76.

[21] Regnier T, Macheix JJ. Changes in wall bound phenolic acids, phenylalanine and tyrosine ammonia-lyases, and peroxidases in developing durum wheat grains (*Triticum turgidum* L. Var. *durum*). J Agric Food Chem 1996; 44: 1727-1730.

[22] Lehmensiek A, Campbell AW, Williamson PM, Michalowitz M, Sutherland MW, Daggard GQTL. For black point resistance in wheat and the identification of potential markers for use in breeding programs. Plant Breed. 2004; 123: 410-6.

[23] Liu JD, He ZH, Wu L, Bai B, Wen WE, Xie CJ, Xia XC. Genome-wide linkage mapping of QTL for black point reaction in bread wheat (*Triticum aestivum* L.). Theor Appl Genet 2016; 129: 2179-2190.

[24] March TJ, Able JA, Willsmore K, Schultz CJ, Able AJ. Comparative mapping of a QTL controlling black point formation in barley. Funct Plant Biol. 2008; 35: 427-37.

[25] Wang S, Wong D, Forrest K, Ailen A, Chao S, Huang BE, Maccaferri M, Salvi S, Milner SG, Cattivelli L, Mastrangelo AM, Stephen S, Barker G, Wieseke R, Plieske J, International Wheat Genome Sequencing Consortium, Lillemo M, Mather D, Appels R, Dulferos R, Brown-Guedira G, Korol A, Akhunova AR, Feuillet C, Salse J, Morgante M, Pozniak C, Luo MC, Dvorak J, Morell M, Dubcovsky J, Ganal M, Tuberosa R, Lawley C, Mikoulitch I, Cavanagh C, Edwards KJ, Hayden M, Akhunov E. Characterization of polyploid wheat genomic diversity using the high-density 90, 000 SNP array. Plant Biotech J. 2014; 12: 787-96.

[26] Winfield MO, Allen AM, Burridge AJ, Barker GL, Benbow HR, Wilkinson PA, Coghill J. High-density

SNP genotyping array for hexaploid wheat and its secondary and tertiary gene pool. Plant Biotechnol J. 2015; 14: 1195-206.

[27] Flint-Garcia SA, Thornsberry JM, Buckler ES. Structure of linkage disequilibrium in plants. Annu Rev Plant Biol. 2003; 54: 357-74.

[28] Breseghello F, Sorrells ME. Association mapping of kernel size and milling quality in wheat (*Triticum aestivum* L.) cultivars. Genetics. 2006; 172: 1165-77.

[29] Zhu CS, Gore M, Buckler ES, Status YJM. Prospects of association mapping in plants. Plant Genome. 2008; 1: 5-20.

[30] Zegeye H, Rasheed A, Makdis F, Badebo A, Ogbonnaya FC. Genome-wide association mapping for seedling and adult plant resistance to stripe rust in synthetic hexaploid wheat. PLoS One. 2014;9: e105593.

[31] Pasam RK, Bansal U, Daetwyler HD, Forrest KL, Wong D, Petkowski J, Willey N. Detection and validation of genomic regions associated with resistance to rust diseases in a worldwide hexaploid wheat landrace collection using BayesR and mixed linear model approaches. Theor Appl Genet. 2017; 130: 777-93.

[32] Marcotuli I, Houston K, Waugh R, Fincher GB, Burton RA, Blanco A, Gadaleta A. Genome-wide association mapping for arabinoxylan content in a collection of tetraploid wheats. PLoS One. 2015; 10: e0132787.

[33] Rasheed A, Xia XC, Ogbonnaya F, Mahmood T, Zhang Z, Mujeeb-Kazi A, He ZH. Genome-wide association for grain morphology in synthetic hexaploid wheats using digital imaging analysis. BMC Plant Biol. 2014; 14: 128.

[34] Ain QU, Rasheed A, Anwar A, Mahmood T, Imtiaz M, Mahmood T, Xia XC, He ZH, Quraishi UM. Genome-wide association for grain yield under rain-fed conditions in historical wheat cultivars from Pakistan. Front Plant Sci. 2015; 6: 743.

[35] Sun CW, Zhang FY, Yan XF, Zhang XF, Dong ZD, Cui DQ, Chen F. Genome-wide association study for 13 agronomic traits reveals distribution of superior alleles in bread wheat from the yellow and Huai Valley of China. Plant Biotechnol J. 2017; doi: 10. 1111/ pbi. 12690.

[36] Bradbury PJ, Zhang Z, Kroon DE, Casstevens TM, Ramdoss Y, Buckler ESTASSEL. Software for association mapping of complex traits in diverse samples. Bioinformatics. 2007; 23: 2633-5.

[37] Liu X, Huang M, Fan B, Buckler ES, Zhang Z. Iterative usage of fixed and random effect models for powerful and efficient genome-wide association studies. PLoS Genet. 2016; 12 (2): e1005767.

[38] Kahl SM, Ulrich A, Kirichenko AA, Müller ME. Phenotypic andphylogenetic segregation of *Alternaria infectoria* from small-spored *Alternaria* species isolated from wheat in Germany and Russia. J Appl Microbiol. 2015; 119: 1637-50.

[39] Chen XJ, Min DH, Tauqeer AY, Genetic HYG. Diversity, population structure and linkage disequilibrium in elite Chinese winter wheat investigated with SSR markers. PLoS One. 2012; 7: e44510.

[40] Lopes M, Dreisigacker S, Peña R, Sukumaran S, Reynolds M. Genetic characterization of the wheat association mapping initiative (WAMI) panel for dissection of complex traits in spring wheat. Theor Appl Genet. 2014; 128: 453-64.

[41] Botstein D, Wlllte RL, Skolinck M. Construction of a genetic linkage map in man using restriction fragment length polymorphisms. Am J Hum Genet. 1980; 32: 314-9.

[42] Chao S, Zhang W, Dubcovsky J, Sorrells M. Evaluation of genetic diversity and genome-wide linkage disequilibrium among US wheat (*Triticum aestivum* L.) germplasm representing different market classes. Crop Sci. 2007; 47: 1018-30.

[43] Pritchard JK, Stephens M, Rosenberg NA, Donnelly P. Association mapping in structured populations. Am J Hum Genet. 2000; 67: 170-81.

[44] Yu J, Buckler ES. Genetic association mapping and genome organization of maize. Curr Opin Biotechnol. 2006; 17: 155-60.

[45] Cormier F, Gouis JL, Dubreuil P, Lafarge S, Praud S. A genome-wide identification of chromosomal regions determining nitrogen use efficiency components in wheat (Triticum aestivum L.). Theor Appl Genet 2014; 127: 2679-2693.

[46] Hao CY, Wang LF, Ge HM, Dong YC, Zhang XY. Genetic diversity and linkage disequilibrium in Chinese bread wheat (*Triticum aestivum* L.) revealed by SSR markers. PLoS One. 2011; 6: e17279.

[47] Sukumaran S, Dreisigacker S, Lopes M, Chavez P, Reynolds MP. Genome-wide association study for grain yield and related traits in an elite spring wheat population grown in temperate irrigated environments.

Theor Appl Genet. 2015; 128: 353-63.

[48] Chao S, Dubcovsky J, Dvorak J, Luo MC, Baenziger SP, Matnyazov R, Clark DR, Talbert LE, Anderson JA, Dreisigacker S, Glover K, Chen J, Campbell K, Bruckner PL, Rudd JC, Haley S, Carver BF, Perry S, Sorrells ME, Akhunov ED. Population and genome specific patterns of linkage disequilibrium and SNP variation in spring and winter wheat (*Triticum aestivum* L.). BMC Genomics 2010; 11: 727.

[49] Edae EA, Byrne PF, Haley SD, Lopes MS, Reynolds MP. Genome-wide association mapping of yield and yield components of spring wheat under contrasting moisture regimes. Theor Appl Genet. 2014; 127: 791-807.

[50] Chen GF, Zhang H, Deng ZY, Wu RG., Li DM, Wang MY, Tian JC. Genome-wide association study for kernel weight-related traits using SNPs in a Chinese winter wheat population. Euphytica 2016; 212: 173-185.

[51] Reif JC, Maurer HP, Korzun V, Ebmeyer E, Miedaner T, Würschum T, Mapping QTL. With main and epistatic effects underlying grain yield and heading time in soft winter wheat. Theor Appl Genet. 2011; 123: 283-2927.

[52] Würschum T, Maurer HP, Kraft T, Janssen G, Nilsson C, Reif JC. Genome-wide association mapping of agronomic traits in sugar beet. Theor Appl Genet. 2011; 123: 1121-31.

[53] Poznial CJ, Clarke JM, Clarke FR. Potential for detection of marker-trait associations in durum wheat using unbalanced, historical phenotypic datasets. Mol Breed. 2012; 30: 1537-50.

[54] Allen AM, Barker GLA, Berry ST, Coghill JA, Gwilliam R, Kirby S, Robinson P, Brenchley RC, D'Amore R, McKenzie N, Waite D, Hall A, Bevan M, Hall N, Edwards KJ. Transcript-specific, single-nucleotide polymorphism discovery and linkage analysis in hexaploid bread wheat (*Triticum aestivum* L.). Plant Biotechnol J 2011; 9: 1086-1099.

[55] Zhai SN, He ZH, Wen WE, Jin H, Liu JD, Zhang Y, Liu ZY, Xia XC. Genome-wide linkage mapping of flour color-related traits and polyphenol oxidase activity in common wheat. Theor Appl Genet. 2016; 129: 377-94.

[56] Sun DJ, He ZH, Xia XC, Zhang LP, Morris CF, Appels R, Ma WJ, Wang HA, Novel STS. Marker for polyphenol oxidase activity in bread wheat. Mol Breed. 2005; 16: 209-18.

[57] Wei JX, Geng HW, Zhang Y, Liu JD, Wen WE, Xia XC, Chen XM, He ZH. Mapping quantitative trait loci for peroxidase activity and developing gene-specific markers for *TaPod-A1* on wheat chromosome 3AL. Theor Appl Genet. 2015; 128: 2067-76.

[58] Kumar S, Kawałek A, van der Klei IJ. Peroxisomal quality control mechanisms. Curr Opin Microbiol. 2014; 22: 30-7.

[59] Sun TP, Gubler F. Molecular mechanism of gibberellin signaling in plants. Annu Rev Plant Biol. 2004; 55: 197-223.

[60] Frigerio M, Alabadí D, Pérez-Gómez J, García-Cárcel L, Phillips AL, Hedden P, Blázquez MA. Transcriptional regulation of gibberellin metabolism genes by auxin signaling in *Arabidopsis*. Plant Physiol. 2006; 142: 553-63.

[61] Kim HS, Delaney TP. *Arabidopsis* SON1 is an F-box protein that regulates a novel induced defense response independent of both salicylic acid and systemic acquired resistance. Plant Cell. 2002; 14: 1469-82.

[62] Xiong L, Yang Y. Disease resistance and abiotic stress tolerance in rice are inversely modulated by an abscisic acid-inducible mitogen-activated protein kinase. Plant Cell. 2003; 15: 745-59.

[63] Li FH, FL F, Sha LN, He L, Li WC. Differential expression of serine/threonine protein phosphatase type-2C under drought stress in maize. Plant Mol Biol Rep. 2009; 27: 29-37.

[64] Hameed U, Pan YB, Iqbal J. Genetic analysis of resistance gene analogues from a sugarcane cultivar resistant to red rot disease. J Phytopathol. 2015; 163: 755-63.

[65] Semagn K, Babu R, Hearne S, Olsen M. Single nucleotide polymorphism genotyping using Kompetitive allele specific PCR (KASP): overview of the technology and its application in crop improvement. Mol Breed 2014; 33: 1-14.

[66] Rasheed A, Hao YF, Xia XC, Khan A, Yb X, Varshney RK, He ZH. Crop breeding chips and genotyping platforms: progress, challenges, and perspectives. Mol Plant. 2017; 10: 1047-64.

[67] Long YM, Chao WS, Ma GJ, SS X, Qi LL. An innovative SNP genotyping method adapting to multiple platforms and throughputs. Theor Appl Genet. 2016; 130: 597-607.

［68］ Veldboom LR，Lee M. Genetic mapping of quantitative trait loci in maize in stress and non-stress environments： I. Grain yield and yield components. Crop Sci. 1996；36： 1310-9.

［69］ Doyle JJ，Doyle JL. A rapid DNA isolation procedure from small quantities of fresh leaf tissues. Phytochem Bull. 1987；19：11-5.

［70］ Liu K，Muse SV. PowerMarker. An integrated analysis environment for genetic marker analysis. Bioinformatics. 2005；21：2128-9.

［71］ Evanno G，Regnaut S，Goudet J. Detecting the number of clusters of individuals using the software STRUCTURE：a simulation study. Mol Ecol. 2005； 14：2611-20.

［72］ Holm SA. Simple sequentially rejective multiple test procedure. Scand J Stat. 1979；6：65-70.

［73］ Houston K，Russell J，Schreiber M，Halpin C， Oakey H，Washington JM，Booth A，Shirley N， Burton RA，Fincher GB，Waugh R. A Genome wide association scan for （1，3；1，4） -β-glucan content in the grain of contemporary 2-row spring and winter barleys. BMC Genomics 2014；15：907.

［74］ Gurung S，Mamidi S，Bonman JM，Xiong M， Brown-Guedira G，Adhikari TB. Genome-wide association study reveals novel quantitative trait loci associated with resistance to multiple leaf spot diseases of spring wheat. PLoS One. 2014；9：e108179.

［75］ Bellucci A，Torp AM，Bruun S，Magid J，Andersen SB，Rasmussen SK. Association mapping in scandinavian winter wheat for yield，plant height， and traits important for second-generation bioethanol production. Front Plant Sci. 2015；6：1046.

［76］ Barrett JC，Fry B，Maller J，Daly MJ. Haploview： analysis and visualization of LD and haplotype maps. Bioinformatics. 2005；21：263-5.

加工品质与营养特性

黄淮冬麦区小麦品种植酸含量与植酸酶活性聚类分析

李颖睿[1]　陈茹梅[2]　朱　伟[3]　阎　俊[4]　何中虎[1,5]　张　勇[1,*]

[1] 中国农业科学院作物科学研究所/国家小麦改良中心，北京 100081；[2] 中国农业科学院生物技术研究所，北京 100081；[3] 河南省商丘市农林科学院，河南商丘 476000；[4] 中国农业科学院棉花研究所，河南安阳 455000；[5] 国际玉米小麦改良中心（CIMMYT）中国办事处，北京 100081；* 通信作者，E-mail：Zhangyong 05@caas cn.

摘要：植酸含量与植酸酶活性是影响铁、锌等微量元素生物有效性的关键因子。2009—2010 和 2010—2011 年度，在河南安阳种植 212 个黄淮麦区代表性小麦品种和高代品系，分析其籽粒植酸含量和植酸酶活性。结果表明，这 2 个指标变异范围较大，植酸含量为 2.18～13.37 g·kg^{-1}，平均 5.72 g·kg^{-1}；植酸酶活性为 10～1759 U·kg^{-1}，平均 657 U·kg^{-1}。品种及品种与年度互作效应显著影响植酸含量和植酸酶活性，以品种效应较大。根据植酸含量与植酸酶活性将参试品种分别聚为 5 类，类间植酸含量和植酸酶活性差异显著。石麦 12、衡 4568、洛麦 21 和济麦 096141 的植酸含量较低，且植酸酶活性较高，可作为进一步改良植酸含量和植酸酶活性的亲本。

关键词：普通小麦；植酸含量；植酸酶活性；营养品质

Variability of Phytate Content and Phytase Activity among Wheat Cultivars from Yellow and Huai River Valleys

LI Ying-Rui[1]，CHEN Ru-Mei[2]，ZHU Wei[3]，YAN Jun[4]，HE Zhong-Hu[1,5]，and ZHANG Yong[1,*]

[1] Institute of Crop Science / National Wheat Improvement Center, Chinese Academy of Agricultural Sciences (CAAS)，Beijing 100081，China；[2] Biotechnology Research Institute，CAAS，Beijing 100081，China；[3] Shangqiu Research Institute of Agriculture and Forestry Science，Shangqiu 476000，China；[4] Cotton Research Institute，CAAS，Anyang 455004，China；[5] CIMMYT-China Office，c/o CAAS，Beijing 100081，China；　* Corresponding author

Abstract：Phytate content and phytase activity are key factors influencing bioavailability of iron and zinc. To understand the status of phytate content and phytase activity in wheats from the Yellow and Huai River Valleys Winter Wheat Region，212 representative cultivars and advanced lines were sown in Anyang，Henan Province，China in 2009—2010 and 2010—2011 cropping seasons. Phytate content and phytase activity varied greatly among these cultivars，ranging from 2.18 to 13.37 g·kg^{-1} of phytate content and from 10 to 1 759 U·kg^{-1} of phytase activity，with the mean values of 5.72 g·kg^{-1} and

原文发表在《作物学报》，2014，40（2）：329-336.

657 U·kg^{-1}, respectively. Both indices were significantly affected by genotype and genotype × season interaction, with genotype effect being predominant. All cultivars were classified into five groups based on the seasonal standardized values of phytate content and phytase activity, with significant difference among groups. Four cultivars, i. e., Shimai 12, Heng 4568, Luomai 21, and Jimai 096141 exhibited low phytate content and high phytase activity, and can be used in wheat breeding program aiming at improving iron and zinc nutritional quality.

Keywords: Bread wheat; Phytate content; Phytase activity; Nutritional quality

微量营养元素缺乏造成的营养不良非常严重。全世界约20亿人患有不同程度的贫血，其中约12%由缺铁所致，而铁缺乏者的数量约为缺铁性贫血患病率的2.5倍[1-2]，超过一半人口摄入的锌和维生素A等营养元素不足[2]。发展中国家超过50%的妇女和儿童及约58%的孕妇存在缺铁性贫血[3]，每年约有800万儿童死于锌缺乏症[4]。我国的缺铁性贫血发病率为20%左右，贫困地区儿童和孕妇则高达45%和35%[1]。

小麦是我国北方地区的主要粮食作物，黄淮麦区是最重要的商品粮基地，提高人体对小麦籽粒中矿质元素的吸收利用，对于解决我国人民由于矿物质元素含量摄入不足造成的健康问题具有重要意义[4-5]。人体对食物中铁、锌等矿物质元素的吸收利用取决于其含量和生物有效性的高低[2-4]，植酸含量和植酸酶活性是影响微量矿物质元素生物有效性的主要因子[6]。影响铁、锌等生物有效性的限制性因子包括植酸、纤维素、丹宁和重金属等，以植酸最为重要[2-4]。植酸是磷在小麦籽粒中的主要贮存形式，通过与铁、锌、钙等二价金属离子结合，形成螯合态植酸盐[7]，降低人对铁、锌等的吸收利用[8]，植酸含量与锌含量呈显著正相关[9]。食物中的植酸盐可以被植酸酶有效分解，但人和猪等单胃动物的消化系统中植酸酶活性非常低[10]，导致铁、锌和磷等矿物质元素无法被有效吸收，同时大量的有机磷还随排泄物排出，引起土壤和水污染[11]。植酸虽然具有抗癌[12]、预防心脏病与糖尿病等代谢类疾病的功能[13]，并可提高种子活力，降低籽粒中黄曲霉毒素的含量[4]，但对微量矿物质元素生物有效性的阻遏作用非常显著，这对婴儿和孕妇等需要大量铁、锌等元素的人群来说非常重要[1,3]。植酸酶可有效分解植酸，通过提高铁、锌和磷等元素的生物有效性，大大改善人体因铁、锌等元素缺乏所导致的营养不良问题[2]。

植酸酶已作为添加剂广泛用于食品和饲料工业[14]，采用微生物方法进行工业化生产成本较高，生产过程中大量消耗能源并严重污染环境[6]。转植酸酶基因玉米已获生产应用安全证书并进行环境释放[15]，但转基因作物受大众认知和接受程度的严重影响，目前还难以大规模应用[16]。因此，有必要对现有小麦品种的植酸酶活性进行筛选，以充分利用籽粒本身所含的植酸酶。

国内外对小麦的铁、锌等元素含量已进行了较深入研究，品种间存在显著差异，并受环境显著影响[17]。铁含量相关基因/主效QTL定位在2A、4A、5A、7A和4D等染色体上，锌含量相关基因/主效QTL定位在1A、2D、3A、4A、4D、5A和7A等染色体上[18-19]。对印度和国际玉米小麦改良中心（CIMMYT）品种的植酸含量和植酸酶活性已进行了初步报道[20]，Liu等[9]和吴澎等[21]曾对我国的地方品种和河南、山东、陕西、江苏和四川等少数地区的品种进行了植酸含量和植酸酶活性分析。但总体来说，有关植酸含量与植酸酶活性分析的资料十分有限，黄淮主产麦区主要品种的植酸含量与植酸酶活性尚不清楚。本研究选用该区212个主栽品种和高代品系，连续两年度种植在中国农业科学院作物科学研究所安阳试验站，检测其植酸含量和植酸酶活性，为通过育种途径改良铁、锌等营养品质提供理论依据。

1 材料与方法

1.1 品种及其田间种植

212份供试材料（表1）中，来自河北80份、河南92份和山东34份，均为当地主栽品种或苗头品系；还包括已被主要育种单位用作优质亲本的中优9507和西农979，面包、面条品质优良的澳大利亚品种Sunco和Sunstate，7D1.7Ag易位系品种Wheatear，以及在国际上广泛用作条锈、叶锈和白粉病抗病亲本的RL6077（Weebill×2/Brambling）。

2009—2010和2010—2011年度，将所有品种种

植于中国农业科学院作物科学研究所安阳试验站。田间采用拉丁方 alpha 格子设计[5,22]，2 次重复，每隔15 个品种加 1 个当地对照。双行播种，行长 1.5m，行间距 0.20m，每行 50 粒，按当地常规进行田间管理。

1.2　性状测定

两年度收获前均无有效降雨，未受穗发芽影响，籽粒样品饱满度均较好。收获后各样品随机数 200 粒种子，3 次重复，称重，换算成千粒重。采用近红外分析仪（Foss 1241，Sweden）测定籽粒蛋白质含量（14%湿基）。在-20℃冷库中保存籽粒样品，用高通量组织研磨机（SPEX GENO 2010 GRINDER，美国）磨粉。按 Chen 等[15]的方法分析植酸含量，并适当改进。将 30mg 全粉置于 1.5mL 离心管中，加入 0.4mol·L^{-1} HCl 1mL 和 15%的 TCA 提取液，室温下振荡 3h，再以 2 000×g 离心 10min，然后取 50μL 上清液，置于 1.5mL 离心管中（内装 36.3mol·L^{-1} NaOH 550μL）；加入 200μL 显色液（含 0.03%氯化铁，0.3%磺基水杨酸），反应后取 200μL 溶液，用酶标仪（SPECTRA max PLUS384）在 500nm 下读数，测定植酸含量。参考 Chen 等[15]的方法，并略加调整，分析植酸酶活性。100mg 全粉置 1.5mL 离心管中，加入 0.25mL 抽提缓冲液（含 50mol·L^{-1} NaAc，1mol·L^{-1} CaCl$_2$、0.5% BSA、0.075% Triton X-100，pH 5.15）。室温振荡 1h，3 000×g 离心 15min 后取上清液，按 Wyss 等[23]的方法，通过测定无机磷的增加量来确定植酸酶的活性。20μL 上清液 50℃预热 5min，加入 6.25mol·L^{-1} 植酸钠溶液 80μL，30min 后用 15%TCA 终止反应。20μL 反应后溶液中加 80μL 去离子水和 100μL 显色液（7.35% FeSO$_4$·7H$_2$O、1%钼酸铵、0.6mol·L^{-1} H$_2$SO$_4$），37℃反应 10min 后，700nm 下测定吸光值。空白对照除先加 TCA 再加植酸钠以外，其余操作均相同。需要注意的是，在小麦植酸酶提取液与植酸钠反应时，需将最佳反应温度调整为 50℃，溶液 pH 值调整为 5.15。

1.3　数据处理

采用合适的空间模型[17]分析试验数据，将两年度所得品种各性状最佳线性无偏预测值用于数据处理和分析[24]。采用 SAS（Statistical Analysis System）8.0 软件，调用 PROC MEANS、PROC MIXED、PROC CLUSTER 和 PROC CORR 分别进行基本统计量、方差、聚类和相关等分析。其中调用 PROC MIXED 进行方差分析时，将基因型类作为固定效应，年度及年度相关互作、类内基因型和年度内重复作为随机效应。调用 PROC CLUSTER 时，将植酸含量和植酸酶活性数据分别按年度进行标准化后，以欧氏距离为标准，按 Ward 类平方和法分别对品种进行聚类[25]。

2　结果与分析

2.1　基本统计量分析

品种和年度间籽粒植酸含量和植酸酶活性均存在较大差异（表2）。植酸含量和植酸酶活性平均值分别为 5.72g·kg^{-1} 和 657U·kg^{-1}，变异范围分别为 2.18～13.37g·kg^{-1} 和 10～1 759U·kg^{-1}，品种间差异远大于年度间差异。千粒重和蛋白质含量平均为 46.0g 和 12.9%，变异范围分别为 30.6～63.1g 和 9.6%～16.7%。由此可见，我国黄淮冬麦区品种间植酸含量与植酸酶活性变异范围广，通过品种筛选降低籽粒植酸含量或提高植酸酶活性的潜力较大。

2.2　聚类分析

将所有品种的植酸含量和植酸酶活性数据分别按年度进行标准化，在决定系数（R^2）为 90.0%和 93.6%水平分别将品种聚为 5 类（表3和表4），不同地区来源品种间植酸含量和植酸酶活性差异不显著（表略）。方差分析表明，植酸含量的品种、年度及其互作效应均达 0.001 显著水平，以品种效应较大，其次为品种和年度互作效应；品种效应中，品种类别效应较大，达 0.001 显著水平，类内品种效应不显著。植酸酶活性的品种及品种和年度互作效应均达 0.001 显著水平，以品种效应较大，年度效应不显著；品种效应中，品种类别效应较大，达 0.001 显著水平，类内品种效应不显著（表5）。

5 类植酸含量存在显著差异的品种中，第 3 和第 5 极端类的品种数量较少（分别包括 19 个和 15 个品种），绝大多数品种表现植酸含量中等。第 3 类品种的平均植酸含量最高，达 10.33g·kg^{-1}，显著高于其他 4 类。第 5 类品种的植酸含量最低，为 2.78g·kg^{-1}；其次是第 4 类，平均为 4.04g·kg^{-1}；这两类品种的植酸含量变异范围都较小，在低植酸含量育种中具有较高的利用价值。从品种来源看，各类型在多个地区均有分布，黄淮麦区广泛应用的重要亲本中，只有周麦 16 属于第 4 类。墨西哥抗病品种 RL6077 也属于第 4 类（表 3），表现植酸含量较低。

5 类植酸酶活性存在显著差异的品种类群中所包含的品种与基于植酸含量的聚类结果不尽相同。其中，第 4 类品种的植酸酶活性最高，达 1 146U·kg⁻¹；其次是第 5 类，平均植酸酶活性为 930U·kg⁻¹；这两类分别包括 6 个和 27 个品种，数量较少，分别来自多个地区，类间植酸酶活性没有显著差异，但都显著高于其他 3 类，可作为重要的高植酸酶活性育种亲本。值得一提的是，澳大利亚品种 Sunstate 和 Sunco 分别属于第 4 和第 5 类（表 4），表现植酸酶活性较高。

综合植酸含量和植酸酶活性的聚类分析结果，发现低植酸含量和高植酸酶活性的共同品种有石麦 12、衡 4568、洛麦 21 和济麦 096141，分别来自河北、河南和山东省。这 4 个品种有可能成为重要的育种亲本，在低植酸含量和高植酸酶活性小麦育种中发挥作用。

表 1　212 份材料代号及其名称和来源

Table 1　Code, name, and origin of the 212 cultivars and advanced lines used in this study

代号 Code	品种 Variety	来源 Origin	代号 Code	品种 Variety	来源 Origin	代号 Code	品种 Variety	来源 Origin	代号 Code	品种 Variety	来源 Origin
1	冀麦 36	中国河北	54	科农 20	中国河北	107	矮抗 58	中国河南	160	曙麦 05-2	中国河南
2	冀麦 38	中国河北	55	科农 2009	中国河北	108	丰德存麦 1 号	中国河南	161	许科 316	中国河南
3	冀麦 112	中国河北	56	科农 2011	中国河北	109	丰德存麦 6 号	中国河南	162	许农 5 号	中国河南
4	冀麦 5265	中国河北	57	冀师 02-1	中国河北	110	丰德存麦 10 号	中国河南	163	许科 99087	中国河南
5	冀矮 1 号	中国河北	58	师栾 08-2	中国河北	111	百农 64	中国河南	164	豫保 2 号	中国河南
6	冀 325	中国河北	59	师栾 08-4	中国河北	112	中育 5 号	中国河南	165	长海大穗	中国河南
7	冀麦 518	中国河北	60	藁城 9415	中国河北	113	中麦 349	中国河南	166	保丰 10-82	中国河南
8	冀麦 585	中国河北	61	藁优 5218	中国河北	114	中麦 895	中国河南	167	内乡 188	中国河南
9	冀麦 602	中国河北	62	藁优 5766	中国河北	115	中麦 875	中国河南	168	豫麦 34	中国河南
10	冀麦 867	中国河北	63	U07-6308	中国河北	116	中麦 871	中国河南	169	豫安 208	中国河南
11	石 4185	中国河北	64	振桥大穗	中国河北	117	中 892	中国河南	170	偃 4110	中国河南
12	石家庄 8 号	中国河北	65	新麦 8 号	中国河北	118	00-62154	中国河南	171	偃展 9998	中国河南
13	石家庄 10 号	中国河北	66	新麦 9 号	中国河北	119	01-52289	中国河南	172	豫教 0520	中国河南
14	石家庄 11	中国河北	67	邢 04-1135	中国河北	120	04 中 36	中国河南	173	济麦 19	中国山东
15	石麦 12	中国河北	68	邢 05-1241	中国河北	121	04 中 70	中国河南	174	济麦 20	中国山东
16	石 06-6136	中国河北	69	邢麦 9 号	中国河北	122	06CA25	中国河南	175	济麦 22	中国山东
17	石 08-4741	中国河北	70	邢麦 10 号	中国河北	123	07CA255	中国河南	176	济麦 6097	中国山东
18	石 09-4276	中国河北	71	邢麦 13	中国河北	124	07CA266	中国河南	177	济麦 8186	中国山东
19	石 5341	中国河北	72	邢台 456	中国河北	125	08CA101	中国河南	178	济麦 0836262	中国山东
20	石 6136	中国河北	73	邯 6172	中国河北	126	08CA062	中国河南	179	济麦 0850187	中国山东
21	石 83H-366	中国河北	74	邯 07-6092	中国河北	127	08CA137	中国河南	180	济麦 0860223	中国山东
22	石 B05-7388	中国河北	75	邯 09-41307	中国河北	128	08CA307	中国河南	181	济麦 096141	中国山东
23	石 B07-4056	中国河北	76	邯 09-41344	中国河北	129	09CA034	中国河南	182	鲁原 502	中国山东
24	石 B07-4179	中国河北	77	邯 5093	中国河北	130	09CA163	中国河南	183	烟农 19	中国山东
25	石新 633	中国河北	78	邯 5672	中国河北	131	09CA170	中国河南	184	烟农 24	中国山东
26	石新 703	中国河北	79	邯 5849	中国河北	132	09CA175	中国河南	185	烟 99102	中国山东
27	石新 733	中国河北	80	邯 9983	中国河北	133	09CA86	中国河南	186	山农 05-066	中国山东
28	石新 811	中国河北	81	郑麦 005	中国河南	134	10CA23	中国河南	187	山农 055843	中国山东
29	石新 828	中国河北	82	郑麦 106	中国河南	135	10CA79	中国河南	188	山农 2149	中国山东

（续）

代号 Code	品种 Variety	来源 Origin	代号 Code	品种 Variety	来源 Origin	代号 Code	品种 Variety	来源 Origin	代号 Code	品种 Variety	来源 Origin
30	衡观 33	中国河北	83	郑 7698	中国河南	136	10CA70	中国河南	189	PH6911	中国山东
31	衡观 35	中国河北	84	郑麦 98	中国河南	137	11CA105	中国河南	190	泰农 2413	中国山东
32	衡 05-4444	中国河北	85	郑丰 4431	中国河南	138	11CA26	中国河南	191	泰农 8681	中国山东
33	衡 09-4061	中国河北	86	郑麦 110	中国河南	139	11CA40	中国河南	192	泰农 8968	中国山东
34	衡 10-5197	中国河北	87	郑麦 366	中国河南	140	9705-0-1-2-3	中国河南	193	泰农 9862	中国山东
35	衡 10S99-2	中国河北	88	郑育麦 9987	中国河南	141	94462-0-30-5	中国河南	194	泰山 064199	中国山东
36	衡 4041	中国河北	89	花培 5 号	中国河南	142	濮麦 9 号	中国河南	195	泰山 6195	中国山东
37	衡 4422	中国河北	90	花培 87-3	中国河南	143	丰优 66	中国河南	196	良星 619	中国山东
38	衡 4568	中国河北	91	新麦 18	中国河南	144	金育麦 2 号	中国河南	197	良星 66	中国山东
39	衡 5114	中国河北	92	新麦 19	中国河南	145	浚 K8	中国河南	198	良星 99	中国山东
40	衡 5218	中国河北	93	新麦 0208	中国河南	146	兰考 15	中国河南	199	临麦 2 号	中国山东
41	衡 5229	中国河北	94	新麦 21	中国河南	147	洛麦 21	中国河南	200	潍麦 8 号	中国山东
42	衡 5364	中国河北	95	新麦 26	中国河南	148	洛麦 22	中国河南	201	汶农 14	中国山东
43	衡 6421	中国河北	96	新选 2039	中国河南	149	洛麦 23	中国河南	202	YB66180	中国山东
44	衡辐 418	中国河北	97	周麦 16	中国河南	150	洛麦 24	中国河南	203	莱州 127	中国山东
45	衡辐 9103	中国河北	98	周麦 18	中国河南	151	洛新 998	中国河南	204	枣 9119	中国山东
46	河农 6049	中国河北	99	周 9823	中国河南	152	洛麦 02133	中国河南	205	淄 995015	中国山东
47	河农 7069	中国河北	100	周 98343	中国河南	153	洛麦 05123	中国河南	206	淄麦 12	中国山东
48	河农 9311	中国河北	101	周麦 22	中国河南	154	平安 8 号	中国河南	207	中优 9507	中国北京
49	金丰 7183	中国河北	102	周麦 24	中国河南	155	平安 10 号	中国河南	208	西农 979	中国陕西
50	金禾 2417	中国河北	103	周麦 25	中国河南	156	平安 11	中国河南	209	Sunco	澳大利亚
51	金禾 9123	中国河北	104	周麦 26	中国河南	157	漯麦 9908	中国河南	210	Sunstate	澳大利亚
52	科奥 08-6	中国河北	105	周麦 27	中国河南	158	漯麦 6082	中国河南	211	RL6077	墨西哥
53	科农 1006	中国河北	106	周麦 28	中国河南	159	汝州 0319	中国河南	212	Wheatear	墨西哥

表 2　212 个小麦品种千粒重、蛋白质含量、植酸含量和植酸酶活性的变异

Table 2　Variations of thousand-kernel weight, protein content, phytate content, and phytase activity in 212 wheat cultivars

性状 Trait	均值±标准差 Mean±SD	变幅 Range		
		样品间 Among samples	品种间 Among cultivars	年度间 Between years
千粒重 1000-kernel weight（g）	46.0±5.7	30.6~63.1	33.9~58.7	45.1~46.9
蛋白质含量 Protein content（%）	12.9±1.2	9.6~16.7	10.0~16.1	12.1~13.7
植酸含量 Phytate content（g·kg⁻¹）	5.72±2.21	2.18~13.37	2.25~13.36	5.14~6.23
植酸酶活性 Phytase activity（U·kg⁻¹）	657±292	10~1759	12~1715	621~687

表 3　两年度 212 份品种植酸含量聚类分析结果

Table 3　Cluster of 212 cultivars based on phytate content across two seasons

类别 Group	植酸含量 Phytate content（g·kg⁻¹）	变幅 Range（g·kg⁻¹）	品种来源 Origin of cultivar	品种代号 Code of cultivar
PC1（90）	6.05±1.49b	4.09~9.72	河北 Hebei（24）	49，78，58，41，12，60，77，1，29，26，68，21，69，7，40，46，32，4，13，28，11，8，52，30

（续）

类别 Group	植酸含量 Phytate content (g·kg⁻¹)	变幅 Range (g·kg⁻¹)	品种来源 Origin of cultivar	品种代号 Code of cultivar
			河南 Henan（47）	137，110，140，90，124，101，84，102，118，151，132，172，158，121，113，105，122，104，165，85，131，81，159，127，144，133，125，117，96，123，89，161，129，160，88，92，93，148，112，141，91，168，150，98，162，157，149
			山东 Shandong（16）	199，201，202，197，196，190，175，174，205，200，188，198，187，195，203，182
			其他 Others（3）	207，208，209
PC2（62）	5.08±1.01c	3.51～6.89	河北 Hebei（26）	45，67，10，50，24，25，37，53，47，36，14，70，63，66，39，19，3，51，22，31，27，16，5，2，72，49
			河南 Henan（27）	134，170，106，139，164，108，145，114，142，100，107，120，95，103，126，153，143，115，152，116，167，128，154，156，163，171，119
			山东 Shandong（7）	206，204，176，189，173，191，184
			其他 Others（2）	210，212
PC3（15）	10.33±4.01a	7.13～13.36	河北 Hebei（9）	65，20，42，75，73，23，64，57，48
			河南 Henan（4）	109，99，111，87
			山东 Shandong（2）	186，183
PC4（26）	4.04±0.88d	2.67～5.11	河北 Hebei（11）	56，71，35，59，76，54，62，55，74，15，79
			河南 Henan（9）	130，169，86，97，135，138，136，83，147
			山东 Shandong（5）	177，192，181，193，194
			其他 Others（1）	211
PC5（19）	2.78±0.35e	2.25～3.24	河北 Hebei（11）	44，18，6，38，34，33，43，61，80，17，9
			河南 Henan（4）	94，155，166，82
			山东 Shandong（4）	178，179，180，185

括号中数字表示品种数。植酸含量为平均值±标准差，数据后不同字母表示类间有显著差异（$P<0.05$）。各类中品种排序按植酸含量从高到低，品种名称见表1。

Thefigure in parentheses is the number of cultivars. Phytate content is shown in mean ± standard deviation, and different letters afterwards indicate significant difference among groups at $P<0.05$. In each group, cultivars were sorted by phytate content with value from high to low. Cultivar names are given in Table 1.

3　讨论

对我国76份小麦地方品种和62份来自黄淮、长江中下游和西南麦区品种的分析表明，小麦植酸含量为5.16～9.87g·kg⁻¹，植酸酶活性为620～2 192 U·kg⁻¹[9]。在我国137份微核心种质资源中，白蚂蚱的植酸含量较低，为9.59g·kg⁻¹，辐w 070261的植酸含量则高达29.63g·kg⁻¹[21]。400份印度及CIMMYT品种和人工合成种植酸含量与植酸酶活性的变异范围分别为11.7～19.3g·kg⁻¹与284～962 U·kg⁻¹[20]。本研究表明，黄淮麦区212份代表性品种的植酸含量和植酸酶活性分别为2.18～13.37g·kg⁻¹和10～1 759U·kg⁻¹，与前人报道结果[9,20,26]基本一致，但本研究中植酸含量和植酸酶活性的变异范围更大，可能与所选取材料的代表性和数量有关。目前还没有对植酸含量和植酸酶活性进行育种选择，因而其变异范围较大。因此，在改良小麦籽粒的微量矿物质营养品质时，首先应对现有品种的目标性状进行筛选，我们在对小麦品种铁、锌等微量元素含量的分析时也有类似结论[5]。本研究发现4个品种的植酸含量较低且植酸酶活性较高，分别是石麦12、衡4568、洛麦21和济麦096141，可作为进一步改良植酸含量和植酸酶活性的亲本。此外，相关分析结果（资料未列出）表明，籽粒中植酸含量和植酸酶活性呈显著正相关（r＝0.31，P＜0.001），说明植酸含量与植酸酶活性间可能存在某种程度的依存关系，要同时提高小麦籽粒中的植酸酶活性并降低其植酸含量存在一定困难。鉴于目前尚未对我国的大多数品种进行植酸含量和植酸酶活性分析，建议在本研究的基础上，进一步筛选我国现有品种特别是各地主栽品种的植酸含量和植酸酶活性，对植酸含量低且植酸酶活性高的主栽品种加大推广力度，同时将其用于育种，为提高我国品种铁、锌等微量矿物质元素的生物有效性奠定基础。

本研究发现籽粒植酸含量与植酸酶活性受品种、年度及品种和年度互作效应的显著影响；植酸含量比植酸酶活性更易受年度效应的影响，这些结果与前人报道一致[26-27]。Kim 等[27]选用 3 个品种在多个环境中种植，结果表明植酸含量受降雨量等环境因素的显著影响，而植酸酶活性的环境效应不显著。因此，有必要进一步分析温度、降水量、日照时数等环境因子对植酸含量与植酸酶活性的影响。

4　结论

植酸含量和植酸酶活性变异范围均较大，受品种、年度及其互作效应的显著影响，其中品种效应较大。石麦 12、衡 4568、洛麦 21 和济麦 096141 的植酸含量低且植酸酶活性较高，可望在小麦营养品质育种中发挥重要作用。

表 4　两年度 212 份品种植酸酶活性聚类分析结果

Table 4　Cluster of 212 cultivars based on phytase activity across two seasons

类别 Group	植酸酶活性 Phytase activity （U·kg⁻¹）	变幅 Range （U·kg⁻¹）	品种来源 Origin of cultivar	品种代号 Code of cultivar
PA1（56）	339±247 e	12～930	河北 Hebei（28）	21，29，52，27，20，24，75，34，79，6，68，44，14，37，3，55，32，77，78，2，26，66，7，19，70，1，41，25
			河南 Henan（18）	108，115，123，106，113，111，92，116，86，127，128，146，156，105，99，100，130，104
			山东 Shandong（9）	198，176，187，189，191，204，185，195，193
			其他 Others（1）	208
PA2（79）	757±135 c	415～978	河北 Hebei（30）	74，71，80，33，45，10，76，51，9，56，64，11，59，4，16，58，18，13，61，62，53，35，50，43，60，49，28，17，47，57
			河南 Henan（36）	139，122，134，160，117，83，137，82，126，169，136，110，88，155，109，135，154，132，170，153，125，166，94，89，157，159，163，112，143，142，168，96，133，98，150，158
			山东 Shandong（10）	174，177，180，179，197，188，192，205，186，200
			其他 Others（3）	212，211，207
PA3（44）	590±160 d	235～772	河北 Hebei（14）	22，5，39，54，42，46，72，23，36，8，12，31，48，30
			河南 Henan（21）	171，118，85，120，145，81，148，97，161，172，164，138，141，150，114，95，144，107，167，162，102
			山东 Shandong（9）	182，201，190，173，178，194，183，203，175
PA4（6）	1146±411 a	859～1 736	河北 Hebei（2）	73，40
			河南 Henan（2）	151，124
			山东 Shandong（1）	199
			其他 Others（1）	210
PA5（27）	930±185 ab	644～1085	河北 Hebei（6）	65，67，69，63，15，38
			河南 Henan（15）	103，121，165，87，131，84，119，129，140，152，147，91，90，93，101

（续）

类别 Group	植酸酶活性 Phytase activity (U·kg^{-1})	变幅 Range (U·kg^{-1})	品种来源 Origin of cultivar	品种代号 Code of cultivar
			山东 Shandong（5）	202，196，206，184，181
			其他 Others（1）	209

括号中数字表示品种数。植酸酶活性为平均值±标准差，数据后不同字母表示类间有显著差异（$P<0.05$）。各类中品种排序按植酸酶活性从高到低，品种名称见表1。

The figure in parentheses is the number of cultivars. Phytase activity is shown in mean $\pm SD$, and different letters afterwards indicate significant difference among groups at $P<0.05$. In each group，cultivars were sorted by phytase activity with value from high to low. Cultivar names are given in Table 1.

表5 两年度212份品种植酸含量与植酸酶活性方差分析

Table 5 Analysis of variance for phytate content and phytase activity of 212 cultivars across two seasons

变异来源 Source	自由度 df	植酸含量 Phytate content			植酸酶活性 Phytase activity		
		SS	F	Pr	SS	F	Pr
年度 Season（S）	1	589	165.6	<0.000 1	151 431	2.1	0.147 7
品种 Cultivar（C）	211	1 892	3.6	<0.000 1	31 963 975	6.2	<0.000 1
类别 Group（G）	4	1 004	59.9	<0.000 1	21 177 268	101.9	<0.000 1
类内品种 C（G）	207	888	1.2	0.135 4	10 786 707	0.7	0.958 6
品种×年度 C×S	211	673	2.7	<0.000 1	11 204 179	4.3	<0.000 1
类别×年度 G×S	4	361	34.0	<0.000 1	5 059 351	17.9	<0.000 1
类内品种×年度 C（G）×S	207	312	962.2	<0.000 1	6 144 828	8.8	<0.000 1
年度内重复 R（S）	2	0.1	0.9	0.391 1	55 428	3.5	0.032 7
误差 Error	398	1.1			2 387 414		

❖ 参考文献
References

[1] 陈春明.中国营养状况十年跟踪（1990—2000）.北京：北京人民卫生出版社，2004.
Chen C M. Ten Year Tracking Nutritional Status in China (1990 — 2000). Beijing：Beijing People's Medical Publishing House，2004. (in Chinese)

[2] Welch R M，Graham R D. A new paradigm for world agriculture：meeting human needs. Productive, sustainable，nutritious. *Field Crops Res*，1999，60：1-10.

[3] Bouis H E，Graham R D，Welch R M. The consultative group on international agriculture research (CGIAR) micronutrients project：justification and objectives. *Food Nutr Bull*，2000，21：374-381.

[4] Ortiz-Monasterio J I，Palacios-Rojas N，Meng E，Pixley K，Trethowan R，Pena R J. Enhancing the mineral and vitamin content of wheat and maize through plant breeding. *J Cereal Sci*，2007，46：293-307.

[5] 张勇，王德森，张艳，何中虎.北方冬麦区小麦品种籽粒主要矿物质元素含量分布及其相关性分析.中国农业科学，2007，40：1871-1876.
Zhang Y，Wang D S，Zhang Y，He Z H. Variation of major mineral elements concentration and their relationships in grain of Chinese wheat. *Sci Agric Sin*，2007，40：1871-1876. (in Chinese with English abstract)

[6] Lei X G，Stahl C H. Biotechnological development of effective phytases for mineral nutrition and environmental protection. *Appl Microbiol Biot*，2001，57：474-481.

[7] Cosgrove D J. The chemistry and biochemistry of inositol poly-phosphates. *Pure Appl Chem*，1966，16：209-224.

[8] Asada K，Tanaka K，Kasai Z. Formation of phytic acid in cereal grains. *Annu New York Acad Sci*，1970，165：801-814.

[9] Liu Z H，Wang H Y，Wang X E，Zhang G P，Chen P D，Liu D J. Genotypic and spike positional difference in grain phytase activity，phytate，inorganic phosphorus，iron，and zinc contents in wheat. *J Cereal Sci*，2006，44：212-219.

[10] Schroder B, BrevesG, Rodehutscord M. Mechanisms of intestinal phosphorus absorption and availability of dietary phosphorus in pigs. *Dtsch Tieraerztl Wochenschr*, 1996, 103：209-214.

[11] Brinch-Pedersen H, Sorensen L D, Holm P B. Engineering crop plants：getting a handle on phosphate. *Trends Plant Sci*, 2002, 7：118-125.

[12] Shamsuddin A M, Vucenik I. Mammary tumor inhibition by IP6：a review. *Anticancer Res*, 1999, 19：36-71.

[13] Jenab M, Thompson L U. Role of phytic acid in cancer and other diseases. In：Reddy N R, Sathe S K, eds. Food Phytates. Boca Raton：The Chemical Rubber Company Press, 2002. pp 225-248.

[14] Oh B C, Choi W C, Park S, Kim Y O, Oh T K. Biochemical properties and substrate specificities of alkaline and histidine acid phytases. *Appl Microbiol Biot*, 2004, 63：362-372.

[15] Chen R M, Xue G X, Chen P, Yao B, Yang W Z, Ma Q L, Fan Y L, Zhao Z Y, Tarczynski M C, Shi J R. Transgenic maize plants express-ing a fungal phytase gene. *Transgenic Res*, 2008, 17：633-643.

[16] 王延锋, 郎志宏, 赵奎军, 黄大昉. 转基因作物的生态安全性问题及其对策. 生物技术通报, 2010,（7）：1-6.
Wang Y F, Lang Z H, Zhao K J, Huang D F. Ecological risks and countermeasures of genetically modified crops. *Biotechnol Bull*, 2010,（7）：1-6.（in Chinese）

[17] Zhang Y, Song Q, Yan J, Tang J, Zhao R, Zhang Y, He Z, Zou C, Ortiz-Monasterio I. Mineral element concentrations in grains of Chinese wheat cultivars. *Euphytica*, 2010, 174：303-313.

[18] Tiwari V K, Rawat N, Chhuneja P, Neelam K, Aggarwal R, Randhawa G S, Dhaliwal H S, Keller B, Singh K. Mapping of quantitative trait loci for grain iron and zinc concentration in dip-loid a genome wheat. *J Hered*, 2009, 100：771-776.

[19] Shi R L, Li H W, Tong Y P, Jing R L, Zhang F S, Zou C Q. Identification of quantitative trait locus of zinc and phosphorus density in wheat（*Triticum aestivum* L.）grain. *Plant Soil*, 2008, 306：95-104.

[20] Ram S, Verma A, Sharma S. Large variability exits in phytase levels among Indian wheat varieties and synthetic hexaploids. *J Cereal Sci*, 2010, 52：486-490.

[21] 吴澎, 陈建省, 田纪春. 137 个微核心种质资源植酸含量的聚类分析. 中国粮油学报, 2010, 25（10）：19-23.
Wu P, Chen J S, Tian J C. Cluster analysis of phytic acid for 137 wheat micro-core collections. *J Chin Cereals Oils Assoc*, 2010, 25（10）：19-23.（in Chinese with English abstract）

[22] Barreto H J, Edmeades G O, Chapman S C, Crossa J. The alphalattice design in plant breeding and agronomy：generation and analysis. In：Edmeades G O, Banziger M, Mickelson H R, Peña-Valdivia C B, eds. Developing Drought-and Low N-Tolerant Maize. Proceedings of a Symposium. Mexico, D.F.：CIMMYT, 1997. pp 544-551.

[23] Wyss M, Pasamontes L, Friedlein A, Rémy R, Tessier M, Kronenberger A, Middendorf A, Lehmann M, Schnoebelen L, Röthlisberger U, Kusznir E, Wahl G, Müller F, Lahm H W, Vogel K, van Loon A P. Biophysical characterization of fungal phytases（myo-inositol hexakisphosphate phosphohydrolases）：molecular size, glycosylation pattern, and engineering of proteolytic resistance. *Appl Environ Microb*, 1999, 65：359-366.

[24] 张勇, 吴振录, 张爱民, van Ginkel M, 何中虎. CIMMYT 小麦在中国春麦区的适应性分析. 中国农业科学, 2006, 39：655-663.
Zhang Y, Wu Z L, Zhang A M, van Ginkel M, He Z H. Adaptation of CIMMYT wheat germplasm in China's spring wheat regions. *Sci Agric Sin*, 2006, 39：655-663.（in Chinese with English abstract）

[25] WardJ H. Hierarchical grouping to optimize an objective function. *J Am Stat Assoc*, 1963, 58：236-244.

[26] Liu Z H, Wang H Y, Wang X E, Zhang G P, Chen P D, Liu D J. Phytase activity, phytate, iron, and zinc contents in wheat pearling fractions and their variation across production locations. *J Cereal Sci*, 2007, 45：319-326.

[27] Kim J C, Mullan BP, Selle P H, Pluske J R. Levels of total phosphorus, phytate phosphorus and phytase activity in three varieties of Western Australian wheats in response to growing region, growing season and storage. *Aust J Agric Res*, 2002, 53：1361-1366.

Phenolic acid profiles of Chinese wheat cultivars

Yong Zhang[a], Lan Wang [a], Yang Yao [a], Jun Yan [a,b], Zhonghu He [a,c,*]

[a] Institute of Crop Science, National Wheat Improvement Centre/The National Key Facility for Crop Gene Resources and Genetic Improvement, Chinese Academy of Agricultural Sciences (CAAS), 12 Zhongguancun South Street, Beijing 100081, China

[b] Cotton Research Institute, Chinese Academy of Agricultural Sciences (CAAS), Huanghedadao, Anyang 455000, Henan Province, China

[c] CIMMYT China Office, C/O CAAS, Beijing 100081, China

[*] Corresponding author

ABSTRACT

Phenolic acid concentrations were determined in 37 Chinese commercial winter wheat cultivars grown at a single site over two seasons, and fractions comprising free and bound types were analyzed using HPLC with measurements of individual phenolic acids in each fraction. Most of the parameters were significantly influenced by cultivar, season, and their interaction effects, with cultivar variance being predominant. Wide ranges of concentration among the 37 cultivars were observed. The average concentration of bound type was 661 μg g^{-1} of dm, making up 97.5% of the phenolic acid determined with ferulic accounting for 70.7% of it, while free type made up only 2.5% of the phenolic acid determined with syringic accounting for 44.7% of it. Bound type was the predominant source to the grain phenolic acid concentrations determined. There were highly significant and positive correlations between bound ferulic concentration and total bound phenolic acid concentration, and between free syringic concentration and total free phenolic acid concentration. Cultivars Liangxing 66 and Zhongmai 895 were stable in concentration of components of phenolic acids across seasons, with high values of free and bound phenolic acids indicating they could be selected as parents in wheat breeding for health beneficial phenolic acid.

1. Introduction

Common wheat (*Triticum aestivum* L.) is a major crop and an important component of the human diet, accounting for approxi-mately 30% of the total grain consumption with an annual production of over 660 million tonnes globally (FAO, 2010). It is used in the production of various food products including bread, noodles, steamed bread, and cakes (He et al., 2004), supplying protein, starch, micronutrients such as Fe and Zn (Bouis et al., 2000), together with dietary fiber, vitamins, and phytochemicals thought to contribute protective effects to health (Ward et al., 2008).

Current interest in the health benefits provided by wheatand other grains has led to an increased focus on variation of phytochemical concentrations among

Published in Journal of Cereal Science, 2012, 56: 629-635.

cultivars (Adom et al., 2003, 2005; Ward et al., 2008). Epidemiological studies have increasingly shown protective roles of wheat grain against the risk of many chronic diseases, especially those related to metabolic syndrome (Meyer et al., 2000; Ward et al., 2008). Grain consumption helps to lower the incidence of cardiovascular disease (Jacobs et al., 1998; Thompson, 1994) and cancer-related deaths (Jacobs et al., 1998; Nicodemus et al., 2001), partly due to unique phytochemical concentrations of a number of chemicals including derivatives of benzoic and cinnamic acids, flavonols, phenolic compounds, tocotrienols, tocopherols, and carotenoids (Adom et al., 2003, 2005; Ward et al., 2008). Grain phytochemicals exert their health benefits through multifactorial physiologic mechanisms including antioxidant activity, mediation of hormones, enhancement of the immune system, and facilitation of substance transit through the digestive tract (Lupton et al., 1995), butyric acid production in the colon, and absorption or dilution of substances in the gut (Gazzaniga and Lupton, 1987). Among the health-promoting phytochemicals residing in cereal grains, phenolic compounds as the most diverse and complex group have gained much attention in scientific research (Dykes and Rooney, 2007; Thompson, 1994). Phenolic acids are the most abundant form of phenolic compounds in cereal grain (Mattila et al., 2005), and play an important role in combating oxidative stress in the human body by maintaining a balance between oxidants and antioxidants (Temple, 2000). Phenolic acids can be divided into two groups (Kim et al., 2006; Li et al., 2008; Ward et al., 2008), with hydroxybenzoic acid derivatives including p-hydroxybenzoic, protocatechuic, vannilic, syringic, and gallic acids, and hydroxycinnamic acid derivatives including p-coumaric, caffeic, ferulic, and sinapic acids (Verma et al., 2009). These acids are present mainly in bound forms, linked to cell wall structural components such as cellulose, lignin, and proteins through ester bonds (Parker et al., 2005). They were reported to possess good antioxidant activities (Miller and Rice-Evans, 1997). The presence of phenolic acids both in free and bound form

attached to various polysaccharides is of significant interest in preventing oxidative stress induced diseases. The free phenolic acids are easily absorbed into the circulation, whereas the bound type are released by the intestinal enzymes as well as by the colonic microflora and can be absorbed into the circulatory system (Andersson et al., 2010). Thus a strategy of breeding cultivars with enhanced abilities to load more phenolic acids into the grains offers sustainable, cost-effective health benefits and has the advantage of requiring no changes in consumer behavior (Adom et al., 2003; Irmak et al., 2008). Exploiting genetic variation in cultivars for increased phenolic acids is likely to be an effective method and is the first step toward enhancement in wheat based foods. Highly significant genotypic differences in phenolic acids among wheat genotypes from Europe, Canada and the United States of American have been reported (Irmak et al., 2008; Li et al., 2008; Verma et al., 2009; Ward et al., 2008).

China is the largest wheat producer and consumer in the world, with an annual production of more than 110 million tonnes from about 24 million hectares, supplying the cheapest source of calories and protein for the local inhabitants. Research has traditionally focused on improving yield, and disease and pest resistance, but more recently also on industrial quality characteristics for food products (He et al., 2001). To date, little attention was given to improving nutritional value, and no information is available on phenolic acid profiles in the leading wheat cultivars. Studies have indicated that the cardiovascular incidence in urban China is about 18% and has almost doubled in the last ten years (Chen, 2004). This has received recent national attention as metabolic syndrome. Therefore, full characterization of the phenolic acid profiles of the leading wheat cultivars could lead to new opportunities for breeding and eventual commercial production of value-added cultivars rich in beneficial components for making nutraceuticals and other functional foods. The objectives of this study were to determine the phenolic acid profiles and concentrations in grain of leading Chinese wheat cultivars and promising advanced lines. The information generated from the study could be important for Chinese wheat breeding programs, and may have potential application in other wheat producing countries.

2. Materials and methods

2.1. Wheat samples and sample preparation

Thirty-seven high-yielding commercial wheat cultivarswere used in this study. All of them are from the Yellow and Huai River Valleys Winter Wheat Region which contributes 70% of Chinese wheat production (He et al., 2001), except for the superior pan bread quality cultivar Zhongyou 9507 which is from the North China Plain Winter Wheat Region. Yannong 19, Jimai 19, and Han 6172 have been the leading cultivars since 2001, each with cumulative areas of cultivation of more than four million ha. The cultivation area of Jimai 22 in the 2010－2011 season exceeded 2 million ha, whereas Zhongmai 155, Zhongmai 875, and Zhongmai 895 were the most promising lines. Xiaoyan 6 was a leading cultivar from 1981 to 1995 and widely used as a quality donor parent. Jishi 02-1, Yumai 34, and Zhengmai 366 displayed superior pan bread quality (contact for detailed information). The tested cultivars were sown in the 2008－2009 and 2009－2010 cropping seasons at the wheat breeding station of the Institute of Crop Science, Chinese Academy of Agricultural Sciences, at Anyang (lat. 36° 06 ' N, long. 114°21' E, 61 m above sea level, with a loam soil type) in Henan province, located in the Yellow and Huai River Valleys Winter Wheat Region. The heading dates of the cultivars ranged from April 19 to 26 in the 2008－2009 season and from April 27 to May 3 in the 2009－2010 season, and the harvest dates varied from June 4 to 7 and from June 7 to 10 in the two seasons, respectively. The total precipitation and sunshine hour, and mean temperature between heading and harvest varied from 113.9 to 19.3 mm, from 326.7 to 257.0 h, and from 19.6 to 21.1℃ in the two seasons, respectively. When the total period of

3 months before heading to harvest was considered, total precipitation and sunshine hours varied from 154.1 to 68.3 mm and from 822.7 to 753.9 h, and mean temperature between 11.4 and 10.0℃ in the two seasons. Fertilization was similar in the two seasons, both with a total of about 300 kg ha^{-1} nitrogen applied. A completely randomized block design with two replications was used, and each plot consisted of 2 ×2 m rows, 0.2 m apart. Seeding rate was about 180 kg ha^{-1} and field management was according to local practices. Grain samples harvested from each replication were cleaned by hand. Thousand kernel weight was obtained as an average of three samples, with each containing 200 seeds. Grain diameter was determined on 300-kernel samples with a Perten 4100 Single Kernel Characterization System (SKCS, Perten Instruments North America Inc., Reno, NV). Grain protein content was obtained with a near-infrared (NIR) analyzer (Instalab 600, New-port Scientific Sales and Services Ltd., Australia) following AACC approved method 39－10 (AACC, 2000). Each sample was milled into fine powder with a 60-mesh screen and thoroughly mixed, and cooled immediately and stored at －20℃ until analyzed.

2.2. Analytical methods

2.2.1. Extraction of free phenolic acids

Free phenolic acids in wheat flour were extracted essentially according to a previously reported method (Adom et al., 2003, 2005). A 1 g sample of flour was extracted with 20 mL of 80% chilled ethanol. Eppendorf tubes containing samples wereshaken on a shaker at room temperature for 10 min. After centrifugation at 2500 g for 10 min, the supernatant was transferred into a new tube, and extraction was repeated once more for the residue. Supernatants were pooled together, evaporated at 45℃ to less than 5 mL, and reconstituted into 10 mL with distilled water. The extracts were stored at － 20℃ until further analysis within a three-month period.

2.2.2. Extraction of bound phenolic acids

Bound phenolic acids were extracted according to the published method of Mattila et al. (2005) with minor modifications: 15 mL of distilled water and 5 mL of 6 M NaOH were added to test tubes with the residue after the extraction of free phenolic compounds, and sealed and stirred overnight (about 16 h) at room temperature (20℃) using a magnetic stirrer. The solution was then adjusted to pH2, and liberated phenolic acids were extracted three times with 15 mL of a mixture of cold diethyl ether (DE) and ethyl acetate (EA, 1: 1 v/v). DE/EA layers were combined, evaporated to dryness, and the residue was dissolved in 1.5 mL of methanol. Acid hydrolysis was then performed by adding 2.5 mL of concentrated 12 M HCl into the test tube and incubating in a water bath at 85℃ for 30 min after completion of the above alkaline hydrolysis. The sample was then cooled, and adjusted to pH 2, with DE/EA extraction performed in the same manner as that for alkaline hydrolysis. The results of the alkaline and acid hydrolyzes were calculated to represent the bound phenolic acid concentrations.

2.2.3. Determination of individual phenolic acids

Individual phenolic acids in the wheat meal extractswere analyzed by an Agilent 1100 Series high-performance liquid chromatogram equipped with UV detector and Agilent TC-C18 (250 × 4.60 mm, 5 mm) column (Irmak et al., 2008). The mobile phase of water with 0.05% trifluoroacetic acid (solvent A) and 30% acetonitrile, 10% methanol, 59.95% water and 0.05% trifluoroacetic acid (solvent B) was used at a flow rate of 1.0 mL/min. Total run time was 50 min and the gradient program was as follows: 10%—12% B for 16 min, 12%—38% for 9 min, 38%—70% B for 7 min, 70%—85% B for 8 min and 85%—10% B for 10 min. The time of post-run for reconditioning was 5 min. The injection volume was 20 mL. Detection was done at 280 nm via the UV detector. Identification and quantification of phenolic acids in samples were performed by comparison with chromatographic retention times and areas of external standards. Calibration curves of phenolic acid standards were constructed using authentic standards that had undergone the same extraction procedure to ensure that losses due to the extraction were accounted

for. The authentic standards used for peak identification of ferulic, syringic, chlorogenic, caffeic, vanillic, p-coumaric and gentisic acids, were purchased from Sigma and Aldrich (Sigma-Aldrich Corporation, St. Louis, MO) and used without further purification (97% and higher purity). Standard compounds were prepared as stock solutions at 2 mg mL^{-1} in 80% ethanol. The stock solutions were stored in darkness at -18℃ and remained stable for at least 3 months. All chemicals used in this study were of analytical grade, and samples were prepared and analyzed at least in duplicate.

2.3. Statistical analysis

Means, standard deviations, and ranges were determinedusing PROC MEANS in the Statistical Analysis System (2000). Analysis of variance was conducted by PROC GLM, and Fisher's F-protected least significant difference (LSD) method was used to separate means of cultivars. PROC CORR was used to analyze the correlations among individual phenolic acids, and between individual phenolic acids and grain properties.

3. Results

The concentration of free phenolic acids in flour samples was 17 μg g^{-1} of dm (dry matter), making up only 2.5% of the total phenolic acids determined (Table 1). Despite the low concentration, syringic, caffeic, and ferulic accounted for 44.7, 30.8, and 24.5% of the free phenolic acids, respectively. There were no detectable amounts of free phenolic acid components chlorogenic, vanillic, p-coumaric, and gentisic. The concentration of bound phenolic acids was 661 μg g^{-1} of dm, making up 97.5% of the total phenolic acids determined, among which ferulic accounted for 70.7% of the bound class, followed by caffeic and p-coumaric, while syringic and gentisic accounted for only 0.3 and 0.8% of the bound class, respectively. Wide ranges of variation in concentrations of phenolic acids among cultivars and between seasons were observed, especially for free syringic, total free phenolic acid, bound ferulic, bound vanillic, bound p-coumaric, bound gentisic, and total bound phenolic acid. The range in concentrations of phenolic acids and all the components among cultivars was much larger than that measured across seasons.

Analysis of variance indicated that all of the parameters were significantly influenced by cultivar, season, and interaction effects except for free ferulic and caffeic, which were not significantly influenced by season (Table 2). The cultivar variance predominated for almost all parameters, while cultivar × season interaction variance was also important for all phenolic acids and their components, and season variance was important for phenolic acids and their components except for free ferulic and caffeic.

3. 1. Distribution of phenolic acids in cultivars

The mean concentrations of free phenolic acids rangedfrom 11 μg g^{-1} of dm to 29 μg g^{-1} of dm, with cultivar Jishi 02-1 having the highest value, followed by Taishan 23, Liangxing 66, and Zhoumai 18 (Table 3). Zhoumai 16 and Xinmai 18 had the highest ferulic concentration. Zhoumai 18 had the highest caffeic concentration. Taishan 23, Jishi 02-1, Liangxing 66, Zhengmai 366, and Yumai 47 had the highest syringic concentration.

The mean concentrations of bound phenolic acids ranged from 530 μg g^{-1} of dm to 855 μg g^{-1} of dm, with Kaimai 18 having the highest value, followed by Han 6172, Shaanyou 225, and Zhengmai 366 (Table 4). Among the bound phenolic acids, ferulic concentration ranged from 353 μg g^{-1} of dm to 629 μg g^{-1} of dm, with Kaimai 18 and Zhengmai 366 having the highest value. Caffeic concentra-tion ranged from 47 μg g^{-1} of dm to 106 μg g^{-1} of dm, with Jimai 22 and Kaimai 18 having the highest value. Chlorogenic concentration ranged from 3 μg g^{-1} of dm to 61 μg g^{-1} of dm, with Shi 4185, Zhoumai 16 and Linyou 145 having the highest value. Syringic concentration ranged from 1 mg g^{-1} of dm to 5 mg g^{-1} of dm, with Yumai 70, Yannong 19, Jimai 20, Jimai 22, and Jinhe 9123 having the highest value. Vanillic concentration ranged from 3 mg g^{-1} of dm to 44μg g^{-1} of dm, with Xingtai 456 having the highest value. P-coumaric concentration ranged from 61 μg g^{-1} of dm to 97 μg g^{-1} of dm, with Aikang 58, Shi 4185, Liangxiang 66, and Yumai 70 having the highest value. Gentisic concentration ranged from 2 μg g^{-1} of dm to 35 μg g^{-1} of dm, with Jishi 02-1 having the highest value, followed by Hengguan 35.

Table 1　Mean and range of phenolic acids concentrations （μg g^{-1} dry matter） in 37 wheat cultivars sown in two seasons

Class	Parameter	Mean	Range		
			Overall	Cultivar	Season
Free	Ferulic	4	0—12	2—8	4—4
	Caffeic	5	0—12	3—9	5—6
	Chlorogenic	—	—	—	—
	Syringic	7	0—41	3—21	5—9
	Vanillic	—	—	—	—
	p-Coumaric	—	—	—	—
	Gentisic	—	—	—	—
	Subtotal	17	6—58	11—29	15—18
Bound	Ferulic	467	235—752	353—629	416—519
	Caffeic	76	21—164	47—106	71—80
	Chlorogenic	25	0—96	3—61	25—26
	Syringic	2	0—10	1—5	1—3
	Vanillic	13	0—99	3—44	10—16
	p-Coumaric	72	51—145	61—97	65—80

(continued)

Class	Parameter	Mean	Range		
			Overall	Cultivar	Season
	Gentisic	6	0—79	2—35	4—8
	Subtotal	661	453—964	530—855	619—705

—, value below the limit of detection (1 μg g^{-1}).

Table 2　Sum of square from a combined analysis of variance for phenolic acid concentrations (μg g^{-1} dry matter) in 37 wheat cultivars sown in two seasons

Class	Source df	Cultivar (C)	Season (S)	C×S	Rep (S)	Error
		37	1	37	2	74
Free	Ferulic	242***	0	277***	0	41
	Caffeic	276***	3	222***	3	95
	Syringic	1 476***	348***	2 098***	0	633
	Subtotal	2 493***	282***	2 614***	1	854
Bound	Ferulic	675 731***	400 489***	431 554***	101	53 379
	Caffeic	37 237***	3 211***	52 908***	12	12 834
	Chlorogenic	38 146***	115***	54 100***	0	186
	Syringic	360***	113***	310***	1	13
	Vanillic	16 036***	1 309***	20 203***	1	2 160
	Gentisic	13 257***	8 649***	8 782**	88	6 929
	p-coumaric	6 421***	649***	6 194***	0	693
	Subtotal	777 301***	286 119***	437 366***	1	66 844

*,**, and *** indicate significant at P=0.05, 0.01, and 0.001, respectively.

Table 3　Free phenolic acid concentrations (μg g^{-1} dry matter) of the 37 wheat cultivars sown in two seasons

Cultivar	Ferulic	Caffeic	Syringic	Free
Xiaoyan 6	4bcd	7ab	7de	18bc
Shaanyou 225	6ab	7ab	7de	20b
Shi 4185	4bcd	6bc	7de	17bc
Yumai 47	3de	4cd	11bc	18bc
Bainong 64	4bcd	7ab	3g	14cd
Yumai 34	3de	6bc	3g	12d
Yumai 70	3de	5bcd	4g	12d
Zhongyou 9507	3de	4cd	5fg	12d
Yannong 19	4bcd	5bcd	7de	16bc
Jimai 19	6ab	7ab	7de	20b
Han 6172	3de	5bcd	8cde	16bc
Zhongyu 5	5abc	7ab	5ef	17bc
Jishi 02-1	6ab	7ab	16ab	29a
Jimai 20	4bcd	6bc	7de	17bc
Linyou 145	4bcd	4cd	6ef	14cd
Zhoumai 16	8a	6bc	6ef	20b

(continued)

Cultivar	Ferulic	Caffeic	Syringic	Free
Yanzhan 4110	6ab	5bcd	9cd	20b
Taishan 23	3de	3d	21a	27a
Xinmai 18	7a	6bc	4g	17bc
Kaimai 18	4bcd	7ab	4g	15bcd
Pumai 9	3de	5bcd	5fg	13cd
Hengguan 35	4bcd	6bc	6ef	16bc
Zhoumai 18	3de	9a	10cd	22b
Aikang 58	4bcd	4cd	4g	12d
Xunong 5	4bcd	5bcd	4g	13cd
Zhengmai 366	2e	5bcd	11bc	18bc
Xinong 979	4bcd	4cd	7de	15bcd
Jimai 22	5abc	3d	3g	11d
04 Zhong36	5abc	5bcd	5ef	15bcd
Zhoumai 22	5abc	4cd	7de	16bc
Xingtai 456	3de	4cd	4g	11d
Liangxing 66	6ab	6bc	11bc	23b
Zhongmai 349	4bcd	5bcd	10cd	19bc
Jinhe 9123	4bcd	6bc	5fg	15bcd
Zhongmai 155	3de	7ab	6ef	16bc
Zhongmai 895	3de	4cd	7de	14cd
Zhongmai 875	4bcd	5bcd	7de	16bc

Values with different letters in the same column are significantly different from each other at $P = 0.05$.

3.2. Impact of growing season on phenolic acids

The mean total amounts of free phenolic acid were 15 μg g^{-1} of dm in cultivars in the 2009—2010 season and 18 μg g^{-1} of dm in cultivars in 2008 — 2009. The samples grown in 2008 — 2009 displayed higher concentration of syringic, but lower caffeic than those in 2009—2010.

The total amounts of bound phenolic acid were 619 μg g^{-1} of dm in cultivars in the 2008—2009 season and 705 μg g^{-1} of dm in cultivars in 2009 — 2010. The samples grown in 2008 — 2009 displayed higher concentrations of syringic, vanillic, p-coumaric and gentisic, but lower concentrations of ferulic, caffeic, and chlorogenic than those in 2009—2010.

3.3. Cultivars with the highest free and bound phenolic acid concentrations in two seasons

Free and bound phenolic acid concentrations of 12 cultivars determined on the basis of classification analysis of the grain free and bound phenolic acid concentrations determined (data not shown), with the highest value in the two seasons shown in Table 5, due to the high significant effects of season and cultivar× season interaction variances. In the 2008 — 2009 season, Jishi 02-1 and Taishan 23 had the highest free phenolic acid concentrations, and Kaimai 18, Linyou 145, Xingtai 456, and Zhoumai 22 had the highest bound phenolic acid concentrations. In 2009 —2010, Jimai 19, Liangxing 66, and Yanzhan 4110 had the highest free phenolic acid concentrations, and Kaimai 18, Jimai 19, and Zhengmai 366 had the highest bound phenolic acid concentrations. Liangxing 66, Zhongmai 349, and Zhongmai 895 had high concentrations of free and bound phenolic acids in the 2008 — 2009 season, and Jimai 19, Liangxing 66, Shaanyou 225, Zhengmai 366, Zhongmai 895, and Zhoumai 22 had high concentrations of free and bound phenolic acids in 2009 — 2010. Liangxing 66, Taishan 23, Zhongmai 349, and Zhongmai 895 had high concentrations of free phenolic acid in both seasons, and Liangxing 66, Han 6172, Shaanyou 225, Kaimai 18, Zhoumai 22, Zhongmai 895, and Zhengmai 366 had high concentrations of bound phenolic acids in both seasons.

3. 4. Relationships among phenolic acids and between phenolic acids and grain properties

As shown in Table 6, bound ferulic concentration was highly significant and positively correlated with total bound phenolic acid ($r = 0.92$, $P < 0.001$). There were significant and positive correlations of free ferulic, caffeic, and syringic concentration with total free phenolic acid concentrations ($r = 0.36$, 0.40, and 0.87, $P < 0.05$, 0.05, and 0.001), of free ferulic concentration with free caffeic concentration ($r = 0.37$, $P < 0.05$), of bound caffeic concentration with total bound phenolic acid concentration ($r = 0.42$, $P < 0.01$), and of bound vanillic concentration with bound p-coumaric ($r = 0.48$, $P < 0.01$) and total bound phenolic acid ($r = 0.54$, $P < 0.001$) concentrations. Significant but negative correlations of free syringic and total free phenolic acid concentrations with bound vanillic concentration were also observed, both with $r = 0.33$ ($P < 0.05$). The thousand kernel weight, grain diameter, and grain protein content varied between 34.5 and 54.1 g, between 2.17 and 2.76 mm, and between 13.1 and 17.5% for the cultivars, respectively. Although phenolic acids are known to be concentrated in the bran, no significant correlations were found between free, bound phenolic acid and their composition concentrations with thousand kernel weight, grain diameter, and grain protein content, except for bound chlorogenic which showed significant but negative correlation with grain diameter ($r = -0.35$, $P < 0.05$).

4. Discussion

China is the largest wheat producer and consumer in the world. Its germplasm is genetically very distinct from that of other countries, and is playing an important role in the wheat breeding program at the International Maize and Wheat Improvement Centre (CIMMYT) and thereby in many breeding programs throughout the world (He et al., 2001; Zhang et

al., 2006). Therefore, information on the phenolic acid profiles in leading Chinese wheat cultivars is of interest not only to Chinese wheat breeders and consumers, but also to the international community.

Many studies have shown that dietary phenolics havehigh antioxidant activities (Fukumoto and Mazza, 2000; Liyana-Pathirana and Shahidi, 2006) with potential health benefits (Handelman et al., 1999). Previous reports have indicated significant influence of wheat cultivars on phenolic acids, including components such as ferulic acid, and on antioxidant activities (Adom et al., 2003, 2005). As in the present work, those studies generally showed that bound type represent the most abundant class of phenolic acids in wheat, contributing about 97.5% of the phenolic acids determined, in agreement with Ward et al. (2008). This is very important when the health benefits of whole grains are considered. Bound phenolics are more likely to survive upper gastrointestinal digestion and can be released from the colon through microflora digestion activity (Adom et al., 2003). Thus, the wheat phenolics are more likely to exert health benefits in the colon where they are released. This may partly explain the reduced incidence of colon cancers and other chronic diseases associated with the consumption of wheat grain products (Jacobs et al., 1998). However, considerable differences among cultivars were shown in both free and bound, as well as their compositions. The free phenolic acids of the cultivars ranged from 9 μg g^{-1} of dm to 27 μg g^{-1} of dm and bound phenolic acids ranged from 530 μg g^{-1} of dm to 854 μg g^{-1} of dm. Contributions from free type to the phenolic acid concentration ranged from 1.5% in Xingtai 456 to 3.6% in Jishi 02-1, whereas that of the bound type ranged from 96.4% in Jishi 02-1 to 98.5% in Xingtai 456, in agreement with the large variations observed in previous studies (Adom et al., 2005; Li et al., 2008). This also indicates that genetic improvement could play a significant role in further improvement of the health aspects of wheat.

Ferulic acid, representing up to 90% of total

polyphenols, is the predominant phenolic acid in wheat, contributing from 62.4% in Pumai 9 to 79.2% in Zhengmai 366, in agreement with previous reports (Adom et al., 2005; Andersson et al., 2010; Kim et al., 2006; Liyana-Pathirana and Shahidi, 2006). It is a precursor for synthesis of other aromatic compounds in plants, and comprises a significant proportion of each of the phenolic acid fractions, with concentrations ranging from 2 to 8 μg g^{-1} of dm for the free form to 353 to 628 μg g^{-1} of dm for the bound form, consistent with the study of Li et al. (2008). The average percentage contributions to the grain ferulic acid by bound type were 99%, much larger than previous reports (Adom et al., 2003, 2005; Li et al., 2008). This may largely be due to the soluble conjugated phenolic acids undetermined in this study. Li et al. (2008) indicated that free phenolic acids make the smallest (typically less than 1%) contribution to the total phenolic acid concentrations in wheats, with the average total free type being of 11 μg g^{-1} of dm; whereas soluble conjugated type undetermined in this study make up a greater proportion (about 22%) of the total phenolic acid concentrations, containing around 162 μg g^{-1} of dm on average; while bound phenolic acids contribute the highest proportion (around 77%) of the total phenolic acids, with mean concentrations being of 492 μg g^{-1} of dm. In addition, phenolic acid components caffeic, vanillic, syringic, chlorogenic, and p-coumaric acids reported in this study are in agreement with previous reports (Kim et al., 2006).

Table 4　Bound phenolic acid concentrations（$\mu g\ g^{-1}$ dry matter）of the 37 wheat cultivars sown in two seasons

Cultivar	Ferulic	Caffeic	Chlorogenic	Syringic	Vanillic	p-Coumaric	Gentisic	Bound
Xiaoyan 6	491cde	73defg	34f	2cde	10def	65ef	3d	678cde
Shaanyou 225	593ab	76def	17klm	1e	3g	73cde	3d	766bc
Shi 4185	408ef	84bcde	61a	2cde	14cde	96a	2d	667def
Yumai 47	450def	87bcd	42e	1e	3g	71de	3d	657def
Bainong 64	399fg	47i	20ij	1e	14cde	76cde	3d	560hi
Yumai 34	405ef	51i	5rs	2cde	19cd	79bcd	4d	565ghi
Yumai 70	456def	60gh	15n	5a	4fg	92ab	3d	635defg
Zhongyou 9507	460def	64gh	46d	1e	3g	71de	5d	650def
Yannong 19	450def	62gh	33f	5a	17cd	75cde	3d	645defg
Jimai 19	528bc	94abc	49c	2cde	3g	66ef	3d	745bc
Han 6172	546bc	80cde	10opq	4abc	38ab	68def	4d	750bc
Zhongyu 5	436ef	73defg	46d	1e	32ab	73cde	3d	664def
Jishi 02-1	531bc	78def	9pq	1e	5fg	73cde	35a	732bc
Jimai 20	409ef	91abcd	12o	5a	9def	67def	4d	597fgh
Linyou 145	498cde	96ab	51bc	2cde	3g	79bcd	3d	732bc
Zhoumai 16	386fg	60gh	52b	1e	5fg	81bcd	4d	589ghi
Yanzhan 4110	447def	51i	27g	2cde	8ef	64ef	4d	603fgh
Taishan 23	442def	70efg	17lmn	3cd	3g	61f	5d	601fgh
Xinmai 18	369g	70efg	6r	1e	16cd	65ef	3d	530i
Kaimai 18	629a	104a	46d	1e	3g	68def	4d	855a
Pumai 9	353g	73defg	23h	1e	28bc	68def	13c	559hi
Hengguan 35	420ef	62gh	18jkl	4abc	24bc	68def	22b	618defg

(continued)

Cultivar	Ferulic	Caffeic	Chlorogenic	Syringic	Vanillic	p-Coumaric	Gentisic	Bound
Zhoumai 18	453def	68fgh	3st	1e	5fg	88abc	2d	620defg
Aikang 58	408ef	92abcd	19jk	1e	13cde	97a	4d	634defg
Xunong 5	474cde	54hi	33f	1e	20cd	64ef	4d	650def
Zhengmai 366	628a	70efg	3t	1e	8ef	64ef	3d	777b
Xinong 979	417ef	75def	42e	3cd	4fg	67def	4d	612efg
Jimai 22	491cde	106a	16lmn	5a	3g	69def	13c	703cde
04 Zhong36	388fg	93abc	11op	4abc	17cd	63ef	4d	580ghi
Zhoumai 22	561bc	62gh	35f	1e	4fg	66ef	3d	732bc
Xingtai 456	478cde	81cde	22hi	2cde	44a	81bcd	17c	725bcd
Liangxing 66	525bc	70efg	28g	2cde	8ef	90ab	3d	726bcd
Zhongmai 349	486cde	99a	3st	1e	17cd	65ef	3d	674def
Jinhe 9123	489cde	56hi	8q	5a	12cde	67def	4d	641defg
Zhongmai 155	513bcd	80cde	33f	2cde	16cd	63ef	4d	711bcd
Zhongmai 895	512bcd	93abc	15lmn	1e	26bc	73cde	14c	734bc
Zhongmai 875	403ef	72efg	33f	1e	8ef	67def	3d	587ghi

Values with different letters in the same column are significantly different from each other at $P<0.05$.

Table 5 Twelve cultivars with the highest free, bound, and total phenolic acid concentrations ($\mu g\ g^{-1}$ dry matter)

Season	Cultivar	Free	Cultivar	Bound
2008—2009	Jishi 02-1	45	Kaimai 18	756
	Taishan 23	39	Linyou 145	725
	Zhoumai 18	28	Xingtai 456	711
	Zhongmai 349	22	Zhoumai 22	705
	Shi 4185	22	Hengguan 35	694
	Yumai 47	21	Han 6172	693
	Xiaoyan 6	21	Zhongmai 895	684
	Liangxing 66	20	Shaanyou 225	683
	Xinong 979	20	Liangxing 66	674
	Zhongmai 895	19	Zhongmai 349	666
	Yannong 19	19	Zhengmai 366	651
	Jimai 20	19	Yanzhan 4110	645
2009—2010	Jimai 19	25	Kaimai 18	951
	Liangxing 66	25	Jimai 19	923
	Yanzhan 4110	25	Zhengmai 366	902
	Zhoumai 16	24	Jishi 02-1	851
	Zhengmai 366	23	Shaanyou 225	846
	Shaanyou 225	23	Xiaoyan 6	844
	Jinhe 9123	20	Jimai 22	839
	Zhongmai 895	19	Zhongmai 155	815

（continued）

Season	Cultivar	Free	Cultivar	Bound
	Zhoumai 22	16	Han 6172	809
	Zhongyu 5	16	Zhongmai 895	783
	Zhongmai 349	16	Liangxing 66	777
	Taishan 23	16	Zhoumai 22	757

Table 6　Correlations among phenolic acid concentrations （µg g^{-1} dry matter） including components among 37 wheat cultivars grown in two seasons

Class	Component	Free			Bound								KW	GD	GPC
		Caffeic	Syringic	Subtotal	Ferulic	Caffeic	Chlorogenic	Syringic	Vanillic	p-Coumaric	Gentisic	Subtotal			
Free	Ferulic	0.37*	0.01	0.36*	−0.02	−0.18	0.13	−0.17	−0.21	−0.09	0.16	−0.07	−0.06	−0.06	0.01
	Caffeic		−0.02	0.40*	0.12	−0.17	0.03	−0.13	−0.03	0.13	0.01	0.09	−0.15	−0.13	0.02
	Syringic			0.87***	0.19	0.01	−0.13	−0.02	−0.33*	−0.12	0.21	0.10	−0.19	0.10	0.10
	Subtotal				0.20	−0.12	−0.10	−0.11	−0.33*	−0.12	0.23	0.09	−0.28	0.07	0.31
Bound	Ferulic					0.17	−0.01	−0.17	−0.18	−0.18	0.01	0.92***	0.07	0.06	0.25
	Caffeic						0.11	−0.01	−0.11	0.07	0.01	0.42**	−0.27	−0.14	−0.21
	Chlorogenic							−0.24	−0.21	0.13	−0.23	0.18	0.07	−0.35*	−0.09
	Syringic								0.06	−0.1	0.03	−0.19	−0.15	−0.04	−0.15
	Vanillic									0.48**	0.21	0.54***	0.05	−0.10	−0.31
	p-Coumaric										0.22	−0.06	−0.17	−0.13	−0.03
	Gentisic											0.04	0.01	0.03	0.18
	Subtotal												0.01	−0.08	0.13

*, **, and ***, significantly different from zero at $P=0.05$, 0.01, and 0.001, respectively. TKW, thousand kernel weight （g）; GD, grain diameter （mm）; GPC, grain protein content （%）.

The extent of variation among components in the overall wheat gene pool is important for breeders, as it demonstrates the extent of genetic potential for exploitation by breeding. The results may also help in developing breeding programs aimed at producing value added cultivars with higher concentrations and better compositions of phenolic acids with regard to human health. However, it will be very important to know more about the stability of phenolic acid concentrations across a wider range of environments. The limited information available indicates that environmental interactions are important （Fernandez-Orozco et al., 2010）. Li et al. （2008） reported Yumai 34 had high concentration of total phenolic content, with value between 800 and 900 µg g^{-1} of dm, much larger than that in this study. The reason could be related to the conjugated phenolic acids undetermined in this study, and due to the difference in growth sites between China and Hungary. There were only two cultivars from China used in Li et al. （2008）, with all wheat samples grown on one site. The season and cultivar × season effects were significant for most phenolic acids and their components, although cultivar variance predominated for almost all parameters. The total precipitation levels and sunshine hour between the heading to harvest dates in the 2008 — 2009 season were much higher than those in the 2009 — 2010 season, while mean temperature between the heading to harvest dates in the 2008 — 2009 season was much lower than those in the 2009 — 2010 season. When taking the data of 3 months before heading to harvest into consideration, the total precipitation levels and sunshine hour, and mean temperature in the 2008 — 2009 season were much larger than those in the 2009 — 2010 season. However, there were 4 times of flooding irrigation in the 2009 — 2010 season including before-winter, elongation, heading, and grain-filling stages, while only 2 times of flooding irrigation in the 2008—2009 season were performed, including elongation and heading stages. There was no significant correlation between phenolic acids including the component concentrations and precipitation, sunshine hour, or mean temperature. Therefore, more research is required to understand the effects of growing location and interaction between cultivar and location on phenolic acid concentrations.

Based on this study, Liangxing 66, Taishan 23, Zhongmai 349, and Zhongmai 895 had high concentrations of free phenolic acid in both seasons, whereas Liangxing 66, Han 6172, Shaanyou 225, Kaimai 18, Zhoumai 22, Zhongmai 895, and Zhengmai 366 had high levels of bound phenolic acid concentrations in both seasons. It is likely that Liangxing 66 and Zhongmai 895 display more consistent performance than others for both free and bound phenolic acids. This suggests that these genotypes with higher and more stable concentrations of phenolic acids could be selected for production and for use as parents in plant breeding.

❖ Acknowledgement

This study was supported by the Core Research Budget of the Non-profit Governmental Research Institutions (ICS, CAAS), National Basic Research Program (2009CB118300), and an inter national collaboration project on wheat improvement from the Chinese Ministry of Agriculture (2011-G3).

❖ References

AACC, 2000. Approved Methods of the AACC, tenth ed. American Association of Cereal Chemists, St. Paul, MN.

Adom, K. K., Sorrells, M. E., Liu, R. H., 2003. Phytochemical profiles and antioxidant activity of wheat varieties. Journal of Agricultural and Food Chemistry 51, 7825-7834.

Adom, K. K., Sorrells, M. E., Liu, R. H., 2005. Phytochemicals and antioxidant activity of milled fractions of different wheat varieties. Journal of Agricultural and Food Chemistry 53, 2297-2306.

Andersson, A. A. M., Kamal-Eldin, A., Åman, P., 2010. Effects of environment and variety on alkylresorcinols in wheat in the HEALTHGRAIN diversity screen. Journal of Agricultural and Food Chemistry 58, 9299-9305.

Bouis, H. E., Graham, R. D., Welch, R. M., 2000. The consultative group on international agriculture research (CGIAR) micronutrients project: justification and objectives. Food Nutrition and Bulletin 21, 374-381.

Chen, C. M., 2004. Ten Year Tracking Nutritional Status in China. People's Medical Publishing House, Beijing.

Dykes, L., Rooney, L. W., 2007. Phenolic compounds in cereal grains and their health benefits. Cereal Foods World 52, 105.

FAO, 2010. Cereal Supply and Demand Brief. In Food and Agriculture Organization of the United Nations.

Fernandez-Orozco, R., Li, L., Harflett, C., Shewry, P. R., Ward, J. L., 2010. Effects of environment and genotype on phenolic acids in wheat in the HEALTHGRAIN diversity screen. Journal of Agricultural and Food Chemistry 58, 9341e9352.

Fukumoto, L. R., Mazza, G., 2000. Assessing antioxidant and prooxidant activities of phenolic compounds. Journal of Agricultural and Food Chemistry 48, 3597-3604.

Gazzaniga, J. M., Lupton, J. R., 1987. Dilution effect of dietary fiber sources: an in vivo study in the rat. Nutrition Research 7, 1261-1268.

Handelman, G. J., Cao, G., Walter, M. F., Nightingale, Z. D., Paul, G. L., Prior, R. L., Blumberg, J. B., 1999. Antioxidant capacity of oat (Avena sativa L.) extracts. 1. Inhibition of low-density lipoprotein oxidation and oxygen radical absorbance capacity. Journal of Agricultural and Food Chemistry 47, 4888-4893.

He, Z. H., Rajaram, S., Xin, Z. Y., Huang, G. Z., 2001. A History of Wheat Breeding in China. CIMMYT, Mexico, D. F., pp. 1-14.

He, Z. H., Yang, J., Zhang, Y., Quail, K., Pena, R. J., 2004. Pan bread and dry white Chinese noodle quality in Chinese winter wheats. Euphytica 139, 257-267.

Irmak, S., Jonnala, R. S., MacRitchie, F., 2008. Effect of genetic variation on phenolic acid and policosanol contents of Pegaso wheat lines. Journal of Cereal Science 48, 20-26.

Jacobs, D. R., Meyer, K. A., Kushi, L. H., Folsom,

A. R.，1998. Whole grain intake may reduce risk of coronary heart disease death in postmenopausal women: the Iowa women's health study. American Journal of Clinical Nutrition 68，248-257.

Kim，K. H.，Tsao，R.，Yang，R. W. S.，Cui，S. W.，2006. Phenolic acid profiles and antioxidant activities of wheat bran extracts and the effect of hydrolysis conditions. Food Chemistry 95，466-473.

Li，L.，Shewry，P.，Ward，J. L.，2008. Phenolic acids in wheat varieties in the HEALTHGRAIN diversity screen. Journal of Agricultural and Food Chemistry 56，9732-9739.

Liyana-Pathirana，C. M.，Shahidi，F.，2006. Importance of insoluble-bound phenolics to antioxidant properties of wheat. Journal of Agricultural and Food Chemistry 54，1256-1264.

Lupton，J. R.，Chen，X. Q.，Frolich，W.，1995. Calcium phosphate supplementation results in lower rat fecal bile acid concentrations and a more quiescent colonic cell proliferation pattern than does calcium lactate. Nutrition and Cancer 23，221-231.

Mattila，P.，Pihlava，J. M.，Hellström，J.，2005. Contents of phenolic acids, alkyl-and alkenylresorcinols, and avenanthramides in commercial grain products. Journal of Agricultural and Food Chemistry 53，8290-8295.

Meyer，K. A.，Kushi，L. H.，Jacob，D. J.，Slavin，J.，Sellers，T. A.，Folsom，A. R.，2000. Carbohydrates, dietary fiber, incident type 2 diabetes mellitus in older women. American Journal of Clinical Nutrition 71，921-930.

Miller，N. J.，Rice-Evans，C. A.，1997. Cinnamates and hydroxybenzoates in the diet: antioxidant activity assessed using the ABTS+ radical cation. British Food Journal 99，57-62.

Nicodemus，K. K.，Jacobs，D. R.，Folsom，A. R.，2001. Whole and refined grain intake and risk of incident postmenopausal breast cancer. Cancer Causes and Control 12，917-925.

Parker，M. L.，Ng，A.，Waldron，K. W.，2005. The phenolic acid and polysaccharide composition of cell walls bran layers of mature wheat (Triticum aestivum L. cv. Avalon) grains. Journal of Agricultural and Food Chemistry 85，2539-2547.

SAS Institute SAS User's Guide，2000. Statistics. SAS Institute，Inc.，Cary，NC. Temple，N. J.，2000. Antioxidants and disease: more questions than answers. Nutrition Research 20，449-459.

Thompson，L. U.，1994. Antioxidants and hormone-mediated health benefits of whole grains. Food Science and Nutrition 34，473-497.

Verma，B.，Hucl，P.，Chibbar，R. N.，2009. Phenolic acid composition and antioxidant capacity of acid and alkali hydrolysed wheat bran fractions. Food Chemistry 116，947-954.

Ward，J. L.，Poutanen，K.，Gebruers，K.，Piironen，V.，Lampi，A. M.，Nyström，L.，Andersson，A. A. M.，Åman，P.，Boros，D.，Rakszegie，M.，Bedö，Z.，Shewry，P. R.，2008. The HEALTHGRAIN cereal diversity screen: concept, results and pros-pects. Journal of Agricultural and Food Chemistry 56，9699-9709.

Zhang，Y.，He，Z. H.，Zhang，A. M.，van Ginkel，M.，Peña，R. J.，Ye，G. Y.，2006. Pattern analysis on protein properties of Chinese and CIMMYT spring wheat cultivars sown in China and CIMMYT. Australian Journal of Agriculture Research 57，811-822.

Effects of heat stress and cultivar on the functional properties of starch in Chinese wheat

Shujun Wang,[1,†] Tiangui Li,[1] Yongjie Miao,[2] Yong Zhang,[2] Zhonghu He,[2,3] and Shuo Wang[1]

† Corresponding author. Phone：+86-22-60912486. E-mail：sjwang@tust. edu. cn

[1] Key Laboratory of Food Nutrition and Safety，Ministry of Education，College of Food Engineering and Biotechnology，Tianjin University of Science and Technology，Tianjin 300457，China.

[2] Institute of Crop Science，Chinese Academy of Agricultural Sciences，Beijing 100081，China.

[3] CIMMYT China Office，c/o Chinese Academy of Agricultural Sciences，Beijing 100081，China.

ABSTRACT

Heat stress during the grain-filling stage is a major limiting factor for improving Chinese wheat production，and its effect on functional properties of flours and starches in 10 leading cultivars from the Yellow and Huai Valleys grown under normal and heat-stress environments was investigated. Heat stress during the grain-filling stage decreased total starch content but increased protein and lipid contents of wheat grains. Amylose content of wheat starch was little altered under a heat-stress environment. Heat stress did not significantly change swelling power and starch solubility of wheat starches but significantly decreased swelling power of wheat flours. Pasting viscosities of wheat starches and flours were affected differentially by heat stress. Heat stress had a significant effect on gelatinization and retrogradation properties of starches. The in vitro enzymatic digestibility of wheat starches was affected slightly by heat stress. Analysis of variance indicated that heat stress had a significant effect on some functional properties of starch and flour，although the largest source of variability in these properties was cultivar.

Wheat is cultivated in many countries around the world. Approximately 21% of the world's food relies on the wheat crop, which is cultivated on over 200 million hectares of farmland (Ortiz et al. 2008). Wheat is an important cereal grain, and the yield and quality of wheat are the most important attributes for breeder and consumer. Genotype and environmental factors, including temperature, water, fertilizer, sunshine intensity, and soil, have a combined effect on yield and quality of wheat grains (Hunt et al. 1991). The fluctuation of temperature, particularly high temperature after anthesis, influences yield and quality through modification of duration and speed of grain filling, and approximately 20-30℃ is considered the optimum temperature during the grain-filling stage in both the temperate zone and subtropics (Keeling et al. 1994). However, wheat often encounters high temperatures (over 30℃) during the grain-filling stage (Randall and Moss 1990)；this has been identified as one of the crucial factors limiting yield and quality (Skylas et

Published in Cereal Chemistry. ，2017，94（3）：443-450.

al. 2002). Resistance or tolerance to heat stress during the grain-filling stage is a major breeding objective under global-warming scenarios. High temperature during the grain-filling stage can negatively impact the milling and breadmaking quality of wheat (Borghi et al. 1995).

Starch, mainly composed of amylose and amylopectin molecules, takes up approximately 60-70% of the grain weight. It is the main storage reserve polysaccharide of higher plants and a biopolymer of considerable significance for humans (Wang and Copeland 2015), contributing approximately 50% of energy intake in the human diet (Rahman et al. 2007). The functional properties of starch are the major determinant of the quality of wheat products such as bread and noodles. Environment, especially heat stress during the grain-filling period, has a marked effect on starch biosynthesis and, therefore, influences the structure of starch granules and their functional characteristics (Beckles and Thitisaksakul 2014). High air and soil temperatures are the primary factors that reduce grain starch production and modify starch structural and functional characteristics (Beckles and Thitisaksakul 2014). High temperature can modify the chain length of amylopectin in endosperm starches, and the extent of modifications varies with species and genotypes (Matsuki et al. 2003; Suzuki et al. 2004; Yamakawa et al. 2007). Thus, it is possible to develop cultivars with consistent performance under both normal and heat-stress conditions. Heat-stress treatment led to more pitting and fissures on the surface of starch granules (Lu et al. 2014) and a greater amount of small granules (Blumenthal et al. 1995; Hurkman et al. 2003). The changes in starch functionality induced by heat stress may have a great influence on the quality of wheat flour and final end products.

There are many studies focusing on how heat stress influences the structure of starch granules; however, little is understood regarding the effect of heat stress during the grain-filling period on starch functionality, especially retrogradation and digestibility. The objective of this study was to investigate the effect of heat stress during grain filling and cultivar on the functional properties of starch in Chinese wheat. Ten leading cultivars from the Yellow and Huai Valleys in China, where 70% of China's wheat is produced, were selected. The information will be useful for breeders to develop cultivars with favorable functionality grown under heat-stress conditions.

MATERIALS AND METHODS

Samples and Field Experiment. Ten leading wheat cultivars from the Yellow and Huai Valleys in China were used in this study. All genotypes were planted at a normal sowing date in four replicates at Anyang Experimental Station, located in Henan Province, at the Institute of Crop Sciences, Chinese Academy of Agricultural Sciences in the 2013-2014 season. Two replicates were grown under a normal environment and the other two under a heat-stress environment during the grain-filling stage. The heat-stress environment was initiated by covering the wheat with polyethylene films (0.12 mm thickness) at 14 days after anthesis, with a 40 cm distance to the ground without film covering. By doing so, the average daytime high temperature under heat stress was 2.2℃ higher than that of the normal environment (Fig. 1). Six hundred plants were grown in five-row plots with a length of 2 m.

Basic Composition of Wheat Grains. Harvested sound grain samples from each of two replicates under normal and heat-stress environments were collected. A plant grinder (BJ-350, Baijie, China) was used to mill the grains into flour. The obtained flour was passed through a 250 μm sieve and used for the determination of total starch, lipid, and protein contents in wheat. A Megazyme total starch assay kit (Megazyme International Ireland, Bray, Ireland) was used for the determination of total starch content. Lipid content was determined gravimetrically after Soxhlet extraction with petroleum ether. Protein content was determined with a Foss-Tecator 1241 near-infrared transmittance analyzer (Foss, Höganas, Sweden), calibrated by the Kjeldahl method (AACC International Approved Method 46-12. 01).

Starch Isolation. Grain (400 g) was steeped in 2 L of 0. 2 M ammonia solution for 24 h at room temperature before starch isolation. Starch was isolated from softened wheat grains according to the method described by Wang et al. (2015b).

Determination of Amylose Content. The iodine binding method of Chrastil (1987) was used to determine the amylose content of isolated wheat starches. Amylose (A0512) and amylopectin (A8515) from potato starch (Sigma-Aldrich, St. Louis, MO, U. S. A.) were used to establish a calibration curve with a set of starches with 0-50% amylose.

Swelling Power and Solubility. Swelling power and solubility of wheat flours and starch granules were measured according to the method of Wang and Copeland (2012a). The suspensions of wheat flours or starch granules (40 mg) and distilled water (1 mL) were heated in a water bath at 92.5℃ for 30 min. Swelling power was calculated as the ratio of the weight of the sedimented materials to the initial dry weight of samples, and the solubility was calculated as the ratio of dissolved materials to the initial dry weight of samples.

Pasting Properties. The pasting properties of wheat flour and starch granules were measured with a Rapid Visco Analyzer (RVA-3, Newport Scientific, Australia). Wheat flour (3. 5 g, dry weight basis) or starch (2. 5 g, dry weight basis) was mixed with distilled water to obtain a suspension with a total weight of 28 g. The slurries were heated according to the procedure described by Wang et al. (2015b). Peak viscosity (PV), final viscosity (FV), trough viscosity (TV), and pasting temperature (PT) can be obtained from the pasting curves. The breakdown (BD) and setback (SB), measured as PV _ TV and FV _ TV, respectively, can be calculated by using the

Thermocline software provided with the instrument.

Differential Scanning Calorimetry (DSC). The properties of gelatinization and retrogradation of wheat starch samples were determined with a DSC (200 F3, Netzsch, Germany). The sample preparation and heating profiles were described by Wang and Copeland (2012b). The onset temperature (T_o), peak temperature (T_p), conclusion temperature (T_c), and enthalpy of gelatinization (ΔH_{gel}) were obtained through the Proteus analysis software provided with the instrument.

The samples after DSC measurements were stored at 4℃ for six days and reheated in the DSC according to the heating profiles for gelatinization tests. The retrogradation enthalpy (ΔH_{ret}) was obtained, and the retrogradation percentage (R_{et}%) was calculated as $R_{et}\% = \Delta H_{ret}/\Delta H_{gel} \times 100$.

In Vitro Starch Digestibility. The in vitro starch digestibility was determined according to the method of Wang et al. (2014), with slight modifications. Native starch (100 mg) was mixed with 1 mL of distilled water and heated at 90℃ for 10 min. Porcine a-amylase (EC 3. 2. 1. 1, type VI-B from porcine pancreas, 28 U/mg, Sigma-Aldrich) and amyloglucosidase (3, 260 U/mL, Megazyme International Ireland) were used to digest the gelatinized starch. At specified time points during digestion, the concentration of glucose in the digestion solution was measured with the Megazyme glucose oxidase/peroxidase kit. The digestograms of starch were obtained by plotting the percentage of starch digestion as a function of hydrolysis time.

Statistical Analysis. All experiments were conducted at least twice, and the results were reported as the mean values and standard deviations. Analysis of variance (ANOVA) by Duncan's test ($P < 0. 05$) was conducted by using the SPSS 19. 0 statistical software program (SPSS, Chicago, IL, U. S. A.).

Table I   Analysis of variance of starch and flour Characteristics of multiple wheat cultivars grown under two different conditions[z]

Factors	Cultivar (C)	Treatment (T)	C×T
Df	9	1	9
Starch content	51. 6***	34. 3***	6. 9***
Protein content	1. 45***	0. 36 *	0. 01
Lipid content	0. 26***	0. 34***	0. 06***
Amylose content	3. 4**	0. 06	1. 07
SP of flour	0. 51***	1. 3***	0. 16
SP of starch	1. 3***	0. 35	0. 28 *
Solubility of starch	9. 9***	0. 93	1. 03
PV of starch	95509. 1***	1113. 0	13644. 6***
TV of starch	47252. 3***	102. 4	9220. 0***
BD of starch	18929. 4***	1890. 6***	652. 5***
FV of starch	94795. 5***	8065. 6 *	20038. 1***
SB of starch	15319. 1***	9985. 6 *	3722. 9 *
PT of starch	9. 1***	0. 13	1. 1**
PV of flour	345839. 8***	10758. 4**	10854. 5***
TV of flour	154825. 0***	0. 40	4321. 1***
BD of flour	106161. 8***	10890. 0***	3114. 2***
FV of flour	457978. 3***	4515. 6	10929. 9***
SB of flour	88813. 9***	4431. 3 *	1837. 6 *
PT of flour	3. 0***	0. 05	0. 27
T_o	1. 7***	0. 91***	0. 12***
T_p	2. 8***	0. 34 *	0. 05
T_c	3. 7***	0. 16	0. 04
Gelatinization enthalpy ΔH_{gel}	2. 5***	2. 42***	1. 43***
Retrogradation enthalpy ΔH_{ret}	0. 23***	0. 63***	0. 11***
Retrogradation percentage $R_{et}\%$	20. 2***	68. 1***	8. 3***

[z] Values are the sum of squares divided by the degrees of freedom (df), and those designated with *, **, and *** indicate significance at $P < 0.05$, 0.01, and 0.001, respectively. SP= swelling power; PV= peak viscosity; TV= trough viscosity; BD= breakdown; FV= final viscosity; SB= setback; PT = pasting temperature; T_o = onset temperature; T_p = peak temperature; T_c = conclusion temperature; ΔH_{gel}=gelatinization enthalpy; ΔH_{ret}=retrogradation enthalpy; and $R_{et}\%$=retrogradation percentage.

RESULTS AND DISCUSSION

ANOVA indicated that cultivar had a significant effect on all parameters of starch and flour at $P < 0.01$ or 0.001 (Table I). However, the effect of treatment (heat stress versus normal) was significantly different for only 15 parameters such as starch content at $P < 0.05$, 0.01, or 0.001, and the effect of cultivar-treatment interaction was significantly different for 18 parameters such as lipid content at $P < 0.05$, 0.01, or 0.001.

Basic Composition of Wheat Grains. Total starch content in wheat under a heat-stress environment was significantly lower than that under normal environment (Table II), although significant differences were also observed between cultivars under each environment. Of the 10 cultivars, total starch contents of Zhongmai 816, ShiU 09-4366, Shimai 15, and Zhongmai 895 were decreased significantly by heat stress. The variance of total starch content was caused by cultivar, treatment, and the cultivar-treatment interaction ($P < 0.001$) (Table I). The reduction in starch content under a heat-stress environment was also reported in previous studies (Keeling et al. 1994; Stone and Nicolas 1998; Liu et al. 2011; Lu et al. 2013). The starch content of wheat grains is mainly determined by the rate and duration of starch accumulation

(Bhullar and Jenner 1986), which are both affected greatly by genotype and growing-season temperature. High temperature after anthesis accelerates the accumulation rate of starch in the endosperm in a short time, although this is not enough to offset the decrease in starch content caused by the shorter accumulation time (Randall and Moss 1990). On the other hand, the activities of starch biosynthetic enzymes can be inhibited under high temperatures after anthesis, resulting in the reduction of starch content (Keeling et al. 1994).

Protein content was increased slightly but significantly by heat stress, and there were significant differences betweencultivars under each environment (Table II). Zhongmai 816 under a heat-stress environment had the highest protein content (14.5%) among 10 tested cultivars, whereas Shiyou 09-4366 had the lowest value (12.8%) under a normal environment. Cultivar contributed the largest component of variance in protein content ($P<0.001$), whereas treatment was a less important factor ($P<0.05$) (Table I).

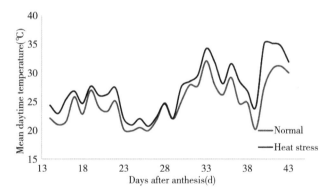

Fig. 1. Change of mean daytime temperature as a function of days after anthesis.

In general, heat stress was reported to increase protein content of wheat grains (Corbellini et al. 1997; Stone and Nicolas 1998; Spiertz et al. 2006), and the increase was more obvious for heat-sensitive cultivars compared with the heat-tolerant ones. However, the protein content of waxy maize was reported to decrease and increase at early and later heat-stress stages of grain filling, respectively (Lu et al. 2013).

There was a slight but statistically significant increase in lipid content after the heat-stress treatment, and

significant differences were observed in lipid content between cultivars under each environment (Table II). Zhongmai 816, Shimai 15, Zhongmai 175, and Shiyou 17 displayed a significant increase in lipid content after heat-stress treatment, whereas Zhongmai 162, ShiU 09-4366, and Zhongmai 875 showed an opposite trend. No significant differences in lipid content were noted for Heng08guan 29, Hengguan 35, and Zhongmai 895 under both environments. Zhongmai 162 showed the highest lipid content among 10 cultivars (2.54 and 2.17% under normal and heat-stress conditions, respectively), and the lowest lipid content was noted for Heng08guan 29 under normal environment (1.07%). Cultivar, treatment, and their interaction made different contributions to variance in lipid content ($P<0.001$) (Table I). Williams et al. (1994) reported that high temperature reduced total lipid, starch lipid, and nonstarch lipid contents of wheat grains. Behl et al. (1996) reported that heat stress increased the content of membrane lipids in wheat, and the increase was more significant in mutant wheat compared with normal cultivar wheat.

Amylose content was not changed significantly by heat-stress treatment and was 27.3-30.2%. The amylose content of starches from Zhongmai 816, Zhongmai 162, Zhongmai 875, Shimai 15, and Zhongmai 895 decreased slightly after heat-stress treatment, whereas it was slightly increased for five other cultivars. Cultivar accounted for the majority of variance in amylose content of starch ($P<0.01$) (Table I). Some studies showed that the amylose content of wheat cultivars was increased or little affected by heat stress (Hurkman et al. 2003; Matsuki et al. 2003), whereas others reported that the amylose content increased slightly with increasing temperature during grain filling (Shi et al. 1994; Stone and Nicolas 1995).

Swelling Power and Solubility. Swelling power and solubility of starch granules in 10 tested cultivars are presented in Table III. Heat stress did not result in significant changes in swelling power and solubility of wheat starch granules but significantly decreased swelling power of wheat flour. Significant differences were noted in swelling power of starch and flour under each environment. Cultivar was the largest source of variance in swelling power and solubility of starch, whereas treatment contributed to the largest variance in swelling power of flour ($P < 0.001$) (Table I). Swelling power of starch and flour was reported to decrease under heat stress during the grain-filling stage (Stone and Nicolas 1994; Tester 1997; Lu et al. 2014). Swelling power of starch granules is considered primarily to be a property of amylopectin

molecules, but it is also influenced by other factors such as amylose content, lipid content, granule size, and surface protein and lipid (Wang et al. 2014; Wang and Copeland 2015). Little change in swelling power of starch granules could be attributed to the small change in amylose content. In contrast, swelling power of wheat flour is affected by both starch granules and other components of flour. The decreased swelling power of wheat flour under heat stress could be attributed to the lower starch content and higher lipid content (Table II).

Table II　Basic Composition of Wheat Grains in 10 Tested Cultivars[z]

Cultivar	Starch Content (%)		Protein Content (%)		Lipid Content (%)		Amylose Content (%)	
	Normal	Heat Stress	Normal	Heat Stress	Normal	Heat Stress	Normal	Heat Stress
Zhongmai 816	56.9±1.3ef	52.3±0.3ab	14.3±0.4d	14.5±0.4d	1.42±0.02bc	1.88±0.04e	28.1±1.1abcde	27.3±0.9a
Zhongmai 162	54.8±0.6cd	54.1±1.3c	13.2±0.3ab	13.3±0.1ab	2.51±0.08g	2.17±0.05f	28.8±0.6abcdef	27.9±1.2abc
Heng08guan 29	60.0±0.8hi	61.0±0.8i	14.2±0.1d	14.4±0.5d	1.07±0.06a	1.11±0.07a	29.5±0.9bcdef	29.5±1.1cdef
Zhongmai 175	60.1±1.3hi	59.8±0.7gh	14.1±0.1cd	14.4±0.6d	1.16±0.05a	1.43±0.01bc	28.9±0.1abcdef	30.2±0.4f
Shiyou 17	53.8±1.1bc	55.6±0.9cde	13.1±0.1ab	13.2±0.0ab	1.11±0.01a	1.52±0.04cd	27.5±1.3ab	28.0±1.4abcd
Zhongmai 875	56.6±1.3def	54.8±0.4cd	14.0±0.2cd	14.1±0.2cd	1.61±0.05d	1.35±0.06b	30.0±1.4ef	29.5±1.2bcdef
Hengguan 35	61.8±0.8i	61.0±0.5hi	13.5±0.3bc	13.9±0.1cd	1.62±0.13d	1.55±0.08cd	29.2±1.3abcdef	29.9±1.1def
ShiU 09-4366	58.1±0.8fg	55.5±0.5cde	12.8±0.1a	13.0±0.4ab	1.66±0.13d	1.34±0.01b	28.2±0.7abcdef	29.0±0.5abcdef
Shimai 15	56.3±1.4def	51.6±1.9a	12.9±0.1ab	13.1±0.1ab	1.43±0.03bc	1.89±0.02e	29.6±0.9cdef	28.8±1.2abcdef
Zhongmai 895	54.9±2.2cde	52.7±0.3ab	14.2±0.4d	14.4±0.2d	1.63±0.05d	1.59±0.03d	29.5±0.9bcdef	28.4±0.7abcdef
Mean value	57.3±2.6b	55.8±3.6a	13.6±0.6a	13.8±0.6b	1.52±0.4a	1.58±0.3b	28.9±0.8a	28.9±0.9a

[z]Values are means ± standard deviations. Values with the same letters in the same column for each of the flours are not significantly different ($P < 0.05$).

Table III　Swelling Power and Solubility of Wheat Starches and Flours[z]

Cultivar	Wheat Starch				Wheat Flour	
	Swelling Power		Solubility (%)		Swelling Power	
	Normal	Heat Stress	Normal	Heat Stress	Normal	Heat Stress
Zhongmai 816	8.8±0.1a	8.8±0.2a	9.3±0.8cdef	9.1±1.0bcdef	8.6±0.1bcd	7.7±0.4a
Zhongmai 162	9.2±0.1ab	9.3±0.0ab	10.3±0.2efg	11.7±1.1g	9.0±0.3cdef	8.2±0.1b
Heng08guan 29	10.5±0.1de	9.9±0.1bcd	9.3±1.1cdef	8.0±1.4abcd	9.3±0.3f	8.8±0.1cdef
Zhongmai 175	10.9±0.6e	10.1±0.1cd	8.5±0.9abcde	7.4±0.9ab	9.2±0.3ef	8.9±0.2cdef
Shiyou 17	9.7±0.3bc	9.6±0.3bc	7.3±0.4ab	6.7±0.7a	8.8±0.4cdef	8.6±0.1bcd
Zhongmai 875	9.6±0.1bc	10.1±0.7cd	7.8±0.1abc	8.0±0.9abcd	8.6±0.1bcd	8.5±0.5bc
Hengguan 35	9.4±0.3abc	9.4±0.4abc	10.2±1.4efg	10.6±1.1fg	9.1±0.5def	9.0±0.5cdef
ShiU 09-4366	9.3±0.4ab	9.7±0.5bc	7.5±0.6abc	7.7±1.3abc	8.7±0.2bcde	8.7±0.2bcde
Shimai 15	9.9±0.4bcd	9.3±0.5ab	10.1±1.6efg	9.8±1.1def	9.0±0.2cdef	9.0±0.3cdef
Zhongmai 895	9.8±0.5bc	9.4±0.3ab	9.2±0.3bcdef	8.1±0.9abcd	8.7±0.3bcd	8.6±0.3bcd
Mean value	9.7±0.6a	9.6±0.4a	9.0±1.1a	8.7±1.6a	8.9±0.3b	8.6±0.4a

[z]Values are means ± standard deviations. Values with the same letters in the same column for each starch or flour are not significantly different ($P < 0.05$).

Table IV Pasting Properties of Starches from Wheat Cultivars[z]

Cultivar	Peak Viscosity		Trough Viscosity		Breakdown		Final Viscosity		Setback		Pasting Temperature	
	Normal	Heat Stress	Normal	Heat Stress	Normal	Heat Stress	Normal	Heat Stress	Normal	Heat Stress	Normal	Heat Stress
Zhongmai 816	1,478±0.0d	1,386±9.9c	1,345±7.1ef	1,265±7.1c	133±7.1de	121±2.8cd	1,964±43.1c	1,745±27.6a	619±50.2b	480±20.5a	89.7±0.0efg	89.7±0.0efg
Zhongmai 162	1,348±7.1b	1,533±2.1ef	1,261±17.0c	1,405±0.7hi	87±9.9ab	128±2.8de	2,000±22.6cd	2,135±4.2fg	739±39.6cdef	731±5.0cdef	87.2±0.0b	86.0±0.7a
Heng08guan 29	1,789±6.4j	1,692±7.1g	1,530±0.0k	1,472±4.2j	259±6.4jk	220±2.8i	2,296±17.7i	2,140±2.1fg	766±17.7f	668±6.4bcde	89.3±0.5def	90.5±0.0ghi
Zhongmai 175	1,776±1.4ij	1,723±18.4gh	1,479±9.9j	1,457±5.7j	297±11.3l	266±24.0k	2,199±21.9gh	2,142±8.1fg	720±31.8def	685±53.7bcdef	87.6±0.6bc	88.8±0.0de
Shiyou 17	1,745±14.1hi	1,766±21.2ij	1,476±0.0j	1,521±14.1k	269±14.1k	245±7.1j	2,243±28.3hi	2,270±50.2hi	767±28.3f	749±36.1ef	89.6±0.1defg	89.3±0.6def
Zhongmai 875	1,483±38.9d	1,293±39.6a	1,378±26.2fgh	1,222±32.5b	105±12.7bc	71±7.1a	2,066±94.8def	1,872±67.2b	689±68.6bcdef	650±34.7bc	90.5±1.1ghi	91.6±0.5ij
Hengguan 35	1,567±4.2f	1,501±12.7de	1,354±5.7efg	1,328±9.9de	213±1.4i	173±2.8h	2,094±2.8ef	2,002±0.7cd	740±2.8def	674±9.2bcde	88.4±0.7cd	88.8±0.0de
ShiU 09-4366	1,478±21.2d	1,563±17.7f	1,226±16.3b	1,306±19.1d	253±5.0jk	257±1.4jk	1,879±29.0b	2,042±69.3cde	653±12.7bcd	737±88.4cdef	92.1±0.1j	91.7±0.6ij
Shimai 15	1,271±1.4a	1,412±11.3c	1,100±5.7a	1,235±5.0bc	171±7.1gh	178±6.4h	1,720±24.8a	1,861±38.9b	620±30.4b	626±33.9b	90.9±0.5hi	89.2±0.7def
Zhongmai 895	1,570±32.5f	1,531±9.2ef	1,417±38.2i	1,388±7.8ghi	153±5.7fg	143±1.4ef	2,152±19.1fg	2,120±8.5efg	735±19.1cdef	733±16.3cdef	90.1±0.6fgh	90.9±0.6hi
Mean value	1,550.5±176.5a	1,540.0±153.1a	1,356.6±131.5a	1,359.9±104.5a	194.0±74.2b	180.2±65.5a	2,061.3±176.2b	2,032.9±162.3a	704.8±56.4b	673.3±79.6a	89.5±1.5a	89.7±1.7a

[z] Values are means± standard deviations. Values with the same letters in the same column for each starch are not significantly different (P<0.05).

Table V Pasting Properties of Wheat Flours[z]

Cultivar	Peak Viscosity		Trough Viscosity		Breakdown		Final Viscosity		Setback		Pasting Temperature	
	Normal	Heat Stress	Normal	Heat Stress	Normal	Heat Stress	Normal	Heat Stress	Normal	Heat Stress	Normal	Heat Stress
Zhongmai 816	1,516±23.3c	1,408±2.8b	1,395±33.9c	1,329±9.9b	121±10.6c	79±12.7ab	2,886±70.0bc	2,806±51.6b	1,491±36.1efghi	1,477±41.7defg	85.2±0.5ef	85.2±0.6ef
Zhongmai 162	1,557±9.2cd	1,626±7.1de	1,494±7.1de	1,519±1.4ef	63±2.1a	107±8.5bc	3,138±7.1f	3,158±2.8f	1,644±0.0k	1,639±1.4k	83.2±0.0ab	82.8±0.6a
Heng08guan 29	2,116±15.6i	1,930±52.3gh	1,628±16.3i	1,565±27.6fgh	489±0.7l	366±24.8jk	3,148±24.8f	3,088±29.7ef	1,520±8.5fghij	1,524±2.1ghij	84.8±0.1def	84.8±0.0def
Zhongmai 175	2,195±73.5j	2,170±70.0ij	1,546±49.5efgh	1,594±37.5hi	649±24.0m	576±32.5m	3,019±57.3e	3,133±69.3f	1,473±7.8defg	1,540±31.8hij	83.6±0.5abc	84.4±0.5cde
Shiyou 17	1,934±29.0gh	1,999±23.3h	1,579±17.0ghi	1,604±19.1hi	355±12.0ij	396±4.2k	3,092±21.9ef	3,165±7.1f	1,513±4.9fghi	1,562±12.0ij	84.8±0.1def	84.7±0.1def
Zhongmai 875	1,722±20.5f	1,611±24.0de	1,530±2.1efg	1,449±12.0cd	192±18.4e	163±12.0de	3,003±17.7de	2,896±8.5bc	1,473±19.8defg	1,448±3.5bcde	85.6±0.1fg	85.6±0.1fg
Hengguan 35	1,727±26.9f	1,641±16.3e	1,399±26.9c	1,403±14.1c	328±0.0hi	238±2.1f	2,806±17.7b	2,863±26.9bc	1,407±9.2bc	1,460±12.7cdef	83.9±0.0bcd	84.8±0.1def
ShiU 09-4366	1,738±50.9f	1,900±33.9g	1,442±36.1cd	1,580±12.7ghi	297±14.9gh	320±21.2hi	2,923±19.8cd	3,155±44.6f	1,482±55.9efgh	1,575±31.8j	86.4±0.1g	85.6±0.1fg
Shimai 15	1,195±7.1a	1,157±43.1a	922±28.9a	930±37.5a	274±21.9g	227±5.7f	1,987±69.3a	2,018±73.5a	1,066±40.3a	1,089±36.1a	84.8±0.0def	84.7±0.0def
Zhongmai 895	1,619±48.1de	1,548±19.8cd	1,443±38.2cd	1,408±6.4c	176±9.9e	141±26.2cd	2,871±24.8bc	2,801±42.4b	1,428±13.4bcd	1,394±36.1b	85.1±0.5ef	85.6±1.2fg
Mean value	1,731.9±294.6b	1,699.0±302.5a	1,437.8±196.9a	1,438.1±201.9a	294.4±175.4b	261.3±154.4a	2,887.3±337.0a	2,908.3±347.8a	1,449.7±149.1a	1,470.8±151.8b	84.7±1.0a	84.8±0.8a

[z] Values are means± standard deviations. Values with the same letters in the same column for each of the flours are not significantly different (P<0.05).

(continued on next page)

Pasting Properties. No significant difference was found in PV, TV, and PT of starches between heat-stress and normal environments. However, BD, FV, and SB were reduced significantly by heat-stress treatment (Table IV). Cultivar made the largest contribution to the variance in PT, PV, BD, TV, FV, and SB of starches ($P < 0.001$) (Table I). Interaction of cultivar and treatment also accounted for a large part of the total variability in most pasting parameters of starches, except SB viscosity. In a previous study (Liu et al. 2011), the pasting properties of wheat starch were not altered when the wheat was grown at 25 and 30℃, whereas the pasting viscosities decreased when the growing temperatures were over 30℃. Lu et al. (2013) also reported that heat stress during grain filling decreased the viscosities but increased the PT of waxy maize starch. Amylose content and amylopectin fine structure are the critical factors that affect starch pasting properties (Copeland et al. 2009).

For wheat flour, no significant differences were observed in TV, FV, and PT under both environments, but heat stress significantly altered PV, BD, and SB (Table V). Similar to starch, the major contributor of variability in pasting properties of wheat flours was cultivar, followed by the interaction of cultivar and treatment ($P < 0.001$) (Table I). Heat-stress treatment was reported to affect the pasting properties of cereal flours in different ways. Chun et al. (2015) reported that PV and TV of rice flour increased significantly with increasing ripening temperature, whereas SB presented an opposite trend. Zhong et al. (2005) reported that the FV of rice flour decreased at a higher growing temperature, whereas the PV, TV, and BD varied with cultivars.

Gelatinization and Retrogradation. Significant differences were observed in most gelatinization and retrogradation parameters, except for T_c (Table VI). Cultivar was the major source of variance in T_o, T_p, T_c, and DH_{gel} ($P < 0.001$) (Table I). Previous studies showed that heat-stress treatment increased gelatinization temperatures of wheat (Shi et al. 1994; Tester et al. 1995; Matsuki et al. 2003), waxy maize (Lu et al. 2014), and rice starches (Suzuki et al. 2004; Chun et al. 2015). The increase in gelatinization temperature after heat-stress treatment was attributed to the increased degree of crystallinity (Matsuki et al. 2003). For waxy maize starches, the enthalpy change was not altered greatly by heat-stress treatment (Lu et al. 2014). In contrast to ΔH_{gel}, ΔH_{ret} and $R_{et}\%$ were found to be affected more greatly by heat-stress treatment ($P < 0.001$) (Table I). ΔH_{ret} is a measure of crystallinity regained by starch gels during retrogradation (Wang et al. 2015a). $R_{et}\%$ can be used to measure the tendency of gelatinized starch to retrograde. The increased ΔH_{ret} and $R_{et}\%$ indicated that wheat starches were liable to retrograde after heat-stress treatment, which may spoil the quality of wheat-based food products during storage. Similar observations were also reported in waxy maize starches (Lu et al. 2014).

In Vitro Enzymatic Digestibility. After 120 min, no significant differences in hydrolysis percentages were found among wheat starches under both environments, except for ShiU 09-4366 and Hengguan 35. Most cultivars displayed a rapid increase in digestion percentage within 60 min followed by a subsequent slow increase. Heat stress altered the in vitro digestion patterns of wheat starch, which were cultivar dependent (Fig. 2).

Table VI Gelatinization and Retrogradation Properties of Wheat Starch[z]

Cultivar	T_o Normal	T_o Heat Stress	T_p Normal	T_p Heat Stress	T_c Normal	T_c Heat Stress	ΔH_{gel} Normal	ΔH_{gel} Heat Stress	ΔH_{ret} Normal	ΔH_{ret} Heat Stress	R_{ret} % Normal	R_{ret} % Heat Stress
Zhongmai 816	56.8±0.1efg	57.0±0.1fgh	60.9±0.2efg	61.0±0.1fg	66.6±0.3cde	66.5±0.1cde	11.3±0.1i	10.0±0.1ab	0.68±0.06a	1.14±0.03def	6.1±0.5a	11.4±0.2gh
Zhongmai 162	55.9±0.1a	56.4±0.1bc	60.5±0.3bc	60.7±0.1cdef	66.3±0.2bc	66.6±0.1cde	11.1±0.2hi	9.9±0.2ab	1.23±0.01ghi	1.38±0.03lm	11.1±0.2fg	13.9±0.1m
Heng08guan 29	57.7±0.2j	57.7±0.1j	62.0±0.3hi	62.0±0.1hi	68.0±0.4f	68.0±0.2f	10.3±0.2cde	10.6±0.1fg	1.18±0.03efgh	1.53±0.04n	11.5±0.3ghi	14.4±0.1m
Zhongmai 175	56.7±0.1cde	56.6±0.1cd	60.9±0.3defg	60.8±0.1cdefg	66.4±0.4cd	66.6±0.1cde	12.8±0.1k	10.6±0.1fg	1.34±0.05kl	1.42±0.01m	10.4±0.4de	13.4±0.0l
Shiyou 17	56.4±0.1bc	57.0±0.1fgh	60.5±0.0bcd	61.0±0.3fg	66.6±0.1cde	66.5±0.3cd	11.2±0.1i	10.6±0.2efg	0.85±0.08b	1.07±0.02d	7.6±0.7b	10.1±0.3d
Zhongmai 875	56.8±0.2def	56.7±0.1cde	60.5±0.4bcde	60.7±0.3cdef	66.2±0.2abc	66.4±0.2cd	10.8±0.2gh	9.8±0.1a	1.16±0.06efg	0.96±0.04c	10.8±0.4ef	9.9±0.4cd
Hengguan 35	57.3±0.2i	57.8±0.1j	61.9±0.2h	62.3±0.2i	67.9±0.3f	68.2±0.3f	10.1±0.4bcd	10.8±0.1gh	1.24±0.01hij	1.33±0.04kl	12.0±0.1ijk	12.3±0.3jk
ShiU 09-4366	57.1±0.1ghi	57.3±0.2i	61.2±0.2g	61.2±0.1g	66.8±0.2de	67.0±0.1e	12.1±0.2j	12.3±0.2j	1.13±0.02de	1.27±0.03ijk	9.4±0.1c	10.3±0.1de
Shimai 15	57.2±0.1hi	57.7±0.3j	61.8±0.2h	61.9±0.3hi	67.9±0.1f	67.9±0.3f	10.2±0.1bcd	10.8±0.3gh	1.20±0.03efghi	2.27±0.01o	11.8±0.2hij	20.7±0.0n
Zhongmai 895	56.2±0.2b	56.3±0.0b	60.1±0.4a	60.2±0.0ab	65.9±0.2ab	65.8±0.3a	10.1±0.1abc	10.4±0.2def	1.21±0.04fghi	1.31±0.01ijkl	12.0±0.3ijk	12.5±0.2k
Mean value	56.8±0.5a	57.1±0.6b	61.0±0.7a	61.2±0.7b	66.9±0.8a	67.0±0.8a	11.0±0.9b	10.6±0.7a	1.1±0.2a	1.4±0.4b	10.3±2.0a	12.9±3.2b

[z] Values are means±standard deviations. Values with the same letters in the same column for each starch are not significantly different ($P<0.05$).

(continued on next page)

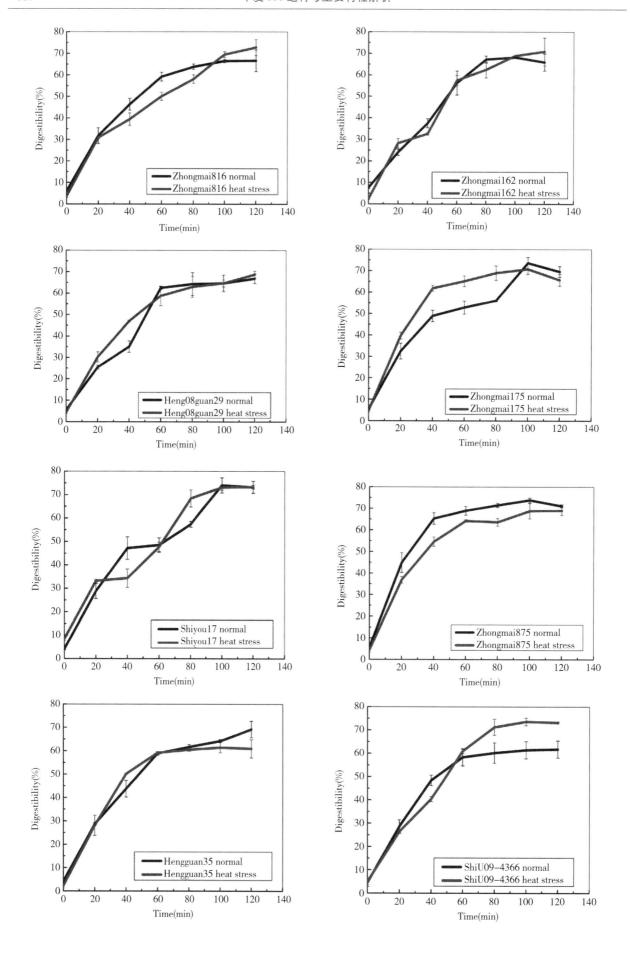

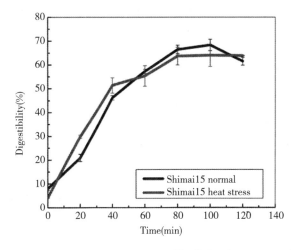
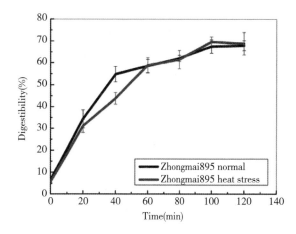

Fig. 2. In vitro enzymatic digestibility of wheat starch.

CONCLUSIONS

Cultivar was the main source of variability in functional properties of wheat starch and flour, although contributions to variability from heat stress or interaction of cultivar and heat stress were also significant in some cases. Heat stress during the grain-filling stage decreased total starch content but increased protein and lipid contents of wheat grains. Amylose content of wheat starches was not altered significantly by heat stress. Heat stress had little effect on swelling power and starch solubility of wheat starch but decreased the swelling power of wheat flour. Heat stress reduced the BD, FV, and SB but had little effect on other pasting parameters of wheat starches. In contrast, heat stress decreased the PV and BD and increased SB but had little effect on other pasting parameters of wheat flours. In addition to T_c, heat stress had a significant effect on gelatinization and retrogradation parameters. The in vitro enzymatic digestibility of wheat starches was affected slightly by heat stress. ShiU 09-4366 was identified with good heat resistance and could serve as crossing parent in breeding for improving heat resistance during the grain-filling stage.

❖ ACKNOWLEDGEMENTS

We thank the National Natural Science Foundation of China (31522043 and 31401651) and the Program for Innovative Research Team in University (IRT IRT_15R49) for their financial support.

❖ LITERATURE CITED

AACC International. Approved Methods of Analysis, 11th Ed. Method 46-12. 01. Crude protein—Kjeldahl method, boric acid modification. Approved November 3, 1999. http://dx. doi. org/10. 1094/AACCIntMethod-46-12. 01. Available online only. AACC International: St. Paul, MN.

Beckles, D. M., and Thitisaksakul, M. 2014. How environmental stress affects starch composition and functionality in cereal endosperm. Starch/Stärke 66: 58-71.

Behl, R. K., Heise, K. P., and Moawad, A. M. 1996. High temperature tolerance in relation to changes in lipids in mutant wheat. Beitr. Trop. Landwirtsch. Veterinaermed. 97: 131-135.

Bhullar, S., and Jenner, C. 1986. Effects of temperature on the conversion of sucrose to starch in the developing wheat endosperm. Funct. Plant Biol. 13: 605-615.

Blumenthal, C., Bekes, F., Gras, P. W., Barlow, E. W. R., and Wrigley, C. W. 1995. Identification of wheat genotypes tolerant to the effects of heat stress on grain quality. Cereal Chem. 72: 539-544.

Borghi, B., Corbellini, M., Ciaffi, M., Lafiandra, D., Stefanis, E., Sgrulletta, D., Boggini, G., Fonzo, N., De, S. E., and Di, F. N. 1995. Effect of heat shock during grain filling on grain quality of bread anddurum wheats. Crop Pasture Sci. 46: 1365-1380.

Chrastil, J. 1987. Improved colorimetric determination of

amylose instarches or flours. Carbohydr. Res. 159: 154-158.

Chun, A., Lee, H.-J., Hamaker, B. R., and Janaswamy, S. 2015. Effects of ripening temperature on starch structure and gelatinization, pasting, and cooking properties in rice (*Oryza sativa*). J. Agric. Food Chem. 63: 3085-3093.

Copeland, L., Blazek, J., Salman, H., and Tang, M. C. 2009. Form and functionality of starch. Food Hydrocolloids 23: 1527-1534.

Corbellini, M., Canevar, M., Mazza, L., Ciaffi, M., Lafiandra, D., and Borghi, B. 1997. Effect of the duration and intensity of heat shock during grain filling on dry matter and protein accumulation, techno logical quality and protein composition in bread and durum wheat. Funct. Plant Biol. 24: 245-260.

Hunt, L., van der Poorten, G., and Pararajasingham, S. 1991. Postanthesis temperature effects on duration and rate of grain filling in some winterand spring wheats. Can. J. Plant Sci. 71: 609-617.

Hurkman, W. J., McCue, K. F., Altenbach, S. B., Korn, A., Tanaka, C. K., Kothari, K. M., Johnson, E. L., Bechtel, D. B., Wilson, J. D., and Anderson, O. D. 2003. Effect of temperature on expression of genes encoding enzymes for starch biosynthesis in developing wheat endo sperm. Plant Sci. 164: 873-881.

Keeling, P., Banisadr, R., Barone, L., Wasserman, B., and Singletary, G. 1994. Effect of temperature on enzymes in the pathway of starch biosynthesis in developing wheat and maize grain. Funct. Plant Biol. 21: 807-827.

Liu, P., Guo, W., Jiang, Z., Pu, H., Feng, C., Zhu, X., Peng, Y., Kuang, A., and Little, C. 2011. Effects of high temperature after anthesis on starch granules in grains of wheat (*Triticum aestivum* L.). J. Agric. Sci. 149: 159-169.

Lu, D., Shen, X., Cai, X., Yan, F., Lu, W., and Shi, Y.-C. 2014. Effects of heat stress during grain filling on the structure and thermal properties of waxy maize starch. Food Chem. 143: 313-318.

Lu, D., Sun, X., Yan, F., Wang, X., Xu, R., and Lu, W. 2013. Effects of high temperature during grain filling under control conditions on the physicochemical properties of waxy maize flour. Carbohydr. Polym. 98: 302-310.

Matsuki, J., Yasui, T., Kohyama, K., and Sasaki, T. 2003. Effects of environmental temperature on structure and gelatinization properties of wheat starch. Cereal Chem. 80: 476-480.

Ortiz, R., Sayre, K. D., Govaerts, B., Gupta, R., Subbarao, G., Ban, T., Hodson, D., Dixon, J. M., Ortiz-Monasterio, J. I., and Reynolds, M. 2008. Climate change: Can wheat beat the heat? Agric. Ecosyst. Environ. 126: 46-58.

Rahman, S., Bird, A., Regina, A., Li, Z., Ral, J. P., McMaugh, S., Topping, D., and Morell, M. 2007. Resistant starch in cereals: Exploiting genetic engineering and genetic variation. Cereal Sci. 46: 251-260.

Randall, P., and Moss, H. 1990. Some effects of temperature regimeduring grain filling on wheat quality. Crop Pasture Sci. 41: 603-617.

Shi, Y.-C., Seib, P. A., and Bernardin, J. E. 1994. Effects of temperature during grain-filling on starches from six wheat cultivars. Cereal Chem. 71: 369-383.

Skylas, D., Cordwell, S., Hains, P., Larsen, M., Basseal, D., Walsh, B., Blumenthal, C., Rathmell, W., Copeland, L., and Wrigley, C. 2002. Heat shock of wheat during grain filling: Proteins associated with heat-tolerance. J. Cereal Sci. 35: 175-188.

Spiertz, J., Hamer, R., Xu, H., Primo-Martin, C., Don, C., and van der Putten, P. 2006. Heat stress in wheat (*Triticum aestivum* L.): Effects on grain growth and quality traits. Eur. J. Agron. 25: 89-95.

Stone, P., and Nicolas, M. 1994. Wheat cultivars vary widely in their responses of grain yield and quality to short periods of post-anthesisheat stress. Funct. Plant Biol. 21: 887-900.

Stone, P., and Nicolas, M. 1995. A survey of the effects of high tem-perature during grain filling on yield and quality of 75 wheat cultivars. Crop Pasture Sci. 46: 475-492.

Stone, P., and Nicolas, M. 1998. The effect of duration of heat stress during grain filling on two wheat varieties differing in heat tolerance: Grain growth and fractional protein accumulation. Funct. Plant Biol. 25: 13-20.

Suzuki, Y., Sano, Y., Ishikawa, T., Sasaki, T., Matsukura, U., and Hirano, H. Y. 2004. Differences in starch characteristics of rice strains having different sensitivities to maturation temperatures. J. Agron. Crop Sci. 190: 218-221.

Tester, R. F. 1997. Influence of growth conditions on barley starch properties. Int. J. Biol. Macromol. 21: 37-45.

Tester, R, F., Morrison, W. R., Ellis, R. H., Piggo, J. R., Batts, G. R., Wheeler, T. R., Morison,

J. I. L. , Hadley, P. , and Ledward, D. A. 1995. Effects of elevated growth temperature and carbon dioxide levels on some physicochemical properties of wheat starch. J. Cereal Sci. 22: 63-71.

Wang, S. , and Copeland, L. 2012a. New insights into loss of swelling power and pasting profiles of acid hydrolyzed starch granules. Starch/Stärke 64: 538-544.

Wang, S. , and Copeland, L. 2012b. Phase transitions of pea starch over awide range of water content. J. Agric. Food Chem. 60: 6439-6446.

Wang, S. , and Copeland, L. 2015. Effect of acid hydrolysis on starch structure and functionality: A review. Crit. Rev. Food Sci. Nutr. 55: 1081-1097.

Wang, S. , Li, C. , Copeland, L. , Niu, Q. , and Wang, S. 2015a. Starch retrogradation: A comprehensive review. Compr. Rev. Food Sci. Food Saf. 14: 568-585.

Wang, S. , Luo, H. , Zhang, J. , Zhang, Y. , He, Z. , and Wang, S. 2014. Alkali-induced changes in functional properties and in vitro digestibility of wheat starch: The role of surface proteins and lipids. J. Agric. Food Chem. 62: 3636-3643.

Wang, S. , Wang, J. , Zhang, W. , Li, C. , Yu, J. , and Wang, S. 2015b. Molecular order and functional properties of starches from three waxywheat varieties grown in China. Food Chem. 181: 43-50.

Williams, M. , Shewry, P. , and Harwood, J. 1994. The influence of the 'greenhouse effect' on wheat (*Triticum aestivum* L.) grain lipids. J. Exp. Bot. 45: 1379-1385.

Yamakawa, H. , Hirose, T. , Kuroda, M. , and Yamaguchi, T. 2007. Com-prehensive expression profiling of rice grain filling-related genes under high temperature using DNA microarray. Plant Physiol. 144: 258-277.

Zhong, L. , Cheng, F. , Wen, X. , Sun, Z. , and Zhang, G. 2005. The deterioration of eating and cooking quality caused by high temperature during grain filling in early-season indica rice cultivars. J. Agron. Crop Sci. 191: 218-225.

Effects of bran hydration and autoclaving on processing quality of Chinese steamed bread and noodles produced from whole grain wheat flour

Yan Zhang[1] Fengmei Gao[1,2] Zhonghu He[1,3]

[1] Institute of Crop Sciences, National Wheat Improvement Center, Chinese Academy of Agricultural Sciences (CAAS), Beijing, China

[2] Crop Breeding Institute, Heilongjiang Academy of Agricultural Sciences, Harbin, China

[3] International Maize and Wheat Improvement Center (CIMMYT) China Office, Beijing, China
Correspondence

Yan Zhang, Institute of Crop Sciences, National Wheat Improvement Center, Chinese Academy of Agricultural Sciences (CAAS), Beijing, China.

Email: zhangyan07@caas.cn

Background and objectives: Consumption of Chinese steamed bread (CSB) and noodles made from whole grain wheat flour (WWF) is encouraged due to their important nutritional elements in the outer seed layers that benefit human health. However, the use of WWF is limited because of the poor processing quality and less attractive product properties of dark color, rough texture and coarse crumb structure. Three Chinese wheat cultivars, Zhongmai 895, Yumai 49, and Zhoumai 27, were used to investigate the potentials of bran hydration and autoclaving treatments for the improvement of WWF functionality in making CSB and noodles.

Findings: Bran particle size, hydration, and autoclaving treatment of WWF significantly affected the quality of dough, CSB, and noodles made from it. Farinograph water absorption and stability of WWFs with fine bran were larger than those with coarse bran. WWFs with coarse bran compared with fine bran exhibited inferior color and higher water retention capacity (WRC), and produced CSB with better texture and crumb structure and noodles with less smoothness and poor appearance. Prehydration of bran decreased dough water absorption and increased dough stability. WWFs containing bran hydrated to 40% moisture content produced CSB with the best texture and crumb structure, and the highest total score. Bran hydration merely induced negative effects on smoothness of noodles made with fine bran. Autoclaving of bran significantly improved texture and crumb structure of CSB made from WWFs with bran hydrated to 40% moisture. Autoclaved bran with 20% moisture contributed to higher total score of CSB. However, bran hydration and autoclaving resulted in darker color of noodle sheet made from WWFs.

Conclusions: Hydration and autoclaving of bran could be an effective way to improve the processing quality of CSB made from WWFs, but there was little improvement in eating quality of noodles made from WWFs.

Published in Cereal Chemistry, 2019, 96: 104-114.

Significance and novelty: This is the first research to show that hydration and autoclaving of bran could be an effective way to improve the processing quality of CSB made from WWFs. These results provide useful information for the incorporation of wheat bran to make fiber-rich CSB.

Keywords: bran granule size, healthy food, *Triticum aestivum*, whole grain flour

1 INTRODUCTION

There is a growing demand for a new generation of healthy cereal products with low calorie and sugar levels, and high fiber contents. Wheat bran is one of the major dietary fiber sources widely used in the food industry in order to reduce risks of obesity, hypertension, diabetes, and colon cancer (Gil, Ortega, & Maldonado, 2011; Liu, 2007; Slavin, Jacobs, Marquart, & Wiemer, 2001; Slavin, Tucker, Harriman, & Jonnalagadda, 2016;). However, the current intake of whole wheat grain products is far below recommended levels, primarily due to the lower attractiveness of products quality and lower sensory acceptance of whole wheat products compared with refined flour products (Adams & Engstrom, 2000; Lang & Jebb, 2003). The presence of bran in whole grain wheat flour (WWF) physically disrupts the starch-gluten matrix and decreases the gas-holding capacity, resulting in reduced bread loaf volume (DeKock, Taylor, & Taylor, 1999; Gan, Galliard, Ellis, Angold, & Vaughan, 1992; Lai, Hosency, & Davis, 1989). Higher bran content in whole wheat pasta increases cooking loss, lowers firmness, and induces strong aromas (Aravind, Sissons, Egan, & Fellows, 2012; West, Seetharaman, & Duizer, 2013). Therefore, it is important to develop strategies to overcome the deleterious effects of bran on whole wheat products.

Various approaches, including manipulation of bran particle sizes (Cai, Choi, Hyun, Jeong, & Baik, 2014; DeKock et al., 1999), bran fermentation and enzymatic treatment (Katina et al., 2012; Salmenkallio-Marttila, Katina, & Autio, 2001), germination (Seguchi, Uozu, Oneda, Murayama, & Okusu, 2010), incorporation of enzymes into WWF formulations (Katina, Salmenkallio-Marttila, Partanen, Forssell, & Autio, 2006), bran hydration and physical treatments such as autoclaving and freezing (Cai, Choi, Park, & Baik, 2015), have been introduced to modify bran and improve dough properties as well as the bread-baking quality of WWF products. Cai et al. (2015) reported that WWFs supplemented with bran preautoclaved at 121° for 2 hr and hydrated to 60% moisture produced bread with significantly higher loaf volume. For pasta, heat-treated bran (Sudha, Ramasarma, & Venkateswara Rao, 2011), optimized relative humidity (Villeneuve & Gelinas, 2007), high-temperature drying (West et al., 2013), and optimization of bran particle size (Steglich, Bernin, Moldin, Topgaard, & Langton, 2015) helped to obtain the desired texture. Smaller particle bran led to a smoother surface and increased hardness and chewiness of cooked noodles (Chen et al., 2011; Niu, Hou, Lee, & Chen, 2014). In addition, the incorporation of wheat bran into wheat flour reduced the extensibility of the dough, decreased Chinese steamed bread (CSB) specific volume, and increased its hardness (Liu, Brennan, Serventi, & Brennan, 2017). However, few studies have reported a reduction in the adverse influence of bran on the processing quality of CSB and noodles made from WWF through bran hydration and autoclaving treatments.

Chinese steamed bread and noodles are the main traditional foods in China and are widely consumed as staple food in other Asian countries, accounting for about 80% of the total wheat flour consumption. They have now become global foods (Chen et al., 2011; Niu, Hou, Wang, & Chen, 2014). Thus, popularization of CSB and noodles made with WWF can be an effective way to promote high-fiber food consumption and increase the health benefits for consumers. Previous research related to use of wheat bran in Asian foods mainly focused on the effects of added levels of wheat bran and its particle size on dough rheological properties and processing qualities of CSB and noodles (Chen et al., 2011; Liu et al., 2017; Niu, Hou, Wang, et al., 2014). The objective of this study was to investigate the potential of bran hydration and autoclaving in improving WWF functionality in production of CSB and noodles.

2 MATERIALS AND METHODS

2.1 Materials

Cultivars Zhongmai 895, Yumai 49, and Zhoumai 27 from the Yellow and Huai Valley Facultative Wheat Region were used for this study. Field trials were conducted at Zhoukou in Henan Province during the 2014-2015 cropping seasons. Grain hardness and protein content (dry basis) were medium and 14.4% in Zhongmai 895, soft and 13.9% in Yumai 49, and medium and 12.9% in Zhoumai 27, respectively.

2.2 Grain milling

Wheat grains tempered to 14.5% of moisture content for Yumai 49 and 15.5% for Zhongmai 895 and Zhoumai 27 for 16 hr was milled by a Buhler MLU 202 laboratory experimental mill (Buhler Bros. Ltd., Uzwil, Switzerland) by AACC method 26-21A, with flour extraction rates of 72.6%, 74.2%, and 73.6% for Zhongmai 895, Yumai 49, and Zhoumai 27, respectively. The separated fractions included flour, bran, and pollard. The bran yield ranged from 12.9% to 15.1%.

Bran of each cultivar was divided into two portions and ground to coarse and fine particle sizes. Coarse and fine particles were prepared by grinding bran with a Cyclotec 1093 Mill (Foss Tecator, Stockholm, Sweden) fitted with a 0.8 mm screen and 0.3 mm openings. Bran particle size was determined using a Laser Diffraction Particle Size Analyzer (HELOS and RODOS, Japan Laser Co, Ltd., Tokyo, Japan). Each sample was assayed in triplicate. The averaged percentage volumes of wheat bran with fine and coarse particles ranged from 96.14% to 100.00% (Table 1).

2.3 Bran processing

Coarse and fine particle brans were hydrated by mixing with distilled water to obtain various moisture contents of as-is, 20%, 40%, and 60%. Hydrated bran was kept in a closed container and stored at 4° for 12 hr for moisture equilibration. For the autoclaving treatment, wheat bran (100 g, as-is moisture) was placed in a 1 L cylindrical glass bottle, covered with aluminum foil, and autoclaved at 121° for 2 hr. Hydration of autoclaved bran was performed by the addition of distilled water to obtain various moisture contents of as-is, 20%, 40%, and 60%. Hydrated autoclaved bran was kept in a sealed container and stored at 4° for 12 hr for moisture equilibration.

2.4 Preparation of WWFs

To prepare WWFs, bran from each cultivar at varying hydrations and levels of autoclaving was blended with the corresponding flour in 10% ratios on a dry basis.

2.5 WWF characterization

The water retention capacity (WRC, 14% moisture basis) of WWF was determined in duplicate according to AACC Method 56-11. The color of the WWF was measured with a chromameter CR 310 (Minolta Camera Co., Ltd., Tokyo, Japan) in duplicate. The CIE-Lab L^* (lightness), a^* (redness-greenness), and b^* (yellowness-blueness) values were recorded. Farinograph parameters of WWF including water absorption, dough development time, and stability were determined according to AACC methods 54-21.

2.6 CSB preparation and quality evaluation

Chinese steamed bread was prepared and evaluated according to Chen et al. (2007) with minor modifications. The optimum water addition was set at 80% of Farinograph water absorption for WWF samples. WWF (200 g) was mixed with yeast slurry and water in a National Mixer (National Mfg. Co., Lincoln, NE, USA) for 1.5-2.5 min. The dough was divided into two parts with similar weight and then sheeted by 10 passes through a pair of rollers set with a gap of 9/32 in. After each pass, the sheeted dough was folded along the side and rotated 90° before the next pass through the rollers. The dough piece was gently shaped by hand to form a rounded shape and then bowl the rounded dough five times in a space of 40 cm diameter, and five more times in a space of 28-30 cm diameter. The rounded dough pieces were proofed for 25 min in a National Fermentation Cabinet (35℃, 85% RH) and steamed for 25 min in a steamer containing cold water. The CSB score included specific volume (volume/weight, weighting, 20), skin color (10), smoothness (10), shape (10), structure (15), and stress relaxation (35). Specific volume was determined by rapeseed displacement and electronic scales. Skin color measurement was carried out using a Minolta color meter (Model CR 310). Smoothness, shape, and structure were scored subjectively by an experienced researcher. Stress relaxation was measured by a Texture Analyzer TA-XT2i (Stable Micro Systems, Ltd., Surrey, England).

TABLE 1   Particle sizes distribution of wheat bran ground to fine (0. 3 mm) and coarse (0. 8 mm) particles

Cultivar	Fine particle size (%)	Coarse particle size (%)
Zhongmai 895	99. 28[a]	99. 17
Yumai 49	98. 23	100. 00
Zhoumai 27	96. 14	99. 12

[a] Values are the averaged percentage volumes of finely (0. 3mm) and coarsely (0. 8mm) ground bran, respectively.

2. 7   Noodles preparation and quality evaluation

Noodles were prepared and evaluated following Zhang et al. (2005). Sensory evaluations were performed by a panel of five well-trained members of a permanent descriptive analysis panel for noodles. Three members of the group were female and two were male ranging in age from 35 to 50. The evaluation parameters included color (weighting 15), appearance (10), firmness (20), viscoelasticity (30), smoothness (15), and taste/flavor (10). Tests performed at room temperature (20°-25°) and 50%-60% RH. A Chinese

commercial flour (Hetao Xuehua flour) was used as a control for sensory evaluation of noodle quality.

2. 8   Color measurements of noodle sheets

Noodle sheet color was assessed with a Chromameter CR 310 (Minolta Camera Co.). The CIE-Lab L^* (lightness), a^* (redness-greenness), and b^* (yellowness-blueness) values were measured. The measurements were made on a Royal Australian Cereal Institute standard backing tile, with three measurements being made on each side of the noodle sheet.

2. 9   Statistical analysis

All tests were replicated at least twice. Data analysis was performed by SAS software 9. 2 (SAS institute, Cary, NC, USA). Least square differences were calculated for each parameter and used to test the significance of differences ($p < 0. 05$) between samples by Student's t tests.

3   RESULTS AND DISCUSSION

3. 1   Effects of bran particle size on CSB and noodle quality of whole wheat flour

Whole grain wheat flours with fine particle size bran exhibited higher Farinograph water absorption than those with coarse particle size bran based on three samples (Table 2). Similar results were reported by Sanz Penella, Collar, and Haros (2008) and suggested that reduced wheat bran size could increase farinograph water absorption by causing more water uptake through hydrogen bonding in the fiber structure. The stabilities of WWFs with fine bran were greater than those with coarse bran in the cases of Yumai 49 and Zhoumai 27. This was mainly due to the fact that fine bran was less destructive to dough mixing and formation than coarse bran (Zhang & Moore, 1997). A more attractive color of WWFs containing fine bran compared with those containing coarse bran was observed based on three samples, with higher L^* value and lower a^* and b^* values. This was possibly caused by decreased visible specks surface area in WWFs with fine particle size bran. WWFs with fine bran showed lower WRC than coarse bran based on three samples. The result agreed with the previous finding that coarse bran retained significantly more water than medium or fine bran when measured by a centrifuge method (Zhang & Moore, 1997).

The total CSB scores made for WWFs with fine bran were lower than those with coarse bran based on three samples (Figure 1). Variation was mainly caused by differences in stress relaxation and structure of CSB made with WWFs supplemented with fine and coarse bran. The CSB from WWFs supplemented with fine bran showed lower stress relaxation and poorer structure than coarse bran for three samples, whereas other parameters such as specific volume, skin color, smoothness, and shape of CSB were not significantly affected by bran particle size, indicating that coarse particle bran might improve the texture and loosen up the inter gas cells in CSB.

TABLE 2 Dough Farinograph parameters, water retention capacity (WRC) and color of whole grain wheat flours (WWFs) containing wheat bran ground to different particle sizes[a]

Bran particle size	Water absorption (%)	Developmenttime (min)	Stability (min)	WRC (%)	Flour		
					L*	a*	b*
Zhongmai895							
FGB	63.7a	3.3a	2.5a	56.2b	79.71a	1.81b	13.37b
CGB	63.2b	3.4a	2.5a	81.5a	76.38b	2.52a	13.87a
Yumai49							
FGB	57.6a	1.5a	5.0a	69.6b	79.34a	2.32b	13.50a
CGB	57.4a	1.5a	4.6b	86.8a	77.11b	2.79a	13.59a
Zhoumai27							
FGB	61.2a	2.1a	5.6a	70.2b	78.58a	2.14b	14.14b
CGB	58.5b	2.1a	4.7b	104.8a	75.92b	2.79a	14.32a

FGB: finely ground bran (0.3 mm); CGB: coarsely ground bran (0.8 mm).

[a]Values with different letters within a column are significantly different ($p < 0.05$).

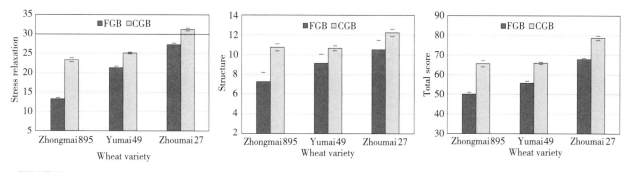

FIGURE 1　Stress relaxation (left), structure (middle) and total score (right) of steamed bread made from whole grain wheat flours supplemented with finely ground bran (FGB) and coarsely ground (CGB) wheat bran

Noodles made with WWFs with fine bran compared with coarse bran showed higher smoothness and appearance, and total scores based on three samples (Figure 2), whereas other sensory evaluation parameters including color, firmness, viscoelasticity, and flavor showed no significant differences. This observation is consistent with those of Steglich et al. (2015) and Chen et al. (2011). Steglich et al. (2015) found that spaghetti surface roughness could be related to the size of larger particles making the particles easily recognizable if they were located close to the surface. Chen et al. (2011) reported that smaller particles improved appearance and smoothness in cooked noodles. The noodle sheet from WWFs of Zhongmai 895 and Yumai 49 containing fine bran compared with coarse bran had slightly higher L^* value and lower a^* and b^* values (Figure 3). Similar results obtained by Niu, Hou, Wang, et al. (2014) showed that the lightness (L^*) of noodle sheets made from Jimai 22 increased at 0 hr as the WWF particle size was reduced. However, the noodle sheets from WWFs containing fine bran compared with coarse bran exhibited lower L^* value and higher a^* and b^* values for Zhoumai 27, in agreement with Chen et al. (2011), who reported that wet dough sheet color from Shannong 2 at 10% added wheat bran had an increased L^* value and decreased a^* and b^* values with the increasing bran particle size from 0.53 mm to 1.72 mm. These results implied different effects of bran particle size on color of noodle sheet made from different wheat cultivars.

3.2 Effects of bran prehydration on CSB and noodle quality of WWF

Water absorptions of WWFs with both fine and coarse bran decreased as the hydration level increased from as-is (5.1%-7.9% moisture) to 60%, except for that of WWFs with fine bran of Zhongmai 895 and coarse bran of Zhoumai 27 with as-is moisture (Table 3). Increases in 0.9%-3.4% and 0.6%-2.3% in water absorptions of WWFs with bran hydrated to 60% were observed for fine and coarse particles respectively. It is possible that prehydration of bran likely lowers competition with protein for moisture during dough mixing as the bran hydration level increased from as-is to 60%. However, Cai et al. (2015) found that the mixograph water absorptions of WWFs increased as hydration moisture contents of bran increased from as-is to 60%. These adverse results can possibly be attributed to the different cultivars having different chemical compositions. The stabilities of WWFs with fine and coarse bran increased as hydration increased from as-is to 60% for Yumai 49 and Zhoumai 27 except for the stability of WWF with fine and coarse bran of Zhoumai 27 with as-is moisture, whereas the stabilities of WWFs with fine and coarse bran of Zhongmai 895 with 40% hydration were the largest (4.4 and 3.6 min). This agrees generally with Cai et al. (2015) who reported that the mixograph mixing times of WWFs increased with hydration of bran up to 60% moisture.

Stress relaxation and total score of CSB from WWFs with coarse bran consistently increased as bran hydration increased from as-is (6.6%-7.9% moisture) to 40% and decreased at 60%, indicating that CSB from WWFs with coarse bran of 40% hydration had the highest stress relaxation and total score, whereas the structure showed inconsistent trends as bran hydration increased from as-is moisture to 60% based on three samples (Table 4). However, stress relaxation, structure, and total score of CSB from WWFs with fine bran increased, albeit inconsistently, as bran hydration increased from as-is to 40%, and consistently decreased with 60% hydration, leading to the largest stress relaxation, structure and total score of CSB from WWFs with fine bran of 40% hydration moisture. Similar observations were reported by Nelles, Randall, and Taylor (1998) and Cai et al. (2015). Nelles et al. (1998) found that prehydration of bran produced a significantly larger and softer bread loaf than untreated bran. Cai et al. (2015) found that loaf volume of WWF bread increased and crumb structure loosened as bran hydration increased to 60%. However, stress relaxation, structure, and total score of CSB from WWFs decreased with 60% bran hydration moisture in this study. It is possible that with hydration to 60%, bran no longer competes with protein for water during dough mixing, resulting in a negative influence of bran on gluten development and texture and structure of CSB. These results suggest that the hydration of bran could be an effective method in improving the processing quality of CSB made from WWF.

Viscoelasticity and total noodle scores with fine and coarse bran showed inconsistent changes as bran hydration increased from as-is moisture to 60% (Table 5). Previous studies showed that larger bran particles induced a more heterogeneous water distribution and thus led to uneven starch swelling in pasta (Steglich et al., 2015) and higher firmness and chewiness in noodles with smaller bran particles during texture analysis (Chen et al., 2011; Niu, Hou, Lee, et al., 2014). Hence, bran particle size might have larger effects on noodle texture than does bran hydration. Differences in effects of bran hydration on the processing quality between CSB and noodles might relate to method of preparation and hence the gluten networks in doughs of CSB and noodles formed by mixing and sheeting respectively. The noodle smoothness from WWFs supplemented with fine bran consistently decreased as bran hydration increased from as-is moisture to 60%, whereas for coarse bran, the noodle smoothness had inconsistent changes as bran hydration increased to 60%. With bran hydration increases from as-is moisture to 60%, increasing numbers of sticky masses were found during noodle dough mixing of WWFs supplemented with fine bran, resulting in rougher noodle sheets during sheeting. In addition, bran coarseness might have a stronger effect than bran hydration on noodle smoothness. Similar results were reported for noodles, whereby decreasing bran particle size significantly improved appearance and smoothness of cooked noodles (Chen et al., 2011).

These results suggest that there were no consistent effects of bran hydration or wheat genotype on the eating quality of noodles made from WWFs.

Bran hydration made a negative contribution to noodle sheet color from WWFs supplemented with fine and coarse bran, with decreased brightness (L^*) and increased redness (a^*) as hydration increased from as-is moisture to 60%, whereas there were inconsistent changes in yellowness (b^*) across genotype (Table 6). One explanation is that bran hydration can promote polyphenol oxidase (PPO) activity in bran creating an increased opportunity for PPO to catalyze substrates, thus resulting in increasing darkness of noodle sheets as bran hydration increased.

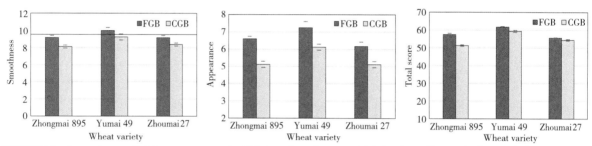

FIGURE 2   Smoothness (left), appearance (middle) and total score (right) of noodles made from whole grain wheat flours supplemented with finely ground bran (FGB) and coarsely ground (CGB) wheat bran

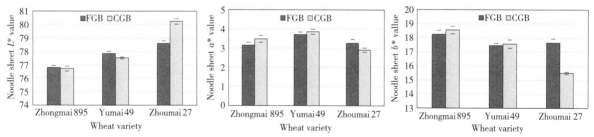

FIGURE 3 L^* (left), a^* (middle) and b^* (right) values of noodle sheets prepared from whole grain wheat flours supplemented with finely ground bran (FGB) and coarsely ground (CGB) wheat bran

TABLE 3   Dough water absorption and stability determined by Farinograph of whole grain wheat flours supplemented wheat finely (0.3 mm) and coarsely (0.8 mm) ground bran with and without autoclaving (AC) and prehydration to various moisture contents[a]

Moisture	Fine bran				Coarse bran			
	Water absorption (%)		Stability (min)		Water absorption (%)		Stability (min)	
	No AC	With AC	No AC	With AC	No AC	With AC	No AC	With AC
Zhongmai895								
6.2% (6.6%) /10.9% (14.0%)[b]	63.7b	62.5b	2.5c	3.5b	63.2a	61.7a, b	2.5c	3.8b

(continued)

Moisture	Fine bran				Coarse bran			
	Water absorption (%)		Stability (min)		Water absorption (%)		Stability (min)	
	No AC	With AC	No AC	With AC	No AC	With AC	No AC	With AC
20%	63.9a	63.1a	2.4c	3.5b	62.5b	61.5b	3.3b	3.8b
40%	63.5c	62.3c	4.4a	4.6a	62.5b	61.4b	3.6a	3.6b
60%	62.2d	62.4b, c	4.0b	4.7a	60.9c	61.9a	3.3b	4.4a
Yumai49								
5.4% (7.8%) /14.3% (13.6%)	57.4a	58.0a	5.0d	3.0b	57.6a	57.7a	4.6d	4.4b
20%	57.5a	58.5a	5.4c	3.0b	57.3b	57.7a	5.2c	4.1c
40%	57.4a	58.2a	6.5b	2.7c	57.2b, c	57.6a	7.6b	3.1d
60%	56.5b	58.0a	9.9a	4.5a	57.0c	57.0b	9.7a	6.6a
Zhoumai27								
5.1% (7.9%) (15.0% (12.4%)	61.2a	61.2a	5.6c	3.2b	58.5b	58.5b	4.7c	4.3b
20%	58.9b	60.1b	4.8d	2.8c	59.4a	59.9a	4.4d	4.1b
40%	58.9b	59.7c	6.0b	2.8c	58.5b	58.5b	5.4b	4.3b
60%	57.8c	59.4d	10a	5.8a	58.1c	58.6b	9.3a	5.6a

[a] Values with different letters within a column are significantly different ($p < 0.05$). [b] Values before and after the slash are the as-is moisture contents of bran without and with autoclaving, respectively. Numbers before and in brackets indicate as-is moisture contents of finely (0.3 mm) and coarsely (0.8 mm) ground bran, respectively.

TABLE 4   Sensory evaluation parameters for Chinese steamed bread produced from whole grain wheat flours supplemented with finely(0.3 mm) and coarsely(0.8 mm) ground wheat bran with and without autoclaving (AC) and prehydration to various moisture levels[a]

Moisture	Fine bran						Coarse bran					
	Stress relaxation (35[b])		Structure (15)		Total score (100)		Stress relaxation (35)		Structure (15)		Total score (100)	
	No AC	With AC	No AC	With AC	No AC	With AC	No AC	With AC	No AC	With AC	No AC	With AC
Zhongmai 895												
6.2% (6.6%) /10.9% (14.0%)[c]	13.0c	27.2a	7.5b	9.0b	49.5c	66.0a	23.0c	25.0c	10.5a	10.5b	64.5b	66.5c
20%	13.1c	25.0b	6.0c	9.1b	46.0d	63.0b	25.1b	27.1b	10.5a	12.0a	66.5a	72.0a
40%	21.0a	25.1b	10.5a	10.5a	59.5a	62.5b	27.2a	29.0a	10.5a	10.5b	66.5a	67.5b
60%	17.2b	19.3c	10.5a	9.0b	55.5b	55.0c	21.0d	21.0d	7.5b	9.0c	60.5c	58.0d
Yumai 49												
5.4% (7.8%) /14.3% (13.6%)	21.0b	31.0a	9.0b	10.5a	55.0d	72.5a	25.0d	33.0a	10.5a	10.5a	66.5c	77.5a
20%	21.2b	31.0a	9.0b	10.5a	56.0c	71.5b	31.0b	33.1a	10.5a	10.5a	72.5b	77.5a
40%	25.1a	29.3b	10.5a	10.5a	64.5a	69.5c	33.2a	33.4a	10.5a	9.0b	75.5a	75.4b
60%	21.0b	29.2b	10.5a	9.0b	60.5b	68.0d	27.1c	27.0b	10.5a	9.0b	66.5c	70.2c
Zhoumai 27												
5.1% (7.9%) /15.0% (12.4%)	27.2b	31.1a	10.5c	10.5b	67.5b	72.5b	31.0a	31.0a	12.0a	9.0b	74.8c	78.0a
20%	21.0d	31.2a	9.0d	10.5b	58.0d	75.5a	31.1a	31.3a	10.5b	9.0b	75.5b	72.0b
40%	31.0a	29.0b	13.5a	13.5a	74.5a	72.5b	31.1a	31.2a	10.5b	10.5a	76.5a	72.5b
60%	25.3c	25.1c	12.0b	9.0c	62.0c	58.0c	27.0b	27.0b	9.0c	9.0b	66.0d	65.0c

[a] Values with different letters within a column are significantly different ($p < 0.05$). [b] The numbers indicate maximum score. [c] Values before and after the slash are the as-is moisture contents of bran without and with autoclaving, respectively. The numbers before and in brackets indicate as-is moisture contents of finely (0.3 mm) and coarsely (0.8 mm) ground bran, respectively.

3.3 Effects of autoclaving and prehydration of bran on CSB and noodle quality of WWF

Water absorption and stability of dough from WWF supplemented with autoclaved bran hydrated to various moisture contents are presented in Table 3. There were significant effects of autoclaving on water absorption and dough stability. However, the autoclaved bran of different cultivars differed significantly in their effects on the water absorption and dough stability. Autoclaved Zhongmai 895 bran tended to give slightly lower water absorption and higher stability of WWFs than non-autoclaved bran except for water absorption of WWF dough with 60% bran hydration, whereas slightly higher water absorption and significantly lower stability were found in WWFs containing autoclaved Yumai 49 and Zhoumai 27 bran, regardless of bran particle size. The results are partly in agreement with Cai et al. (2015) who reported that mixograph and bread-baking dough mixing times of WWFs were reduced by autoclaved bran. The differences in water absorption and stability between as-is bran and bran hydrated to 60% were smaller for autoclaved bran than non-autoclaved bran. By autoclaving bran prior to hydration, the mixing time required for the development of dough with bran hydration was shortened, and saved energy consumed during CSB making. The reduced mixing time of WWFs containing autoclaved bran is possibly attributable to inactivation of enzymes or proteins that interfere with gluten development (Cai et al., 2015).

Stress relaxation, structure, and total score for CSB from WWFs containing autoclaved and hydrated wheat bran are summarized in Table 4. Stress relaxation and total scores were significantly enhanced by autoclaving fine bran with up to 40% bran hydration, whereas autoclaved fine bran with 60% hydration generally reduced the stress relaxation and total score for CSB. For autoclaved coarse bran, small changes in stress relaxation and total score for CSB were observed with up to 60% bran hydration. In addition, the autoclaved bran with 20% hydration contributed to a higher CSB score, regardless of bran particle size. Autoclaved fine bran produced a higher CSB structure with up to 40% bran hydration, but a lower structure at 60% hydration than non-autoclaved bran. Autoclaved coarse bran had inconsistent effects on CSB structure when bran was hydrated to 60% for all three varieties. This is generally consistent with the report of Cai et al. (2015), showing a positive effect of autoclaving of bran on WWF bread baking. It is probable that autoclaving the bran increases the soluble fiber content (Kabel et al., 2002), resulting in improved texture and structure of CSB. These results suggest that autoclaving treatment of bran could be an effective way to modify the physical characteristics of bran in order to improve the eating quality of CSB.

TABLE 5 Sensory evaluation parameters for noodles produced from whole grain wheat flours supplemented with finely (0.3 mm) and coarsely (0.8 mm) ground wheat bran with and without autoclaving (AC) and prehydration to various moisture levels[a]

Moisture	Fine bran						Coarse bran					
	Viscoelasticity (30[b])		Smoothness (15)		Total score (100)		Viscoelasticity (30)		Smoothness (15)		Total score (100)	
	No AC	With AC	No AC	With AC	No AC	With AC	No AC	With AC	No AC	With AC	No AC	With AC
Zhongmai 895												
6.2% (6.6%) /10.9% (14.0%)[c]	18.0b	21.0b	9.0a	9.8b	57.0b	61.0b	15.0c	21.0a	8.3b	9.0b	51.0b	59.8a
20%	18.1b	22.5a	7.5b	10.5a	54.0c	62.3a	15.0c	19.5b	7.5c	9.0b	49.7c	56.3c
40%	19.5a	22.5a	9.0a	9.8b	58.3a	62.5a	16.5b	19.5b	9.8a	9.8a	54.3a	58.0b
60%	16.5c	22.5a	6.8c	8.3c	54.0c	58.0c	18.0a	16.5c	6.8d	7.5c	54.0a	49.8d
Yumai 49												

(continued)

Moisture	Fine bran						Coarse bran					
	Viscoelasticity (30[b])		Smoothness (15)		Total score (100)		Viscoelasticity (30)		Smoothness (15)		Total score (100)	
	No AC	With AC	No AC	With AC	No AC	With AC	No AC	With AC	No AC	With AC	No AC	With AC
5.4% (7.8%) /14.3% (13.6%)	19.5a	19.5a	9.8a	9.0a	61.5b	57.5a, b	19.5a	19.5a	9.0a	8.3b	59.0a	59.5a
20%	19.5a	19.5a	9.0b	9.0a	63.3a	58.0a	18.0b	18.0b	8.3b	7.5c	56.3c	56.3c
40%	18.0b	18.0b	8.3c	9.0a	59.3c	57.0b	19.5a	19.5a	7.5c	9.0a	57.0b	57.8b
60%	18.0b	16.5c	7.5d	8.3b	58.8c	53.5c	16.5c	16.5 c	6.8d	6.8d	53.5d	51.0d
Zhoumai 27												
5.1% (7.9%) /15.0% (12.4%)	16.5b	19.5b	9.0a	9.0b	55.5b	58.0c	16.5c	18.0b	8.3c	8.3a	54.8c	56.8b
20%	16.5b	21.0a	8.3b	9.0b	56.8a	60.5b	18.0b	19.5a	9.0b	8.3a	57.5b	58.3a
40%	16.5b	21.0a	7.5c	9.0b	56.8a	61.5a	18.0b	19.5a	9.8a	8.3a	58.8a	58.3a
60%	18.0a	15.0c	6.8d	10.5a	56.3ab	55.8d	19.5a	15.0c	7.5d	7.5b	57.3b	45.8c

[a] Values with different letters within a column are significantly different ($p < 0.05$). [b] The numbers indicate maximum score. [c] Values before and after the slash are the as-is moisture contents of bran without and with autoclaving, respectively. The numbers before and in brackets indicate as-is moisture contents of finely (0.3 mm) and coarsely (0.8 mm) ground bran, respectively.

Viscoelasticity, smoothness, and total scores of noodles made from WWFs containing autoclaved and hydrated wheat bran are summarized in Table 5. The viscoelasticity and total scores of noodles from Zhongmai 895 and Zhoumai 27 WWFs supplemented with fine and coarse bran autoclaved and hydrated up to 40% were higher than non-autoclaved bran, whereas brans autoclaved and hydrated to 60% gave lower values than non-autoclaved bran, except for fine bran of Zhongmai 895. For Yumai 49, there were no significant differences in the viscoelasticity of noodles from WWFs supplemented with fine and coarse bran hydrated to various moisture levels between autoclaved and non-autoclaved bran except for fine bran hydrated to 60%. The total score for noodles made from WWFs supplemented with autoclaved bran compared with non-autoclaved bran were lower and not significant for fine or coarse bran hydrated to various moisture levels These results indicated that differential influences of bran autoclaving on viscoelasticity and total score of noodles from WWFs supplemented with brans of different particle size could be responsible for the lack of further improvement in eating quality of WWF noodles. The smoothness of noodles from WWFs with fine bran autoclaved and hydrated to various moisture were higher than those with non-autoclaved bran, whereas the smoothness of noodles from WWFs with coarse bran had inconsistent changes by autoclaving as bran hydration to various moisture. It is probable that the high pressure and temperature during autoclaving loosened up the structure of bran, making the bran fibers more hydrophilic and soluble in the finely ground bran (Kabel et al., 2002), leading to a smoother surface of the noodles.

Autoclaved fine and coarse bran had strongly negative effects on the color of noodle sheets prepared from WWFs, causing decreased brightness (L^*) of 3.20-9.98 and 3.91-7.59, and increased redness (a^*) of 0.89-2.72 and 0.75-2.12, and increased yellowness (b^*) of 1.07-5.36 and 0.46-3.39, respectively. The decrease in L^* value and increase in a^* value of noodle sheets prepared from WWFs containing autoclaved bran and subsequent hydration up to 60% was similar to that of WWFs containing non-autoclaved bran, whereas inconsistent changes in yellowness (b^*) were observed among the three genotypes (Table 6). The changes in color of noodle sheets were likely driven by the

Maillard browning that occurred during autoclaving (Kim, Chung, & Lim, 2014). This is strongly supported by Cai et al. (2015) who reported that autoclaving of wheat bran caused darkening of noodles.

TABLE 6 Color of noodle sheets prepared from whole grain wheat flours supplemented finely (0.3 mm) and coarsely (0.8 mm) ground wheat bran with and without autoclaving (AC) and prehydrated to various moisture levels[a]

Moisture	Fine bran						Coarse bran					
	L*		a*		b*		L*		a*		b*	
	No AC	With AC	No AC	With AC	No AC	With AC	No AC	With AC	No AC	With AC	No AC	With AC
Zhongmai 895												
6.2% (6.6%) /10.9% (14.0%)[b]	76.70a	71.07b	3.26c	5.00d	18.04b	20.61c	76.69b	70.90b	3.35b	4.91b	18.36a	19.01b
20%	76.38b	71.67a	3.39b	5.09c	18.12a	20.48c	77.61a	71.56a	3.09c	4.62c	17.59c	18.68c
40%	75.52c	70.42c	3.36b	5.22b	17.32c	21.38a	75.71c	70.39c	3.32b	4.89b	18.09b	19.03b
60%	73.92d	67.99d	3.98a	5.83a	17.20d	21.17b	73.98d	66.54d	3.70a	5.82a	16.50d	19.75a
Yumai 49												
5.4% (7.8%) /14.3% (13.6%)	77.73a	72.54a	3.61d	4.89c	17.34d	19.48c	77.57a	73.25a	3.76c	4.51c	17.35a	17.81c
20%	76.39b	72.43b	3.83c	4.91bc	18.95a	20.02b	77.05b	72.14c	3.82c	4.60c	17.20b	18.09b
40%	76.02c	71.93c	3.98b	4.97b	17.85c	19.53c	76.58c	72.67b	3.90b	4.74b	16.69c	17.82c
60%	72.68d	69.48d	4.66a	5.55a	18.42b	20.58a	73.53d	67.58d	4.62a	5.99a	16.66c	20.05a
Zhoumai 27												
5.1% (7.9%) /15.0% (12.4%)	78.74a	73.89a	3.03d	4.42c	17.45b	19.69b	80.37a	74.24a	2.83d	4.13c	15.44d	17.02c
20%	78.41b	72.86b	3.11c	4.64b	17.58a	19.24c	79.35b	73.45b	3.03c	4.35b	16.46c	17.80b
40%	78.23c	72.78b	3.18b	4.62b	16.16d	18.98d	78.22c	73.31c	3.16b	4.36b	16.55b	17.80b
60%	75.50d	65.52c	3.63a	6.35a	16.32c	21.68a	73.07d	66.83d	4.48a	5.89a	18.16a	20.10a

[a] Values with different letters within a column are significantly different ($p < 0.05$). [b] Values before and after the slash are the as-is moisture contents of bran without and with autoclaving, respectively. The numbers before and in brackets indicate as-is moisture contents of finely (0.3 mm) and coarsely (0.8 mm) ground bran, respectively.

4 CONCLUSIONS

In this study bran particle size, hydration and autoclaving significantly affected quality of doughs, destined to produce CSB and noodles made from WWFs. Farinograph water absorption and stability of WWFs supplemented with fine bran were superior to those supplemented with coarse bran. The WWFs with amount of coarse bran com-parable to fine bran exhibited inferior color and higher WRC, and produced CSB with better texture and crumb structure and noodles with inferior smoothness and appearance. Prehydration of bran decreased WWF dough water absorption and increased dough stability. WWFs containing bran hydrated to 40% produced CSB with the best texture and crumb structure, and highest total score. Bran hydration had negative effects on smoothness of noodles made with WWFs supplemented with fine bran. Autoclaving of bran significantly improved texture and crumb structure of CSB made from WWFs supplemented with bran hydrated to 40%. Autoclaved bran hydrated to 20% contributed to the highest total score for CSB. Auto-claving the bran had negligible effects on eating quality of noodles prepared from WWFs. Bran hydration and autoclaving led to a darker color of noodle sheets made from WWFs.

❖ ACKNOWLEDGEMENTS

We are grateful to Prof. R. A. McIntosh, Plant Breeding Institute, University of Sydney, for critical review of this manuscript. This work was supported by CAAS Science and Technology Innovation Program (CAAS-XTCX2016009), National Key R&D Program of China (2016YFD0100502), National Key Technology R&D Program of China (2014BAD01B05).

❖ REFERENCES

Adams, J. F., & Engstrom, A. (2000). Helping consumers achieve recommended intake of whole grain foods. *Journal of the American College of Nutrition*, *19*, 339S-344S. https://doi.org/10. 1080/07315724. 2000. 10718970

Aravind, N., Sissons, M. J., Egan, N. E., & Fellows, C. M. (2012). Effect of insoluble dietary fibre addition on technological, sensory, and structural properties of durum wheat spaghetti. *Food Chemistry*, *130*, 299-309. https://doi. org/10. 1016/j. foodchem. 2011. 07. 042

Cai, L., Choi, I., Hyun, J. N., Jeong, Y. K., & Baik, B. K. (2014). Influence of bran particle size on bread-baking quality of whole grain wheat flour and starch retrogradation. *Cereal Chemistry*, *91*, 65-71. https://doi. org/10. 1094/CCHEM-02-13-0026-R

Cai, L., Choi, I., Park, C. S., & Baik, B. K. (2015). Bran hydration and physical treatments improve the bread-baking quality of whole grain wheat flour. *Cereal Chemistry*, *92*, 557-564. https://doi. org/10. 1094/CCHEM-04-15-0064-R

Chen, J. S., Fei, M. J., Shi, C. L., Tian, J. C., Sun, C. L., Zhang, H., ⋯ Dong, H. X. (2011). Effect of particle size and addition level of wheat bran on quality of dry white Chinese noodles. *Journal of Cereal Science*, *53*, 217-224. https://doi. org/10. 1016/j. jcs. 2010. 12. 005

Chen, F., He, Z. H., Chen, D. S., Zhang, C. L., Zhang, Y., & Xia, X. C. (2007). Influence of puroindoline alleles on milling performance and qualities of Chinese noodles, steamed bread and pan bread in spring wheats. Journal of Cereal Science, 45, 59-66. https://doi. org/10. 1016/j. jcs. 2006. 06. 006

DeKock, S., Taylor, J., & Taylor, J. R. N. (1999).

Effect of heat treatment and particle size of different brans on loaf volume of brown bread. *LWT-Food Science Technology*, *32*, 349-356. https://doi. org/ 10. 1006/fstl. 1999. 0564

Gan, Z., Galliard, T., Ellis, P. R., Angold, R. E., & Vaughan, J. G. (1992). Effect of the outer bran layers on the loaf volume of wheat bread. *Journal of Cereal Science*, *15*, 151-163. https://doi. org/10. 1016/S0733-5210 (09) 80066-0

Gil, A., Ortega, R. M., & Maldonado, J. (2011). Wholegrain cereals and bread: a duet of the Mediterranean diet for the prevention of chronic diseases. *Public Health Nutrition*, *14*, 2316-2322. https://doi. org/10. 1017/S1368980011002576

Kabel, M. A., Carvalheiro, F., Garrote, G., Avgerinos, E., Koukios, E., Parajo, J. C., ⋯ Voragen, A. G. J. (2002). Hydrothermally treated xylan rich by-products yield different classes of xylo-oligo-saccharides. *Carbohydrate Polymers*, *50*, 47-56. https://doi. org/10. 1016/S0144-8617 (02) 00045-0

Katina, K., Juvonen, R., Laitila, A., Flander, L., Nordlund, E., Karilu-oto, S., ⋯ Poutanen, K. (2012). Fermented wheat bran as a functional ingredient in baking. *Cereal Chemistry*, *89*, 126-134. https://doi. org/10. 1094/CCHEM-08-11-0106

Katina, K., Salmenkallio-Marttila, M., Partanen, R., Forssell, P., & Autio, K. (2006). Effect of sourdough and enzymes on staling of high fibre wheat bread. *LWT-Food Science Technology*, *39*, 479-491. https://doi. org/10. 1016/j. lwt. 2005. 03. 013

Kim, S. M., Chung, H. J., & Lim, S. T. (2014). Effect of various heat treatments on rancidity and some bioactive compounds of rice bran. *Journal of Cereal Science*, *60*, 243-248. https://doi. org/ 10. 1016/j. jcs. 2014. 04. 001

Lai, C. S., Hosency, R. C., & Davis, A. B. (1989). Effects of wheat bran in breadmaking. *Cereal Chemistry*, *66*, 217-219.

Lang, R., & Jebb, S. A. (2003). Who consumes whole grains, and how much? Proceeding of the

Nutrition Society, 62, 123-127. https：//doi. org/ 10. 1079/PNS2002219

Liu, R. H. (2007). Whole grain phytochemicals and health. *Journal of Cereal Science*, 46, 207-219. https：//doi. org/10. 1016/j. jcs. 2007. 06. 010

Liu, W. J., Brennan, M., Serventi, L., & Brennan, C. (2017). Effect of wheat bran on dough rheology and final quality of Chinese steamed bread. *Cereal Chemistry*, 94, 581-587. https：// doi. org/10. 1094/CCHEM-09-16-0234-R

Nelles, E. M., Randall, P. G., & Taylor, J. R. N. (1998). Improvement of brown bread quality by prehydration treatment and cultivar selection of bran. *Cereal Chemistry*, 75, 536-540. https：// doi. org/10. 1094/CCHEM. 1998. 75. 4. 536

Niu, M., Hou, G. G., Lee, B., & Chen, Z. (2014). Effects of fine grinding of mill feeds on the quality attributes of reconstituted whole-wheat flour and its raw noodle products. *LWT-Food Science Technology*, 57, 58-64. https：//doi. org/10. 1016/ j. lwt. 2014. 01. 021

Niu, M., Hou, G. G., Wang, L., & Chen, Z. (2014). Effects of super-fine grinding on the quality characteristics of whole-wheat flour and its raw noodle product. *Journal of Cereal Science*, 60, 382-388. https：//doi. org/10. 1016/j. jcs. 2014. 05. 007

Salmenkallio-Marttila, M., Katina, K., & Autio, K. (2001). Effects of bran fermentation on quality and microstructure of high-fiber wheat bread. *Cereal Chemistry*, 78, 429-435. https：//doi. org/10. 1094/ CCHEM. 2001. 78. 4. 429

Sanz Penella, J. M., Collar, C., & Haros, M. (2008). Effect of wheat bran and enzyme addition on dough functional performance and phytic acid levels in bread. *Journal of Cereal Science*, 48, 715-721. https：//doi. org/10. 1016/j. jcs. 2008. 03. 006

Seguchi, M., Uozu, M., Oneda, H., Murayama, R., & Okusu, H. (2010). Effect of outer bran layers from germinated wheat grains on bread-making

properties. *Cereal Chemistry*, 87, 231-236. https：// doi. org/10. 1094/CCHEM-87-3-0231

Slavin, J. L., Jacobs, D., Marquart, L., & Wiemer, K. (2001). The role of whole grains in disease prevention. *Journal of the American Dietetic Association*, 101, 780-785. https：// doi. org/10. 1016/S0002-8223 (01) 00194-8

Slavin, J., Tucker, M., Harriman, C., & Jonnalagadda, S. S. (2016). Whole grains：definition, dietary recommendations, and health benefits. *Cereal Chemistry*, 93, 209-216.

Steglich, T., Bernin, D., Moldin, A., Topgaard, D., & Langton, M. (2015). Bran particle size influence on pasta microstructure, water distribution, and sensory properties. *Cereal Chemistry*, 92, 617-623. https：//doi. org/10. 1094/CCHEM-03-15-0038-R

Sudha, M. L., Ramasarma, P. R., & Venkateswara Rao, G. (2011). Wheat bran stabilization and its use in the preparation of high-fiber pasta. *International Journal of Food Science and Technology*, 17, 47-53. https：//doi. org/10. 1177/1082013210368463

Villeneuve, S., & Gelinas, P. (2007). Drying kinetics of whole durum wheat pasta according to temperature and relative humidity. *LWT-Food Science Technology*, 17, 47-53.

West, R., Seetharaman, K., & Duizer, L. M. (2013). Effect of drying profile and whole grain content on flavor and texture of pasta. *Journal of Cereal Science*, 58, 82-88. https：//doi. org/10. 1016/j. jcs. 2013. 03. 018

Zhang, D., & Moore, R. W. (1997). Effect of wheat bran particle size on dough rheological properties. *Journal of the Science of Food and Agriculture*, 74, 490-496. https：//doi. org/10. 1002/ (ISSN) 1097-0010

Zhang, Y., Nagamine, T., He, Z. H., Ge, X. X., Yoshida, H., & Peña, R. J. (2005). Variation in quality traits in common wheat as related to Chinese fresh white noodle quality. *Euphytica*, 141, 113-120. https：//doi. org/10. 1007/s10681-005-6335-0

Genome-wide association mapping of starch granule size distribution in common wheat

Jieyun Li[a], Awais Rasheed [a,b], Qi Guo [a], Yan Dong [a], Jindong Liu [a], Xianchun Xia [a], Yan Zhang[a,*], Zhonghu He [a,b,**]

[a] Institute of Crop Science, National Wheat Improvement Center, Chinese Academy of Agricultural Sciences (CAAS), 12 Zhongguancun South Street, Beijing 100081, China

[b] International Maize and Wheat Improvement Center (CIMMYT) China Office, c/o CAAS, 12 Zhongguancun South Street, Beijing 100081, China

* Corresponding author.

** Corresponding author. Institute of Crop Science, National Wheat Improvement Center, Chinese Academy of Agricultural Sciences (CAAS), 12 Zhongguancun South Street, Beijing 100081, China.

E-mail addresses: zhangyan07@caas.cn (Y. Zhang), zhhecaas@163.com (Z. He).

Abstract: Starch is a crucial component in wheat endosperm and plays an important role in processing quality. Endosperm of matured wheat grains contains two distinct starch granules (SG), referred to as larger A- and smaller B-granules. In the present study, 166 Chinese bread wheat cultivars planted in four environments were characterized for variation in SG size. A genome-wide association study (GWAS) using the 90 K SNP assay identified 23 loci for percentage volumes of A- and B-granules, and 25 loci for the ratio of A-/B-granules volumes, distributing on 15 chromosomes. Fifteen MTAs were associated with both the percentage volumes of A-, B-granules and the ratio of A-/B-granules volumes. MTAs *IWB34623* and *IWA3693* on chromosome 7A and *IWB22624* and *IWA4574* on chromosome 7B associated with the per-centage volumes of A- and B-granules consistently identified in multiple environments were considered to be stable. Linear regression analysis showed a significantly negative correlation of the number of favorable alleles with the percentage volumes of A-granules and a significantly positive correlation between the number of favorable alleles and the percentage volumes of B-granules, respectively. The loci identified in this study and associated markers could provide basis for manipulating SG size to obtain superior noodle quality in wheat.

Keywords: 90K SNP assay, GWAS, SG, *Triticum aestivum*

Abbreviations: GWAS, genome-wide association study; MAF, minor allele frequency; MAS, marker-assisted selection; MTA, marker-trait association; PIC, poly-morphism information content; QTL, quantitative trait loci; SG, starch granule; SGP, starch granule bound protein; SNP, single nucleotide polymorphism.

Published in Journal of Cereal Science, 2017, 77: 211-218.

1　Introduction

Starch is a main component of wheat endosperm, accounting for 65-75% of dry weight and serving as a multifunctional ingredient for the food industries (Zhang et al., 2010). Starch granules (SG) are usually classified into two classes according to size, viz. larger A-granules (10-35 μm) and smaller B-granules (<10 μm) (Stoddard, 2003). Bechtel et al. (1990) proposed a third type, C-granules (<5 μm), which was controversially classified as B-granules because of the difficulty in defining a boundary between them. A-granules are disk-shaped, accounting for 70% of the wheat endosperm starch by weight and less than 10% by number, whereas B-granules are spherical or irregular, making up 30% by weight and up to 90% by number (Peng et al., 1999).

Different SGs have different physical, chemical, and functional properties. Starch amylose (Peterson and Fulcher, 2001; Li et al., 2008), gelatinization and swelling properties (Soh et al., 2006; Balmeet et al., 2007), rapid visco-anaylzer parameters (Geera et al., 2006; Kim and Huber, 2010) and rheological properties (Tang et al., 2001; Barrera et al., 2013) were all affected by granules size. B-granules have higher water adsorption than A-granules due to a less ordered arrangement of the polysaccharide chains in the smaller granules (Chiotelli and Meste, 2002). The differences of SG affect the quality of many final products. Park et al. (2005) showed that the bread with 30% small granules and 70% large granules had the highest crumb grain score and peak fineness value through reconstituted flour. Guo et al. (2014) reported that the content of B-granules was positively related to color, elasticity and smoothness of raw white noodles.

SG size is largely controlled by genetic factors (Peterson and Fulcher, 2001; Li et al., 2008), and to some extent, it is also influenced by environment (Xiong et al., 2014). Stoddard (2003) reported that percentage volumes of A- and B-granules in wheat starch are controlled by major and minor genes, respectively. Boém and Mather (1999) found three QTL on barley chromosomes 2H, 4H and 5H affecting SG. Batey et al. (2001) identified a QTL for percentage volume of B-granules on wheat chromosome 4B. Igrejas et al. (2002) confirmed one QTL on wheat chromosome 1B explaining 34% of the variation in B-granules volume percentage and two QTL on chromosomes 7A and 4D explaining 20% and 27% of variation in A-granule volume percentage, respectively. Howard et al. (2011) mapped a major QTL on chromosome 4S in *Ae. peregrina*, accounting for 44.4% of the variation in percentage volume of B-granules. Feng et al. (2013) detected three QTL on chromosomes 1D, 4A and 7B for percentage volume of A-granules.

All instances of QTL mapping for SG size were performed in biparental mapping populations where only two allelic effects can be evaluated in any single population. In contrast, genome-wide as-sociation (GWAS) is an efficient method to discover significant as-sociation between genotypes and phenotypes in germplasm (Hamblin et al., 2011). Such studies have been carried out in agronomic traits in wheat and were successful in identifying loci determining genetic factors affecting complex traits (Li et al., 2015; Rasheed et al., 2015). To date, no GWAS on wheat SG size has been reported. In the present study, a GWAS analysis of SG size was performed using a panel of 166 Chinese bread wheat cultivars and 18, 207 mapped SNP markers present in the wheat 90 K iSelect chip. The aim was to identify loci associated with SG size and molecular markers for noodle quality improvement in bread wheat.

2　Materials and methods

2.1　*Plant materials*

A collection of 166 cultivars and advanced lines from the Yellow and Huai Valley Facultative Wheat Region was used for this study (Table S1). Field trials were conducted in randomized complete blocks with three replicates in Anyang (Henan province) and Suixi (Anhui province) during the 2012-2013 (recorded as Anyang, 2013 and Suixi, 2013) and 2013-2014 (recorded as Anyang, 2014 and Suixi, 2014) cropping seasons, providing data for four environments. Each plot contained three 2 m rows spaced 20 cm apart. Details on the experimental layout and agronomic practices were described earlier (Dong et al., 2016).

2. 2　Genotyping and quality control

Genomic DNA was extracted by a modified method according to Lagudah et al. (1991), then samples were sent to Capital Bio Corporation (Beijing, China; www. capitalbio. com) for genotyping with the high-density illumina 90 K infinium SNP array (Wang et al., 2014). PowerMarker V3. 2. 5 was used to calculate gene diversity, minor allele frequency (MAF) and polymorphism information content (PIC). Markers were removed if they either had no position information on chromosomes, exhibiting more than 30% missing values, showing MAF of less than 5% or containing more than 10% of heterozygosis.

2. 3　Milling

Grain hardness and water content were measured on 300-kernel samples with a Perten 4100 Single Kernel Characterization System (SKCS, Perten Instruments, Springfield, IL, USA) and a Near Infrared Reflectance (Foss, Hoganas, Sweden) instrument, respectively. Soft, medium and hard wheats were tempered to 14.5%, 15.5% and 16.5% moisture overnight. Selected 100 g samples were milled using a Brabender Quadrumat Junior Mill (Brabender Inc., Duisberg, Germany), following American Association of Cereal Chemists (AACC) approved method 26-50. The ground flour passed through a 60-mesh screen, cooled immediately and stored at -20 ℃ until analyzed.

2. 4　Starch extraction and SG size determination

Starch was extracted following Liu et al. (2007) with minor modifications, in which the tailings were centrifuged twice and all the starch was pooled. Dough was made with 6 g flour and 4 g distilled water, and allowed to stand for 10 min before being washed and kneaded with 60 ml water. The liquid component containing the starch was collected. The gluten component was washed, kneaded twice with 20 ml of distilled water until no more starch was extracted, and the liquid component was pooled with the earlier liquid extract. This starch suspension was filtered through a nylon cloth (75 μm openings) to remove impurities, centrifuged at 2 500 × g for 15 min; supernatant was discarded and the residue moved into a new centrifuge tube. Twenty ml water was added into the lower lighter-colored portions, and stirred to a uniform mixture. These steps were repeated until there were no gray-colored tailings on the top of the starch. The extract portions were combined, frozen, lyophilized and ground lightly with a mortar and pestle to pass a 100-mesh screen.

SG sizes were determined using a Laser Diffraction Particle Size Analyzer (HELOS and RODOS, Japan Laser Co, Ltd., Tokyo, Japan). Particles of 10. 0-35. 0 μm and < 10. 0 μm are defined as A -and B-granules, respectively (Peng et al., 1999), and those of > 35. 0 μm are considered to be aggregate fraction. Each sample was assayed twice, and further tested if differences between two repeats were more than 0. 5%. The contents of A- and B-granules were calculated according to the formula %A = 100 * A/ (A+B) and %B=100 * B/ (A+B), respectively.

2. 5　Statistical analysis

Analysis of variance and correlation coefficients among the four environments were performed using PROC GLM and CORR in SAS software version 9. 2 (SAS Institute Inc, Cary, NC, USA). Least square means were calculated for each parameter and used to test the significance of differences ($P < 0.001$) between samples. The broad-sense heritability (h^2) was calculated following Lin and Allaire (1977).

2.6 Population structure analysis

Population structure was described earlier (Dong et al., 2016). Briefly, Structure v.2.3.4 was used to estimate population structure based on 5624 SNP markers distributed across the entire genome using Bayesian cluster analysis (Pritchard et al., 2000). To ensure the sampling variance of inferred population structure, each K value was run repeatedly and independently. A range of K from 1 to 10 was based on admixture and correlated allele frequencies models. Each run was carried out with 10,000 replicates for the burn-in period and 100,000 replicates during analysis. Then the optimum value of K was chosen by the highest DK (Evanno et al., 2005).

2.7 Association analysis

SG size, genotype, population structure (Q-matrix) and relative kinship matrix (K-matrix) were implemented in TASSEL software version 5.0 using the mixed linear model (MLM) for association analysis. The significances of SNP markers were determined by a threshold P-value of 0.001 (Maccaferri et al., 2015; Dong et al., 2016), and MTAs with less than 5 cM distance were declared to be the same locus (Wang et al., 2014). The Quantile-Quantile plot showed the distribution of observed and expected P values. Manhattan plots were then used to map SNP markers significantly associated with SG size. Both the Quantile-Qantile and Manhattan plots were drawn in R Language (R version 3.1.2; http://www.r-project.org/).

3 Results

3.1 Phenotypic variation

Phenotypic differences and continuous variation in four environments for the percentage volumes of A-, B-granules and the ratio of A-/B-granules volumes are shown in Figs. S2, S3 and S4, respectively. The averaged percentage volumes of A-, B-granules and the ratio of A-/B-granules volumes among cultivars ranged from 70.4 to 83.7%, 16.3-29.6% and 2.4 to 5.1, respectively. Cultivars Nidera Baguette 10, Nidera Baguette 20, Huaimai 21, Luohan 2, Zhoumai 26 had the highest percentage volume of A-granules, whereas Xinmai 19, Xinmai 9408, Wanmai 52, Taishan 1, Jinhe 9123 had the highest percentage volume of B-granules, and Nidera Baguette 10, Nidera Baguette 20, Huaimai 21, Luohan 2, Zhoumai 26 had the highest ratio of A-/B-granules volumes (Table S1). Pearson correlation coefficients of the percentage volumes of A-, B-granules and the ratio of A-/B-granules volumes were significant across all environments, with the ranges of 0.52-0.70, 0.52-0.70 and 0.50-0.70, respectively (Tables S2, S3). Analysis of variance showed significant differences for the percentage volumes of A-, B-granules and the ratio of A-/B-granules volumes ($P < 0.000\ 1$) among genotypes, environments and G E interactions (Table 1). The h^2 of the percentage volumes of A-, B-granules and the ratio of A-/B-granules volumes were 0.79, 0.79 and 0.78, respectively.

3.2 Analysis of SNP markers and population structure

Among the 81,587 SNP markers in the 90 K array, 40,267 (49.4%) were mapped to individual chromosomes. Finally, 18,207 (22.3%) markers were selected after strict quality control in our association panel and were integrated into a linkage map involving all 21 wheat chromosomes. These markers covered genetic distance of 3,700 cM, with an average density of one marker per 0.2 cM. The marker density was much lower for the D genome (254.4 markers per chromosome) compared to the A (1,007.7 markers per chromosome) and B (1,338.9 markers per chromosome) genomes. Among D genome chromosomes, the markers on chromosome 4D were the lowest (50). The average SNP diversity (H) and PIC values were 0.35 and 0.29, respectively (Table S4). Here, all 166 cultivars were divided into three groups (Fig. S1). The population 1 contains 62 cultivars primarily come from Shandong province, and the population 2 has 54 cultivars mainly derived from Henan and Anhui provinces, while the population 3 comprises 50 cultivars primarily from Henan province.

3.3 *Marker-trait associations (MTAs)*

Considering the criteria (P-value $<$ 0.001), 23 loci were associated with the percentage volumes of A- and B-granules and 25 were associated with the ratio of A-/B-granules volumes, distributing on 15 different chromosomes (Tables 2 and 3). Twenty-four loci were associated with the aggregates (Table S5). The maximum number of MTAs was for chromosome 7A, followed by 7B and no MTA was detected on chromosomes 1A, 1B, 1D, 3B, 5B and 6D (Figs. 1 and 2). MTAs consistently identified in more than one environment were considered to be stable. These were located on chromosomes 7AS, 7AL, and 7BS, explaining 9.3-12.1% of the phenotypic variation (R^2) for percentage volumes of A-and B-granules (Table 2), and 8.5-9.2% for the ratio of A-/B-granules volumes (Table 3). There were multiple SNP markers associated with percentage volumes of A-and B-granules on chromosomes 7AS (128 cM) and 7AL (136 cM), identified in four environments. Fifteen loci were associated with both the percentage volumes of A-, B-granules and the ratio of A-/B-granules volumes, viz. *IWB8363*, *IWB71648*, *IWB69831*, *IWB35441*, *IWB30485*, *IWB72476*, *IWB35259*, *IWB14529*, *IWB8050*, *IWB22035*, *IWB60249*, *IWB34623*, *IWB3693*, *IWB22624*, and *IWB4574*. QQ plots for the distribution of expected and observed P values of associated SNP markers were shown in Fig. S5.

3.4 *Effects of favorable alleles on SG size*

Alleles with positive effects to increase the percentage volume of B-granules and those with negative effects on A-granules were considered to be favorable, because a higher content of B-granules is favorable to the water adsorption and noodle quality. There was a significantly negative correlation ($r = -0.98$, $P <$ 0.001) between percentage volumes of A-granules and the number of favorable alleles, and a significantly positive correlation ($r = 0.98$, $P < 0.001$) between percentage volumes of B-granules and the number of favorable alleles (Fig. S7). The number of favorable alleles present in a cultivar ranged from 1 to 17 for the percentage volume of A-granules, and 5 to 20 for the percentage volume of B-granules (Table S1).

4 Discussion

4.1 *Marker-trait associations for SG size*

SG size is controlled mainly by genetic factors (Peterson and Fulcher, 2001; Li et al., 2008) and has high heritability. Therefore, identification of stable QTL associated with SG size should be helpful in selecting for SG size in wheat genetic improvement. Previously, genetic analysis of SG size was performed primarily by QTL mapping which could not identify broad allelic variation. In this study, we used a GWAS approach by assaying 18, 207 SNP markers in a panel of 166 bread wheat cultivars to identify chromosomal regions associated with SG size. This is the first study on identification of SG size loci by GWAS.

Markers *IWB34623* (7AS, 128 cM), *IWA3693* (7AL, 136 cM), *IWB22624* (7BS, 3 cM) and *IWA4574* (7BS, 72 cM) were significantly associated with percentage volumes of A-and B-granules; these QTL were consistently identified in more than one environment and reported for the first time in the present study, being considered as stable and new. Among them, *IWB34623* (10.7%) and *IWA3693* (12.1%) loci explained the highest phenotypic variations across four environments. MTAs at *IWB22624* and *IWA4574* positions were identified in two environments, explaining 9.3% and 9.4% of the phenotypic variation for percentage volumes of A-and B-granules, respectively. *IWA3693* (7AL, 136 cM) and *IWB22624* (7BS, 3 cM) were significantly associated with the ratio of A-/B-granules volumes in two environments. *IWB35259* (3DL, 143 cM), *IWB5734* (5DS, 103 cM) and *IWB51603* (6BL, 105 cM) were significantly associated with aggregate fraction; these QTL were stable and new. Notably,

IWB34623 （7AS，128 cM） and *IWA3693* （7AL，136 cM） simultaneously affected both the percentage volumes of A-，B-granules，aggregate and the ratio of A-/B-granules volumes.

Igrejas et al. （2002） reported three QTL related to SG diameter $<10~\mu m$ on chromosome 4D and 10-15 μm on chromosomes 1B and 7A，with QTL on chromosome 7A accounting for 27.0% of the variation in A-granules percentage volume. In our study，MTAs for percentage volumes of B- and A-granules were detected on chromosomes 4D and 7A，respectively，in agreement with Igrejas et al. （2002） study. Feng et al. （2013） found QTLs on chromosomes 1DL，4AL and 7BL；all of them had minor effects on percentage volume of A-granules. The QTL on chromosome 7BL explained 5.2% of the phenotypic variation，while one locus on chromosome 7BL accounted for 8.2% of the phenotypic variation in the present study. Previously，QTL mapping on SG was re-ported using low-density SSR markers （Feng et al.，2013），whereas GWAS was performed in the present study using high-density SNP markers. It is difficult to make a comparison of QTL locations between these studies because of different kinds of markers used.

Table 1　Analysis of variance of A-，B-granule volume percentages and the ratio of A-/B-granules volumes in 166 cultivars grown in four environments

Trait	Mean squares				
	Genotypes （$df=165$）	Environment （$df=3$）	Replicate （$df=2$）	G×E （$df=495$）	Error （$df=635$）
A- or B-granules	42.41***	1 178.02***	0.90	6.01***	4.10
A-/B-granules	1.76***	48.64***	0.01	0.27***	0.19

***，Significant at $P<0.0001$.

Table 2　SNP markers significantly associated with percentage volume of A- or B-granules in the association panel

SNP ID	Environment	Chromosome	Position[a]	SNP[b]	MAF[c]	P-value	R^2 （%）
IWB8363	Anyang 2014	2AS	82	G/A	0.35	5.29E-04	7.6
IWB75252	Suixi 2013	2AL	184	G/A	0.07	8.71E-04	7.6
IWB71648	Anyang 2013	2BL	107	G/A	0.39	8.25E-04	7.1
IWB69831	Suixi 2013	2DS	8	A/C	0.06	8.80E-04	7.3
IWB35441	Suixi 2014	2DS	47	G/A	0.16	1.13E-04	9.8
IWB30485	Anyang 2013	3AL	130	A/G	0.47	9.62E-04	6.9
IWB72476	Anyang 2014	3AL	154	A/G	0.11	9.79E-04	6.8
IWB35259	Anyang 2014	3DL	143	A/G	0.17	9.64E-04	6.9
IWB58408	Anyang 2014	4AL	60	A/G	0.11	9.90E-04	6.8
IWB14529	Suixi 2013	4BL	74	A/C	0.23	5.55E-05	10.5
IWB8050	Anyang 2014	4DS	74	A/G	0.12	3.56E-04	8.2
IWB22035	Suixi 2014	5AL	62	A/C	0.36	7.31E-04	7.3
IWB48748	Anyang 2013	5DL	143	G/A	0.06	6.70E-04	7.3
IWB60249	Anyang 2013	6AS	13	A/C	0.18	6.17E-04	7.6
IWB59637	Suixi 2013	6BS	34	A/C	0.40	8.87E-04	7.2
IWA3831	Anyang 2013	7AS	115	G/A	0.43	4.66E-04	7.8
IWB34623	Anyang 2013，Anyang 2014，Suixi 2013，Suixi 2014	7AS	128	A/G	0.35	4.45E-05	10.7
IWA3693	Anyang 2013，Anyang 2014，Suixi 2013，Suixi 2014	7AL	136	G/A	0.11	1.87E-05	12.1
IWB1800	Suixi 2013	7AL	210	C/A	0.10	6.09E-04	7.4

(continued)

SNP ID	Environment	Chromosome	Position[a]	SNP[b]	MAF[c]	P-value	R^2 (%)
IWB22624	Anyang 2013,	7BS	3	$\underline{G}$/A	0.48	1.65E-04	9.3
	Suixi 2013						
IWA4574	Suixi 2014,	7BS	72	$\underline{G}$/A	0.08	1.50E-04	9.4
	Anyang 2014						
IWB45006	Suixi 2013	7BL	158	G/$\underline{A}$	0.45	5.38E-04	8.2
IWB10645	Suixi 2014	7DL	147	G/$\underline{A}$	0.07	1.43E-04	9.3

[a] Position from the wheat 90 K SNP consensus map (Wang et al., 2014).

[b] Favorable alleles for increasing noodle quality are underlined.

[c] Minor allele frequency.

Table 3   SNP markers significantly associated with the ratio of A-/B-granules volumes in the association panel

SNP ID	Environment	Chromosome	Position[a]	SNP[b]	MAF[c]	P-value	R^2 (%)
IWA6477	Anyang 2014	2AS	82	$\underline{A}$/C	0.45	7.98E-04	7.1
IWB7897	Anyang 2014	2AL	102	$\underline{A}$/G	0.25	7.20E-04	7.2
IWB55858	Suixi 2013	2AL	140	$\underline{A}$/G	0.32	5.47E-04	7.7
IWB71648	Anyang 2013	2BL	107	$\underline{A}$/G	0.39	4.81E-04	7.8
IWB32143	Anyang 2014	2BL	157	$\underline{A}$/G	0.07	3.84E-04	8.2
IWB69831	Suixi 2013	2DS	8	$\underline{A}$/G	0.06	3.08E-04	8.6
IWB35441	Suixi 2014	2DS	47	$\underline{A}$/C	0.10	7.19E-04	7.4
IWB30485	Anyang 2013	3AL	130	$\underline{A}$/G	0.47	9.42E-04	6.9
IWB72476	Anyang 2014	3AL	154	$\underline{A}$/G	0.11	6.50E-04	7.3
IWB35259	Anyang 2014	3DL	143	$\underline{A}$/G	0.17	4.19E-04	7.9
IWB48433	Anyang 2013,	4AS	62	$\underline{A}$/G	0.10	6.91E-04	7.5
	Anyang 2014						
IWB14529	Suixi 2013	4BL	74	$\underline{A}$/C	0.25	1.21E-04	9.5
IWB8050	Anyang 2014	4DL	74	$\underline{A}$/G	0.12	1.19E-04	9.5
IWA5381	Anyang 2014	4DS	81	$\underline{A}$/G	0.22	7.21E-04	7.3
IWB22035	Suixi 2014	5AL	62	$\underline{A}$/C	0.36	7.86E-04	7.2
IWB43809	Anyang 2013	6AS	0	$\underline{A}$/G	0.17	4.26E-04	8.1
IWB60249	Anyang 2013	6AS	13	$\underline{A}$/C	0.18	1.90E-04	9.0
IWB5710	Anyang 2014	6AL	99	$\underline{A}$/G	0.06	7.79E-04	7.1
IWA233	Suixi 2013	6AS	66	$\underline{A}$/G	0.16	7.74E-04	7.1
IWB56290	Suixi 2014	6BL	123	A/$\underline{G}$	0.07	2.55E-04	8.7
IWB34623	Anyang 2013,	7AS	128	$\underline{A}$/G	0.35	6.93E-05	10.1
	Suixi 2013						
IWA3693	Anyang 2014,	7AL	136	$\underline{G}$/A	0.11	1.61E-04	9.2
	Suixi 2014						
IWB22624	Anyang 2013,	7BS	3	$\underline{A}$/G	0.48	2.83E-04	8.5
	Suixi 2013						
IWA4574	Suixi 2014	7BS	72	A/$\underline{G}$	0.08	9.85E-04	7.0
IWB41869	Anyang 2013	7BL	151	A/$\underline{G}$	0.06	7.27E-04	7.4

[a] Position from the wheat 90 K SNP consensus map (Wang et al., 2014).

[b] Favorable alleles for reducing the ratio of A-/B-granules volumes are underlined.

[c] Minor allele frequency.

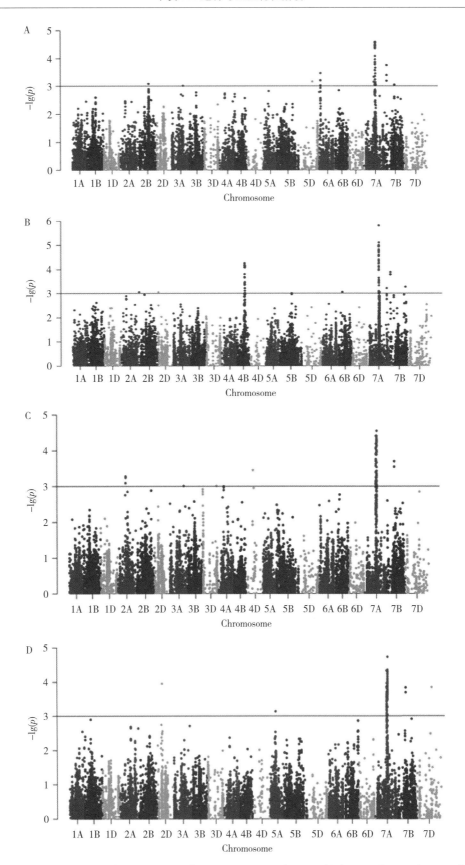

Fig. 1. Manhattan plots from GWAS for percentage volume of A- or B-granules in four environments. The horizontal line depicts the 1E-03 threshold for significant association. A，Anyang 2013；B，Suixi 2013；C，Anyang 2014；D，Suixi 2014.

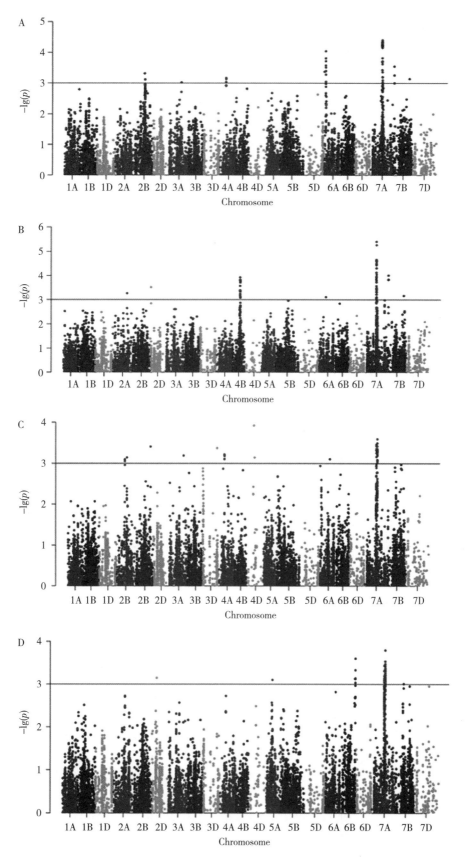

Fig. 2. Manhattan plots from GWAS for the ratio of A-/B-granules volumes in four environments. The horizontal line depicts the 1E-03 threshold for significant association. A Anyang 2013; B, Suixi 2013; C, Anyang 2014; D, Suixi 2014.

4.2 *Putative candidate genes*

Peng et al. (2000) reported two starch granule bound proteins, SGP-145 and SGP-140 that were different variants of a wheat starch branching enzyme. These were associated with SG size in developing and mature wheat kernels (Rahman et al., 1999). In the present study, markers *IWB45006* and *IWB71648* were also on chromosomes 7BL and 2BL may influence SG size. Granule bound starch synthase (*GBSS*) present in both waxy and non-waxy wheat cultivars is a key enzyme in improving amylose synthesis and SG development (Zhang et al., 2010; Yan et al., 2015). The *GBSS* genes were located on chromosomes 4AL, 7AS and 7DS (Murai et al., 1999). In the present study, we identified MTAs on chromosomes 4AL (1) and 7AS (2), but none was at the same position as *GBSS* gene based on the physical map from IWGSC (http://www.wheatgenome.org/), indicating that these may be new QTL. Moreover, marker *IWB5734* at *Pinb-D1* locus associated with aggregate was identified in multiple environments, explaining the highest phenotypic variation in the present study. Previous studies found that grain hardness was correlated with starch and protein properties. A highly significant correlation was found between starch granule surface protein and SG size (Oda et al., 1992). However, the precise molecular basis of the effect of hardness on SG size requires further study.

4.3 *Potential implications for wheat breeding*

Markers for MTAs explaining high phenotypic variation, suchas *IWB14529*, *IWB34623* and *IWA3693* identified in this study could be used for improvement of the percentage volume of B-granules and the ratio of A-/B-granules volumes in marker-assisted selection (MAS). The cultivars Xinmai 19, Xinmai 9408, Wanmai 52, Taishan 1, Jinhe 9123, Yumai 34 and Bainong 64 had higher percentage volume of B-granules (over 27%), which can be used as parents in breeding programs. There are also multiple MTAs associated with percentage volume of B-granules on chromosomes 7AS and 7BS, implying that these regions may provide novel alleles affecting SG size. Favorable alleles identified in this study will facilitate selection of parents for wheat breeding aimed at increasing noodle quality. B-granules have higher affinity for water adsorption and were positively related to color, elasticity and smoothness of raw white noodles (Chiotelli and Meste, 2002; Guo et al., 2014). By increasing the percentage volume of B-granules and decreasing the percentage volume of A-granules should improve the water adsorption and quality of noodle.

5 Conclusions

Twenty-three loci for the percentage volumes of A- and B-granules, and 25 for the ratio of A-/B-granules volumes were identified through GWAS. The loci on chromosomes 7AS, 7AL and 7BS associated with the percentage volumes of A-, B-granules and the ratio of A-/B-granules volumes are new stable QTL. The cultivars Xinmai 19, Xinmai 9408, Wanmai 52, Taishan 1 and Jinhe 9123 with a higher percentage volume of B-granules can be used as parents in wheat breeding. These results provide a basis for fine mapping, gene discovery and MAS of SG size in wheat.

❖ Acknowledgements

We are grateful to Prof. R. A. McIntosh, Plant Breeding Institute, University of Sydney, for critical review of this manuscript. This work was supported by National Basic Research Program (2014CB138105), the National Natural Science Foundation of China (31461143021), Gene Transformation Projects (2016ZX08009003-004), and National Key Technology R&D Program of China (2014BAD01B05).

❖ References

Barrera, G. N., Bustos, M. C., Iturriaga, L., Flores,

S. K. , Leon, A. E. , Ribotta, P. D. , 2013. Effect of damaged starch on the rheological properties of wheat starch sus-pensions. J. Food Eng. 116, 233-239.

Balmeet, S. G. , Narpinder, S. , Saxena, S. K. , 2007. The impact of starch properties on noodle making properties of Indian wheat flours. Int. J. Food Prop. 7, 59-74.

Batey, I. L. , Hayden, M. J. , Cai, S. , Sharp, P. J. , Cornish, G. B. , Morell, M. K. , Appels, R. , 2001. Genetic mapping of commercially significant starch characteristics in wheat crosses. Aust. J. Agric. Res. 52, 1287-1296.

Bechtel, D. B. , Zayas, I. , Kaleikau, L. , Pomeranz, Y. , 1990. Size-distribution of wheat starch granules during endosperm development. Cereal Chem. 67, 59-63.

Borém, A. , Mather, D. E. , 1999. Mapping quantitative trait loci for starch granule traits in barley. J. Cereal Sci. 29, 153-160.

Chiotelli, E. , Meste, M. L. , 2002. Effect of small and large wheat starch granules on thermo mechanical behavior of starch. Cereal Chem. 79, 286-293.

Dong, Y. , Zhang, Y. , Rasheed, A. , Xiao, Y. G. , Fu, L. P. , Yan, J. , Liu, J. D. , Wen, W. E. , Zhang, Y. , Jing, R. L. , Xia, X. C. , He, Z. H. , 2016. Genome-wide association for stem water soluble carbohydrates in bread wheat. PLoS One 11, e0164293.

Evanno, G. , Regnaut, S. , Goudet, J. , 2005. Detecting the number of clusters of in-dividuals using the software STRUCTURE: a simulation study. Mol. Ecol. 14, 2611-2620.

Feng, N. , He, Z. H. , Zhang, Y. , Xia, X. C. , Zhang, Y. , 2013. QTL mapping of starch granule size in common wheat using recombinant inbred derived from a PH82-2/ Ningxiang 188 cross. Crop J. 1, 166-171.

Geera, B. P. , Nelson, J. E. , Souza, E. , Huber, K. C. , 2006. Composition and properties of A-and B-type starch granules of wild-type, partial waxy, and waxy soft wheat. Cereal Chem. 83, 551-557.

Guo, Q. , He, Z. H. , Xia, X. C. , Qu, Y. Y. , Zhang, Y. , 2014. Effects of wheat starch granule size distribution on qualities of Chinese steamed bread and raw white noodles. Cereal Chem. 91, 623-630.

Hamblin, M. T. , Buckler, E. S. , Jannink, J. K. , 2011. Population genetics of genomics-based crop improvement methods. Trends Genet. 27, 98-106.

Howard, T. , Rejab, N. A. , Grif fiths, S. , Leigh, F. , Leverington-Waite, M. , Simmonds, J. , Uauy, C. ,

Trafford, K. , 2011. Identification of a major QTL controlling the content of B-type starch granules in Aegilops. J. Exp. Bot. 62, 2217-2228.

Igrejas, G. , Faucher, B. , Bertrand, D. , Guilbert, D. , Leroy, P. , Branlard, G. , 2002. Genetic analysis of the size of endosperm starch granules in a mapped segregating wheat population. J. Cereal Sci. 35, 103-107.

Kim, H. S. , Huber, K. C. , 2010. Physicochemical properties and amylopectin fine structures of A-and B-type granules of waxy and normal soft wheat starch. J. Cereal Sci. 51, 256-264.

Lagudah, E. S. , Appels, R. , Brown, A. H. D. , 1991. The Nor-D3 locus of Triticum tauschii: natural variation and linkage to markers in chromosome 5. Genome 34, 387-395.

Li, W. Y. , Yan, S. H. , Yin, Y. P. , Li, Y. , Liang, T. B. , Gu, F. , Dai, Z. M. , Wang, Z. L. , 2008. Comparison of starch granule size distribution between hard and soft wheat cultivars in eastern China. J. Agric. Sci. 7, 907-914.

Li, W. Y. , Zhang, B. , Li, R. Z. , Chang, X. P. , Jing, R. L. , 2015. Favorable alleles for stem water-soluble carbohydrates identified by association analysis contribute to grain weight under drought stress conditions in wheat. PLoS One 10, e0119438. Lin, C. Y. , Allaire, F. R. , 1977. Heritability of a linear combination of traits. Theor. Appl. Genet. 51, 1-3.

Liu, Q. , Gu, Z. , Donner, E. , Tetlow, I. , Emes, M. , 2007. Investigation of digestibility in vitro and physicochemical properties of A-and B-type starch from soft and hard wheat flour. Cereal Chem. 84, 15-21.

Maccaferri, M. , Zhang, J. L. , Bulli, P. , Abate, Z. , Chao, S. , Cantu, D. , Bossolini, E. , Chen, X. M. , Pumphrey, M. , Dubcovsky, J. , 2015. A genome-wide association study of resistance to stripe rust (Puccinia striiformis f. sp. tritici) in a worldwide collection of hexaploid spring wheat (Triticum aestivum L.). G3-Genes Genomes Genet. 5, 449-465.

Murai, J. , Taira, T. , Ohta, D. , 1999. Isolation and characterization of the three Waxy genes granule-bound starch synthase in hexaploid wheat. Gene 234, 71-79.

Oda, S. , Komae, K. , Yasui, T. , 1992. Relation between starch granule protein and endosperm softness in Japanese wheat (Triticum aestivum L.) cultivars. Jpn. J. Breed. 42, 161-165.

Park, S. H. , Chung, O. K. , Seib, P. A. , 2005. Effects of varying weight ratios of large and small wheat starch granules on experimental straight-dough bread. Cereal Chem. 82, 166-172.

Peng, M., Gao, M., Abdel-Aal, E. S. M., Hucl, P., Chibbar, R. N., 1999. Separation and characterization of A- and B-type starch granules in wheat endosperm. Cereal Chem. 76, 375-379.

Peng, M. S., Gao, M., Baga, M., Hucl, P., Chibbar, R. N., 2000. Starch-branching enzymes preferentially associated with A-type starch granules in wheat endo-sperm. Plant Physiol. 124, 265-272.

Peterson, D. G., Fulcher, R. G., 2001. Variation in Minnesota HRS wheat: starch granule size distribution. Food Res. Int. 34, 357-363.

Pritchard, J. K., Stephens, M., Donnelly, P., 2000. Inference of population structure using multilocus genotype data. Genetics 155, 945-959.

Rahman, S., Li, Z., Abrahams, S., Abbott, D., Appels, R., Morell, M. K., 1999. Charac-teristics of a gene encoding wheat endosperm starch branching enzymes-I. Theor. Appl. Genet. 98, 156-163.

Rasheed, A., Ain, Q., Anwar, A., Mahmood, T., Mahmood, T., Xia, X. C., He, Z. H., 2015. Genome-wide association for grain yield under rainfed conditions in historical wheat cultivars from Pakistan. Front. Plant Sci. 6, 743.

Soh, H. N., Sissons, M. J., Turner, M. A., 2006. Effect of starch granule size distribution and elevated amylose content on durum dough rheology and spaghetti cooking quality. Cereal Chem. 83, 513-519.

Stoddard, F. L., 2003. Genetics of starch granule size distribution in tetraploid and hexaploid wheat. Aust. J. Agric. Res. 54, 637-648.

Tang, H. J., Ando, H., Watanabe, K., Takeda, Y., Mitsunaga, T., 2001. Physicochemical properties and structure of large, medium and small granule starches in frac-tions of normal barley endosperm. J. Carbohydr. Res. 330, 241-248.

Wang, S. C., Wong, D. B., Forrest, K., Allen, A., Chao, S. M., Huang, B. E., 2014. Charac-terization of polyploid wheat genomic diversity using a high-density 90, 000 single nucleotide polymorphism array. Plant Biotechnol. J. 12, 787-796.

Xiong, F., Yu, X., Zhou, L., Zhang, J., Jin, Y. P., Li, D. L., Wang, Z., 2014. Effect of nitrogen fertilizer on distribution of starch granules in different regions of wheat endosperm. Crop J. 2, 46-54.

Yan, Y. M., Cao, M., Chen, G. X., Wang, C., Zhen, S. M., Li, X. H., Zhang, W. Y., Zeller, F. J., Hsam, S. L. K., Hu, Y. K., 2015. 1S (1B) chromosome substitution in Chinese Spring wheat promotes starch granule development and starch biosynthesis. Crop Pasture Sci. 66, 894-903.

Zhang, C. H., Jiang, D., Liu, F. L., Cai, J., Dai, T. B., Cao, W. X., 2010. Starch granules size distribution in superior and inferior grains of wheat is related to enzyme activities and their gene expressions during grain filling. J. Cereal Sci. 51, 226-233.

A genome-wide association study reveals a rich genetic architecture of flour color-related traits in bread wheat

Shengnan Zhai[1,2], Jindong Liu [2], Dengan Xu [2], Weie Wen [2], Jun Yan [2,3],
Pingzhi Zhang [4], Yingxiu Wan [4], Shuanghe Cao [2], Yuanfeng Hao [2],
Xianchun Xia [2], Wujun Ma [5] and Zhonghu He [2,6]*

[1] Crop Research Institute, National Engineering Laboratory for Wheat and Maize, Key Laboratory of Wheat Biology and Genetic Improvement in the Northern Yellow-Huai Rivers Valley of Ministry of Agriculture, Shandong Academy of Agricultural Sciences, Jinan, China, [2] National Wheat Improvement Center, Institute of Crop Sciences, Chinese Academy of Agricultural Sciences, Beijing, China, [3] Institute of Cotton Research, Chinese Academy of Agricultural Sciences, Anyang, China, [4] Crop Research Institute, Anhui Academy of Agricultural Sciences, Hefei, China, [5] School of Veterinary and Life Sciences, Murdoch University and Australian Export Grains Innovation Centre, Perth, WA, Australia, [6] International Maize and Wheat Improvement Center, Beijing, China

* Correspondence: zhhecaas@163.com

Abstract: Flour color-related traits, including brightness (L^*), redness (a^*), yellowness (b^*) and yellow pigment content (YPC), are very important for end-use quality of wheat. Uncovering the genetic architecture of these traits is necessary for improving wheat quality by marker-assisted selection (MAS). In the present study, a genome-wide association study (GWAS) was performed on a collection of 166 bread wheat cultivars to better understand the genetic architecture of flour color-related traits using the wheat 90 and 660 K SNP arrays, and 10 allele-specific markers for known genes influencing these traits. Fifteen, 28, 25, and 32 marker-trait associations (MTAs) for L^*, a^*, b^*, and YPC, respectively, were detected, explaining 6.5-20.9% phenotypic variation. Seventy-eight loci were consistent across all four environments. Compared with previous studies, *Psy-A1*, *Psy-B1*, *Pinb-D1*, and the 1B · 1R translocation controlling flour color-related traits were confirmed, and four loci were novel. Two and 11 loci explained much more phenotypic variation of a^* and YPC than phytoene synthase 1 gene (*Psy1*), respectively. Sixteen candidate genes were predicted based on biochemical information and bioinformatics analyses, mainly related to carotenoid biosynthesis and degradation, terpenoid backbone biosynthesis and glycolysis/gluconeogenesis. The results largely enrich our knowledge of the genetic basis of flour color-related traits in bread wheat and provide valuable markers for wheat quality improvement. The study also indicated that GWAS was a powerful strategy for dissecting flour color-related traits and identifying candidate genes based on diverse genotypes and high-throughput SNP arrays.

Keywords: brightness (L^*) candidate gene, GWAS, redness (a^*), yellowness (b^*), yellow pigment content (YPC)

Published in Frontiers in Plant Science, 2018, 9: 1136.

Abbreviations: a*, Redness; b*, Yellowness; *BCH2*, Carotenoid β-ring hydroxylase 2; BLUP, Best linear unbiased predictor; *DLD*, Dihydrolipoyl dehydrogenase; *DXR*, 1-deoxy-D-xylulose 5-phosphate reductoisomerase; *FPPS2*, Farnesyl pyrophosphate synthase 2; GWAS, Genome-wide association study; h^2, Broad-sense heritability; *IPPS*, Isopentenyl pyrophosphate isomerase; IWGSC, International Wheat Genome Sequencing Consortium; K, Genetic relatedness; L*, Brightness; *LBR*, Leghemoglobin reductase; *LCYB*, Lycopene β-cyclase; LD, Linkage disequilibrium; LOESS, Locally weighed polynomial regression; *LOX2*, Lipoxygenase 2; MAF, Minor allele frequency; MAS, Marker-assisted selection; *MDPS*, 4-hydroxy-3-methylbut-2-en-1-yl diphosphate synthase; MLM, Mixed linear model; *MK*, Mevalonate kinase; MTAs, Marker-trait associations; *NCED4*, 9-cis-epoxycarotenoid dioxygenase 4; *PMM*, Phosphomannomutase; *PO*, Premnaspirodiene oxygenase; *POD*, Peroxidase; *Psy1*, Phytoene synthase 1; Q, Population structure; Q-Q, quantile-quantile; QTL, Quantitative trait loci; *r*, Correlation coefficient; R^2, Percentage of variation explained by each locus; YPC, Yellow pigment content; *ZISO*, Cis-zeta-carotene isomerase.

INTRODUCTION

Bread wheat (*Triticum aestivum* L.) is among the most important food crops and is one of the most traded commodities in world markets (Curtis and Halford, 2014). Developing cultivars with appropriate end-use quality is the primary objective of all wheat breeding programs. Flour color plays a significant role in the end-use quality of wheat, particularly for Asian noodles and steamed bread, since it affects consumer acceptance, market value and human nutrition (Zhai et al., 2016a). Color measurements were expressed as tristimulus parameters (L*, a*, and b*). L* is a measure of flour brightness, ranging from 0 (black) to 100 (white); a* measures redness when positive or greenness when negative; and b* describes the yellow-blue color value, and is positive for yellowness and negative for blueness (Hutchings, 1999). Yellow pigment content (YPC) is the most important determinant of flour yellowness caused mainly by accumulation of carotenoids in the grain (Mares and Campbell, 2001). It is also a very important quality criterion for pasta made from durum wheat. Understanding the genetic basis of flour color-related traits (L*, a*, b*, and YPC) is necessary for improving wheat quality by marker-assisted selection (MAS). During recent decades, numerous quantitative trait loci (QTL) for flour color-related traits have been identified using bi-parental populations (Mares and Campbell, 2001; Patil et al., 2008; Zhang and Dubcovsky, 2008; Zhang et al., 2009; Blanco et al., 2011; Roncallo et al., 2012; Crawford and Francki, 2013a; Colasuonno et al., 2014; Zhai et al., 2016b). Cloning genes relevant to flour color and developing functional markers have become major research focus (Mares and Campbell, 2001; He et al., 2008, 2009; Howitt et al., 2009; Zhang et al., 2011; Crawford and Francki, 2013b). As observed in previous studies flour color is a complex trait, and knowledge of the genetic control is still limited due to use of low density marker platforms and low resolution in bi-parental mapping studies. Phytoene synthase 1 (PSY1) catalyzes the first committed step in carotenoid biosynthesis, and it is generally accepted as the most important regulatory node, significantly correlated with carotenoid accumulation and determined flour color ($r = 0.8$) (Zhai et al., 2016c). Thus, more effort is needed to further dissect the genetics of flour color-related traits.

Genome-wide association study (GWAS), based on linkage disequilibrium (LD), is an effective complementary strategy to QTL mapping in dissecting associations between genotype and phenotype in germplasm collections (Yu and Buckler, 2006; Yu et al., 2006; Zhu et al., 2008). GWAS has a number of advantages over traditional linkage mapping, including the use of germplasm populations, potential for increased QTL resolution, and a wide sampling of molecular variation (Buckler and Thornsberry, 2002; Flint-Garcia et al., 2005; Waugh et al., 2009). With the development of high-throughput genotyping platforms, GWAS has been increasingly used to identify loci responsible for complex traits in plants, including wheat (Brachi et al., 2010; Zhao et al., 2011; Wang et al., 2012; Marcotuli et al., 2015; Tadesse et al., 2015). A potential disadvantage is identification of spurious associations. The mixed linear model (MLM) method is an effective method to avoid spurious associations as it simultaneously accounts for population structure (Q) and genetic relatedness (K) between individuals (Yu and Buckler, 2006; Yu et al., 2006). The wheat Illumina 90 K SNP array made a dramatic improvement in the number of gene-based markers, and widely used to detect QTL for important traits and identify candidate genes (Ain et al., 2015; Sun et al., 2017). Nevertheless, the genetic distances between markers were large on wheat chromosomes using the 90 K SNP array, particularly a low coverage in the D genome, reducing the power of marker identification and the precision of QTL mapping (Liu et al., 2017). Recently, the wheat Axiom 660 K SNP array was developed, providing higher marker density, higher resolution and better coverage of wheat genome (Cui et al., 2017).

The objectives of the present study were to: (1) identify the genetic basis underlying flour color-related traits in a collection of 166 bread wheat accessions using the wheat 90 and 660 K SNP arrays, and 10 allele-specific markers for known relevant genes; and (2) propose candidate genes affecting flour color-related traits based on biochemical information and bioinformatics analyses, with the ultimate aim of facilitating molecular breeding. Finally, the genetic determinants of flour color-related traits could be identified more accurately by GWAS using diverse genotypes and high-throughput SNP arrays.

MATERIALS AND METHODS

Plant Material and Field Trials

The association panel used in this study was a genetically diverse collection, comprising 166 elite bread wheat cultivars mainly from Yellow and Huai Valley of China, but also Italy, Argentina, Japan, Australia, and Turkey. Information about the accessions, including cultivar names, origins and subpopulation identity, was described in **Table S1**.

Briefly, the materials were grown in the 2012-2013 and 2013-2014 in randomized complete blocks with three replications at Anyang (AY, Henan province, 35°12′N, 113°37′E) and Suixi (SX, Anhui province, 33°17′N, 116°23′E), providing data for four environments. Each plot comprised three 2 m rows in 2012-2013 and four rows in 2013-2014, with 20 cm between rows. Agronomic management was performed according to local practices.

Phenotypic Evaluation and Statistical Analysis

For phenotypic evaluation 500 g clean grain from each plot was ground in a Brabender Quadrumat Junior Mill (Brabender Inc., Duisberg, Germany; http://www.cwbrabender.com) with a 0.15 mm sieve. Kernel hardness was determined with a Single Kernel Characterization System (SKCS 4100, Perten, Sweden). Moisture and protein contents were measured using a Near Infrared Transmittance (NIT; Foss-Tecator 1241, Foss, Höganäs, Sweden). The data of kernel hardness, moisture and protein content were only used to calculate how much water is required to tempering, so we did not describe these in this manuscript. Before milling, samples were conditioned to 14, 15, and 16% moisture overnight for soft (SKCS hardness index, HI<40), medium (HI, 40-59), and hard (HI>60) types, respectively, as mentioned by Jin et al. (2016). Due to time constraints and a laborious milling method only two replications were made for each environment.

Flour color parameters [brightness (L*), redness (a*) and yellowness (b*)] were measured with a Minolta colorimeter (CR-310, Minolta Camera Co., Ltd., Osaka, Japan) using the Commission Internationale de l' Éclairage (CIE) L* a* b* color system, respectively (Oliver et al., 1992). YPC was assessed according to Zhai et al. (2016b). Briefly, 1 g of flour sample was extracted with 5 ml of water-saturated n-butanol, along with shaking on an orbital incubator for 1 h at room temperature. After centrifuging at 5,000 rpm for 10 min, the absorbance was measured at 436.5 nm. YPC was expressed as $\mu g \cdot g^{-1}$ using a correction coefficient of 0.301 (American Association for Cereal Chemistry, 2000). Each sample was assayed twice, and a third assay was performed if the difference between two repeats was more than 10%. The mean values were used for statistical analysis.

Statistical analysis was carried out using SAS 9.2 software (SAS Institute, Inc., Cary, NC, USA, http://www.sas.com). Briefly, basic statistics such as means, standard deviation, and minimum and maximum values were calculated with the UNIVARIATE procedure. Analysis of variance was performed using the PROC MIXED procedure, where environments were treated as fixed effects, and genotypes, genotype × environment interaction and replicates nested in environments as random effects. Broad-sense heritability (h^2) was calculated following Zhai et al. (2016b). For each trait, a best linear unbiased predictor (BLUP) for each accession of each trait was calculated across four environments and used for further analyses. Pearson's correlation coefficient (r) among the flour color-related traits was calculated using the PROC CORR procedure based on BLUP values.

Genotypic Characterization

Genomic DNA was extracted from seedling leaves using the CTAB (cetyltrimethylammonium bromide) method (Doyle and Doyle, 1987). All 166 accessions were genotyped by both the Illumina wheat 90 K (comprising 81,587 SNPs) and Affymetrix wheat 660 K (containing 630,517 SNPs) SNP arrays by Capital Bio Corporation,

Beijing, China (http://www.capitalbiotech.com/). SNP allele clustering and genotype calling were performed using the polyploid version of Genome Studio software (Illumina, http://www.illumina.com). The default clustering algorithm was initially used to classify each SNP call into three allele clusters. Manual curation was then performed for more accurate genotyping. Markers with a minor allele frequency (MAF) < 5%, and more than 20% missing data were removed from the data matrix. After stringent filtration 259, 922 SNP markers from the wheat 90 and 660 K SNP arrays were integrated into a common physical map for GWAS. The physical positions of SNP markers were obtained from the International Wheat Genome Sequencing Consortium (IWGSC RefSeq v1.0, http://www.wheatgenome.org/) and used in association analysis.

Given that many of the genes influencing flour color-related traits have been well characterized, allele-specific markers for *Psy-A1* (He et al., 2008), *Psy-B1* (He et al., 2009), *TaPds-B1* (Dong, 2011), *e-Lcy3A* (Crawford and Francki, 2013b), *Talcye-B1* (Dong, 2011), *Lox-B1* (Geng et al., 2012), *Pinb-D1* (Giroux and Morris, 1997), and 1B·1R translocation (Liu et al., 2008) were used to genotype the association panel following the indications reported in the previous works to assess whether marker-trait associations (MTAs) identified in this study were co-located with these known genes. Detailed information of these markers is provided in **Table S2**.

Population Structure and Linkage Disequilibrium Analysis

A subset of 2,000 polymorphic SNP markers, distributed evenly across all the wheat chromosomes, was selected for population structure analysis using STRUCTURE (Pritchard et al., 2000). We applied the admixture model with correlated allele frequencies to assess numbers of hypothetical subpopulations ranging from $K = 2$ to 12 using 10,000 burn-in iterations followed by 100,000 MCMC (Markov Chain Monte Carlo) replicates. For each *K*, five independent runs were carried out. According to Evanno et al. (2005), the 166 accessions were structured into three subpopulations as described in Liu et al. (2017).

Pairwise LD was calculated as squared correlation coefficients (r^2) among alleles, and significance was computed by 1,000 permutations using TASSEL (Bradbury et al., 2007). LD was calculated separately for loci on the same chromosome (intra-chromosomal pairs) and unlinked loci (inter-chromosomal pairs). The extent and distribution of LD were graphically displayed by plotting intra-chromosomal r^2-values for loci in significant LD at $P \leqslant 0.001$ against the physical distance, and a locally weighed polynomial regression (LOESS) curve was fitted using XLSTAT (Addinsoft, Paris, France). The critical r^2-values beyond which LD is due to true physical linkage was calculated by taking the 95th percentile of the square root transformed r^2-values of unlinked markers (Breseghello and Sorrells, 2006). The intersection of the LOESS curve with the line of the critical r^2 was estimated to see how fast LD decay occurs.

Marker-Trait Association Analysis

Marker-trait association analysis was performed separately for each environment and BLUP values across four environments using the MLM in TASSEL. Association mapping model evaluations were based on visual observations of the quantile-quantile (Q-Q) plots, which are the plots of observed-$\log_{10}$ (P-value) vs. expected-$\log_{10}$ (P-value) under the null hypothesis that there was no association between marker and phenotype. A false discovery rate (FDR) of 0.05 was used as a threshold for significant association (Benjamini and Hochberg, 1995). The association of a marker with a trait was represented by its R^2-value, an estimate of the percentage of variance explained by the marker. To provide a complimentary summary of declared putative MTAs, Manhattan plots were generated using a script written in R software (http://www.r-project.org/).

To assess the pyramiding effect of favorable alleles of MTAs for flour color-related traits identified in this study, the BLUP value for each trait in each accession was regressed against the number of favorable alleles

using the line chart function in Microsoft Excel 2016.

Prediction of Candidate Genes for Flour Color-Related Traits

To assign putative biological functions of significant SNP markers associated with flour color-related traits, the flanking sequences of SNPs were blasted against the NCBI (http://www.ncbi.nlm.nih.gov/), European Molecular Biology Laboratory (http://www.ebi.ac.uk/) and European Nucleotide Archive (http://www.ebi.ac.uk/ena) public databases following Zhai et al. (2016b).

In addition, chromosomal locations and physical distances of 40 known genes influencing flour color-related traits, mainly involved in terpenoid backbone biosynthesis and carotenoid biosynthesis and degradation, were evaluated using IWGSC RefSeq v1.0. If the physical distances between MTAs identified in this study and known genes were less than LD decay, the MTAs were considered to be co-located with these known genes. Following Colasuonno et al. (2017), these 40 known genes were blasted against the available dataset of SNP marker sequences (Wang et al., 2014). Markers aligned with more than 80% similarity were considered to be within the coding sequences of known genes. Gene ontology annotation for the candidate genes was also conducted using EnsemblePlants (http://plants.ensembl.org/index.html).

RESULTS

Phenotypic Variation

There was a wide range of phenotypic variation for flour color-related traits among the 166 accessions, particularly for a*, b* and YPC where nearly 3 to 16-fold differences were observed (**Table 1**). Across all four environments, the mean L* was 90.29, ranging from 87.29 to 92.06; a* averaged -0.86, ranging from -1.75 to -0.11; b* varied from 5.29 to 14.56 and averaged 8.83; and YPC averaged 1.18 $\mu g \cdot g^{-1}$, with a range of 0.58-2.95 $\mu g \cdot g^{-1}$. The frequency distribution of each trait in each environment is provided in（**Figure S1**）.

High broad-sense heritabilities（$h^2 \geqslant 0.89$）were observed for these traits, indicating that genetic effects played a determinant role for each trait（**Table 1**）. Analysis of variance indicated that the effects of genotypes, environments, and their interactions were significant for flour color-related traits, and genotype had a larger effect（**Table S3**）. Pearson's correlation coefficients among all flour color-related traits were mostly significant（$P < 0.001$）, except for L* × a* （**Table S4**）. b* was negatively correlated with L* （$r = -0.71$）and a* （$r = -0.68$）. YPC was negatively correlated with L* （$r = -0.28$）and a* （$r = -0.89$）, and positively correlated with b* （$r = 0.83$）.

TABLE 1 Phenotypic variation of flour color-related traits in 166 bread wheat cultivars across four environments

Trait	Mean	SD	Min	Max	h^2
L*	90.29	0.92	87.29	92.06	0.89
a*	−0.86	0.34	−1.75	−0.11	0.91
b*	8.83	2.00	5.29	14.56	0.92
YPC	1.18	0.42	0.58	2.95	0.93

L*, *flour brightness*; *a**, *flour redness*; *b**, *flour yellowness*; *YPC*, *yellow pigment content*（μg. g-1 ）.

SD, *standard deviation*.

h^2, *broad-sense heritability*.

Population Structure and Linkage Disequilibrium

According to Evanno et al. （2005）, ΔK was plotted against the number of subgroups K. The maximum value of ΔK occurred at $K = 3$, indicating that a K-value of 3 was the most probable prediction for the number of subpopulations（Liu et al. 2017）. In general, subgroup 1 comprised 62 cultivars primarily derived from Shandong province and abroad; subgroup 2 consisted of 54 varieties predominantly from Henan, Shaanxi and Anhui provinces; and subgroup 3 included 50 cultivars mainly from Henan and Hebei provinces.

For the whole genome the threshold r^2, calculated as the 95th percentile of the distribution of r^2 of unlinked markers, was 0.082, and thus all values of $r^2 > 0.082$

were probably due to physical linkage. The intersection between the threshold r^2 and the LOESS curve was at 8 Mb, which was considered as the LD decay rate of the population. Similarly, the LD decay was 6, 4 and 11 Mb for A, B and D sub-genomes, respectively.

Marker-Trait Association

As shown in the Q-Q plots（**Figure S2**）, the observed-$\log_{10}$（P）-values were closed to the expected distribution, suggesting that the MLM model（Q + K）was appropriate for association analysis of flour color-related traits in this study. One hundred MTAs for flour color-related traits were detected, and 78 were identified in all four environments（**Table 2**）. The highest number of MTAs was detected for YPC（32）, followed by a* （28）, b* （25）and L* （15）. The associations between markers and flour color-related traits are shown by Manhattan plots in **Figure 1** and **Figure S3**.

MTAs of brightness were detected on chromosomes 1A, 1B, 1D, 2A, 2B, 3A, 4A（2）, 4B, 5A, 5B, 5D（2）, and 7A（2）, each explaining 7.1-17.5% of the phenotypic variation across environments（**Table 2**）. The MTAs on chromosomes 1D, 2B, 4A（2）, 4B, 5A and 5D（*AX _ 108930866*）were identified in three environments, and the remaining MTAs were detected in all four environments. The two most significant markers were *AX _ 111612184* on chromosome 2B and *Pinb-D1* on chromosome 5D, explaining 11.0-16.4% and 12.7-17.5% of phenotypic variation, respectively.

For redness, the MTAs were distributed on chromosomes 1A（2）, 1B（2）, 1D, 2A（2）, 2B（2）, 2D, 3D, 4A（2）, 4B, 4D, 5A, 5B（2）, 5D, 6A（2）, 6B（2）, 6D, 7A, 7B（2）and 7D, individually explaining 6.5-20.0% of the phenotypic variation（**Table 2**）. The loci on chromosomes 1A （*AX _ 111611571*）, 4A（*AX _ 109058420*）and 4B were identified in three environments, and the other 25 MTAs were significantly associated with a* across all four environments. The MTA on chromosome 1A （*AX _ 108727598*）showed the strongest association with a*, explaining phenotypic variation ranging from 14.6 to 20.0%.

Yellowness MTAs, located on chromosomes 1A (2), 1B (2), 1D, 2A (2), 2B (2), 2D, 3A, 3B, 4A, 5B (2), 5D (2), 6A, 6B (2), 7A (2), 7B (2) and 7D, explained phenotypic variation ranging from 6.6 to 20.7% (**Table 2**). The loci on chromosomes 2A (*AX _ 109848219*), 2B (2), 3A, 3B, 5B (*AX _ 94508455*), 5D (*AX _ 108930866*), 6B (2) and 7A (*AX _ 111616453*) were detected in three environments, whereas the others were significant for b* in all four environments. The MTAs on chromosomes 2A (2), 2D and 7D, and *Pinb-D1*, *Psy-A1* and *Psy-B1* showed stronger association with b*, explaining more than 15% phenotypic variation in some environments.

For yellow pigment content, we detected MTAs on all chromosomes, explaining 6.9-20.9% of the total variation (**Table 2**). MTAs on chromosomes 1A (*AX _ 111611571*) and 3B (*AX _ 108968661*) were identified in three environments, while the remaining were observed in all four environments. The MTAs on chromosomes 1A (*AX _ 108727598*), 1D, 2B (*AX _ 89310598*), 5B (*AX _ 110577474*) and 1B • 1R translocation showed stronger association with YPC than other MTAs.

In Silico Prediction of Candidate Genes

The putative biological functions of 1,543 significant SNP markers associated with flour color-related traits identified in the present study were assigned (data not shown). Briefly, *AX _ 109030196*, associated with a*, b* and YPC, corresponds to premnaspirodiene oxygenase (*PO*) gene involved in sesquiterpenoid and triterpenoid biosynthesis (**Table 3**). For the MTA on chromosome 1D (97, 314, 998-97, 315, 098 Mb), two candidate genes for a* and YPC were identified; one is a putative gene (*IWB35120*) encoding the dihydrolipoyl dehydrogenase (DLD) enzyme, and the other (*IWB31766*) corresponds to leghemoglobin reductase (LBR), both involved in glycolysis/

gluconeogenesis. Either one or even both genes are potential candidates for the observed MTA in this genomic region. *AX _ 111471334*, corresponding to phosphomannomutase (*PMM*) gene, is a potential candidate gene for the MTA on chromosome 3B (146, 971, 117-146, 971, 187 Mb) for YPC. *IWB36240*, significant for a* and YPC, corresponds to 4-hydroxy-3-methylbut-2-en-1-yl diphosphate synthase (*MDPS*) gene involved in terpenoid backbone biosynthesis.

Based on the reference wheat genome sequences from the IWGSC RefSeq v1.0, the 40 known genes influencing flour color-related traits were mapped on the physical map (**Table S5**). Briefly, the farnesyl pyrophosphate synthase 2 gene (*FPPS2*) was located about 4.19 Mb from *AX _ 111611571* significant for L*, a*, b* and YPC (**Table 2**). *AX _ 94596570* (2D), *AX _ 109058420* (4A) and *AX _ 95195654* (7D) associated with a*, b* and YPC was located about 8.65, 1.31 and 1.13 Mb from isopentenyl pyrophosphate isomerase (*IPPS*), 1-deoxy-D-xylulose 5-phosphate reductoisomerase (*DXR*) and *PSY1* genes, respectively. The mevalonate kinase (*MK*) gene on chromosome 2A was located in close physical proximity with *AX _ 110044577* (0.82 Mb) associate with both L* and YPC. *AX _ 108749277* (4B) and *AX _ 108940832* (6D) significant for a* and YPC was located 1.72 and 1.69 Mb from carotenoid β-ring hydroxylase 2 (*BCH2*) and 9-cis-epoxycarotenoid dioxygenase 4 (*NCED4*) genes, respectively. The lipoxygenase 2 gene (*LOX2*) on chromosome 5A was 5.12 Mb from *AX _ 111736921* controlling L*. *AX _ 111736921* significant for L* on chromosome 7A was close to peroxidase gene (*POD*) (1.85 Mb). The *cis*-zeta-carotene isomerase gene (*ZISO*) on chromosome 5B was detected at a distance of 1.44 Mb from *AX _ 110952518* associated with b*. *AX _ 109996966* significant for b* was located about 1.07 Mb proximal to the lycopene β-cyclase gene (*LYCB*). Notably, most of candidate genes, except for *LOX2*, *POD*, *ZISO*, *LCYB* and *PMM*, controlled more than one flour color-related traits.

TABLE 2　Significant marker-trait associations for flour color-related traits using mixed linear model approach

Trait	Marker[a]	Chr[b]	Physical position[c] (bp)	Environments[d]	P-value	R^{2e} (%)	Candidate gene[f]	GenBank No.	Physical position[c] (bp)	Distance (Mb)
L*	AX_111611571	1A	548,493,456-548,493,526	E1,E3,E4,E5	$1.40\text{-}9.83E^{-04}$	7.1-9.6	FPPS2	JX235715	544,198,065-544,198,296	4.19
	AX_94634405	1B	483,114,668-483,114,738	E1,E2,E3,E4,E5	$2.95E^{-05}\text{-}3.71E^{-04}$	10.1-11.8				
	AX_94562220	1D	18,073,435-18,073,505	E1,E3,E4,E5	$1.55\text{-}7.53E^{-04}$	7.4-9.5				
	AX_110044577	2A	778,128,072-778,128,142	E1,E2,E3,E4,E5	$7.29E^{-06}\text{-}8.74E^{-04}$	8.1-14.0	MK	AFV51837	777,287,545-777,287,970	0.82
	AX_111612184	2B	745,494,701-745,494,771	E2,E3,E4,E5	$7.52E^{-06}\text{-}3.30E^{-04}$	11.0-16.4				
	IWB26070	3A	710,831,963-710,832,063	E1,E2,E3,E4,E5	$1.25\text{-}7.84E^{04}$	7.6-10.9				
	AX_109848219	4A	29,058,615-29,058,685	E1,E2,E3,E5	$2.03\text{-}8.72E^{-04}$	7.5-11.5				
	AX_111533733	4A	607,266,755-607,266,825	E1,E2,E4,E5	$9.93E^{-05}\text{-}3.69E^{-04}$	10.3-12.2				
	AX_111138030	4B	524,670,446-524,670,516	E1,E2,E4,E5	$1.24\text{-}9.80E^{-04}$	7.2-9.7				
	AX_111736921	5A	580,948,231-580,948,301	E1,E3,E4,E5	$1.39\text{-}3.04E^{-04}$	9.5-10.7	LOX2	GU167921	575,706,041-575,708,873	5.12
	IWB71821	5B	669,675,280-669,675,380	E1,E2,E3,E5	$1.12\text{-}9.81E^{-04}$	7.9-12.1				
	Pinb-D1	5D	3,031,551-303,2419	E1,E2,E3,E4,E5	$2.97E^{-06}\text{-}8.29E^{-05}$	12.7-17.5				
	AX_108930866	5D	417,084,071-417,084,141	E1,E2,E4	$3.60\text{-}8.47E^{-04}$	8.6-10.4				
	AX_109342544	7A	19,556,781-19,556,851	E1,E2,E3,E4,E5	$1.70\text{-}9.28E^{-04}$	7.3-11.5				
	IWB56095	7A	644,611,038-644,611,138	E1,E2,E3,E4,E5	$2.90\text{-}8.01E^{-04}$	7.3-10.8	POD	EU725470	646509653-646510291	1.85
a*	AX_108727598	1A	7,653,153-7,653,223	E1,E2,E3,E4,E5	$7.74E^{-08}\text{-}2.96E^{-06}$	14.6-20.0				
	AX_111611571	1A	548,493,456-548,493,526	E1,E2,E3,E5	$3.01\text{-}5.84E^{-04}$	7.1-7.9	FPPS2	JX235715	544,198,065-544,198,296	4.19
	1B·1R	1B	—	E1,E2,E3,E4,E5	$3.82E^{-07}\text{-}9.32E^{-06}$	12.5-16.9				
	AX_95217104	1B	540,848,004-540,848,074	E1,E2,E3,E4,E5	$1.40\text{-}E^{-06}\text{-}2.30E^{-04}$	8.3-14.7				
	IWB9456	1D	97,314,998-97,315,098	E1,E2,E3,E4,E5	$3.56E^{-06}\text{-}1.44E^{-05}$	12.0-13.8				
	AX_95151551	2A	79,303,812-79,303,882	E1,E2,E3,E4,E5	$2.65E^{-07}\text{-}9.14E^{-05}$	9.5-17.2				
	AX_109848219	2A	629,632,627-629,632,697	E1,E2,E3,E4,E5	$2.71E^{-06}\text{-}1.84E^{-04}$	8.9-14.9				
	AX_89310598	2B	418,276,653-418,276,723	E1,E2,E3,E4,E5	$2.27E^{-07}\text{-}1.22E^{-05}$	12.1-17.0				
	AX_111612184	2B	745,494,701-745,494,771	E1,E2,E3,E4,E5	$2.04\text{-}9.91E^{-04}$	7.9-9.9				
	AX_94596570	2D	131,312,023-131,312,093	E1,E2,E3,E4,E5	$8.82E^{-07}\text{-}1.62E^{-04}$	10.7-17.9	IPPS	EU783965	187,126,553-187,126,809	8.65
	AX_108794699	3D	10,644,700-10,644,770	E1,E2,E3,E4,E5	$1.40E^{-05}\text{-}4.98E^{-04}$	7.5-11.9				

(continued)

Trait	Marker[a]	Chr[b]	Physical position[c] (bp)	Environments[d]	P-value	R[2e] (%)	Candidate gene[f]	GenBank No.	Physical position[c] (bp)	Distance (Mb)
	AX_109848219	4A	29,058,615-29,058,685	E1,E2,E3,E4,E5	$1.72E^{-06}-6.21E^{-05}$	10.0-14.3				
	AX_109058420	4A	707,784,994-707,785,064	E1,E2,E4	$6.78-9.72E^{-04}$	6.5-11.2		EMS62178	709,122,545-709,122,730	1.31
	AX_108749277	4B	606,865,366-606,865,436	E1,E2,E4,E5	$6.19-9.72E^{-04}$	6.7-9.0	BCH2	JX171673	608624963-608625486	1.72
	IWB17540	4D	26,108,065-26,108,266	E1,E2,E3,E4,E5	$9.35E^{-07}-3.39E^{-05}$	10.6-15.0				
	AX_110042198	5A	74,535,559-74,535,629	E1,E2,E3,E4,E5	$6.37E^{-05}-7.32E^{-04}$	8.9-12.2				
	AX_110577474	5B	379,722,302-379,722,372	E1,E2,E3,E4,E5	$6.21E^{-07}-3.11E^{-05}$	11.2-16.2				
	IWB1821	5B	669,675,280-669,675,380	E1,E2,E3,E4,E5	$3.56E^{-06}-1.50E^{-05}$	11.8-13.7				
	AX_108968610	5D	321,600,750-321,600,820	E1,E2,E3,E4,E5	$2.02E^{-06}-1.34E^{-04}$	9.1-14.5				
	AX_94533255	6A	42,879,926-42,879,996	E1,E2,E3,E4,E5	$4.07E^{-05}-1.59E^{-04}$	8.7-10.3				
	AX_111451223	6A	594,420,538-594,420,608	E1,E2,E3,E4,E5	$1.77E^{-06}-1.14E^{-04}$	9.3-14.6				
	AX_110572276	6B	99,080,197-99,080,267	E1,E2,E3,E4,E5	$7.00E^{-06}-1.57E^{-04}$	9.6-14.1				
	IWB36240	6B	437,483,918-437,484,033	E1,E2,E3,E4,E5	$9.73E^{-07}-1.68E^{-05}$	11.8-15.6				
	AX_108940832	6D	463,914,887-463,914,957	E1,E2,E3,E4,E5	$1.22E^{-06}-1.76E^{-05}$	11.5-15.0	NCED4	KP099105	355857551-355859437	1.69
	Psy-A1	7A	729,328,109-729,323,941	E1,E2,E3,E4,E5	$4.02E^{-07}-1.31E^{-04}$	8.9-16.2				
	AX_110044711	7B	481,839,758-481,839,828	E1,E2,E3,E4,E5	$1.33E^{-06}-1.99E^{-05}$	11.3-14.8				
	Psy-B1	7B	739,105,007-739,107,002	E1,E2,E3,E4,E5	$1.88E^{-07}-5.88E^{-05}$	10.3-18.2				
	AX_95195654	7D	635,560,079-635,560,149	E1,E2,E3,E4,E5	$9.74E^{-06}-2.37E^{-04}$	10.5-15.2	PSY1	EF600063	636721676-636723811	1.13
b*	AX_108727598	1A	7,653,153-7,653,223	E1,E2,E3,E4,E5	$6.70E^{-06}-3.64E^{-05}$	12.1-14.9				
	AX_111611571	1A	548,493,456-548,493,526	E1,E2,E3,E4,E5	$1.65-8.07E^{-04}$	6.9-11.2	FPPS2	JX235715	544198065-544198296	4.19
	1B·1R	1B	—	E1,E2,E3,E4,E5	1.43E-05-1.36E-04	8.8-11.8				
	AX_95217104	1B	540,848,004-540,848,074	E1,E2,E3,E4,E5	$4.22E^{-06}-9.00E^{-05}$	9.6-14.6				
	AX_94562220	1D	18,073,435-18,073,505	E1,E2,E3,E4,E5	$2.53-8.95E^{-04}$	9.1-10.5				
	AX_95151551	2A	79,303,812-79,303,882	E1,E2,E3,E4,E5	$9.60E^{-07}-1.47E^{-05}$	11.9-15.0				
	AX_109848219	2A	629,632,627-629,632,697	E1,E2,E3	$2.04E^{-06}-7.21E^{-04}$	7.7-15.3				
	AX_89310598	2B	418,276,653-418,276,723	E2,E3,E4,E5	$1.02-9.19E^{-04}$	6.7-9.6				
	AX_111612184	2B	745,494,701-745,494,771	E2,E3,E4,E5	$2.13-5.83E^{-04}$	9.2-10.5				

(continued)

Trait	Marker[a]	Chr[b]	Physical position[c] (bp)	Environments[d]	P-value	R^{2e} (%)	Candidate gene[f]	GenBank No.	Physical position[c] (bp)	Distance (Mb)
	AX_94596570	2D	131,312,023-131,312,093	E1,E2,E3,E4,E5	$4.62E^{-06}$-$1.14E^{-04}$	11.5-15.4	IPPS	EU783965	187126553-187126809	8.65
	IWB9681	3A	37,825,997-37,826,097	E1,E2,E3,E5	5.30-$9.97E^{-04}$	6.6-7.3				
	IWB38921	3B	747,712,577-747,712,677	E1,E2,E3	7.07-$8.79E^{-04}$	7.3-7.8				
	AX_109058420	4A	707,784,994-707,785,064	E1,E2,E3,E4,E5	2.45-$7.05E^{-04}$	8.9-10.4	DXR	EMS82178	709122545-709122730	1.31
	AX_110952518	5B	35,290,739-35,290,809	E1,E2,E3,E4,E5	$5.92E^{-06}$-$5.58E^{-04}$	9.5-13.8	ZISO	CV770956	36761350-36762789	1.44
	AX_94508455	5B	531,581,931-531,582,001	E2,E3,E4,E5	5.17-$7.64E^{-04}$	7.1-7.8				
	Pinb-D1	5D	3,031,551-3,032,419	E1,E2,E3,E4,E5	$3.36E^{-07}$-$7.21E^{-05}$	11.7-19.3				
	AX_108930866	5D	417,084,071-417,084,141	E2,E3,E4,E5	$5.62E^{-05}$-$3.49E^{-04}$	9.9-12.4				
	AX_111109507	6A	17,899,493-17,899,563	E1,E2,E3,E4,E5	2.49-$7.15E{-04}$	7.5-9.5				
	AX_109996966	6B	240,379,787-240,379,857	E1,E2,E3	5.30-$8.47E^{-04}$	7.2-8.5	LYCB	JN62196	239283250-239284710	1.07
	AX_111547031	6B	689,671,458-689,671,528	E2,E3,E4,E5	2.14-$6.33E^{-04}$	9.2-10.6				
	AX_111616453	7A	449,976,248-449,976,318	E1,E2,E3	2.16-$6.29E^{-04}$	8.4-9.1				
	Psy-A1	7A	729,328,109-729,323,941	E1,E2,E3,E4,E5	$4.07E^{-08}$-$1.74E^{-05}$	11.7-19.4				
	AX_94457966	7B	11,253,471-11,253,541	E1,E2,E3,E4	3.10-$8.15E^{-04}$	7.5-9.0				
	Psy-B1	7B	739,105,007-739,107,002	E1,E2,E3,E4,E5	$6.47E^{-07}$-$4.34E^{-06}$	13.5-16.1				
	AX_95195654	7D	635,560,079-635,560,149	E1,E2,E3,E4,E5	$1.27E{-07}$-$1.97E{-04}$	11.1-20.7	PSY1	EF600063	636721676-636723811	1.13
YPC	AX_108727598	1A	7,653,153-7,653,223	E1,E2,E3,E4,E5	$2.50E^{-08}$-$1.42E^{-06}$	16.3-20.9				
	AX_111611571	1A	548,493,456-548,493,526	E1,E3,E4,E5	$1.19E^{-05}$-$7.24E^{-04}$	8.7-14.1	FPPS2	JX233715	544198065-544198296	4.19
	1B·1R	1B	-	E1,E2,E3,E4,E5	$3.81E^{-08}$-$3.27E^{-06}$	15.5-20.5				
	AX_95217104	1B	540,848,004-540,848,074	E1,E2,E3,E4,E5	$1.39E^{-05}$-$2.93E^{-04}$	7.9-11.6				
	IWB9456	1D	97,314,998-97,315,098	E1,E2,E3,E4,E5	$2.75E^{-08}$-$3.10E^{-06}$	15.4-20.7				
	AX_95151551	2A	79,303,812-79,303,882	E1,E2,E3,E4	$8.66E{-07}$-$2.36E{-05}$	11.1-15.2				
	AX_110044577	2A	778,128,072-778,128,142	E1,E2,E3,E4,E5	$1.15E{-05}$-$9.19E{-04}$	8.2-14.5	MK	AFV51837	777287545-777287970	0.82
	AX_89310598	2B	418,276,653-418,276,723	E1,E2,E3,E4,E5	$8.34E{-09}$-$1.14E{-06}$	15.0-20.8				
	AX_111612184	2B	745,494,701-745,494,771	E1,E2,E3,E4,E5	$1.08E{-05}$-$1.90E{-04}$	11.0-15.0				
	AX_94596570	2D	131,312,023-131,312,093	E1,E2,E3,E4,E5	$2.29E{-07}$-$1.55E{-05}$	11.6-16.3	IPPS	EU783965	187126553-187126809	8.65
	AX_109399477	3A	549,330,506-549,330,576	E1,E2,E3,E4,E5	$1.99E^{-05}$-$2.63E^{-04}$	9.9-13.9				
	AX_108968661	3B	146,971,117-146971187	E1,E3,E4,E5	3.25-$9.15E^{-04}$	6.9-9.9				
	IWB38921	3B	747,712,577-747,712,677	E1,E2,E3,E4,E5	1.79-$6.92E^{-04}$	7.7-9.1				

(continued)

Trait	Marker[a]	Chr[b]	Physical position[c] (bp)	Environments[d]	P-value	R²[e] (%)	Candidate gene[f]	GenBank No.	Physical position[c] (bp)	Distance (Mb)
	AX_108794699	3D	10,644,700-10,644,770	E1,E2,E3,E4,E5	5.35E-07-6.92E-05	9.7-15.2				
	AX_109848219	4A	29,058,615-29,058,685	E1,E2,E3,E4,E5	2.53E-07-2.51E-05	11.0-16.2				
	AX_109058420	4A	707,784,994-707,785,064	E1,E2,E3,E4,E5	1.06-6.79E-04	10.9-13.7	DXR	EMS62178	709122545-709122730	1.31
	AX_108749277	4B	606,865,366-606,865,436	E1,E2,E3,E4,E5	1.33-8.21E-04	8.6-10.4	BCH2	JX171673	608624963-60862548 6	1.72
	IWB17540	4D	26,108,065-26,108,266	E1,E2,E3,E4,E5	3.92E-08-6.03E-06	12.6-18.4				
	AX_110042198	5A	74,535,559-74,535,629	E1,E2,E3,E4,E5	6.88E-05-2.00E-04	10.9-12.0				
	AX_110577474	5B	379,722,302-379,722,372	E1,E2,E3,E4,E5	2.82E-08-2.69E-06	15.4-20.5				
	IWB71821	5B	669,675,280-669,675,380	E1,E2,E3,E4,E5	1.57E-07-8.57E-06	12.8-17.2				
	AX_108968610	5D	321,600,750-321,600,820	E1,E2,E3,E4,E5	1.10E-07-2.04E-05	11.3-17.4				
	AX_94533255	6A	42,879,926-42,879,996	E1,E2,E3,E4,E5	4.70E-06-1.24E-04	8.9-12.4				
	AX_111451223	6A	594,420,538-594,420,608	E1,E2,E3,E4,E5	8.85E-08-1.69E-05	11.5-17.6				
	AX_110572276	6B	99,080,197-99,080,267	E1,E2,E3,E4,E5	1.99E-07-2.52E-05	11.9-18.1				
	IWB36240	6B	437,483,918-437,484,033	E1,E2,E3,E4,E5	3.18E-08-3.43E-06	14.3-19.8				
	AX_108940832	6D	463,914,887-463,914,957	E1,E2,E3,E4,E5	3.65E-08-4.87E-06	13.1-18.8	NCED4	KP099105	355857551-35585943 7	1.69
	AX_111136917	7A	548,316,594-548,316,664	E1,E2,E3,E4,E5	5.33E-06-6.75E-05	9.2-12.9				
	Psy-A1	7A	729,328,109-729,323,941	E1,E2,E3,E4,E5	9.43E-07-1.98E-05	11.2-15.0				
	AX_11004471	7B	481,839,758-481,839,828	E1,E2,E3,E4,E5	4.11E-08-5.55E-06	12.7-18.5				
	Psy-B1	7B	739,105,007-739,107,002	E1,E2,E3,E4,E5	9.89E-07-1.43E-05	12.0-15.7				
	AX_95195654	7D	635,560,079-635,560,149	E1,E2,E3,E4,E5	3.12E-05-2.12E-04	10.8-13.5	PSY1	EF600063	636721676-636723811	1.13

[a] Representative marker at the MTA.

[b] Chromosome.

[c] The physical positions of SNP markers based on wheat genome sequences from the International Wheat Genome Sequencing Consortium (IWGSC RefSeq v1.0, http: // www. wheatgenome. org /).

[d] E1, 2012-2013 AY; E2, 2012-2013 SX; E3, 2013-2014 AY; E4, 2013-2014 SX; E5, BLUP, a best linear unbiased predictor of flour color-related traits in 166 common wheat cultivars across four environments.

[e] The percentage of variation explained by each locus.

[f] FPPS2, Farnesyl pyrophosphate synthase 2; MK, mevalonate kinase; LOX2, lipoxygenase 2; POD, peroxidase; IPPS, Isopentenyl pyrophosphate isomerase; DXR, 1-deoxy-D-xylulose 5-phosphate reductoisomerase; BCH2, Carotenoid β-ring hydroxylase 2; NCED4, 9-cis-epoxycarotenoid dioxygenase 4; PSY1, Phytoene synthase 1; ZISO, Cis-zeta-carotene isomerase; LCYB, Lycopene β-cyclase.

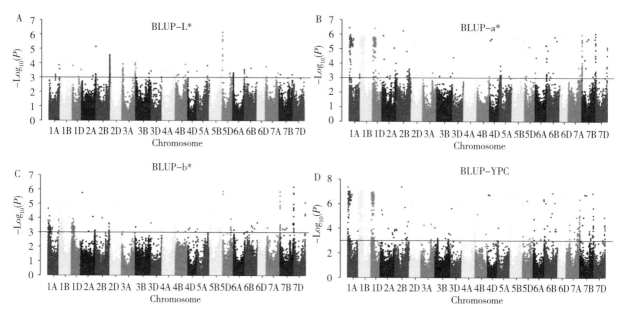

FIGURE 1 Manhattan plots indicating genomic regions associated with flour color-related traits based on BLUP values. (A) L*, (B) a*, (C) b*, and (D) yellow pigment content (YPC). Negative $\log_{10}$-transformed P-values from a genome-wide scan are plotted against SNP marker positions on each wheat chromosome. The blue horizontal line designificant the significant association threshold ($-\log_{10} P \geqslant 3$).

DISCUSSION

Detection of QTL by GWAS

GWAS has become an efficient tool for genetic dissection of complex traits. Among the diverse accessions analyzed in this study, substantial phenotypic variation in flour color-related traits indicated a wide range of genetic diversity (**Table 1, Figure S1**). Based on diverse genotypes and high-throughput SNP arrays, GWAS was a powerful strategy for dissecting flour color-related traits and identifying candidate genes.

Spurious associations are often the result of structured relationships within the population and may be reduced by taking population structure into account (Pritchard et al., 2000; Yu et al., 2006). Therefore, assessment of population structure is important prior to conducting a GWAS. As described in Liu et al. (2017) the core collection in this study was structured into three subpopulations, largely in agreement with their geographical origin. This pattern can be caused by several factors, including disproportional usage of a limited number of founders in developing regional populations and enrichment of alleles associated with regional adaptation by local breeding programs.

As shown in Q-Q plots(**Figure S2**), MLM greatly reduced spurious associations. Consistency across environments was used as an additional criterion for MTAs significant at FDR < 0.05 to reduce the risk of false marker-trait associations. Finally, the 100 MTAs for flour color-related traits were identified in at least three environments, which will be useful for metabolomic studies of flour color-related traits. The more robust markers can then be implemented in breeding programs to ensure that increasingly stringent color requirements imposed by industry are met through early screening of breeding lines. Importantly, the effects of these MTAs and consequent values for selection in breeding programs require validation in bi-parental populations.

Correlations Among Flour Color-Related Traits

Pearson correlation coefficients among all flour color-related traits showed that b* was negatively correlated with brightness ($r = -0.71$) and redness ($r = -0.68$) and yellow pigment content was negatively correlated with redness ($r = -0.89$) and positively correlated with yellowness ($r = 0.83$). As expected the strong phenotypic correlations among flour color-related traits were generally based on a large number of shared MTAs (**Table 2**). b* had five and 13 MTAs in common with L* and a*, respectively. YPC shared 27 and 13 same genetic regions with a* and b*, respectively. Of course, some loci were also identified specific for L*, a*, b* and YPC, respectively, indicating that flour color-related traits could be controlled by common major-effect loci, but also modified by numerous trait-specific loci.

To improve flour color-related traits, using multi-trait markers in MAS may increase QTL pyramiding efficiency. Most Chinese foods, such as steamed buns and noodles, typically require a white or creamy flour type (higher L*) with lower b* and YPC. In this situation, these multi-trait regions could be important targets to reduce b* and YPC and increase L* simultaneously.

TABLE 3 Candidate genes identified by putative biological functions of significant SNP markers

| | MTAs | | | | Candidate genes | | | Distance |
Trait	Marker[a]	Chr[b]	Physical position[c] (bp)	Marker[d]	Physical position[c] (bp)	Candidate gene (bp)[e]	Pathway	(Mb)
a*, b*, YPC	AX_108727598	1A	7, 653, 153- 7, 653, 223	AX_109030196	10, 126, 611- 10, 126, 681	PO	Sesquiterpenoid and triterpenoid biosynthesis	2.42
a*, YPC	IWB9456	1D	97, 314, 998- 97, 315, 098	IWB35120	100, 325, 418- 100, 325, 536	DLD	Glycolysis/ Gluconeogenesis	2.94
a*, YPC	IWB9456	1D	97, 314, 998- 97, 315, 098	IWB31766	103, 350, 946- 103, 351, 043	LBR	Glycolysis/ Gluconeogenesis	5.89
YPC	AX_108968661	3B	146, 971, 117- 146, 971, 187	AX_111471334	146, 551, 873- 146, 551, 943	PMM	Glycolysis/ Gluconeogenesis	0.41
a*, YPC	IWB36240	6B	437, 483, 918- 437, 484, 033	IWB36240	437, 483, 918- 437, 484, 033	MDPS	Terpenoid backbone biosynthesis	0.00

[a] *Representative marker at the MTA.*

[b] *Chromosome.*

[c] *The physical positions of SNP markers based on wheat genome sequences from the International Wheat Genome Sequencing Consortium (IWGSC RefSeq v1.0, http://www.wheatgenome.org/).*

[d] *significant SNP marker.*

[e] *PO, Premnaspirodiene oxygenase; DLD, Dihydrolipoyl dehydrogenase; LBR, Leghemoglobin reductase; PMM, Phosphomannomutase; MDPS, 4-hydroxy-3-methylbut-2-en-1-yl diphosphate synthase.*

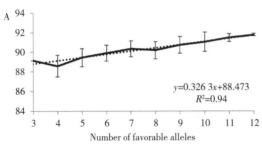

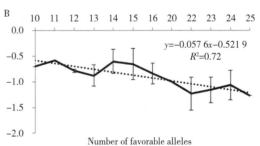

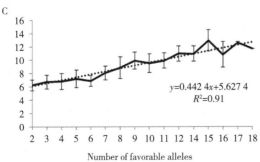

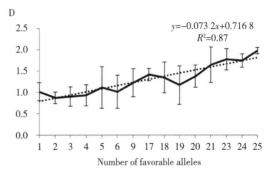

FIGURE 2 Linear regressions between number of favorable alleles and BLUP values of flour color-related traits (A) L*, (B) a*, (C) b*, and (D) yellow pigment content (YPC)

Comparison With Previous Studies

As shown in **Table 2**, known major genes controlling flour color-related traits and previously identified in bi-parental mapping populations were confirmed in the present study, such as *Psy-A1*, *Psy-B1*, *Pinb-D1* and the 1B • 1R translocation (Nagamine et al., 2003; He et al., 2008, 2009; Tsilo et al., 2011). On average, cultivars carrying the *Psy-A1b* allele produced significantly lower b* (7.48 ± 1.53) and YPC (0.87±0.30 μg • g^{-1}) than lines carrying the *Psy-A1a* allele (9.56 ± 1.84 and 1.34 ± 0.39 μg • g^{-1}), while higher a* was associated with the *Psy-A1b* allele (−0.66±0.28 vs. −0.97±0.32). b* (11.52±1.55) and YPC (1.77 ± 0.34 μg • g^{-1}) values for 19 cultivars with the *Psy-B1a* allele were significantly higher than those of 146 lines with *Psy-B1b* (8.46 ± 1.75 and 1.09 ± 0.36 μg • g^{-1}), whereas lines carrying the *Psy-B1b* allele generated higher a* (−0.80±0.30 vs. −1.32±0.27). For *Pinb-D1*, lines with the *Pinb-D1b* allele produced significantly lower L* (89.95±0.56) and higher b* (9.64±1.76) than those with Pinb-D1a (90.75 ± 1.12 and 7.67±1.70). Cultivars with the 1B • 1R translocation had significantly higher b* (9.65 ± 1.03) and YPC (1.39±0.44 μg • g^{-1}) than those without (8.05±0.91) and (0.97±0.29 μg • g^{-1}), while non-1B • 1R translocation genotypes were associated with higher a* (−1.02±0.34 vs. −0.72 ±0.28) (data not shown). These results confirmed that *Psy-A1*, *Psy-B1*, *Pinb-D1* and the 1B • 1R translocation were major-effect loci for flour color-related traits, and can be considered as stable and useful for MAS in breeding programs. However, *e-Lcy3A*, *Talcye-B1*, *TaPds-B1* and *Lox-B1* were not significantly associated with flour color-related traits as they have been reported previously (Dong, 2011; Geng et al., 2012; Crawford and Francki, 2013b). This might be due to the genetic background of the germplasm panel used in this study. More accessions with wider genetic variation will be collected and analyzed in further research.

Because no QTL for L* was previously mapped to chromosome 7A (2), the two MTAs in this work may be new loci associated with this trait. Similarly, MTA for a* located on chromosome 2A (629, 632, 627-629, 632, 697 Mb) and MTA for YPC on chromosome 7D (635, 560, 079-635, 560, 149 Mb) were not matched to any previously identified QTL. Hence these loci are likely to be new and should be attractive candidate regions on which to focus further in dissecting the genetic architecture of flour color-related traits in wheat. Notably, MTAs on chromosomes 1A (7, 653, 153-7, 653, 223 Mb) and 2B (418, 276, 653-418, 276, 723 Mb) showed stronger association with a* than *Psy-A1* and *Psy-B1* in at least three environments (data not shown). Similarly, 11 MTAs on chromosomes 1A (7, 653, 153-7, 653, 223 Mb), 1D, 2B (418, 276, 653-418, 276, 723 Mb), 4D, 5B (379722302-379722372 Mb), 5D, 6A (594, 420, 538-594, 420, 608 Mb), 6B (2), 6D, and 7B (481, 839, 758-481, 839, 828 Mb) explained much more phenotypic variation of YPC than *Psy-A1* and *Psy-B1* in at least three environments (data not shown). These demonstrated that GWAS is a powerful mapping approach for identifying genomic regions underlying variation in flour color based on diverse genotypes and high-throughput SNP arrays; they also confirmed previously detected QTL and revealed novel QTL that were not found in bi-parental populations previously.

Candidate Genes

Knowledge of the functions of significant SNP markers provides a very useful tool for identification of candidate genes for traits under investigation. Based on a comparative genomics approach, annotations of flanking sequences of significant SNP markers predicted that some genomic regions encode proteins that are important components of pathways linked to carotenoid biosynthesis and degradation (**Table 3**). For example, PO (*AX _ 109030196*) participated in sesquiterpenoid and triterpenoid biosynthesis; DLD (*IWB35120*), LBR (*IWB31766*) and PMM (*AX _ 111471334*) were involved in glycolysis/gluconeogenesis; and MDPS

(*IWB36240*) played a role in terpenoid backbone biosynthesis. These genes can be considered potential candidate genes for flour color-related traits.

Over the past decades, the gene discovery in bread wheat has been largely limited due to the absence of reference genome sequences. However, recent advances in high-throughput genotyping platforms and publicly available wheat genome sequences offer researchers new opportunities to achieve their goals. Based on the reference wheat genome sequences from the IWGSC RefSeq v1.0, 40 known genes influencing flour color-related traits were assigned to chromosome locations (**Table S5**). Eleven of the 40 known genes were located close to 23 MTAs for flour color-related traits (**Table 2**), including four terpenoid backbone biosynthesis genes (*FPPS2*, *IPPS*, *DXR* and *MK*), five genes involved in carotenoid biosynthesis (*ZISO*, *LYCB*, *PSY-D1*, *BCH2* and *NCED4*), and two related to carotenoid degradation (*LOX2* and POD). Therefore, the co-linearity of 11 known genes and 23 MTAs suggested that these might be candidate genes for flour color-related traits. In fact, all these candidate genes should be confirmed in bi-parental population mapping or by reverse genetic approaches.

Out of 40 known genes blasted against the available dataset of SNP marker sequences (Wang et al., 2014), SNP markers corresponded to nine genes, including *FPPS1*, *POD*, *FPPS2*, *MK*, aldehyde oxidase 3, carotenoid β-ring hydroxylase, geranylgeranyl transferase I β-subunit, geranylgeranyl transferase I α-subunit and catalase 3 (**Table S6**). However, only *FPPS2*, *POD* and *MK* were significantly associated with the flour color-related traits, based on chromosome locations and physical distances. Briefly, cultivars carrying the "A" SNP in *FPPS2* produced significantly lower L^* (90.20±0.90), a^* (−0.91±0.34) and higher b^* (9.17±1.95) and YPC (1.25±0.42) than those with the "G" (90.44±0.93, −0.78±0.33, 8.24± 1.88 and 1.05±0.38). Seventy-four cultivars with the "A" SNP in *POD* showed significantly lower L^*

(90.05±0.90) than those with "G" (90.48±0.88). For *MK*, lines with the "C" SNP had significantly higher L^* (90.75±1.02) and lower YPC (1.09± 0.26) than those with the "G" (90.25±0.90 and 1.18±0.43).

Based on gene ontology analysis, all these 16 candidate genes are related to four main biological processes. Briefly, *PMM* is involved in GDP-mannose biosynthetic process; *MDPS*, *FPPS2*, *IPPS*, *DXR* and *MK* participate in the isoprenoid/terpenoid biosynthetic process; *ZISO*, *LYCB*, *PSY* and *BCH2* involve in the carotenoid biosynthetic/carotene process; and *PO*, *DLD*, *LBR*, *NCED4*, *LOX2* and *POD* are related to the oxidation-reduction process; these are corresponding to carotenoid precursor supply, carotenoid biosynthesis and carotenoid degradation, respectively (Rodríguez-Concepción et al., 2001; Qin et al., 2012; Zeng et al., 2015).

Effects of Favorable Alleles on Flour Color-Related Traits

To assess the pyramiding effect of favorable alleles of MTAs for flour color-related traits, we examined the number of favorable alleles in each accession, and the BLUP value for each trait was regressed against the number of favorable alleles. The favorable alleles of significantly associated SNPs showed additive effects on brightness, redness, yellowness and yellow pigment content (**Figure 2**). As the number of favorable alleles increased, L^*, b^* and YPC values also increased and a^* decreased, and linear regressions (R^2) between numbers of favorable alleles and phenotype were 0.94, 0.72, 0.91 and 0.87, respectively. Higher b^* and YPC are desirable for yellow alkaline noodles and human health, while most Chinese foods typically require a white or creamy flour type with lower b^* and YPC. In this situation, cultivars with minimum (Wennong 5, Lumai 21, Jinan 17 and Funo) or maximum (Huaimai 18, Zhongmai 875, Shan 715, Lumai 11 and Jinmai 61) numbers of favorable alleles for b^* and YPC can be used as breeding parents to achieve defined color and nutritional properties of end-use products.

CONCLUSIONS

We performed a genome-wide association analysis on 166 bread wheat cultivars using the wheat 90 and 660 K SNP arrays and 10 allele-specific markers, and identified 100 MTAs for flour color-related traits. Broad comparison of MTAs identified in this study with QTL in previous reports indicated many common loci conditioning flour color-related traits, and four MTAs detected were new, including MTAs on chromosome 7A（2）for L*, chromosome 2A（629，632，627-629，632，697 Mb）for a* and chromosome 7D（635，560，079-635，560，149 Mb）for YPC. Two and 11 loci explained much more phenotypic variation of a* and YPC than phytoene synthase 1 gene（*Psy1*）, respectively. Based on biochemical information and bioinformatics analyses 16 predicted candidate genes were related to carotenoid biosynthesis and degradation, terpenoid backbone biosynthesis and glycolysis/gluconeogenesis. We will confirm these candidate genes in bi-parental populations and do some gene function analysis in the future.

The genomic regions associated with flour color-related traits identified in this study bring new insights to understanding the genetic basis of these traits, and new markers are useful for wheat quality improvement by MAS. Moreover, the candidate genes may serve as promising targets for study of the molecular mechanisms underlying flour color-related traits in wheat. This study also confirmed that GWAS is a powerful approach to validate known genes for complex traits and identify novel loci.

✦ ACKNOWLEDGEMENTS

The authors are grateful to Prof. R. A. McIntosh, Plant Breeding Institute, University of Sydney, for critical review of this manuscript. The authors would like to thank the International Wheat Genome Sequencing Consortium for pre-publication access to IWGSC RefSeq v. 1. 0.

✦ REFERENCES

American Association for Cereal Chemistry （2000）. *Approved Methods of the AACC-Method*, 10th Edn. St Paul, MN: AACC, 14-50.

Ain, Q. U., Rasheed, A., Anwar, A., Mahmood, T., Imtiaz, M., Mahmood, T., et al. （2015）. Genome-wide association for grain yield under rain-fed conditions in historical wheat cultivars from Pakistan. *Front. Plant Sci.* 6: 743. doi: 10. 3389/fpls. 2015. 00743

Benjamini, Y., and Hochberg, Y. （1995）. Controlling the false discovery rate: a practical and powerful approach to multiple testing. *J. Roy. Stat. Soc.* 57, 289-300.

Blanco, A., Colasuonno, P., Gadaleta, A., Mangini, G., Schiavulli, A., Simeone, R., et al. （2011）. Quantitative trait loci for yellow pigment concentration and individual carotenoid compounds in durum wheat. *J. Cereal Sci.* 54, 255-264. doi: 10. 1016/j. jcs. 2011. 07. 002

Brachi, B., Faure, N., Horton, M., Flahauw, E., Vazquez, A., Nordborg, M., et al. （2010）. Linkage and association mapping of *Arabidopsis thaliana* flowering time in nature. *PLoS Genet.* 6: e1000940. doi: 10. 1371/journal. pgen. 1000940

Bradbury, P. J., Zhang, Z., Kroon, D. E., Casstevens, T. M., Ramdoss, Y., and Buckler, E. S. （2007）. TASSEL: software for association mapping of complex traits in diverse samples. *Bioinformatics* 23, 2633-2635. doi: 10. 1093/bioinformatics/btm308

Breseghello, F., and Sorrells, M. E. （2006）. Association mapping of kernel size and milling quality in wheat （*Triticum aestivum* L.）cultivars. *Genetics* 172, 1165-1177. doi: 10. 1534/genetics. 105. 044586

Buckler, E. S., and Thornsberry, J. M. （2002）. Plant molecular diversity and applications to genomics. *Curr. Opin. Plant Biol.* 5, 107-111. doi: 10. 1016/S1369-5266 （02）00238-8

Colasuonno, P., Gadaleta, A., Giancaspro, A., Nigro, D., Giove, S., Incerti, O., et al. （2014）. Development of a high-density SNP-based linkage map and detection of yellow pigment content QTLs in durum wheat. *Mol. Breed.* 34, 1563-1578. doi: 10. 1007/s11032-014-0183-3

Colasuonno, P., Lozito, M. L., Marcotuli, I., Nigro, D., Giancaspro, A., Mangini, G., et al. （2017）. The carotenoid biosynthetic and catabolic genes in wheat and their association with yellow pigments. *BMC Genomics* 18: 122. doi: 10. 1186/s12864-016-3395-6

Crawford, A. C., and Francki, M. G. (2013a). Chromosomal location of wheat genes of the carotenoid biosynthetic pathway and evidence for a catalase gene on chromosome 7A functionally associated with flour b* colour variation. *Mol. Genet. Genomics* 288, 483-493. doi: 10. 1007/s00438-013-0767-3

Crawford, A. C., and Francki, M. G. (2013b). *Lycopene-ε-cyclase* (*e-LCY3A*) is functionally associated with QTL for flour b* colour on chromosome 3A in wheat (*Triticum aestivum* L.). *Mol. Breed.* 31, 737-741. doi: 10. 1007/s11032-012-9812-x

Cui, F., Zhang, N., Fan, X. L., Zhang, W., Zhao, C. H., Yang, L. J., et al. (2017). Utilization of a Wheat 660K SNP array-derived high-density genetic map for high-resolution mapping of a major QTL for kernel number. *Sci. Rep.* 7: 3788. doi: 10. 1038/s41598-017-04028-6

Curtis, T., and Halford, N. G. (2014). Food security: the challenge of increasing wheat yield and the importance of not compromising food safety. *Ann. Appl. Biol.* 164, 354-372. doi: 10. 1111/aab. 12108

Dong, C. H. (2011). *Cloning of Genes Associated with Grain Yellow Pigment Content in Common Wheat and Development of Functional Markers*. [Master's Thesis]. [Baoding]: Agricultural University of Hebei Province.

Doyle, J. J., and Doyle, J. L. (1987). A rapid DNA isolation procedure from small quantities of fresh leaf tissues. *Phytochem. Bull.* 19, 11-15.

Evanno, G., Regnaut, S., and Goudet, J. (2005). Detecting the number of clusters of individuals using the software STRUCTURE: a simulation study. *Mol. Ecol.* 14, 2611-2620. doi: 10. 1111/j. 1365-294X. 2005. 02553. x

Flint-Garcia, S. A., Thuillet, A. C., Yu, J., Pressoir, G., Romero, S. M., Mitchell, S. E., et al. (2005). Maize association population: a high-resolution platform for quantitative trait locus dissection. *Plant J.* 44, 1054-1064. doi: 10. 1111/j. 1365-313X. 2005. 02591. x

Geng, H. W., He, Z. H., Zhang, L. P., Qu, Y. Y., and Xia, X. C. (2012). Development of functional markers for a lipoxygenase gene *TaLox-B1* on chromosome 4BS in common wheat. *Crop Sci.* 52, 568-576. doi: 10. 2135/cropsci2011. 07. 0365

Giroux, M. J., and Morris, C. F. (1997). A glycine to serine change in puroindoline b is associated with wheat grain hardness and low levels of starch-surface friabilin. *Theor.*

Appl. Genet. 95, 857-864. doi: 10. 1007/s00122 0050636

He, X. Y., He, Z. H., Ma, W. J., Appels, R., and Xia, X. C. (2009). Allelic variants of phytoene synthase 1 (*Psy1*) genes in Chinese and CIMMYT wheat cultivars and development of functional markers for flour colour. *Mol. Breed.* 23, 553-563. doi: 10. 1007/s11032-009-9255-1

He, X. Y., Zhang, Y. L., He, Z. H., Wu, Y. P., Xiao, Y. G., Ma, C. X., et al. (2008). Characterization of phytoene synthase 1 gene (*Psy1*) located on common wheat chromosome 7A and development of a functional marker. *Theor. Appl. Genet.* 116, 213-221. doi: 10. 1007/s00122-007-0660-8

Howitt, C. A., Cavanagh, C. R., Bowerman, A. F., Cazzonelli, C., Rampling, L., Mimica, J. L., et al. (2009). Alternative splicing, activation of cryptic exons and amino acid substitutions in carotenoid biosynthetic genes are associated with lutein accumulation in wheat endosperm. *Funct. Integr. Genomics* 9, 363-376. doi: 10. 1007/s10142-009-0 121-3

Hutchings, J. B. (1999). *Food Color and Appearance*. Gaithesburg, MD: Aspen Publishers, Inc.

Jin, H., Wen, W., Liu, J., Zhai, S., Zhang, Y., Yan, J., et al. (2016). Genome-wide QTL mapping for wheat processing quality parameters in a Gaocheng 8901/Zhoumai 16 recombinant inbred line population. *Front. Plant Sci.* 7: 1032. doi: 10. 3389/fpls. 2016. 01032

Liu, C., Yang, Z. J., Li, G. R., Zeng, Z. X., Zhang, Y., Zhou, J. P., et al. L (2008). Isolation of a new repetitive DNA sequence from *Secale africanum* enables targeting of *Secale* chromatin in wheat background. *Euphytica* 159, 249-258. doi: 10. 1007/s10681-007-9484-5

Liu, J., He, Z., Rasheed, A., Wen, W., Yan, J., Zhang, P., et al. (2017). Genome-wide association mapping of black point reaction in common wheat (*Triticum aestivum* L.). *BMC Plant Biol.* 17: 220. doi: 10. 1186/s12870-017-1167-3

Marcotuli, I., Houston, K., Waugh, R., Fincher, G. B., Burton, R. A., Blanco, A., et al. (2015). Genome-wide association mapping for arabinoxylan content in a collection of tetraploid wheats. *PloS ONE* 10: e0132787. doi: 10. 1371/journal. pone. 0132787

Mares, D., and Campbell, A. (2001). Mapping components of flour and noodle color in Australian wheat. *Aust. J. Agric. Res.* 52, 1297-1309. doi: 10. 1071/AR01048

Nagamine, T., Ikeda, T. M., Yanagisawa, T.,

Yanaka, M., and Ishikawa, N. (2003). The effects of hardness allele *Pinb-D1b* on the flour quality of wheat for Japanese white salty noodles. *J. Cereal Sci.* 37, 337-342. doi: 10.1006/jcrs. 200 2.0505

Oliver, J. R., Blakeney, A. B., and Allen, H. M. (1992). Measurement of flour color in color space parameters. *Cereal Chem.* 69, 546-551.

Patil, R. M., Oak, M. D., Tamhankar, S. A., Sourdille, P., and Rao, V. S. (2008). Mapping and validation of a major QTL for yellow pigment content on 7AL in durum wheat (*Triticum turgidum L. ssp. durum*). *Mol. Breed.* 21, 485-496. doi: 10.1007/s11032-007-9147-1

Pritchard, J. K., Stephens, M., Rosenberg, N. A., and Donnelly, P. (2000). Association mapping in structured populations. *Am. J. Hum.* Genet. 67, 170-181. doi: 10.1086/302959

Qin, X., Zhang, W., Dubcovsky, J., and Tian, L. (2012). Cloning and comparative analysis of carotenoid β-hydroxylase genes provides new insights into carotenoid metabolism in tetraploid (*Triticum turgidum* ssp. durum) and hexaploid (*Triticum aestivum*) wheat grains. *Plant Mol. Biol.* 80, 631-646. doi: 10.1007/s11103-012-9972-4

Rodríguez-Concepción, M., Ahumada, I., Diez-Juez, E., Sauret-Güeto, S., Lois, L. M., Gallego, F., et al. (2001). 1-Deoxy-D-xylulose 5-phosphate reductoisomerase and plastid isoprenoid biosynthesis during tomato fruit ripening. *Plant J.* 27, 213-222. doi: 10.1046/j. 1365-313x. 2001. 01089. x

Roncallo, P. F., Cervigni, G. L., Jensen, C., Miranda, R., Carrera, A. D., Helguera, M., et al. (2012). QTL analysis of main and epistatic effects for flour color traits in durum wheat. *Euphytica* 185, 77-92. doi: 10.1007/s10681-012-0628-x

Sun, C., Zhang, F., Yan, X., Zhang, X., Dong, Z., Cui, D., et al. (2017). Genome-wide association study for 13 agronomic traits reveals distribution of superior alleles in bread wheat from the Yellow andHuai Valley of China. *Plant Biotechnol. J.* 15, 953-969. doi: 10.1111/pbi. 12690

Tadesse, W., Ogbonnaya, F. C., Jighly, A., Sanchez-Garcia, M., Sohail, Q., Rajaram, S., et al. (2015). Genome-wide association mapping of yield and grain quality traits in winter wheat genotypes. *PLoS ONE* 10: e0141339. doi: 10.1371/journal. pone. 0141339

Tsilo, T. J., Hareland, G. A., Chao, S., and Anderson, J. A. (2011). Genetic mapping and QTL analysis of flour color and milling yield related traits using recombinant inbred lines in hard red spring wheat. *Crop Sci.* 51, 237-246. doi: 10.2135/cropsci2009. 12. 0711

Wang, M., Yan, J, Zhao, J., Song, W., Zhang, X., Xiao, Y., et al. (2012). Genome-wide association study (GWAS) of resistance to head smut in maize. *Plant Sci.* 196, 125-131. doi: 10.1016/j. plantsci. 2012. 08. 004

Wang, S. C., Wong, D., Forrest, K., Allen, A., Chao, S., Huang, B. E., et al. and Akhunov, E. (2014). Characterization of polyploid wheat genomic diversity using the high-density 90, 000 SNP array. *Plant Biotech. J.* 12, 787-796. doi: 10.1111/pbi. 12183

Waugh, R. J., Jannink, L., Muehlbauer, G. J., and Ramsey, L. (2009). The emergence of whole genome association scans in barley. *Curr. Opin. Plant Biol.* 12, 218-222. doi: 10.1016/j. pbi. 2008. 12. 007

Yu, J., and Buckler, E. S. (2006). Genetic association mapping and genome organization of maize. *Curr. Opin. Biotechnol.* 17, 155-160. doi: 10.1016/j. copbio. 2006. 02. 003

Yu, J., Pressoir, G., Briggs, W. H., Vroh Bi, I., Yamasaki, M., Doebley, J. F., et al. (2006). A unified mixed-model method for association mapping that accounts for multiple levels of relatedness. *Nat. Genet.* 38, 203-208. doi: 10.1038/ng1702

Zeng, J., Wang, C., Chen, X., Zang, M. L., Yuan, C. H., Wang, X. T., et al. (2015). The lycopene β-cyclase plays a significant role in provitamin A biosynthesis in wheat endosperm. *BMC Plant Biol.* 15: 112. doi: 10.1186/s12870-015-0514-5

Zhai, S., Xia, X., and He, Z. (2016a). Carotenoids in staple cereals: metabolism, regulation, and genetic manipulation. *Front. Plant Sci.* 7: 1197. doi: 10.3389/fpls. 2016. 01197

Zhai, S. N., He, Z. H., Wen, W. E., Jin, H., Liu, J. D., Zhang, Y., et al. (2016b). Genome-wide linkage mapping of flour color-related traits and polyphenol oxidase activity in common wheat. *Theor. Appl. Genet.* 129, 377-394. doi: 10.1007/s00122-015-2634-6

Zhai, S. N., Li, G. Y., Sun, Y. W., Song, J. M., Li, J. L., Song, G. Q., et al. (2016c). Genetic analysis of phytoene synthase 1 (*Psy1*) gene function and regulation in common wheat. *BMC Plant Biol.* 16: 228. doi: 10.1186/s12870-016-0916-z

Zhang, C. Y., Dong, C. H., He, X. Y., Zhang, L. P., Xia, X. C., and He, Z. H. (2011). Allelic variants at the *TaZds-D1* locus on wheat chromosome 2DL and their association with yellow pigment content. *Crop Sci. 51*, 1580-1590. doi: 10. 2135/cropsci2010. 12. 0689

Zhang, W. , and Dubcovsky, J. (2008). Association between allelic variation at the *phytoene synthase 1* gene and yellow pigment content in the wheat grain. *Theor. Appl. Genet.* 116, 635-645. doi: 10. 1007/s00122-007-0697-8

Zhang, Y. L. , Wu, Y. P. , Xiao, Y. G. , He, Z. H. , Zhang, Y. , Yan, J. , et al. (2009). QTL mapping for flour and noodle colour components and yellow pigment

content in common wheat. *Euphytica* 165, 435-444. doi: 10. 1007/s10681-008-9744-z

Zhao, K. Y. , Tung, C. W. , Eizenga, G. C. , Wright, M. H. , Ali, M. L. , Price, A. H. , et al. (2011). Genome-wide association mapping reveals a rich genetic architecture of complex traits in *Oryza sativa*. *Nat. Commun.* 2: 467. doi: 10. 1038/ncomms1467

Zhu, C. S. , Gore, M. , Buckler, E. S. , and Yu, J. M. (2008). Status and prospects of association mapping in plants. *Plant Genome* 1, 5-20. doi: 10. 3835/plantgenome 2008. 02. 0089

Genetic architecture of polyphenol oxidase activity in wheat flour by genome-wide association study

Shengnan Zhai[1,2] Zhonghu He[2,3] Weie Wen[2] Jindong Liu[2] Hui Jin[2]
Jun Yan[4] Yong Zhang[2] Pingzhi Zhang[5] Yingxiu Wan[5] Xianchun Xia[2]

[1] Crop Research Institute, National Engineering Lab. for Wheat and Maize, Key Lab. of Wheat Biology and Genetic Improvement in the Northern Yellow-Huai Rivers Valley of Ministry of Agriculture, Shandong Academy of Agricultural Sciences, Jinan, Shandong, 250100, China

[2] Institute of Crop Sciences, National Wheat Improvement Center, Chinese Academy of Agricultural Sciences (CAAS), 12 Zhongguancun South St., Beijing, 100081, China

[3] International Maize and Wheat Improvement Center (CIMMYT) China Office, c/o CAAS, 12 Zhongguancun South St., Beijing, 100081, China

[4] Institute of Cotton Research, Chinese Academy of Agricultural Sciences (CAAS), 38 Huanghe St., Anyang, Henan, 455000, China

[5] Crop Research Institute, Anhui Academy of Agricultural Sciences, 40 Nongke South St., Hefei, Anhui, 230001, China

Abstract: The primary cause of time-dependent discoloration in wheat (*Triticum aestivum* L.) - based products is polyphenol oxidase (PPO) activity. A comprehensive genetic characterization of PPO activity in wheat could expedite the development of wheat varieties with low PPO activity. To dissect the genetic architecture of PPO activity in wheat flour, a panel of 166 diverse bread wheat cultivars was phenotyped in four different environments and genotyped by high-density wheat 90K and 660K arrays. A genome-wide association study (GWAS) was performed by a mixed linear model that incorporated population structure and relative kinship. Four hundred and sixty-five stable marker-trait associations representative of 43 quantitative trait loci (QTL) for PPO activity identified in at least two environments were located on all wheat chromosomes except 5A, with each explaining 6.6-32.4% of the phenotypic variance. Based on IWGSC RefSeq v1.0, we found that *QPPO2A.3*, *QPPO2B.1*, *QPPO2B.2*, and *QPPO2D.2* were consistent with the previously reported QTL, and 12 QTL located on homoeologous group 1 chromosomes (6 QTL), chromosomes 4B, 4D, and 7A (2 QTL), and chromosome 7B (2 QTL) are likely to be new PPO loci. Based on physical positions of QTL and previous studies, *Ppo-A1*, *Ppo-A2*, *Ppo-B2*, *Ppo-D2*, *Pinb-D1*, and genes encoding a Cu ion-binding protein and a Cu-transporting ATPase RAN1-like protein were considered as candidates for *QPPO2A.3*, *QPPO2B.1*, *QPPO2D.2*, *QPPO5D.1*, *QPPO4A.2*, and *QPPO7D.2*, respectively. This study provided insights into the molecular basis of wheat PPO activity, and associated markers can be used for selection of low PPO breeding lines as a means of improving the marketing quality of wheat products.

Published in Crop Science, 2020, 1-13.

Abbreviations: AY, Anyang; BLUP, best linear unbiased predictor; GWAS, genome-wide association study; IWGSC, International Wheat Genome Sequencing Consortium; L-DOPA, L-3, 4-dihydroxyphenylalanine; LD, linkage disequilibrium; MTA, marker-trait association; PIC, polymorphism information content; PPO, polyphenol oxidase; QTL, quantitative trait locus/loci; SNP, single nucleotide polymorphism; SX, Suixi.

1 INTRODUCTION

Polyphenol oxidase (PPO) activity in flour is considered a major contributor to time-dependent discoloration anddarkening of wheat (*Triticum aestivum* L.) products (Anderson & Morris, 2003; Baik, Czuchajowska, & Pomeranz, 1995; Fuerst, Anderson, & Morris, 2006; Mares & Campbell, 2001; Morris, Jeffers, & Engle, 2000). Polyphenol oxidase is a Cu-containing metalloenzyme that catalyzes phenolic compounds to quinones, which in turn react with thiol and amine groups or by self-polymerization to generate melanin. The discoloration greatly reduces the quality of wheat end-use products and affects consumer acceptance (Anderson & Morris, 2001; Baik & Ullrich, 2008; Feillet, Autran, & Vernière, 2000; Jukanti, Bruckner, & Fischer, 2004; Morris et al. , 2000; Simeone, Pasqualone, Clodoveo, & Blanco, 2002). Therefore, one of the priorities in wheat breeding is the development of wheat varieties with low-level PPO activity (Ransom, Berzonsky, & Sorenson, 2006).

A thorough understanding of the genetics of PPO activity in wheat flour and derivatives could provide better strategies to resolve the issue of the undesirable discoloration in wheat breeding programs for quality improvement. To date, many attempts have been made to determine the molecular basis of PPO activity in wheat. Several quantitative trait loci (QTL) studies have shown that the main genes influencing PPO activity are located on homoeologous group 2 chromosomes (Anderson & Morris, 2001; He et al. , 2007; He, He, Morris, & Xia, 2009; Jimenez & Dubcovsky, 1999; Raman et al. , 2005; Sadeque & Turner, 2010; Watanabe, Akond, &

Nachit, 2006; Zhang et al. , 2005). Other QTL with minor effects were mapped to homoeologous groups 3, 5, and 6 and on chromosomes 4A, 4B, 7B, and 7D (Demeke et al. , 2001; Fuerst, Xu, & Beecher, 2008; Li et al. , 1999; Sadeque & Turner, 2010; Simeone et al. , 2002; Udall, 1997).

The PPO genes on homoeologous group 2 chromosomes, named *Ppo-A1*, *Ppo-B1*, and *Ppo-D1*, have been cloned and shown to co-segregate with wheat PPO activity (Demeke & Morris, 2002; He et al. , 2007; Jukanti et al. , 2004; Si, Zhou, Wang, & Ma, 2012; Sun et al. , 2005). In addition, three paralogous *Ppo* genes, *Ppo-A2*, *Ppo-B2*, and *Ppo-D2*, are also located to homoeologous group 2 chromosomes. They have also been cloned and mapped 8. 9, 11. 4, and 10. 7 cM from their respective *Ppo1* counterparts (Beecher & Skinner, 2011; Beecher, Carter, & See, 2012; Taranto, Mangini, Pasqualone, Gadaleta, & Blanco, 2015). In one study, the expression level of *Ppo2* genes accounted for 72% of the total *Ppo* gene transcription level in developing grains, indicating that *Ppo2* genes play an important role in overall *Ppo* gene expression (Beecher & Skinner, 2011). However, in another study, *Ppo-A1* and *Ppo-D1* contributed much more substantial genotypic effects on PPO activity than *Ppo2* (Hystad, Martin, Graybosch, & Giroux, 2015). Thus, the relative effects of *Ppo1* and *Ppo2* genes on wheat PPO activity remain inconclusive.

Although the foregoing research has provided a deeper understanding of PPO activity and the means to select wheat lines with lower PPO activity by pyramiding favorable alleles at multiple PPO loci, the discoloration remains to be observed in noodles (Morris, 2018). Therefore, there might be other

QTL for PPO activity that have not been detected. The disadvantage of traditional QTL analysis using biparental populations is that the number and resolution of alleles is limited in each cross. In contrast, genome-wide association studies (GWAS) can increase the resolution to identify multiple alleles at known loci or new QTL by use of germplasms with undefined molecular variation (Buckler & Thornsberry, 2002; Flint-Garcia et al., 2005; Waugh, Jannink, Muehlbauer, & Ramsey, 2009). As an effective, alternative means of genetic analysis, GWAS has been widely applied in the genetic studies of complex traits in wheat (Liu et al., 2017; Tadesse et al., 2015; Zhai et al., 2018). Based on the high-throughput genotyping assays by wheat 90K and 660K single nucleotide polymorphism (SNP) arrays in the present study, GWAS will be used to further investigate the molecular basis of PPO activity.

To further investigate the regulation of PPO activity, it was considered necessary to map loci for PPO activity using diverse germplasm and high-throughput SNP genotyping arrays. The aims of our study were (a) to dissect genomic regions controlling PPO activity by GWAS using the wheat 90K and 660K arrays and an association panel of 166 bread wheat cultivars, and (b) to predict candidate genes involved in PPO activity. The study has potential to provide information on the genetic architecture of PPO activity and to provide markers for developing wheat varieties with low PPO activity and thus improve end-use and marketing quality of wheat-based products.

2　MATERIALS AND METHODS

2.1　Mapping population and field trials

A total of 166 elite bread wheat cultivars representing a wide range of genetic variation were used to identify genomic regions controlling PPO activity. Briefly, all the 166 cultivars were winter wheat, including 144 accessions from the Yellow and Huai River Valley Facultative Wheat Region of China, nine from Italy, seven from Argentina, four from Japan, and one each from Australia and Turkey. Details relating to the panel are provided in Supplemental **Table S1**.

All 166 accessions were grown during the 2012-2013 and 2013—2014 growing seasons in complete randomized blocks with three replications at Anyang in Henan Province and Suixi in Anhui Province (hereafter referred to as 2013 AY, 2013 SX, 2014 AY, and 2014 SX, respectively). Each plot comprised three 2-m rows in 2012—2013 and four 2-m rows in 2013—2014, with 0.2-m spacing. Fertilizer, irrigation water, and herbicide applications followed the local agricultural practices.

2.2　Polyphenol oxidase activity assays

Grains of each cultivar were milledby a Brabender Quadrumat Junior mill fitted with a 0.25-mm screen. Prior to milling, cultivars with soft, medium, and hard grain texture were conditioned to 14, 15, and 16% moisture content, respectively. The PPO activities of the flour were determined by a modified protocol of Anderson and Morris (2001). Briefly, 1 g flour was incubated in 7.5 ml of 10 mM L-DOPA (L-3, 4-dihydroxyphenylalanine) in 50 mM MOPS [3-(N-morpholino) propane sulfonic acid] buffer (pH 6.5), and constantly rotated on a reciprocating shaker for 30 min at room temperature. The reaction solution was centrifuged at 5,000 rpm for 10 min, and absorbance of 200 μl of supernatant was measured at 475 nm in a SpectraMax Plus 384 Microplate Reader (Molecular Devices). The assay was standardized by including a bulk flour sample from cv. Zhongmai 175 with each batch. Each sample was assayed twice and three individual absorbance measurements per extracted sample were recorded. Absorbance values were averaged and converted to PPO activity units (U g^{-1} min^{-1}). The mean values for each environment, shown in Supplemental **Table S1**, were used in further analyses.

2.3　Statistical analysis

Statistical analyses were done using the SAS software version 9.2 (SAS Institute). Briefly, ANOVA was performed with the PROC MIXED procedure, where environments were treated as fixed effects, and lines, line×environment interaction, and replicates nested in environments were all treated as random effects. Broad-sense heritability ($h2$) of PPO activity was calculated as $h^2 = \sigma_g^2/ (\sigma_g^2 + \sigma_{ge}^2/r + \sigma_\varepsilon^2/re)$, where σ_g^2, σ_{ge}^2, and σ_ε^2 are variances for genotype, genotype×environment interaction, and residual error, respectively, whereas r and e are the numbers of replications and environments. A best linear unbiased predictor (BLUP) for each cultivar was calculated across four environments using the PROC MIXED procedure and used for subsequent analyses. Pearson's correlation coefficients of PPO activity among different environments were calculated using the PROC CORR procedure based on BLUP values. Scatterplots were assessed in Microsoft Excel 2016.

2.4 Phenotypic characterization

Total genomic DNA was isolated from seedling leaves of each accession following the CTAB (cetyl trimethylammonium bromide) protocol (Doyle & Doyle, 1987). All 166 accessions were genotyped using the high-density Illumina wheat 90K (consisting of 81,587 SNPs) and Axiom wheat 660K (630,517 SNPs) SNP arrays at the Capital Bio Corporation in Beijing (http: //www. capitalbio. com/). The SNP allele clustering was performed with GenomeStudio software (Illumina) using the default filtering and clustering algorithm, and the accuracy was then validated by manual curation.

After removing redundant markers with minor allele frequencies＜5％ and missing rate＞20％, 259,922 SNP markers were used for further analyses. Based on the International Wheat Genome Sequencing Consortium (IWGSC, 2018), the physical positions of these SNP markers were identified and used for GWAS. Genetic diversity and polymorphism infor-mation content (PIC) were computed by PowerMarker version 3.25 (http: // statgen. ncsu. edu/powermarker/).

2.5 Population structure and linkage disequilibrium analyses

Population structure was analyzed by STRUCTURE version 2.3.3 (Pritchard, Stephens, Rosenberg, & Donnelly, 2000) using 2,000 markers evenly distributed across the entire genome. Hypothetical subpopulations (k) in a range of 2-12 were assessed with admixture and correlated allele frequencies models, and five independent runs for each k were conducted using a burn-in of 10,000 iterations followed by 100,000 Monte Carlo Markov chain iterations. The optimum number of subpopulations was determined following Evanno, Regnaut, and Goudet (2005). For the selected k, five runs were performed again, and the Q matrix was estimated whereby most cultivars were assigned to a subpopulation at a probability of ＞0.5. Subpopulation identity of each accession is listed in Supplemental **Table S1**.

Linkage disequilibrium (LD) was measured in TASSEL version 5.0 (Bradbury et al., 2007) using 259,922 markers for the whole genome and 89,519, 146,270, and 24,133 markers for the A, B, and D subgenomes, respectively. Pairwise LD was measured using squared allele frequency correlation coefficients (r^2), and significance was calculated with 1,000 permutations. The LD was calculated separately for unlinked loci and for loci on the same chromosome. Intrachromosomal r^2 values at $P＜0.001$ were plotted against the physical distance, and a LOESS (locally weighed polynomial regression) curve was fitted to the plot to check how fast LD decay occurs. The critical value for r^2, beyond which LD is likely to be caused by true physical linkage, was measured by taking the 95th percentile of the square-root-transformed r^2 values of unlinked markers as reported by Breseghello and Sorrells (2006). The physical distance where LD fell below the critical r^2 value was also used to obtain the confidence intervals of marker-trait associations (MTAs) identified in the present study as mentioned in previous studies (Khatkar et al., 2008; Lawrence et al., 2009; Shifman, Kuypers, Kokoris, Yakir, & Darvasi, 2003).

2.6 Marker-trait association analysis

To avoid spurious associations, MTA analysis was

performed by TASSEL version 5. 0 (Bradbury et al., 2007) with the mixed linear model, which incorporates kinship and population structure. The association mapping model was evaluated by quantile-quantile plots constructed by plotting the observed P value against the expected P value in a negative $\log_{10}$ scale under the null hypothesis that there was no correlation between markers and phenotypes. The false discovery rate < 0.05 was used as a threshold to determine whether SNP were significantly associated with PPO activity (Benjamini & Hochberg, 1995), and R^2 value was used as an estimate to indicate the percentage of PPO activity variation explained by the significant marker. To provide a complimentary summary of significant SNPs, Manhattan plots were drawn by the R language (http: //www. r-project. org/) to provide a comprehensive display of significant SNPs.

To compare QTL identified in the present study with those reported previously, we first searched sequences of restriction fragment length polymorphism (RFLP), simple sequence repeat (SSR), or Diversity Arrays Technology (DArT) markers underlying PPO activity in previous studies (Beecher et al., 2012; Demeke et al., 2001; He et al., 2007; Li et al., 1999; Raman et al., 2005; Raman, Raman, & Martin, 2007; Sun et al., 2005; Taranto et al., 2015; Watanabe et al., 2006; Zhang et al., 2005), based on the database in Grain-Genes (https: //wheat. pw. usda. gov/GG3/). Then, the marker sequences were blasted against IWGSC RefSeq v1. 0 to find out their regions on the physical map. The physical positions of markers linked to PPO activity reported in previous studies were listed in Supplemental Table S2. The QTL located within 8 Mb were considered to be the same.

2. 7　The effect of favorable alleles on polyphenol oxidase activity

The alleles with positive effects leading to lower PPO activity were defined as "favorable alleles," whereas those resulting in higher PPO activity were referred to

as "unfavorable alleles." The pyramiding effect of each favorable allele of QTL identified in this study was assessed by regression analysis between the BLUP value and number of accumulated favorable alleles in each accession using the line chart function in Microsoft Excel 2016.

2. 8　Prediction of candidate genes

To find the candidate genes linked to significant markers, we used the IWGSC RefSeq v1. 0 database to retrieve high-confidence annotated genes located $\pm$ 250 kb proximal to each MTA, as described by Beyer et al. (2019). Survey sequences to which the SNPs were best hits were screened in a BLASTN (Nucleotide Basic Local Alignment Search Tool) search against the National Center for Biotechnology Information (NCBI, http: //www. ncbi. nlm. nih. gov/) and European Nucleotide Archive (ENA) (http: //www. ebi. ac. uk/ena) to confirm the gene annotations by evaluating orthologous genes in related species, such as *Triticum urartu* Thumanian ex Gandilyan, *Aegilops tauschii* Coss., and *Hordeum vulgare* L. The BLAST hits were filtered with an e value threshold of 10^{-5} and sequence similarities $>$ 75%. Specific candidate genes of interest were selected based on the criteria of close proximity to the significant SNPs, and possible involvement in PPO activity reported in literature. All gene descriptions were shown in Supplemental Table S3. Additionally, significantly associated markers on homoeologous group 2 chromosomes were also searched for correspondence with genomic regions of the *Ppo1* and *Ppo2* gene series based on IWGSC RefSeq v1. 0.

3　RESULTS

3. 1　Phenotypic variance of polyphenol oxidase activity

There was substantial variation of flour PPO activity among 166 wheat cultivars, ranging from 0. 95 to 10. 95 U g^{-1} min^{-1} across four environments (Table 1, Supplemental Table S1). The frequency distribution among accessions was skewed towards low PPO activity (Supplemental Figure S1). The PPO activities from different environments were highly correlated, ranging from 0. 81 ($P<10^{-4}$) between 2013 AY and 2014 AY to 0. 92 ($P<10^{-4}$) between 2014 AY and 2014 SX (Supplemental Table S4). Scatterplots also indicated a high repeatability of PPO activity among different environments (Supple-mental Figure S2).

TABLE 1 **Phenotypic variation in polyphenol oxidase activity in 166 common wheat cultivars across four environments**

Environment[a]	Mean[b]	SD	Min.	Max.
	U g^{-1} min^{-1}			
2013 AY	5. 82	2. 00	1. 46	10. 95
2013 SX	5. 35	1. 94	1. 87	10. 42
2014 AY	3. 26	1. 15	0. 95	6. 57
2014 SX	4. 47	1. 78	1. 27	9. 52
BLUP	4. 72	1. 62	1. 41	9. 37

[a] AY, Anyang; SX, Suixi; BLUP, best linear unbiased predictor.

[b] Polyphenol oxidase activity.

The ANOVA revealed highly significant effects of genotypes, environments, and their interactions on PPO activity, and the magnitude of the genotypic contribution was much larger than others (Supplemental Table S5). A high broad-sense heritability (0. 91) was estimated, indicating that the phenotypic variance of PPO activity was largely influenced by genetic factors rather than environmental effects.

3. 2 Population structure and linkage disequilibrium

Based on the second order rate of change in the log probability, a k value of 3, at a clear peak, was the optimum number of subpopulations following the ad-hoc quantity statistic ΔK described by Evanno et al. (2005). Generally, the subpopulations were in agreement with geographical groups, as indicated in our previous study (Zhai et al., 2018). Subpopulation 1 comprised a majority of cultivars from Shandong Province and foreign countries, whereas the Subpopulation 2 included lines from Anhui, Henan, and Shaanxi Provinces. Subpopulation 3 consisted of varieties predominantly bred in Henan and Hebei

Provinces. There were significant differences in PPO activity between subpopulations, except for Sub-populations 1 and 2. Subpopulation 3 had a significantly higher mean PPO activity than the other two subpopulations ($P<10^{-5}$).

The 95th percentile of r^2 calculated based on unlinked markers on different chromosomes was 0. 082. Then any r^2 value larger than this was considered as significant LD based on linkage. Based on this threshold, the extent of elevated LD for subgenomes A, B, and D were 6, 4, and 11 Mb, respectively. On average, the entire genome showed an extent LD of 8 Mb. This was consistent with Kidane et al. (2019), where LD decay was 7. 4 Mb on average, ranging from 3. 7 (chromosome 4A) to 18. 7 Mb (chromosome 3B).

3. 3 Marker-trait associations for polyphenol oxidase activity

As demonstrated in the quantile-quantile plots, the mixed linear model greatly reduced the occurrence of false positives andwas suitable for the present association analysis of PPO activity (Supplemental Figure S3). A total of 465 SNP markers were significantly associated with flour PPO activity in multiple environments (Figure 1, Supplemental Table S3). To determine QTL, all significant SNPs located in close physical proximity were clustered at $r^2>0. 082$, and only the most significant trait-associated SNP in each LD block is reported. Finally, 43 significantly associated loci were mapped on almost all chromosomes except chromosome 5A, individually explaining 6. 6-32. 4% phenotypic variances (Table 2). Thirteen QTL were identified in all four environments, whereas nine were detected in three environments, and the remaining 21 were found in two environments. The three most significant QTL were mapped on chromosomes 2A (*QPPO2A. 3*), 2B (*QPPO2B. 1*), and 2D (*QPPO2D. 2*), explaining 17. 5-32. 4, 15. 0-25. 5, and 13. 6-24. 9% of the phenotypic variances, respectively. Their effects on PPO activity were shown in Figure 2.

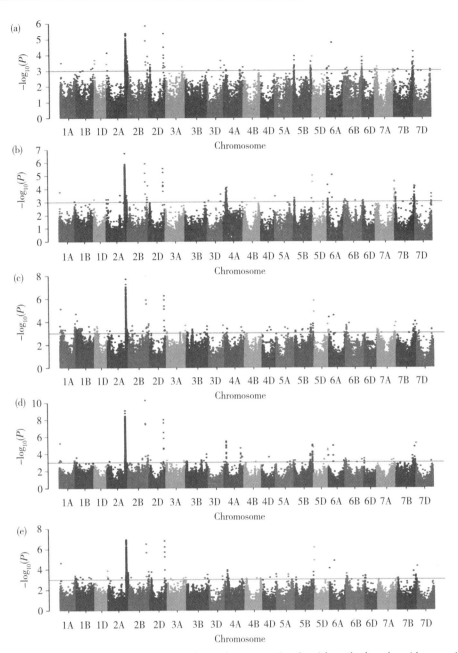

FIGURE 1　Manhattan plots indicating genomic regions associated with polyphenol oxidase activity in four environments: (a) 2013 Anyang (AY), (b) 2013 Suixi (SX), (c) 2014 AY, (d) 2014 SX, and (e) best linear unbiased predictor (BLUP). Negative $\log_{10}$-transformed P values from a genome-wide scan were plotted against single nucleotide polymorphism marker positions on each wheat chromosome. Blue horizontal line indicates the significant association threshold of $-\log_{10}$ (P) ⩾3

Based on IWGSC RefSeq v1.0, we further found that *QPPO2A.3*, *QPPO2B.1*, *QPPO2B.2*, and *QPPO2D.2* were consistent with the previously reported QTL (Beecher et al., 2012; He et al., 2007; Raman et al., 2005, 2007; Zhai et al., 2016), confirming that these QTL have key roles in PPO activity (Table 2, Supplemental Table S2). *QPPO4B.1*, *QPPO7B.1*, and *QPPO7B.2* were different from *QPPO. caas-4BL*

(Zhai et al., 2016) and *QPPO. dmh45-7B* (Sadeque & Turne, 2010), respectively, with physical distances of >23.7 Mb (Supplemental Table S2). It is not clear whether QTL mapped on homoeologous groups 2, 3, 5, and 6 and chromosomes 4A and 7D in the present and previous studies (Li et al., 1999; Sadeque & Turne, 2010; Udall, 1997) are the same or not, due to lack of marker information or unavailable sequences of some RFLP markers. Notably, six QTL on homoeologous group 1 chromosomes, *QPPO4D. 1*, *QPPO7A. 1*, and *QPPO7A. 2* are likely to be new, considering that no QTL for PPO activity were previously mapped on these chromosomes.

3. 4　Effects of favorable alleles on polyphenol oxidase activity

The favorable alleles of strongly associated SNPs showed additive effects on PPO activity (Figure 3). Polyphenol oxidase activity decreased with the number of favorable alleles, and linear regressions ($R2$) were as high as 0. 86. Cultivars with lower PPO activity can potentially produce Asian noodles and other wheat end-use products with desirable color. Therefore, cultivars with maximum numbers of favorable alleles, such as Funo, Kanto 107, Zhengzhou 3, and Barra, can be used as parents to breed wheat varieties with low PPO activity and desirable color of end-use products.

3. 5　In silico functional annotation of SNPs associated with polyphenol oxidase activity

A total of 398 high-confidence protein-coding genes closely linked (± 250 kb) to 465 significant SNP markers underlying PPO activity were described based on IWGSC RefSeq v1. 0 (Supplemental Table S3). Seven potential candidate genes were identified for PPO activity according to the physical positions of QTL and previous studies. Briefly, *AX _ 109465981* (2A: 712, 239, 612-712, 239, 682 bp) was 0. 05 Mb away from *Ppo-A1* (2A: 712, 187, 200-712, 188, 721 bp) and 0. 1 Mb from *Ppo-A2* (712, 344, 578-712, 346, 518bp); *wsnp _ Ra _ c10658 _ 17500498* (2B: 689, 871, 339-689, 871, 539 bp) and *AX _ 95205011* (2D: 573, 969, 879-573, 969, 949 bp) was 0. 1 and 0. 06 Mb away from *Ppo-B2* (2B: 689, 765, 730-689, 766, 587 bp) and *Ppo-D2* (2D: 573, 904, 022-573, 905, 141 bp), respectively. *Kukri _ c77040 _ 87* (4A: 625, 862, 284-625, 862, 384 bp), corresponding to a copper ion-binding protein, was 1. 37 Mb from *QPPO4A. 2*; *BS00000020 _ 51* (5D: 3, 609, 844-3, 609, 944 bp), the most significantly associated SNP at *QPPO5D. 1*, corresponds to puroindoline b (*Pinb-D1*); *AX _ 94749119* (7D: 580, 574, 591-580, 574, 661 bp), closely linked to *QPPO7D. 2*, encodes a Cu-transporting ATPase RAN1-like protein. They are likely candidate genes for the QTL in these genomic regions.

TABLE 2　**Quantitative trait loci (QTL) for polyphenol oxidase activity in 166 wheat cultivars using a mixed linear model**

Name	Marker	Chr[a]	Physical position[b]	Environment[c]	P value	R[2d]	QTL[e]
			bp			%	
QPPO1A. 1	*AX _ 94858130*	1A	32, 281, 136—32, 281, 206	E1, E2, E3, E4, E5	$5.73 \times 10^{-6} - 3.20 \times 10^{-4}$	8. 4—14. 5	
QPPO1A. 2	*AX _ 111093292*	1A	558, 292, 426—558, 292, 496	E2, E3, E4, E5	$1.40 \times 10^{-4} - 8.91 \times 10^{-4}$	6. 6—11. 7	
QPPO1B. 1	*AX _ 94921536*	1B	50, 186, 199—50, 186, 269	E3, E4, E5	$7.83 \times 10^{-5} - 8.57 \times 10^{-4}$	7. 2—12. 8	
QPPO1B. 2	*AX _ 95152096*	1B	248, 236, 014—248, 236, 084	E1, E3, E5	$3.67 \times 10^{-4} - 9.64 \times 10^{-4}$	7. 3—10. 4	
QPPO1D. 1	*BS00079462 _ 51*	1D	27, 576, 013—27, 576, 113	E1, E2	$2.10 \times 10^{-4} - 6.44 \times 10^{-4}$	7. 4—9. 2	
QPPO1D. 2	*AX _ 110914190*	1D	482, 835, 286—482, 835, 356	E1, E3, E5	$1.39 \times 10^{-4} - 5.57 \times 10^{-4}$	9. 8—11. 5	
QPPO2A. 1	*AX _ 110038506*	2A	5, 108, 253—5, 108, 323	E1, E2, E5	$7.26 \times 10^{-5} - 7.77 \times 10^{-4}$	7. 0—10. 1	
QPPO2A. 2	*AX _ 94634529*	2A	93, 190, 895—93, 190, 965	E2, E3, E4	$6.22 \times 10^{-4} - 7.45 \times 10^{-4}$	7. 2—7. 8	

（continued）

Name	Marker	Chr[a]	Physical position[b]	Environment[c]	P value	R[2d]	QTL[e]
QPPO2A. 3	AX_109465981	2A	712, 239, 612— 712, 239, 682	E1, E2, E3, E4, E5	$2.49 \times 10^{-8} - 1.24 \times 10^{-5}$	17.5—32.4	He et al., 2007; Raman et al., 2005, 2007; Zhai et al., 2016
QPPO2A. 4	AX_109378069	2A	753, 566, 703— 753, 566, 773	E1, E2, E3, E5	$2.22 \times 10^{-4} - 6.90 \times 10^{-4}$	9.3—11.3	
QPPO2B. 1	wsnp_Ra_c10658_17500498	2B	689, 871, 339— 689, 871, 539	E1, E2, E3, E4, E5	$2.05 \times 10^{-9} - 1.41 \times 10^{-6}$	15.0—25.5	He et al., 2007
QPPO2B. 2	AX_110403106	2B	786, 051, 908— 786, 051, 978	E1, E2, E3, E4, E5	$5.64 \times 10^{-5} - 6.36 \times 10^{-4}$	9.3—11.4	He et al., 2007
QPPO2D. 1	AX_111593514	2D	69, 244, 597— 69, 244, 667	E1, E4, E5	$5.14 \times 10^{-4} - 6.76 \times 10^{-4}$	7.4—9.6	
QPPO2D. 2	AX_95205011	2D	573, 969, 879— 573, 969, 949	E1, E2, E3, E4, E5	$9.46 \times 10^{-9} - 4.37 \times 10^{-6}$	13.6—24.9	Beecher et al., 2012; He et al., 2007
QPPO3A. 1	AX_110057689	3A	711, 090, 705— 711, 090, 775	E3, E4	$2.15 \times 10^{-4} - 6.85 \times 10^{-4}$	7.6—11.2	
QPPO3B. 1	AX_110939127	3B	50, 543, 628— 50, 543, 698	E3, E4	$7.73 \times 10^{-4} - 8.85 \times 10^{-4}$	7.3—7.5	
QPPO3B. 2	AX_110922129	3B	779, 569, 728— 779, 569, 798	E3, E4, E5	$2.79 \times 10^{-4} - 8.08 \times 10^{-4}$	9.1—10.9	
QPPO3D. 1	AX_109461736	3D	58, 392, 177— 58, 392, 247	E1, E2	$7.89 \times 10^{-4} - 9.00 \times 10^{-4}$	7.1—7.3	
QPPO3D. 2	Kukri_c8913_655	3D	512, 675, 071— 512, 675, 171	E1, E2, E5	$2.34 \times 10^{-4} - 4.86 \times 10^{-4}$	7.8—8.7	
QPPO4A. 1	AX_109980263	4A	104, 862, 727— 104, 862, 797	E1, E2, E3, E4, E5	$2.06 \times 10^{-5} - 8.85 \times 10^{-4}$	9.1—14.6	
QPPO4A. 2	AX_111505789	4A	627, 304, 625— 627, 304, 695	E1, E3, E4, E5	$8.83 \times 10^{-5} - 9.15 \times 10^{-4}$	7.0—12.6	
QPPO4B. 1	AX_94869270	4B	446, 520, 186— 446, 520, 256	E1, E2, E4, E5	$5.25 \times 10^{-4} - 9.31 \times 10^{-4}$	7.5—8.3	
QPPO4D. 1	AX_94417710	4D	271, 271, 977— 271, 272, 047	E3, E4, E5	$2.52 \times 10^{-4} - 6.45 \times 10^{-4}$	9.5—11.1	
QPPO5B. 1	AX_94713620	5B	43, 860, 785— 43, 860, 855	E1, E2, E3, E4, E5	$1.18 \times 10^{-4} - 6.78 \times 10^{-4}$	7.4—11.5	
QPPO5B. 2	AX_110639926	5B	601, 419, 594— 601, 419, 664	E3, E4	$1.22 \times 10^{-4} - 3.52 \times 10^{-4}$	11.8—13.3	
QPPO5B. 3	AX_109431771	5B	684, 087, 987— 684, 088, 057	E1, E2, E3, E4, E5	$1.41 \times 10^{-5} - 8.16 \times 10^{-4}$	7.1—15.6	
QPPO5D. 1	BS00000020_51	5D	3, 609, 844— 3, 609, 944	E1, E2, E3, E4, E5	$7.11 \times 10^{-7} - 1.28 \times 10^{-4}$	9.6—16.9	
QPPO5D. 2	AX_110379055	5D	542, 315, 674— 542, 315, 744	E3, E4	$2.95 \times 10^{-5} - 7.80 \times 10^{-4}$	9.6—14.6	
QPPO6A. 1	AX_94620967	6A	1, 496, 950— 1, 497, 020	E1, E2, E3, E4, E5	$4.18 \times 10^{-5} - 8.32 \times 10^{-4}$	9.1—13.7	
QPPO6A. 2	AX_94819835	6A	184, 823, 634— 184, 823, 704	E1, E2, E3, E4, E5	$8.95 \times 10^{-6} - 2.93 \times 10^{-5}$	12.1—13.4	
QPPO6A. 3	AX_111696637	6A	581, 715, 403— 581, 715, 473	E3, E4	$6.85 \times 10^{-4} - 9.13 \times 10^{-4}$	9.3—9.9	

(continued)

Name	Marker	Chr[a]	Physical position[b]	Environment[c]	P value	R[2d]	QTL[e]
QPPO6B. 1	AX _ 109475521	6B	10, 684, 647— 10, 684, 717	E2, E4	$9.54×10^{-4}—9.83×10^{-4}$	8. 3—9. 2	
QPPO6B. 2	AX _ 110541841	6B	51, 025, 146— 51, 025, 216	E2, E3, E4, E5	$2.16×10^{-4}—8.62×10^{-4}$	7. 1—9. 1	
QPPO6B. 3	AX _ 111820514	6B	145, 950, 886— 145, 950, 956	E1, E3, E4	$1.40×10^{-4}—8.83×10^{-4}$	7. 3—11. 8	
QPPO6B. 4	AX _ 110081827	6B	712, 235, 402— 712, 235, 472	E1, E2	$1.44×10^{-4}—8.78×10^{-4}$	6. 7—10. 4	
QPPO6D. 1	AX _ 111649730	6D	13, 616, 741— 13, 616, 811	E2, E4, E5	$3.86×10^{-4}—8.67×10^{-4}$	6. 8—9. 3	
QPPO7A. 1	AX _ 108824499	7A	16, 293, 405— 16, 293, 475	E1, E3	$4.90×10^{-4}—8.41×10^{-4}$	9. 0—10. 0	
QPPO7A. 2	AX _ 94738780	7A	709, 405, 155— 709, 405, 225	E1, E2, E3, E5	$3.09×10^{-5}—9.81×10^{-4}$	7. 2—13. 9	
QPPO7B. 1	AX _ 94571186	7B	626, 068, 866— 626, 068, 936	E2, E3, E4, E5	$4.56×10^{-4}—7.25×10^{-4}$	6. 9—10. 2	
QPPO7B. 2	AX _ 110906405	7B	709, 469, 379— 709, 469, 449	E1, E2, E3, E4, E5	$1.51×10^{-5}—3.97×10^{-4}$	10. 4—15. 3	
QPPO7D. 1	AX _ 109886696	7D	15, 222, 380— 15, 222, 450	E1, E2, E3, E4, E5	$6.07×10^{-6}—5.84×10^{-4}$	9. 9—17. 2	
QPPO7D. 2	AX _ 94749119	7D	580, 574, 591— 580, 574, 661	E3, E4	$5.24×10^{-4}—7.04×10^{-4}$	7. 9—8. 2	
QPPO7D. 3	AX _ 94483979	7D	633, 033, 169— 63, 3033, 239	E1, E2	$4.55×10^{-4}—8.76×10^{-4}$	9. 0—9. 7	

[a] Chromosome.

[b] Physical positions of SNP markers were based on wheat genome sequences from the International Wheat Genome Sequencing Consortium (IWGSC, http：//www. wheatgenome. org/).

[c] E1, 2013 Anyang (AY); E2, 2013 Suixi (SX); E3, 2014 AY; E4, 2014 SX; E5, BLUP, a best linear unbiased predictor of flour color-related traits in 166 wheat cultivars across four environments.

[d] Percentage of variance explained by each locus.

[e] The previously reported QTL within the same chromosomal regions.

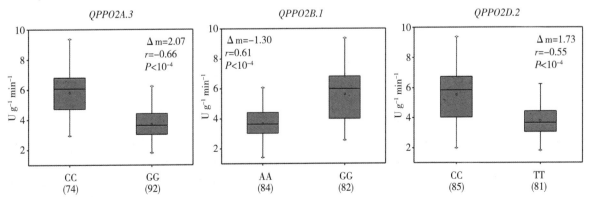

FIGURE 2   Phenotypic differences between lines carrying different alleles in the three most significant quantitative trait loci for polyphenol oxidase activity based on the most significant trait-associated single nucleotide polymorphisms. The box shows the first quartile, median, and third quartile. The number of individuals for each allele is given in the parentheses. The difference of mean (Δm), the Pearson's correlation coefficients (r) between genotypes and phenotypic values, and the P values of correlations are also shown

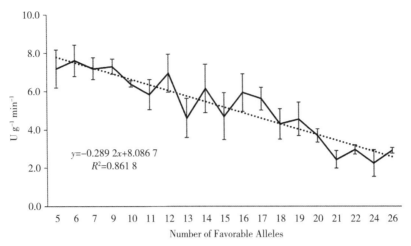

FIGURE 3   Linear regressions between the number of favorable alleles and best linear unbiased predictor values of polyphenol oxidase activity

4   DISCUSSION

4. 1   QTL for polyphenol oxidase activity

Polyphenol oxidase activity in flour is the main cause of undesirable darkening of wheat end-use products, especially Asian noodles. Characterization of the genetic architecture of PPO activity is an important prerequisite for developing wheat varieties with low PPO activity, and it is assessed by high levels of color stability in end-use products. Therefore, the identification of genes or genomic regions significantly associated with PPO activity facilitates further understanding of the molecular basis of PPO activity and makes it feasible to select genotypes with low PPO activity. The present GWAS was performed to identify genomic regions controlling PPO activity in 166 wheat accessions using wheat 90K and 660K arrays. The average genetic diversity and PIC were 0. 36 and 0. 29, respectively, indicating higher polymorphisms in this panel of wheat accessions that are suitable for GWAS (Chen, Min, Tauqeer, & Hu, 2012; Lopes, Dreisigacker, Peña, Sukumaran, & Reynolds, 2014). The large phenotypic variance of PPO activity also indicated that the wheat collection used in this work was an appropriate panel to study the genetic architecture of flour PPO activity (Table 1, Supplemental Table S1).

Many studies showed that homoeologous group 2

chromosomes were the main genomic areas influencing PPO activity (Anderson, Fuerst, Hurkman, Vensel, & Morris, 2006; He et al. , 2007, 2009; Martin et al. , 2011; Nilthong, Graybosch, & Baenziger, 2012; Raman et al. , 2005; Sadeque & Turner, 2010; Sun et al. , 2005). In the present study, the three QTL(*QPPO2A. 3*, *QPPO2B. 1*, and *QPPO2D. 2*) with largest effects on PPO activity were again located on chromosomes 2A, 2B, and 2D, respectively. The contribution from the *QPPO2A. 3* (17. 5-32. 4%) was greater than that from the *QPPO2B. 1* (15. 0-25. 5%) and *QPPO2D. 2* (13. 6-24. 9%). This was in agreement with previous studies indicating that PPO activity was primarily controlled by genes located on homoeologous group 2 chromosomes, and effects from chromosome 2A were much larger than those from chromosomes 2B or 2D (Demeke et al. , 2001; He et al. , 2007; Jukanti, Bruckner, & Fischer, 2006; Nilthong et al. , 2012; Si et al. , 2012; Wang, Ma, Si, Qiao, & He, 2009; Zhang et al. , 2005).

Polyphenol oxidase activity was also controlled by genes on homoeologous groups 3, 5, and 6, chromosomes 4A, 4B, 7B, and 7D, in agreement with previous studies (He et al. , 2007; Sadeque & Turner, 2010; Zhai et al. , 2016). Based on IWGSC RefSeq v1. 0, *QPPO4B. 1* was different from *QPPO. caas-4BL* (Zhai et al. , 2016), and *QPPO7B. 1* and *QPPO7B. 2* were different from

QPPO. dmh45-7B (Sadeque & Turne, 2010) due to large physical distances (Supplemental Table S2). Nevertheless, it is not clear if QTL mapped on homoeologous groups 2, 3, 5 and 6 chromosomes, and chromosomes 4A and 7D were the same as those in previous studies (Li et al., 1999; Sadeque & Turne, 2010; Udall, 1997), due to lack of marker information or unavailable sequences of some RFLP markers. In addition, QTL on homoeologous group 1 chromosomes (6 QTL) and chromosomes 4D and 7A (2 QTL) identified in this study are likely to be new, because no QTL for PPO activity was previously mapped on these chromosomes. These new genomic loci add to the genetic complexity of flour PPO activity.

4. 2　Prediction of candidate genes

Candidate genesfor PPO activity on chromosomes 2A (*QPPO2A. 3*), 2B (*QPPO2B. 1*), 2D (*QPPO2D. 2*), 4A (*QPPO4A. 2*), 5D (*QPPO5D. 1*), and 7D (*QPPO7D. 2*) were predicted following the physical positions of QTL and literatures. Based on IWGSC RefSeq v1. 0, we further found that *AX _ 109465981* (*QPPO2A. 3*) was 0. 05 Mb away from *Ppo-A1* and 0. 1 Mb from *Ppo-A2*; *wsnp _ Ra _ c10658 _ 17500498* (*QPPO2B. 1*) and *AX _ 95205011* (*QPPO2D. 2*) were 0. 1 and 0. 06 Mb from *Ppo-B2* and *Ppo-D2*, respectively. These three genes were cloned and co-segregated with PPO activity in previous studies (Beecher & Skinner, 2011; Sun et al., 2005). However, the relative effects of *Ppo1* and *Ppo2* genes should be further validated in biparental populations with parents having different alleles at both the *Ppo1* and *Ppo2* loci.

Polyphenol oxidase is a Cu-containing metalloenzyme that catalyzes phenolic compounds to quinones. Two catalytic Cu-binding sites as well-conserved functional domains are present in all known PPO genes (Cary, Lax, & Flurkey, 1992; Sommer, Ne'eman, Steffens, Mayer, & Harel, 1994; van Gelder, Flurkey, & Wichers, 1997). *AX _ 94749119* (*QPPO7D. 2*) encoding a Cu-transporting ATPase RAN1-like protein and

Kukri _ c77040 _ 87 (1. 37 Mb from *QPPO4A. 2*) corresponding to a Cu ionbinding protein were considered as candidate genes for participation in PPO activity.

In addition, *BS00000020 _ 51* corresponding to puroindoline b (*Pinb-D1*) was also proposed as a potential candidate gene for QPPO5D. 1. The PPO activity of cultivars from Sub-population 3 was significantly higher than that from the other two subpopulations, whereas there was no significant difference in kernel hardness between them (52. 14 vs 52. 49, $P = .91$; data not shown). Therefore, we could exclude the effect of kernel hardness on population structures. One interpretation might be that *Pinb* influences kernel texture and thus milling characteristics and flour granularity (Edwards, Osborne, & Henry, 2010). Considering that PPO is localized in plastids, damaged starch and flour granularity may influence PPO release and reaction with L-DOPA, causing variation in PPO activity.

4. 3　Marker-assisted selection for low polyphenol oxidase activity

Knowledgeof the frequency of favorable alleles within a breeding population would allow breeders to determine whether sufficient variation is available within a breeding population or whether further introgression of additional favorable alleles is required. The use of molecular markers to aid increases in the number of favorable alleles is becoming less expensive and could substantially reduce the overall cost of genetic improvement of wheat for specific end uses. An important finding in this study was that cultivars with higher frequencies of favorable alleles can be selected as parental resources for developing wheat cultivars with lower PPO activity, such as Funo, Kanto 107, Zhengzhou 3, and Barra. Marker-assisted selection is an efficient way to pyramid or maintain favorable alleles for low PPO activity in wheat breeding programs.

5　CONCLUSION

The 43 QTL for PPO activity identified in this study

included 12 new loci that provide a wider molecular base for further lowering PPO activity in wheat end-use products. The three most significant QTL (*QPPO2A. 3*, *QPPO2B. 1*, and *QPPO2D. 2*) will enable marker-assisted selection in developing wheat varieties with low level PPO activity and hence a reduced likelihood of browning of wheat flour-based products. *Ppo-A1*, *Ppo-A2*, *Ppo-B2*, *Ppo-D2*, *Pinb-D1*, and genes encoding a Cu ion-binding protein and a Cu-transporting ATPase RAN1-like protein were suggested as candidate genes for *QPPO2A. 3*, *QPPO2B. 1*, *QPPO2D. 2*, *QPPO5D. 1*, *QPPO4A. 2*, and *QPPO7D. 2*, respectively. Favorable alleles of strongly associated SNPs showed cumulative effects in decreasing PPO activity. Characterizing the genetic architecture of PPO activity and distribution of favorable alleles within natural wheat germplasm will provide the foundation for designing breeding strategies to achieve high flour quality.

❖ ACKNOWLEDGEMENTS

The authors are grateful to Prof. R. A. McIntosh at the Plant Breeding Institute, University of Sydney, for review of this manuscript. This work was funded by the Natural Science Foundation of Shandong Province (ZR2017BC038), National Natural Science Foundation of China (31701420), Young Elite Scientists Sponsorship Program by the China Association for Science and Technology (CAST) (2017QNRC001), and the Agricultural Scientific and Technological Innovation Project of Shandong Academy of Agricultural Sciences (CXGC2018E01).

❖ REFERENCES

Anderson, J. V., Fuerst, E. P., Hurkman, W. J., Vensel, W. H., & Morris, C. F. (2006). Biochemical and genetic characterization of wheat (*Triticum* spp.) kernel polyphenol oxidases. *Journal of Cereal Science*, 44, 353-367. https://doi.org/10.1016/j.jcs.2006.06.008

Anderson, J. V., & Morris, C. F. (2001). An improved whole-seed assay for screening wheat germplasm for polyphenol oxidase activity. *Crop Science*, 41, 1697-1705. https://doi.org/10.2135/cropsci2001.1697

Anderson, J. V., & Morris, C. F. (2003). Purification and analysis of wheat grain polyphenol oxidase (PPO) protein. *Cereal Chemistry*, 80, 135-143. https://doi.org/10.1094/cchem.2003.80.2.135

Baik, B. K., Czuchajowska, Z., & Pomeranz, Y. (1995). Discoloration of dough for oriental noodles. *Cereal Chemistry*, 72, 198-204.

Baik, B. K., & Ullrich, S. E. (2008). Barley for food: Characteristics, improvement, and renewed interest. *Journal of Cereal Science*, 48, 233-242. https://doi.org/10.1016/j.jcs.2008.02.002

Beecher, B. S., Carter, A. H., & See, D. R. (2012). Genetic mapping of new seed-expressed polyphenol oxidase genes in wheat (*Triticum aestivum* L.). *Theoretical and Applied Genetics*, 124, 1463-1473. https://doi.org/10.1007/s00122-012-1801-2

Beecher, B., & Skinner, D. Z. (2011). Molecular cloning and expression analysis of multiple polyphenol oxidase genes in developing wheat (*Triticum aestivum*) kernels. *Journal of Cereal Science*, 53, 371-378. https://doi.org/10.1016/j.jcs.2011.01.015

Benjamini, Y., & Hochberg, Y. (1995). Controlling the false discovery rate: A practical and powerful approach to multiple testing. *Journal of the Royal Statistical Society*, 57, 289-300. https://doi.org/10.1111/j.2517-6161.1995.tb02031.x

Beyer, S., Daba, S., Tyagi, P., Bockelman, H., Brown-Guedira, G., & Mohammadi, M. (2019). Loci and candidate genes controlling root traits in wheat seedlings-a wheat root GWAS. *Functional & Integrative Genomics*, 19, 91-107. https://doi.org/10.1007/s10142-018-0630-z

Bradbury, P. J., Zhang, Z., Kroon, D. E., Casstevens, T. M., Ramdoss, Y., & Buckler, E. S. (2007). TASSEL: Software for association mapping of complex traits in diverse samples. *Bioinformatics*, 23, 2633-2635. https://doi.org/10.1093/bioinformatics/btm308

Breseghello, F. , & Sorrells, M. E. (2006). Association mapping of kernel size and milling quality in wheat (*Triticum aestivum* L.) cultivars. *Genetics*, *172*, 1165-1177. https: //doi. org/10. 1534/ genetics. 105. 044586

Buckler, E. S. , & Thornsberry, J. M. (2002). Plant molecular diversity and applications to genomics. *Current Opinion in Plant Biology*, *5*, 107-111. https://doi. org/10. 1016/s1369-5266 (02) 00238-8

Cary, J. W. , Lax, A. R. , & Flurkey, W. H. (1992). Cloning and characterization of cDNAs coding for *Vicia faba* polyphenol oxidase. *Plant Molecular Biology*, *20*, 245-253. https: //doi. org/10. 1007/bf00014492

Chen, X. J. , Min, D. H. , Tauqeer, A. Y. , & Hu, Y. G. (2012). Genetic diversity, population structure and linkage disequilibrium in elite Chinese winter wheat investigated with SSR markers. *PLOS ONE*, *7* (*9*). https: //doi. org/10. 1371/journal. pone. 0044510

Demeke, T. , & Morris, C. F. (2002). Molecular characterization of wheat polyphenol oxidase (PPO). *Theoretical and Applied Genetics*, *104*, 813-818. https: //doi. org/10. 1007/s00122-001-0847-3

Demeke, T. , Morris, C. F. , Campbell, K. G. , King, G. E. , Anderson, J. A. , & Chang, H. G. (2001). Wheat polyphenol oxidase: Distribution and genetic mapping in three inbred line populations. *Crop Science*, *41*, 1750-1757. https: //doi. org/ 10. 2135/cropsci2001. 1750

Doyle, J. J. , & Doyle, J. L. (1987). A rapid DNA isolation procedure from small quantities of fresh leaf tissues. *Phytochem Bulletin*, *19*, 11-15

Edwards, M. A. , Osborne, B. G. , & Henry, R. J. (2010). Puroindoline genotype, starch granule size distribution and milling quality of wheat. *Journal of Cereal Science*, *52*, 314-320. https: //doi. org/ 10. 1016/j. jcs. 2010. 05. 015

Evanno, G. , Regnaut, S. , & Goudet, J. (2005). Detecting the number of clusters of individuals using the software STRUC-TURE: A simulation study. *Molecular Ecology*, *14*, 2611-2620. https: // doi. org/10. 1111/j. 1365-294x. 2005. 02553. x

Feillet, P. , Autran, J. C. , & Vernière, C. L. (2000). Pasta brownness: An assessment. *Journal of Cereal Science*, *32*, 215-233. https://doi. org/10. 1006/jcrs. 2000. 0326

Flint-Garcia, S. A. , Thuillet, A. C. , Yu, J. , Pressoir, G. , Romero, S. M. , Mitchell, S. E. , … Buckler, E. S. (2005). Maize association population: A high-resolution platform for quantitative trait locus dissection. *Plant Journal*, *44*, 1054-1064. https: // doi. org/10. 1111/j. 1365-313x. 2005. 02591. x

Fuerst, E. P. , Anderson, J. V. , & Morris, C. F. (2006). Delineating the role of polyphenol oxidase in the darkening of alkaline wheat noodles. *Journal of Agricultural and Food Chemistry*, *54*, 2378-2384. https: //doi. org/10. 1021/jf0526386

Fuerst, E. P. , Xu, S. S. , & Beecher, B. (2008). Genetic characterization of kernel polyphenol oxidases in wheat and related species. *Journal of Cereal Science*, *48*, 359-368. https: //doi. org/ 10. 1016/j. jcs. 2007. 10. 003

He, X. Y. , He, Z. H. , Morris, C. F. , & Xia, X. C. (2009). Cloning and phylogenetic analysis of polyphenol oxidase genes in common wheat and related species. *Genetic Resources and Crop Evolution*, *56*. https: //doi. org/10. 1007/s10722-008-9365-3

He, X. Y. , He, Z. H. , Zhang, L. P. , Sun, D. J. , Morris, C. F. , Fuerst, E. P. , & Xia, X. C. (2007). Allelic variation of polyphenol oxidase (PPO) genes located on chromosomes 2A and 2D and development of functional markers for the PPO genes in common wheat. *Theoretical and Applied Genetics*, *115*, 47-58. https: //doi. org/ 10. 1007/s00122-007-0539-8

Hystad, S. M. , Martin, J. M. , Graybosch, R. A. , & Giroux, M. J. (2015). Genetic characterization and expression analysis of wheat (*Triticum aestivum*) line 07OR1074 exhibiting very low polyphenol oxidase (PPO) activity. *Theoretical and Applied Genetics*, *128*, 1605-1615. https: // doi. org/10. 1007/s00122-015-2535-8

International Wheat Genome Sequencing Consortium (IWGSC). (2018). Shifting the limits in wheat research and breeding using a fully annotated reference genome. *Science*, *361* (6403). https：//doi. org/10. 1126/science. aar7191

Jimenez, M., &Dubcovsky, J. (1999). Chromosome location of genes affecting polyphenol oxidase activity in seeds of common and durum wheat. *Plant Breeding*, *118*, 395-398. https：//doi. org/10. 1046/j. 1439-0523. 1999. 00393. x

Jukanti, A. K., Bruckner, P. L., & Fischer, A. M. (2004). Evaluation of wheat polyphenol oxidase genes. *Cereal Chemistry*, *81*, 481-485. https：//doi. org/10. 1094/cchem. 2004. 81. 4. 481

Jukanti, A. K., Bruckner, P. L., & Fischer, A. M. (2006). Molecular and biochemical characterisation of polyphenol oxidases in developing kernels and senescing leaves of wheat (*Triticum aestivum*). *Functional Plant Biology*, *33*, 685-696. https：//doi. org/10. 1071/fp06050

Khatkar, M. S., Nicholas, F. W., Collins, A. R., Zenger, K. R., Cavanagh, J. A. L., Barris, W. ,... Raadsma, H. W. (2008). Extent of genome-wide linkage disequilibrium in Australian Holstein-Friesian cattle based on a high-density SNP panel. *BMC Genomics*, *9*. https：//doi. org/10. 1186/1471-2164-9-187

Kidane, Y. G., Gesesse, C. A., Hailemariam, B. N., Abate, E. A., Mengistu, D. K., Fadda, C. ,... Dell' Acqua, M. (2019). A large nested association mapping (NAM) population for breeding and QTL mapping in Ethiopian durum wheat. *Plant Biotechnology Journal*, *17*, 1380-1393. https：//doi. org/10. 1111/pbi. 13062

Lawrence, R., Day-Williams, A. G., Mott, R., Broxholme, J., Cardon, L. R., & Zeggini, E. (2009). GLIDERS-A web-based search engine for genome-wide linkage disequilibrium between HapMap SNPs. *BMC Bioinformatics*, *10*. https：//doi. org/10. 1186/1471-2105-10-367

Li, W. L., Farris, J. D., Chittoor, J., Leach, J. F., Liu, D. J., Chen, P. D., & Gill, B. S. (1999). Genomic mapping of defense response genes in wheat. *Theoretical and Applied Genetics*, *98*, 226-233. https：//doi. org/10. 1007/s00122005

1062

Liu, J. D., He, Z. H., Rasheed, A., Wen, W. E., Yan, J., Zhang, P. Z. ,... Xia, X. C. (2017). Genome-wide association mapping of black point reaction in common wheat (*Triticum aestivum* L.). *BMC Plant Biology*, *17*. https：//doi. org/10. 1186/s12870-017-1167-3

Lopes, M., Dreisigacker, S., Peña, R. J., Sukumaran, S., & Reynolds, M. P. (2014). Genetic characterization of the wheat association mapping initiative (WAMI) panel for dissection of complex traits in spring wheat. *Theoretical and Applied Genetics*, *128*, 453-464. https：//doi. rg/10. 1007/s00122-014-2444-2

Mares, D., & Campbell, A. (2001). Mapping components of flour and noodle color in Australian wheat. *Australian Journal of Agricultural Research*, *52*, 1297-1309. https：//doi. org/10. 1071/ar01048

Martin, J. M., Berg, J. E., Hofer, P., Kephart, K. D., Nash, D., & Bruckner, P. L. (2011). Allelic variation of polyphenol oxidase genes impacts on Chinese raw noodle color. *Journal of Cereal Science*, *54*, 387-394. https：//doi. org/10. 1016/j. jcs. 2011. 08. 003

Morris, C. F. (2018). Determinants of wheat noodle color. *Journal of the Science of Food and Agriculture*, *98*, 5171-5180. https：//doi. org/10. 1002/jsfa. 9134

Morris, C. F., Jeffers, H. C., & Engle, D. A. (2000). Effect of processing, formula and measurement variables on alkaline noodle color-toward an optimized laboratory system. *Cereal Chemistry*, *77*, 77-85. https：//doi. org/10. 1094/cchem. 2000. 77. 1. 77

Nilthong, S., Graybosch, R. A., & Baenziger, P. S. (2012). Inheritance of grain polyphenol oxidase (PPO) activity in multiple wheat (*Triticum aestivum* L.) genetic backgrounds. *Theoretical and Applied Genetics*, *125*, 1705-1715. https：//doi. org/10. 1007/s00122-012-1947-y

Pritchard, J. K., Stephens, M., Rosenberg, N. A., & Donnelly, P. (2000). Association mapping in structured populations. *American Journal of Human*

Genetics, *67*, 170-181. https: //doi. org/10. 1086/ 302959

Raman, R. , Raman, H. , Johnstone, K. , Lisle, C. , Smith, A. , Matin, P. , & Allen, H. (2005). Genetic and in silico comparative mapping of the polyphenol oxidase gene in bread wheat (*Triticum aestivum* L.). *Functional & Integrative Genomics*, *5*, 185-200. https: //doi. org/10. 1007/ s10142-005-0144-3

Raman, R. , Raman, H. , & Martin, P. (2007). Functional gene markers for polyphenol oxidase locus in bread wheat (*Triticum aestivum* L.). *Molecular Breeding*, *19*, 315-328. https: //doi. org/10. 1007/s11032-006-9064-8

Ransom, J. K. , Berzonsky, W. A. , & Sorenson, B. K. (2006). *Hard white wheat: Producing North Dakota' s next market opportunity*. Fargo: North Dakota State University Extension Service.

Sadeque, A. , & Turner, M. A. (2010). QTL mapping of polyphenol oxidase (PPO) activity and yellow alkaline noodle (YAN) color components in an Australian hexaploid wheat population. *Thai Journal of Agricultural Science*, *43*, 109-118.

Shifman, S. , Kuypers, J. , Kokoris, M. , Yakir, B. , & Darvasi, A. (2003). Linkage disequilibrium patterns of the human genome across populations. *Human Molecular Genetics*, *12*, 771-776. https: //doi. org/10. 1093/hmg/ddg088

Si, H. Q. , Zhou, Z. L. , Wang, X. B. , & Ma, C. X. (2012). A novel molecular marker for the polyphenol oxidase gene located on chromosome 2B in common wheat. *Molecular Breeding*, *30*, 1371-1378. https: //doi. org/10. 1007/s11032-012-9723-x

Simeone, R. , Pasqualone, A. , Clodoveo, M. L. , & Blanco, A. (2002). Genetic mapping of polyphenol oxidase in tetraploid wheat. *Cellular & Molecular Biology Letters*, *7*, 763-770.

Sommer, A. , Ne' eman, E. , Steffens, J. C. , Mayer, A. M. , & Harel, E. (1994). Import, targeting, and processing of a plant polyphenol oxidase. *Plant Physiology*, *105*, 1301-1311. https: //doi. org/10. 1104/pp. 105. 4. 1301

Sun, D. J. , He, Z. H. , Xia, X. C. , Zhang, L. P. ,

Morris, C. F. , Appels, R. ,... Wang, H. (2005). A novel STS marker for polyphenol oxidase activity in bread wheat. *Molecular Breeding*, *16*, 209-218. https: //doi. org/10. 1007/s11032-005-6618-0

Tadesse, W. , Ogbonnaya, F. C. , Jighly, A. , Sanchez-Garcia, M. , Sohail, Q. , Rajaram, S. , & Baum, M. (2015). Genome-wide association mapping of yield and grain quality traits in winter wheat genotypes. *PLOS ONE*, *10* (10). https: //doi. org/ 10. 1371/journal. pone. 0141339

Taranto, F. , Mangini, G. , Pasqualone, A. , Gadaleta, A. , & Blanco, A. (2015). Mapping and allelic variations of *Ppo-B1* and *Ppo-B2* gene-related polyphenol oxidase activity in durum wheat. *Molecular Breeding*, *35*. https: //doi. org/ 10. 1007/s11032-015-0272-y

Udall, J. (1997). *Important alleles for noodle quality in winter wheat as identified by molecular markers* (Master' s thesis). Moscow: University of Idaho.

van Gelder, C. W. G. , Flurkey, W. H. , & Wichers, H. J. (1997). Sequence and structural features of plant and fungal tyrosinases. *Phytochemistry*, *45*, 1309-1323. https: //doi. org/10. 1016/　s0031-9422 (97) 00186-6

Wang, X. B. , Ma, C. X. , Si, H. Q. , Qiao, Y. Q. , & He, X. F. (2009). Allelic variation of PPO genes in Chinese wheat micro-core collections. *Scientia Agricultura Sinica*, *42*, 28-35.

Watanabe, N. , Akond, A. M. , & Nachit, M. M. (2006). Genetic mapping of the gene affecting polyphenol oxidase activity in tetraploid durum wheat. *Journal of Applied Genetics*, *47*, 201-205. https: //doi. org/10. 1007/bf03194624

Waugh, R. J. , Jannink, L. , Muehlbauer, G. J. , & Ramsey, L. (2009). The emergence of whole genome association scans in barley. *Current Opinion in Plant Biology*, *12*, 218-222. https: //doi. org/ 10. 1016/j. pbi. 2008. 12. 007

Zhai, S. N. , He, Z. H. , Wen, W. E. , Jin, H. , Liu, J. D. , Zhang, Y. ,... Xia, X. C. (2016). Genome-wide linkage mapping of flour color-related traits and polyphenol oxidase activity in common wheat. *Theoretical and Applied Genetics*, *129*, 377-394. https: //doi. org/10. 1007/s00122-015-2634-6

Zhai, S. N. , Liu, J. D. , Xu, D. A. , Wen, W. E. , Yan, J. , Zhang, P. Z. ,... He, Z. H. (2018). A genome-wide association study reveals a rich genetic architecture of flour color-related traits in bread wheat. *Frontiers in Plant Science*, *9*. https: //

doi. org/10. 3389/fpls. 2018. 01136

Zhang, L. P. , Ge, X. X. , He, Z. H. , Wang, D. S. , Yan, J. , Xia, X. C. , & Sutherland, M. W. (2005). Mapping QTLs for polyphenol oxidase activity in a DH population from common wheat. *Acta Agronomica Sinica*, *31*, 7-10.

图书在版编目（CIP）数据

中麦 895 选育与主要特性解析 / 何中虎主编 . —北京 : 中国农业出版社，2023.11
ISBN 978-7-109-31243-2

Ⅰ.①中…　Ⅱ.①何…　Ⅲ.①小麦－选择育种－研究
Ⅳ.①S512.103

中国国家版本馆 CIP 数据核字（2023）第 196441 号

中麦 895 选育与主要特性解析
ZHONGMAI895 XUANYU YU ZHUYAO TEXING JIEXI

中国农业出版社出版
地址：北京市朝阳区麦子店街 18 号楼
邮编：100125
责任编辑：李　梅
版式设计：王　晨　　责任校对：吴丽婷
印刷：北京通州皇家印刷厂
版次：2023 年 11 月第 1 版
印次：2023 年 11 月北京第 1 次印刷
发行：新华书店北京发行所
开本：889mm×1168mm　1/16
印张：25.25
字数：1200 千字
定价：298.00 元

版权所有·侵权必究
凡购买本社图书，如有印装质量问题，我社负责调换。
服务电话：010-59195115　010-59194918